Lecture Notes in Computer Science 5868

Commenced Publication in 1973
Founding and Former Series Editors:
Gerhard Goos, Juris Hartmanis, and Jan van Leeuwen

W0081960

Ravindra K. Ahuja   Rolf H. Möhring
Christos D. Zaroliagis (Eds.)

Robust and Online Large-Scale Optimization

Models and Techniques
for Transportation Systems

 Springer

Volume Editors

Ravindra K. Ahuja
University of Florida, Department of Industrial & Systems Engineering
303 Weil Hall, Gainesville, FL 32611-6595, USA
E-mail: ahuja@ufl.edu

Rolf H. Möhring
Technische Universität Berlin, Institut für Mathematik
Straße des 17. Juni 136, 10623 Berlin, Germany
E-mail: Rolf.Moehring@TU-Berlin.de

Christos D. Zaroliagis
University of Patras, Department of Computer Engineering & Informatics
26500 Patras, Greece
E-mail: zaro@ceid.upatras.gr

Library of Congress Control Number: 2009937876

CR Subject Classification (1998): B.8, C.4, G.3, G.4, I.2.8, E.1

LNCS Sublibrary: SL 1 – Theoretical Computer Science and General Issues

ISSN   0302-9743
ISBN-10   3-642-05464-1 Springer Berlin Heidelberg New York
ISBN-13   978-3-642-05464-8 Springer Berlin Heidelberg New York

springer.com

© Springer-Verlag Berlin Heidelberg 2009
Printed in Germany

Typesetting: Camera-ready by author, data conversion by Scientific Publishing Services, Chennai, India
Printed on acid-free paper   SPIN: 12772070   06/3180   5 4 3 2 1 0

Preface

Scheduled transportation networks give rise to very complex and large-scale network optimization problems requiring innovative solution techniques and ideas from mathematical optimization and theoretical computer science. Examples of scheduled transportation include bus, ferry, airline, and railway networks, with the latter being a prime application domain that provides a fair amount of the most complex and largest instances of such optimization problems. Scheduled transport optimization deals with planning and scheduling problems over several time horizons, and quite some progress has been made for strategic planning and scheduling problems in all transportation domains.

In this volume, we focus on two important facets of scheduled transportation planning that pose even harder optimization questions: robust planning and online (real-time) planning. These two, tightly coupled, facets constitute a proactive and a reactive approach, respectively, to deal with disruptions to the normal operation. *Robust planning* is concerned with the development of an a priori plan that allows the absorption of disruptions to the best possible extent. *Online planning* is concerned with real-time decision making when, typically unpredictable, disruptions in daily operations occur, and before the entire sequence of disruptions is known. Since railway systems provide the largest, most complex and hence most challenging problems, we have put a special emphasis in this volume on robust and online railway optimization.

The papers appearing in the volume have been selected after an open call for contributions asking for either research papers or state-of-the-art survey articles. We received 24 submissions that underwent two rounds of the standard peer-review process, out of which 18 were finally accepted for publication.

The selected papers cover several aspects of robust and online large-scale optimization. With respect to the former, they cover issues of robust timetabling and route planning, as well as robust planning under scarce resources. With respect to the latter, they cover issues of delay and disruption management. Moreover, a fair amount of papers introduce new concepts of robustness and recoverability (to the normal operation) that turn out to be particularly useful when dealing with problems in railway optimization. The volume is organized in four parts reflecting the above areas.

The first part, **Robustness and Recoverability: New Concepts**, consists of five papers that introduce new concepts of robustness and recoverability and exemplify their usefulness on various applications. More specifically:

- In *The Concept of Recoverable Robustness, Linear Programming Recovery, and Railway Applications*, Christian Liebchen, Marco Lübbecke, Rolf Möhring, and Sebastian Stiller introduce a new concept of robustness that does not only help to achieve robust plans but also allows recovery to a feasible solution under certain circumstances. The new concept is exemplified in the railway optimization problems of delay resistant, periodic and aperiodic timetabling, and train platforming.

- In *Recoverable Robustness in Shunting and Timetabling*, Serafino Cicerone, Gianlorenzo D'Angelo, Gabriele Di Stefano, Daniele Frigioni, Alfredo Navarra, Michael Schachtebeck, and Anita Schöebel apply the concept of recoverable robustness to the shunting problem and also extend the concept to situations where multiple stages of recovery are required.
- In *Light Robustness*, Matteo Fischetti and Michele Monaci introduce the concept of light robustness, which couples robust optimization with a simplified two-stage stochastic programming approach, and constitutes a flexible counterpart of (classical) robust models.
- In *Incentive-Compatible Robust Line Planning*, Apostolos Bessas, Spyros Kontogiannis, and Christos Zaroliagis introduce the concept of incentive-compatible robustness and demonstrate its application on robust line planning when several competing operators demand line frequencies over a transportation network.
- In *A Bicriteria Approach for Robust Timetabling*, Anita Schöbel and Albrecht Kratz introduce a bicriteria approach for studying the trade-off between an optimal and a robust solution, by adding the robustness of the problem's solution as an additional objective function. They demonstrate their approach on the aperiodic timetabling problem in which a timetable is sought that is robust against delays.

The second part, **Robust Timetabling and Route Planning**, consists of five papers that present new approaches for robust timetabling, route planning, route re-planning, and timetable information updating in case of delays. More specifically:

- In *Meta-Heuristic and Constraint-Based Approaches for Singe-Line Railway Timetabling*, Federico Barber, Laura Ingolotti, Antonio Lova, Pilar Tormos, and Miguel A. Salido study the single-line railway timetabling problem (which is NP-hard) under several heuristic approaches, which are based on constraint techniques (distributed constraint satisfaction and topological constraint optimization) and on meta-heuristic techniques (GRASP-based variable ordering and genetic algorithms).
- In *Engineering Time-Expanded Graphs for Faster Timetable Information*, Daniel Delling, Thomas Pajor, and Dorothea Wagner present an extension of the time-expanded model for computing timetable information that results in faster query times using less space than the original one. They also show how known query speed-up techniques can be adapted to the extended model in order to gain further performance speed-up.
- In *Time-Dependent Route Planning*, Daniel Delling and Dorothea Wagner survey query speed-up techniques for route planning under the time-dependent model, and identify the most important ingredients along with their augmentations that make some techniques superior to others.
- In *The Exact Subgraph Recoverable Robust Shortest Path Problem*, Christina Büsing presents approximate approaches for route re-planning on a small

subnetwork when delays occur, and demonstrates that the achieved approximation ratio is the best possible.

– In *Efficient Timetable Information in the Presence of Delays*, Matthias Müller-Hannemann and Mathias Schnee present an efficient method for updating timetable information when a stream of delay information and schedule changes arise, and demonstrate its applicability on a real-world scenario.

The third part, **Robust Planning Under Scarce Resources**, consists of four papers that deal with several problems that demand scarce resources. More specifically:

– In *Integrating Robust Network Design and Line Planning Under Failures*, Angel Marin, Juan A. Mesa, and Federico Perea present a heuristic approach for robust network design and line planning that integrates these two phases, and consider two new notions for measuring robustness.

– In *Effective Allocation of Fleet Frequencies by Reducing Intermediate Stops and Short Turning in Transit Systems*, Juan A. Mesa, Francisco A. Ortega, and Miguel A. Pozo develop an effective model for allocating rolling-stock frequencies at stops along a line, and develop a heuristic approach for its solution.

– In *Shunting for Dummies: An Introductory Survey with an Algorithmic Focus*, Michael Gatto, Jens Maue, Matus Mihalak, and Peter Widmayer survey several commonly used as well as new train classification (or shunting) methods from an algorithmic perspective.

– In *Integrated Gate and Bus Assignment at Amsterdam Airport Schiphol*, Guido Diepen, Marjan van den Akker, and Han Hoogeveen present a column generation approach for achieving a robust model that integrates the phases of gate and bus assignment at an airport, and show that it is acceptable in practice.

The fourth part, **Online Planning: Delay and Disruption Management**, consists of four papers that deal with several aspects of delay and disruption management including detection of delay dependencies and conflict resolution among complex train routes. More specifically:

– In *Mining Railway Delay Dependencies in Large-Scale Real-World Delay Data*, Holger Flier, Rati Gelashvili, Thomas Graffagnino, and Marc Nunkesser present efficient algorithms to detect important types of systematic delay dependencies (that are one of the main sources of delay propagation), and demonstrate their practical applicability on real-world data.

– In *Rescheduling Dense Train Traffic over Complex Station Interlocking Areas*, Francesco Corman, Rob M.P. Goverde, and Andrea D'Ariano present two graph-theoretic approaches for modeling multiple conflicting train routes in busy stations along with their solution methods and their experimental comparison.

– In *Online Train Disposition: To Wait or not to Wait?*, Luzi Anderegg, Paolo Penna, and Peter Widmayer present deterministic polynomial-time optimal algorithms and matching lower bounds for several variants of an online delay management problem, where the delay is unknown and the vehicle can only wait in a station so as to minimize the passengers waiting time.

– In *Disruption Management in Passenger Railway Transportation*, J. Jespersen-Groth, D. Potthoff, J. Clausen, D. Huisman, L. Kroon, G. Maroti, and M.N. Nielsen give a comprehensive description of the problems arising in railway disruption management (timetable adjustment, rolling stock and crew rescheduling) along with the actors involved, and also describe the challenges confronted by railway companies in order to improve their operational performance.

Overall, the volume comprises a blend of state-of-the-art surveys and original research contributions. It is addressed to students, researchers, and practitioners who are interested in robust and online optimization of large-scale systems. We hope that they will find it useful.

We would like to thank all those who submitted papers for consideration, as well as the referees for their invaluable contribution. We also thank Apostolos Bessas for helping with several technical issues during the whole process of this volume production.

We gracefully acknowledge the support of the Future and Emerging Technologies Unit of the European Commission, under contract no. FP6-021235-2 (FP6 IST/FET Open/Project ARRIVAL). The ARRIVAL project not only supported part of the work presented in this volume, but most importantly it provided the means to stimulate a new line of research on robust and online railway optimization and also to create a critical mass of researchers, who are now able to deal with challenging problems in this area.

July 2009

Ravindra K. Ahuja
Rolf H. Möhring
Christos D. Zaroliagis

List of Contributors

Luzi Anderegg
Institute of Theoretical
Computer Science
ETH Zürich
Switzerland
anderegg@inf.ethz.ch

Federico Barber
Instituto de Automática
e Informática Industrial
Universidad Politécnica de Valencia
Spain
fbarber@dsic.upv.es

Apostolos Bessas
Department of Computer Engineering
and Informatics
University of Patras
26500 Patras
Greece
and
R.A. Computer Technology Institute
N. Kazantzaki Str.
Patras University Campus
26504 Patras
Greece
mpessas@ceid.upatras.gr

Christina Büsing
Institut für Mathematik
Technische Univeristät Berlin
Strasse des 17 Juni 136
10623 Berlin
Germany
cbuesing@math.tu-berlin.de

Serafino Cicerone
Department of Electrical
and Information Engineering
University of L'Aquila
Italy
serafino.cicerone@univaq.it

Jens Clausen
DSB S-tog, Denmark
and
Department of Informatics
and Mathematical Modelling
Technical University of Denmark
2800 Kongens Lyngby
Denmark
jencl@man.dtu.dk

Francesco Corman
Transport & Planning Department
Delft University of Technology
Stevinweg 1
2628 CN Delft
The Netherlands
f.corman@tudelft.nl

Gianlorenzo D'Angelo
Department of Electrical
and Information Engineering
University of L'Aquila
Italy
gianlorenzo.dangelo@univaq.it

Andrea D'Ariano
Dipartimento di Informatica
e Automazione
Università degli Studi Roma Tre
via della Vasca Navale 79
00146 Rome
Italy
and
Transport & Planning Department
Delft University of Technology
Stevinweg 1, 2628 CN Delft
The Netherlands
a.dariano@dia.uniroma3.it

Daniel Delling
Department of Computer Science
University of Karlsruhe
P.O. Box 6980
76128 Karlsruhe
Germany
delling@ira.uka.de

Guido Diepen
Paragon Decision Technology
Schipholweg 1
2034 LS Haarlem
The Netherlands
Guido.Diepen@aimms.com

Gabriele Di Stefano
Department of Electrical
and Information Engineering
University of L'Aquila
Italy
gabriele.distefano@univaq.it

Matteo Fischetti
DEI, University of Padova,
Via Gradenigo 6/A
35131 Padova
Italy
matteo.fischetti@unipd.it

Holger Flier
Institute of Theoretical
Computer Science
ETH Zürich
Switzerland
holger.flier@inf.ethz.ch

Daniele Frigioni
Department of Electrical
and Information Engineering
University of L'Aquila
Italy
daniele.frigioni@univaq.it

Michael Gatto
Institute of Theoretical
Computer Science
ETH Zürich
Switzerland
michael.gatto@inf.ethz.ch

Rati Gelashvili
Tbilisi State University
Georgia
gelash@gmail.com

Rob M.P. Goverde
Transport & Planning Department
Delft University of Technology
Stevinweg 1
2628 CN Delft
The Netherlands
r.m.p.goverde@tudelft.nl

Thomas Graffagnino
SBB AG Bern
Infrastruktur/Trassenmanagement
Switzerland
thomas.graffagnino@sbb.ch

J.A. Hoogeveen
Department for Information
and Computing Sciences
Utrecht University
P.O. Box 80089
3508 TB Utrecht
The Netherlands
slam@cs.uu.nl

Dennis Huisman
Econometric Institute
Erasmus University Rotterdam
P.O. Box 1738,
3000 DR Rotterdam
The Netherlands
and
Erasmus Center for Optimization
in Public Transport (ECOPT)
and

Department of Logistics
Netherlands Railways
P.O. Box 2025,
3500 HA Utrecht
The Netherlands
huisman@ese.eur.nl

Laura Ingolotti
Instituto de Automática
e Informática Industrial
Universidad Politécnica de Valencia
Spain
lingolotti@dsic.upv.es

Julie Jespersen-Groth
DSB S-tog, Denmark
and
Department of Informatics
and Mathematical Modelling
Technical University of Denmark
2800 Kongens Lyngby
Denmark
jjespersen@s-tog.dsb.dk

Spyros Kontogiannis
Computer Science Department
University of Ioannina
45110 Ioannina
Greece
and
R.A. Computer Technology Institute
N. Kazantzaki Str.
Patras University Campus
26504 Patras
Greece
kontog@cs.uoi.gr

Albrecht Kratz
Institut für Numerische
und Angewandte Mathematik
Georg-August Universität Göttingen
Germany
Albrecht.Kratz@gmx.de

Leo Kroon
Rotterdam School of Management
Erasmus University Rotterdam
P.O. Box 1738,
3000 DR Rotterdam
The Netherlands
and
Erasmus Center for Optimization
in Public Transport (ECOPT)
and
Department of Logistics
Netherlands Railways
P.O. Box 2025,
3500 HA Utrecht
The Netherlands
LKroon@rsm.nl

Christian Liebchen
Institut für Mathematik
Technische Univeristät Berlin
Strasse des 17 Juni 136
10623 Berlin
Germany
liebchen@math.tu-berlin.de

Antonio Lova
Instituto de Automática
e Informática Industrial
Universidad Politécnica de Valencia
Spain
allova@eio.upv.es

Marco Lübbecke
Institut für Mathematik
Technische Univeristät Berlin
Strasse des 17 Juni 136
10623 Berlin
Germany
m.luebbecke@math.tu-berlin.de

Angel Marín
Department of Applied Mathematics
and Statistics
Madrid Polytechnic University
Spain
angel.marin@upm.es

Gábor Maróti
Rotterdam School of Management
Erasmus University Rotterdam
P.O. Box 1738
3000 DR Rotterdam
The Netherlands
and
Erasmus Center for Optimization
in Public Transport (ECOPT)
marotig@gmail.com

Jens Maue
Institute of Theoretical
Computer Science
ETH Zürich
Switzerland
jens.maue@inf.ethz.ch

Juan A. Mesa
Department of Applied Mathematics II
University of Seville
Spain
jmesa@us.es

Matúš Mihalák
Institute of Theoretical
Computer Science
ETH Zürich
Switzerland
matus.mihalak@inf.ethz.ch

Rolf Möhring
Institut für Mathematik
Technische Univeristät Berlin
Strasse des 17 Juni 136
10623 Berlin
Germany
Rolf.Moehring@TU-Berlin.DE

Michele Monaci
DEI, University of Padova,
Via Gradenigo 6/A
35131 Padova
Italy
michele.monaci@unipd.it

Matthias Müller–Hannemann
Martin-Luther-University Halle
Department of Computer Science
Von-Seckendorff-Platz 1
06120 Halle
Germany
muellerh@informatik.uni-halle.de

Alfredo Navarra
Department of Mathematics
and Computer Science
University of Perugia, Italy
navarra@dmi.unipg.it

Morten Nyhave Nielsen
DSB S-tog, Denmark
and
Department of Informatics
and Mathematical Modelling
Technical University of Denmark
2800 Kongens Lyngby
Denmark

Marc Nunkesser
Institute of Theoretical
Computer Science
ETH Zürich
Switzerland
marc.nunkesser@inf.ethz.ch

Francisco A. Ortega
Department of Applied Mathematics I
University of Seville
Spain
riejos@us.es

Thomas Pajor
Department of Computer Science
University of Karlsruhe
P.O. Box 6980
76128 Karlsruhe
Germany
pajor@ira.uka.de

Paolo Penna
Dipartimento di Informatica
ed Applicazioni
"R.M. Capocelli", Università di Salerno
via S. Allende 2
84081 Baronissi (SA)
Italy
penna@dia.unisa.it

Federico Perea
Department of Applied Mathematics II
University of Seville
Spain
perea@us.es

Daniel Potthoff
Econometric Institute
Erasmus University Rotterdam
P.O. Box 1738
3000 DR Rotterdam
The Netherlands
and
Erasmus Center for Optimization
in Public Transport (ECOPT)
potthoff@ese.eur.nl

Miguel A. Pozo
Department of Applied Mathematics I
University of Seville
Spain
miguelpozo@us.es

Miguel A. Salido
Instituto de Automática
e Informática Industrial
Universidad Politécnica de Valencia
Spain
msalido@dsic.upv.es

Michael Schachtebeck
Institut für Numerische
und Angewandte Mathematik
Georg-August Universität Göttingen
37083 Göttingen
Germany
schachte@math.uni-goettingen.de

Mathias Schnee
Darmstadt University of Technology
Department of Computer Science
Hochschulstraße 10
64289 Darmstadt
Germany
schnee@algo.informatik.
tu-darmstadt.de

Anita Schöbel
Institut für Numerische
und Angewandte Mathematik
Georg-August Universität Göttingen
37083 Göttingen
Germany
schoebel@math.uni-goettingen.de

Sebastian Stiller
Institut für Mathematik
Technische Univeristät Berlin
Strasse des 17 Juni 136
10623 Berlin
Germany
stiller@math.tu-berlin.de

Pilar Tormos
Instituto de Automática
e Informática Industrial
Universidad Politécnica de Valencia
Spain
ptormos@eio.upv.es

J.M. van den Akker
Department for Information
and Computing Sciences
Utrecht University
P.O. Box 80089
3508 TB Utrecht
The Netherlands
marjan@cs.uu.nl

Dorothea Wagner
Department of Computer Science
University of Karlsruhe
P.O. Box 6980
76128 Karlsruhe
Germany
wagner@ira.uka.de

Peter Widmayer
Institute of Theoretical
Computer Science
ETH Zürich
Switzerland
widmayer@inf.ethz.ch

Christos Zaroliagis
Department of Computer Engineering
and Informatics
University of Patras
26500 Patras
Greece
and
R.A. Computer Technology Institute
N. Kazantzaki Str.
Patras University Campus
26504 Patras
Greece
zaro@ceid.upatras.gr

Referees

Reinhard Bauer
Apostolos Bessas
Suat Bog
Amy Cohn
Gianlorenzo D'Angelo
Daniel Delling
Camil Demetrescu
Maged Dessouky
Gabriele Di Stefano
Matteo Fischetti
Holger Flier
Dimitris Fotakis
Daniele Frigioni
Michael Gatto
Kalliopi Giannakopoulou
Tobias Harks

Dennis Huisman
Spyros Kontogiannis
Leo Kroon
Janny May-Yee Leung
Christian Liebchen
Marco Lübbecke
Matthias Müller-Hahnemann
Alfredo Navarra
Ashish Nemani
Marco Pranzo
Evangelia Pyrga
Christina Puhl
Anita Schöbel
Sebastian Stiller
Bala Vaidyanathan
Arrigo Zanette

Table of Contents

Robust Planning under Scarce Resources

Online Planning: Delay and Disruption Management

The Concept of Recoverable Robustness, Linear Programming Recovery, and Railway Applications*

Christian Liebchen, Marco Lübbecke, Rolf Möhring, and Sebastian Stiller

Technische Universität Berlin, Strasse des 17. Juni 136, 10623 Berlin, Germany
{liebchen,m.luebbecke,moehring,stiller}@math.tu-berlin.de

Abstract. We present a new concept for optimization under uncertainty: recoverable robustness. A solution is recovery robust if it can be recovered by limited means in all likely scenarios. Specializing the general concept to linear programming we can show that recoverable robustness combines the flexibility of stochastic programming with the tractability and performances guarantee of the classical robust approach. We exemplify recoverable robustness in delay resistant, periodic and aperiodic timetabling problems, and train platforming.

1 Introduction

Solutions for real-world problems found by mathematical optimization can hardly enter into praxis unless they possess a certain robustness. In applications robustness is not an additional feature but a conditio sine qua non. Usually, robustness is achieved ex post or by rules of thumb, i.e., heuristically. As systems work closer to capacity shortcomings of these suboptimal approaches become apparent. The classical two exact methods to deal with uncertain or fluctuating data in linear programming and combinatorial optimization are (2-Stage) Stochastic Programming and Robust Optimization.

A 2-stage stochastic program is a linear program, where part of the input data are random variables of some distribution. The distribution is either known, or partly known, or can at least be sampled. The decision variables split into *first stage decisions* and *second stage decisions*. The first stage decisions must be chosen fixed for all scenarios. The second stage variables can be chosen after the actual value of the random variables is revealed, i.e., the second stage decision can be different for each scenario. Thus, strictly speaking the second stage variables form a vector of random variables. But it is natural to interpret this vector of random variables as a large (deterministic) vector containing for each scenario one copy of the second stage variables. To be feasible in each scenario first and second stage variables together must form a feasible vector for the data realized in the scenario. Usually, the objective function of a 2-stage stochastic program

---

* This work was partially supported by the Future and Emerging Technologies Unit of EC, under contract no. FP6-021235-2 (FP6 IST/FET Open/Project ARRIVAL).

R.K. Ahuja et al. (Eds.): Robust and Online Large-Scale Optimization, LNCS 5868, pp. 1–27, 2009.

comprises a cost function for the first stage variables and the *expected* cost of the second stage variables according to the given distribution. Assuming a discretized scenario set there is an obvious way to interpret a 2-stage stochastic linear program as a (very large) usual linear program: The random variables in the original linear program are resolved by adding for each scenario a copy of the original linear program. In this copy the random variable is replaced by its realization in the scenario. This is called the scenario expansion of a 2-stage stochastic program.

Classical robust optimization considers a quite similar situation, except that one abstains from second stage actions. Again for a linear program a certain part of the data is subject to uncertainty. But as the robust program features no second stage variables, the variables—which are fixed before the actual data is revealed—must form a feasible solution for *every* scenario. A robust solution fits for all scenarios. Likewise, the objective function of a robust program contains no expectation or other stochastic component. The objective is a deterministic, linear function of the solution. Obviously, a robust model avoids the use of probability distributions. It suffices to know the range, in which the uncertain data can fluctuate. Usually, one models this range smaller than given in reality, thus excluding extremely unlikely scenarios.

Both methods have their merits for different types of applications. Still, for a number of applications none of the two provides a suitable method. One of these applications is delay resistant timetabling, e.g., for a railway system. Here instances are usually too large for stochastic programming approaches. Whereas robust optimization appeals for two reasons: A robust solution comes with a guarantee to be feasible in all scenarios of a certain restricted scenario set. We call this the set of *likely scenarios*. In addition, robust optimization yields compact mathematical models which are likely to be solvable on a scale relevant for practical purposes like delay resistant timetabling. But it turns out that the classical robust approach [1,15]—which we call *strict robustness* and which is a special case of the broader concept we present here—is necessarily over-conservative in the context of timetabling. The strict robust model yields a timetable where *each* train ride is scheduled to take its technically minimal travel time plus at least the total time of disturbances that is likely in the *whole* network. Unfortunately, this over-conservativism is intrinsic to the classical robust approach.

In practice, one often encounters a way of handling disturbances that is different from both methods mentioned above. First, the plan is furnished with some slack that shall allow to compensate disturbances. Second, the plan can be recovered during operations. Third, these recoveries are limited. The limits apply to the actions that can be taken and the computational means by which the recovery is calculated. For example, changes to the plan may be restricted to be local or to only affect a certain subset of the planning variables. It is a promising, practical concept that the plan has to be recoverable by limited means in every likely scenario. Still, in practice the plans, their slack and the simple means of recovery are currently not optimized together.

Related work. The concept of *recoverable robustness* has first been formalized together with concepts for the price of robustness in [11]. In the meantime it has attracted several applications to optimization problems particularly in the railway context. In [5,7] the concept has successfully been applied to shunting problems. Specific cases of recoverable robust timetabling are treated in [4,6,7]. In particular, they can identify some types of scenario sets for which finding a recoverable robust timetable becomes NP-hard and other types of scenario sets for which an efficient algorithm exists. In [9] the concept of recoverable robustness is spelled out specifically for the case of multi-stage recovery. Dynamic algorithms for this case have been proposed in [8]. In case the recovery is a linear program [16] provides for efficient algorithms and a stochastic analysis of quality for recovery robust solutions both for the case of right-hand side disturbances and for the case of matrix disturbances. The first application of recoverable robustness in a study on real-world data was carried out in [3]. It uses the techniques developed in this work and will be described in some detail in Section 5.

Our contribution and Outline. In this work we present the concept of *Recoverable Robustness*. The goal of recoverable robustness is to *jointly optimize* the plan and the strategy for limited recovery. This will combine the flexibility of stochastic programming with the performance guarantee and the compactness of models found in robust optimization.

In Section 2 we develop the concept of recoverable robustness formally and in full generality. As delay resistant timetabling has sparked its development, we exemplify the modeling power of recoverable robustness by the case of timetabling in Section 3. To solve recovery robust models we specify to *Linear Recovery Robust Programs* in Section 4, for which we provide an efficient algorithm. Finally, we demonstrate the power of this method by citing a real-world application of the method to the train platforming problem (Section 5).

2   The Concept of Recoverable Robustness

We are looking for solutions to an optimization problem which in a limited set of scenarios can be made feasible, or *recovered*, by a limited effort. Therefore, we need to define

- the <u>o</u>riginal optimization problem (Step O),
- the imperfection of information, that is, the <u>s</u>cenarios (Step S), and
- the limited <u>r</u>ecovery possibilities (Step R).

For Step O and Step S a large toolbox for modeling can be borrowed from classical approaches to optimization respectively optimization with imperfect information. Step R is a little less obvious, and we choose to formalize it via a *class $\mathcal{A}$ of admissible recovery algorithms*.

A solution x for the optimization problem defined in Step O is *recovery-robust*

- *against* the imperfection of information (Step S) and
- *for* the recovery possibilities (Step R),

if in all situations that may occur according to Step S, we can recover from x a feasible solution by means of one of the algorithms given in Step R.

Computations in recovery-robust optimization naturally decompose into a *planning phase* and a *recovery phase*. In the planning phase,

- we compute a solution x which may become infeasible in the realized scenario,
- and we choose $A \in \mathcal{A}$, i.e., one of the admissible recovery algorithms.

Such a pair (x, A) hedges for data uncertainty in the sense that in the recovery phase

- algorithm A is used to turn x into a feasible solution in the realized scenario.

The output (x, A) of the planning phase is more than a solution, it is a *precaution*. It does not only state that recovery is possible for x, but explicitly specifies *how* this recovery can be found, namely by the algorithm A.

The formal definition of recoverable robustness [11] we give next is very broad. The theorems in this paper only apply to strong specializations of that concept.

We introduce some terminology. Let $\mathcal{F}$ denote the original optimization problem. An instance $O = (P, f)$ of $\mathcal{F}$ consists of a set P of feasible solutions, and an objective function $f : P \to \mathbb{R}$ which is to be minimized.

By $\mathcal{R} = \mathcal{R}_{\mathcal{F}}$ we denote a model of imperfect information for $\mathcal{F}$ in the sense that for every instance O we specify a set $S = S_O \in \mathcal{R}_{\mathcal{F}}$ of possible scenarios. Let P_s denote the set of feasible solutions in scenario $s \in S$.

We denote by $\mathcal{A}$ a class of algorithms called *admissible recovery algorithms*. A recovery algorithm $A \in \mathcal{A}$ solves the *recovery problem*, which is a feasibility problem. Its input is $x \in P$ and $s \in S$. In case of a *feasible recovery*, $A(x, s) \in P_s$.

Definition 1. *The triple* $(\mathcal{F}, \mathcal{R}, \mathcal{A})$ *is called a* recovery robust optimization problem, *abbreviated RROP.*

Definition 2. *A pair* $(x, A) \in P \times \mathcal{A}$ *consisting of a* planning solution x *and an* admissible algorithm A *is called a* precaution.

Definition 3. *A precaution is* recovery robust, *iff for every scenario* $s \in S$ *the recovery algorithm* A *finds a feasible solution to the recovery problem, i.e., for all* $s \in S$ *we have* $A(x, s) \in P_s$.

Definition 4. *An optimal precaution is a recovery robust precaution* (x, A) *for which* $f(x)$ *is minimal.*

Thus, we can quite compactly write an RROP instance as

$$\inf_{(x,A)\in P\times\mathcal{A}} f(x)$$

$$\text{s.t.} \quad \forall s \in S : A(x,s) \in P_s \quad .$$

The objective function value of an RROP is infinity, if no recovery is possible for some scenario with the algorithms given in the class $\mathcal{A}$ of admissible recovery algorithms.

It is a distinguished feature of this notion that the planning solution is explicitly accompanied by the recovery algorithm. In some specializations the choice of the algorithm is self-understood. For example, for linear recovery robust programs, to which we will devote our main attention, the algorithm is some solver of a linear program or a simpler algorithm that solves the specific type of linear program that arises as the recovery problem of the specific RROP. Then we will simply speak of the planning solution x, tacitly combining it with the obvious algorithm to form a precaution.

2.1   Restricting the Recovery Algorithms

The class of admissible recovery algorithms serves as a very broad wildcard for different modeling intentions. Here we summarize some important types of restrictions that can be expressed by means of that class.

The definition of the algorithm class $\mathcal{A}$ also determines the computational balance between the planning and the recovery phase. For all practical purposes, one must impose sensible limits on the recovery algorithms (otherwise, the entire original optimization problem could be solved in the recovery phase, when the realized scenario is known). In very bold term, these limits fall into two categories:

– limits on the actions of recovery;
– limits on the computational power to find those actions of recovery.

We mention two important subclasses of the first category:

Strict Robustness. We can forbid recovery entirely by letting $\mathcal{A}$ consist of the single recovery algorithm A with $A(x,s) = x$ for all $s \in S$. This is called *strict robustness*. Note that by strict robustness the classical notion of robust programming is contained in the definition of recoverable robustness.

Recovery Close to Planning. An important type of restrictions for the class of admissible recovery algorithms is, that the recovery solution $A(x,s)$ must not deviate too far from the original solution x according to some measure of distance defined for the specific problem. For some distance measures one can define subsets $P_{s,x} \subseteq P_s$ depending on the scenario s and the original solution x, such that the restriction to the recovery algorithm that $A(x,s)$ will not deviate too far from x, can be expressed equivalently by requiring $A(x,s) \in P_{s,x}$. As an example, think of a railway timetable that must be recovered, such that the difference between the actual and the planned arrival times is not too big, i.e., that the delay is limited.

2.2 Passing Information to the Recovery

If (as it ought to be) the recovery algorithms in $\mathcal{A}$ are allowed substantially less computational power than the precaution algorithms in $\mathcal{B}$, we may want to pass some additional information $z \in Z$ (for some set Z) about the instance to the recovery algorithm. That is, we may compute an extended precaution $B(P, f, S) = (x, A, z)$, and in the recovery phase we require $A(x, s, z) \in P_s$.

As a simple example, consider a class of admissible recovery algorithms $\mathcal{A}$ that is restricted to computational effort linear in the size of a certain finite set of weights, which is part of the input of the RROP instance. Then it might be helpful to pass an ordered list of those weights on to the recovery algorithm, because the recovery algorithm will not have the means to calculate the ordered list itself, but could make use of it.

In Section 3 we present another example, namely rule based delay management policies, which shows that it is a perfectly natural idea to preprocess some values depending on the instance, with which the recovery algorithm becomes a very simple procedure.

2.3 Limited Recovery Cost

The recovery algorithm A solves a feasibility problem, and we did not consider any cost incurred by the recovery so far. There are at least two ways to do so in the framework of recoverable robustness. Let $d(y^s)$ be some (possibly vector valued) function measuring the cost of recovery $y^s := A(x, s)$.

- **Fixed Limit:** Impose a fixed limit λ to $d(y^s)$ for all scenarios s.
- **Planned Limit:** Let λ be a (vector of) variable(s) and part of the planning solution. Require $\lambda \geq d(y^s)$ for every scenario s, and let $\lambda \in \Lambda$ influence the objective function by some function $g : \Lambda \to \mathbb{R}$.

In the second setting, the planned limit λ to the cost of recovery is a variable chosen in the planning phase and then passed to the recovery algorithm A. It is the task of A to respect the constraint $\lambda \geq d(y)$, and it is the task of the planning phase to choose (x, A, λ), such that A will find a recovery for x with cost less or equal to λ. Therefore, and to be consistent with previous notation we formulate the cost bound slightly different. Let P'_s denote the set of feasible recoveries for scenario s. Then we define P_s by:

$$A(x, s, \lambda) \in P_s :\Leftrightarrow d(A(x, s)) \leq \lambda \wedge A(x, s) \in P'_s$$

We obtain the following recovery robust optimization problem with recovery cost:

$$\min_{(x,A,\lambda)\in P\times\mathcal{A}\times\Lambda} f(x) + g(\lambda)$$

$$\text{s.t.} \quad \forall s \in S : A(x,s,\lambda) \in P_s \quad .$$

Including the possibility to pass some extra information $y \in Y$ to A we obtain:

$$\min_{(x,A,z,\lambda)\in P\times\mathcal{A}\times Z\times\Lambda} f(x) + g(\lambda)$$

$$\text{s.t.} \quad \forall s \in S : A(x,s,z,\lambda) \in P_s \quad .$$

These recovery cost aware variants allow for computing an optimal trade-off between higher flexibility for recovery by a looser upper bound on the recovery cost, against higher cost in the planning phase. This is conceptually close to two-stage stochastic programming, however, we do not calculate an expectation of the second stage cost, but adjust a common upper bound on the recovery cost. This type of problem still has a purely deterministic objective. The linear recovery robust programs discussed later are an example of this type of RROP.

3   Recovery Robust Timetabling

Punctual trains are probably the first thing a layman will expect from robustness in railways. Reliable technology and well trained staff highly contribute to increased punctuality. Nevertheless, modern railway systems still feature small disturbances in every-day operations.

A typical example for a disturbance is a prolonged stop at a station because of a jammed door. A disturbance is a seminal event in the sense that the disturbance may cause several delays in the system but is not itself caused by other delays. Informing passengers about the reason for a delay affecting them, railway service providers sometimes do not distinguish between disturbances, i.e., seminal events, and delays that are themselves consequences of some initial disturbance. We will use the term *disturbance* exclusively for initial changes of planning data. A *delay* is any difference between the planned point in time for an event and the time the event actually takes place. We also speak of *negative delay*, when an event takes place earlier than planned.

A good timetable is furnished with *buffers* to absorb small disturbances, such that they do not affect the planned arrival times at all, or that they cause only few delays in the whole system. Those buffer times come at the expense of longer, planned travel times. Hence they must not be introduced excessively. Delay resistant timetabling is about increasing the planned travel times as little as possible, while guaranteeing the consequences of small disturbances to be limited.

We will now show how delay resistant timetabling can be formulated as a recovery robust optimization problem. We actually show that a robust version of timetabling is only reasonable, if it is understood as a recovery robust optimization problem. Moreover, we show how recoverable robustness integrates timetabling and the so-called delay management. Delay management is the term coined for the set of operational decisions reacting to concrete disturbances, i.e., the recovery actions. Its integration with timetabling is an important step forward for delay resistant timetabling, which can be formalized by the notion of recoverable robustness.

Step 0. The original problem is the deterministic timetabling problem. It exists in many versions that differ in the level of modeling detail, the objective function, or whether periodic or aperiodic plans are desired. The virtues of recovery robust timetables can already be shown for a simple version.

A Simple Timetabling Problem. The basic mathematical model that stands to reason for timetabling problems is the so-called Event-Activity Model or Feasible Differential Problem [13]. A timetable assigns points in time for certain events, i.e., arrivals and departures of trains. This assignment is feasible, if the differences in time between two related events are large enough, to allow for the activities relating them. For example, the arrival of a train must be scheduled sufficiently after its departure at the previous station. Likewise, transfers of passengers require the arrival of the feeder train and the departure of the transfer train to take place in the right time order and with a time difference at least large enough to allow for the passengers to change trains. We now describe a basic version of this model.

The input for our version of timetabling is a directed graph $G = (V, E)$ together with a non-negative function $t : E \to \mathbb{R}_+$ on the arc set. The nodes of the graph $V = V_{AR} \cup V_{DP}$ model arrival events (V_{AR}) and departure events (V_{DP}) of trains at stations. The arc set can be partitioned into three sets representing traveling of a train from one station to the next, E_{DR}, stopping of a train at a station, E_{ST}, and transfers of passengers from one train to another at the same station, E_{TF}. For travel arcs $e = (i, j) \in E_{DR}$ we have $i \in V_{DP}$ and $j \in V_{AR}$, for the two other types $e = (i, j) \in E_{ST} \cup E_{TF}$ the contrary holds: $i \in V_{AR}$ and $j \in V_{DP}$. The function $t(e)$ expresses the minimum time required for the action corresponding to $e = (i, j)$, in other words the minimum time between event i and event j. For example, for a travel arc e the value of the function $t(e)$ expresses the technical travel time between the two stations.

A feasible timetable is a non-negative vector $\pi \in \mathbb{R}_+^{|V|}$ such that $t(e) \leq \pi_j - \pi_i$ for all $e = (i, j) \in E$. W.l.o.g. we can assume that G is acyclic.

For the objective function we are given a non-negative weight function $w : E \to \mathbb{R}_+$, where $w_e = w(e)$ states how many passengers travel along arc e, i.e., are in the train during the execution of that action, or change trains according to that transfer arc. An optimal timetable is a feasible timetable that minimizes the total planned (or *nominal*) travel time of the passengers:

$$\sum_{e=(i,j)\in E} w_e(\pi_j - \pi_i).$$

Thus the data for the original problem can be encoded in a triple (G, t, w), containing the event-activity graph G, the arc length function t, and a cost function on the arcs w. The original problem can formulated as a linear program:

$$\min \sum_{e=(i,j)\in E} w_a(\pi_j - \pi_i)$$

$$\text{s.t.} \quad \pi_j - \pi_i \geq t(e) \quad \forall e = (i,j) \in E$$

$$\pi \geq 0$$

Step S. We assume uncertainty in the time needed for traveling and stopping. Those actions typically produce small disturbances. For a scenario s we are given a function $t^s : E \to \mathbb{R}_+$, with the properties $t^s(e) \geq t(e)$ for all $e \in E$, and $t^s(e) = t(e)$ for all $e \in E_{\text{TF}}$. As we only want to consider scenarios with small disturbances, we restrict to those scenarios where $t^s(e) - t(e) \leq \Delta_e$, for some small, scenario independent constant Δ_e. In a linear program one can scale each row, i.e., multiply all matrix entries of the row and the corresponding component of the right-hand side vector by a positive scalar, without changing the set of feasible solutions. Therefore, we can assume w.l.o.g. $\Delta_e = \Delta$ for all $e \in E$. Additionally, we require that not too many disturbances occur at the same time, i.e., in every scenario for all but k arcs $e \in E$ we have $t^s(e) = t(e)$.

Of course, there are situations in practice where larger disturbances occur. But it is not reasonable to prepare for such catastrophic events in the published timetable.

Strict Robustness. The above restrictions to the scenario set can be very strong, in particular, if we choose $k = 1$. But even for such a strongly limited scenario set strict robustness leads to unacceptably conservative timetables. Namely, the strict robust problem can be formulated as the following linear program:

$$\min \sum_{e=(i,j)\in E} w_e(\pi_j - \pi_i)$$

$$\text{s.t.} \quad \pi_j - \pi_i \geq t(e) + \Delta \quad \forall e = (i,j) \in E$$

$$\pi \geq 0$$

In other words, even if we assume that in every scenario at most one arc takes Δ time units longer, we have to construct a timetable as if all (traveling and stopping) arcs were Δ time units longer. This phenomenon yields solutions so conservative, that classical robust programming is ruled out for timetabling. Indeed, delay resistant timetabling has so far been addressed by stochastic programming [12,17] only. These approaches suffer from strong limitations to the size of solvable problems.

The real world expectation towards delay resistant timetables includes that the timetable can be adjusted slightly during its operation. But a strict robust program looks for timetables that can be operated unchanged despite disturbances. This makes the plans too conservative even for very restricted scenario sets. Robust timetabling is naturally *recovery robust timetabling* as we defined it. Naturally, a railway timetable has to be robust against *small* disturbances and for *limited* recovery.

Step R. The recovery of a timetable is called *delay management*. The two central means of delay management are delaying events and canceling transfers. Delaying an event means to propagate the delay through the network. Canceling a transfer means to inhibit this propagation at the expense of some passengers loosing their connection.

Pure delay propagation seems not deserve the name recovery at all. But recall that if delay propagation is not captured in the model, as in the strict robust model, the solutions become necessarily over-conservative. Delay is a form of recovery, and though it is a basic, it is a very important.

Actually, delay management has several other possibilities for recovery. For example, one may cancel train trips, re-route the trains, or re-route the passengers by advising them to use an alternative connection, or hope that they will figure such a possibility themselves. Moreover, delay management has to pay respect to several other aspects of the transportation system. For example, the shifts of the on-board crews are affected by delays. These in turn may be subject to subtle regulations by law or contracts and general terms of employment.

We initially adopt a quite simple perspective to delay management gradually increasing the complexity of the model. First we concentrate on delay, later on delay and broken transfers. The latter means plan with respect to delay management decisions, i.e., decision whether a train shall wait for delayed transferring passengers, or not in order to remain itself on time. Even basic delay management decision lead to PSPACE-hard models.

Roughly speaking, for a PSPACE-hard problem we cannot even recognize an optimal solution, when it is given to us, nor can we compare two solutions suggested to us. (See below for details on the complexity class PSPACE.) Therefore, we describe a variant that yields simpler models and is useful in railway practice.

Simple Recovery Robust Timetabling. First, we describe a model where the recovery can only delay the events but cannot cancel transfers. This is not a recovery in the ordinary understanding of the word. The recovery is simply the delay propagation. But this simple recovery already rids us from the conservatism trap of strict robustness.

In the recovery phase, when the scenario s and its actual traveling and stopping times t^s are known, we construct a *disposition timetable* $\pi^s \in \mathbb{R}_+^{|V|}$ fulfilling the following feasibility condition:

– The disposition timetable π^s of scenario s must be feasible for t^s, i.e.,

$$\forall e = (i,j) \in E : \pi_j^s - \pi_i^s \geq t^s(e).$$

These inequalities define the set (actually, the polytope) P_s of feasible recoveries in scenario s.

If this was the complete set of restrictions to the recovery, every timetable would be recoverable. We set up limits to the recovery algorithms:

TTC. The disposition timetable is bounded by the original timetable in a very strict manner: Trains must not depart earlier than scheduled, i.e.,

$$\forall e \in E_{\mathrm{DP}} : \pi^s(e) \geq \pi(e).$$

This is what we call the *timetabling condition*.

L1. We want the sum of the delays of all arrival events to be limited. Therefore assume we are also given a weight function $\ell : V_{\mathrm{AR}} \to \mathbb{R}_+$ that states how many passengers reach their final destination by the arrival event i. We fix a limit $\lambda_1 \geq 0$ and require:

$$\sum_{i \in V_{\mathrm{AR}}} \ell(i) \left(\pi_i^s - \pi_i \right) \leq \lambda_1.$$

L2. One may additionally want to limit the delay for each arrival separately, ensuring that no passenger will experience an extreme delay exceeding some fixed $\lambda_2 \geq 0$, i.e.:

$$\forall i \in V_{\mathrm{AR}} : \pi_i^s - \pi_i \leq \lambda_2.$$

In our model a recovery algorithm $A \in \mathcal{A}$ must respect all three limits. The bounds λ_1 and λ_2 can be fixed a priori, or made part of the objective function. In this way upper bounds on the *recovery cost* can be incorporated into the optimization process. For a timetabling problem (G, t, w) and a function $\ell : V_{\mathrm{AR}} \to \mathbb{R}_+$ and constants $g_1, g_2 \geq 0$ and an integer k we can describe the first timetabling RROP by the following linear program:

$$\min \sum_{e=(i,j)\in E} w_e(\pi_j - \pi_i) + g_1 \cdot \lambda_1 + g_2 \cdot \lambda_2$$

$$\text{s.t.} \quad \pi_j - \pi_i \geq t(e) \ \ \forall e = (i,j) \in E \tag{1}$$

$$\pi_j^s - \pi_i^s \geq t^s(e) \ \ \forall s \in S, \forall e = (i,j) \in E \tag{2}$$

$$\pi_i^s \geq \pi_i \ \ \forall s \in S, \forall i \in V_{\mathrm{DP}} \tag{3}$$

$$\sum_{i \in V_{\mathrm{AR}}} \ell(i) \left(\pi_i^s - \pi_i \right) \leq \lambda_1 \ \ \forall s \in S \tag{4}$$

$$\pi_i^s - \pi_i \leq \lambda_2 \ \ \forall s \in S, \forall i \in V_{\mathrm{AR}} \tag{5}$$

$$\lambda_{\{1,2\}}, \pi^s, \pi \geq 0$$

The set of scenarios S in this description is defined via the set of all functions $t^s : E \to \mathbb{R}_+$ which fulfill the following four conditions from Step S:

$$t^s(e) \geq t(e) \ \ \forall e \in E$$

$$t^s(e) \leq t(e) + \Delta \quad \forall e \in E$$
$$t^s(e) = t(e) \quad \forall e \in E_{\mathrm{TF}}$$
$$|\{e \in E : t^s(e) \neq t(e)\}| \leq k$$

In our terminology Inequality (1) defines P, Inequality (2) defines P_s, Inequalities (3) to (5) express limits to the action of the algorithm, namely, that the recovery may not deviate to much from the original solution. In detail (3) models the TTC, (4) ensures condition L1 and (5) condition L2.

Here and in the remainder of the example we use mathematical programs to *express* concisely the problems under consideration. These programs are not necessarily the right approach to *solve* the problems. Note that the above linear program is a scenario expansion and therefore too large to be solved for instances of relevant scale. In Section 4, we will devise a general result that allows us to reformulate such scenario expansions in a compact way. Thereby, the recovery robust timetabling problem becomes efficiently solvable.

Breaking Connections. In practice delay management allows for a second kind of recovery. It is possible to cancel transfers in order to stop the propagation of delay through the network. We now include the possibility to cancel transfers into the recovery of our model.

Again we consider a simple version for explanatory purposes. A transfer arc e can be removed from the graph G at a fixed cost $g_3 \geq 0$ multiplied with the weight w_e. With a sufficiently large constant M we obtain a mixed integer linear program representing this model:

$$\min \sum_{e=(i,j) \in E} w_e(\pi_j - \pi_i) + g_1 \cdot \lambda_1 + g_2 \cdot \lambda_2 + g_3 \cdot \lambda_3$$

$$\text{s.t.} \quad \pi_j - \pi_i \geq t(e) \quad \forall e = (i,j) \in E \tag{6}$$
$$\pi_j^s - \pi_i^s \geq t^s(e) \quad \forall s \in S, \forall e = (i,j) \in E_{\mathrm{DR}} \cup E_{\mathrm{ST}} \tag{7}$$
$$\pi_j^s - \pi_i^s + M x_e^s \geq t^s(e) \quad \forall s \in S, \forall e = (i,j) \in E_{\mathrm{TF}} \tag{8}$$
$$\pi_i^s \geq \pi_i \quad \forall s \in S, \forall i \in V_{\mathrm{DP}} \tag{9}$$
$$\sum_{i \in V_{\mathrm{AR}}} \ell(i)\,(\pi_i^s - \pi_i) \leq \lambda_1 \quad \forall s \in S \tag{10}$$
$$\pi_i^s - \pi_i \leq \lambda_2 \quad \forall s \in S, \forall i \in V_{\mathrm{AR}} \tag{11}$$
$$\sum_{e \in E_{\mathrm{TF}}} w_e x_e^s \leq \lambda_3 \quad \forall s \in S \tag{12}$$

$$\lambda_{\{1,2,3\}}, \pi^s, \pi \geq 0$$

$$x^s \in \{0,1\}^{|E_{\mathrm{TF}}|}$$

In our terminology Inequality 6 defines P. Inequalities 7 and 8 define P_s for every s. Again Inequalities 9 to 11 express limits to the actions that can be taken by the recovery algorithms. These are limits to the deviation of the recovered solution π^s from the original solution π.

3.1 Computationally Limited Recovery Algorithms

So far we imposed limits on the *actions* of the recovery algorithms. But delay management is a real-time task. Decisions must be taken in very short time. Thus it makes sense to impose further restrictions on the computational power of the recovery algorithm. Note that in general such restrictions cannot be expressed by a mathematical program as above. We now give two examples for computationally restricted classes of recovery algorithms.

The Online Character of Delay Management. In fact the above model has a fundamental weakness. It assumes that the recovered solution, i.e., the disposition timetable $\pi^s = A(\pi, s)$ can be chosen after s is known completely. This is of course not the case for real-world delay management: The disturbances evolve over time, and delay management must take decisions before the whole scenario is known. This means that the algorithms in A must be non-anticipative[1].

PSPACE-hardness of Delay Management. The multistage structure of some delay management models, namely that uncertain events and dispatching decisions alternate, makes these problems extraordinarily hard. Even quite restricted models have been shown to be PSPACE-hard [2].

The complexity class PSPACE contains those decision problems that can be decided with the use of memory space limited by a polynomial in the input size. The class NP is contained in PSPACE, because in polynomial time only polynomial space can be used. It is widely assumed that NP is a proper subset of PSPACE. Given this, one cannot decide in polynomial time that a given solution to a PSPACE-hard problem is feasible, because else the solution would be a certificate and therefore the problem in NP. (Note, that the complexity terminology is formulated for decision problems. Feasibility in this context means, that the delay management solution is feasible in the usual sense and in addition has cost less or equal to some constant.) Thereby it becomes even difficult to assess the quality of a solution for delay management, or to compare the quality of two competing delay management strategies.

Rule Based Delay Management. The previous observation is quite discouraging. How shall one design a recovery robust timetable, if the recovery itself is already PSPACE-hard? We now dicuss a special restriction to the delay management

---

[1] Given a mapping of the random variables in the input and of the decision variables to some partially ordered set (i.e., a timeline). Then a (deterministic) algorithm is *non-anticipative*, if for any pair of scenarios s and s' and every decision variable x, the algorithm in both scenarios chooses the same value for x, whenever s and s' are equal in all data entries that are mapped to elements less or equal to the image of x. This means that the algorithm can at no time anticipate and react to data revealed later.

that is motivated by the real-world railway application and turns each decision whether to wait or not to wait into a constant time solvable question. The model keeps the multistage character, but the resulting recovery robust timetabling problem is solvable by a mixed integer program.

Delay management decisions must be taken very quickly. Moreover, as delay management is a very sensitive topic for passengers' satisfaction the transparency of delay management decisions can be very valuable. A passenger might be more willing to accept a decision, that is based on explicit rules, about how long, e.g., a local train waits for a high-speed train, than to accept the outcome of some non-transparent heuristic or optimization procedure. For these two reasons, computational limits for real-time decisions and transparency for the passenger, one may want to restrict the class $\mathcal{A}$ of admissible recovery algorithms to *rule based delay management*. The idea is that trains will wait for at most a certain time for the trains connecting to them. These maximal waiting times depend on the type of involved trains. For example, a local train might wait 10 minutes for a high-speed train, but vice versa the waiting time could be zero. Fixing the maximal waiting times determines the delay management (within the assumed modeling precision). But, which waiting times are best? Does the asymmetry in the example make sense? We want to optimize the waiting rules, i.e., the delay management together with the timetable.

Assume we distinguish between m types of trains in the system, i.e., we have a mapping $\mu : V \rightarrow \{1, \ldots, m\}$ of the events onto the train types. A rule based delay management policy A is specified by a matrix $\mathcal{M} = \mathcal{M}_A \in \mathbb{R}_+^{m \times m}$. The y-th entry in the x-th row m_{xy} is the maximum time a departure event of train type y will be postponed in order to ensure transfer from a type x train. Formally, a rule based delay management policy schedules a departure event j at the earliest time π_j^s satisfying

$$\pi_j^s \geq \pi_i^s + t^s(i,j) \quad \forall (i,j) \in E_{ST}$$
$$\pi_j^s \geq \min\{\pi_i^s + t^s(i,j), \pi_j + m_{\mu(i)\mu(j)}\} \quad \forall (i,j) \in E_{TF}$$
$$\pi_j^s \geq \pi_j.$$

Arrival events are scheduled as early as possible respecting TTC and the traveling times in scenario s:

$$\pi_j^s = \max(\{\pi_j\} \cup \{\pi_i^s + t^s(i,j) | (i,j) \in E_{DR}\}) \quad \forall j \in V_{AR}$$

Actually, the maximum is taken over two elements, as only one traveling arc (i,j) leads to each arrival event j.

Moreover, for a transfer arc $(i,j) \in E_{TF}$ the canceling variable $x_{(i,j)}$ is set to 1 if and only if the result of the above rule gives $\pi_i^s + t(i,j) > \pi_j^s$.

It is easy to see that such a recovery algorithm gives a feasible recovery for every (even non-restricted) scenario s and every solution $\pi \in P$. If we restrict $\mathcal{A}$ to the class of rule based delay management policies, the RROP consists in finding a $m \times m$ matrix $\mathcal{M}$ and a schedule π that minimizes an objective function like those in the models we presented earlier:

$$\min_{\mathcal{M},\pi,\lambda_{\{1,2,3\}}} \sum_{e=(i,j)\in E} w_e(\pi_j - \pi_i) + g_1 \cdot \lambda_1 + g_2 \cdot \lambda_2 + g_3 \cdot \lambda_3$$

$$\text{s.t.} \quad \pi_j - \pi_i \geq t(e) \quad \forall e = (i,j) \in E \tag{13}$$

$$\forall s \in S, \forall j \in V_{\mathrm{DP}}: \tag{14}$$
$$\pi_j^s = \max(\{\pi_i^s + t^s(i,j) \,|\, (i,j) \in E_{\mathrm{ST}}\}$$
$$\cup \max\{\min\{\pi_i^s + t^s(i,j), \pi_j + m_{\mu(i)\mu(j)}\} | (i,j) \in E_{\mathrm{TF}}\}$$
$$\cup \{\pi_j\})$$

$$\forall s \in S, \forall j \in V_{\mathrm{AR}}: \tag{15}$$
$$\pi_j^s = \max(\{\pi_j\}$$
$$\cup \{\pi_i^s + t^s(i,j) | (i,j) \in E_{\mathrm{DR}}\}$$

$$\pi_j^s - \pi_i^s + M x_e^s \geq t^s(e) \quad \forall s \in S, \forall e = (i,j) \in E_{\mathrm{TF}} \tag{16}$$

$$\sum_{i \in V_{\mathrm{AR}}} \ell(i)\,(\pi_i^s - \pi_i) \leq \lambda_1 \quad \forall s \in S \tag{17}$$

$$\pi_i^s - \pi_i \leq \lambda_2 \quad \forall s \in S, \forall i \in V_{\mathrm{AR}} \tag{18}$$

$$\sum_{e \in E_{\mathrm{TF}}} w_e x_e^s \leq \lambda_3 \quad \forall s \in S \tag{19}$$

$$\lambda_{\{1,2,3\}}, \pi^s, \pi \geq 0$$

$$x^s \in \{0,1\}^{|E_{\mathrm{TF}}|}$$

The timetabling condition is ensured automatically by the rule based delay management described in Equations (14) and (15).

Rule based delay management algorithms are non-anticipative. The formulation we give even enforces the following behavior: The departure π_i of a train A will be delayed for transferring passengers from train B (with arrival π_j) for the maximal waiting time $m_{\mu(i)\mu(j)}$, even if before time $\pi_i + m_{\mu(i)\mu(j)}$ it becomes known that train B will arrive too late for its passengers to reach train A at time $\pi_i + m_{\mu(i)\mu(j)}$. As formulated, a train will wait the due time, even if the awaited train is hopelessly delayed. In practice, delay managers might handle such a situation a little less short minded.

Rule based delay management is a good example for the idea of integrating robust planning and simple recovery. Consider the following example of two local trains, A and B, and one high-speed train C. Passengers transfer from A to B, and from B to C. Assume local trains wait 7 minutes for each other, but high-speed trains wait at most 2 minutes for local trains. Then train A being late could force train B to loose its important connection to the high-speed train C. Indeed, this could happen, if the timetable and the waiting times are not attuned. In the planning, we might not be willing to increase the time a high-speed train waits, but instead plan a sufficient buffer for the transfer from B to C. This example illustrates that buffer times and waiting rules must be constructed jointly in order to attain optimal delay resistance.

4   Linear Programming Recovery

In this section we specialize to RROPs linear programs as recovery. We call such an RROP a Linear Recovery Robust Problem (LRP). We show how LRPs can be solved for certain scenario sets. This leads us to a special case, namely robust network buffering, which entails the robust timetabling problem. Towards the end of this section we turn to a variant of LRPs, where the planning problem is an *integer* linear program.

4.1   Linear Recovery Robust Programs

Given a linear program $(\min c'x, \text{ s.t. } A^0x \geq b^0)$ with m rows and n variables. We seek solutions to this problem that can be recovered by limited means in a certain limited set of disturbance scenarios. The situation in a disturbance scenario s is described by a set of linear inequalities, notably, by a matrix A^s and a right-hand side b^s. We slightly abuse notation when we say that the scenario set S contains a scenario (A^s, b^s), which, strictly speaking, is the image of scenario s under the random variable (A, b). We will discuss later more precisely the scenario sets considered in this analysis. For the linear programming case the limited possibility to recover is defined via a recovery matrix $\hat{A}$, a recovery cost d, and a recovery budget D. A vector x is recovery robust, if for all (A^s, b^s) in the scenario set S exists y such that $A^sx + \hat{A}y \geq b^s$, and $d'y \leq D$. Further, we require that x is feasible for the original problem without recovery, i.e., $A^0x \geq b^0$. The problem then reads:

$$\inf_{x} c'x$$
$$\text{s.t.} \qquad A^0x \geq b^0$$
$$\forall(A, b) \in S \ \exists y \in \mathbb{R}^{\hat{n}} :$$
$$Ax + \hat{A}y \geq b$$
$$d'y \leq D$$

When S is a closed set in the vector space $\mathbb{R}^{(m \times n + m)}$ we know that either the infimum is attained, or the problem is unbounded. This case constitutes the principal object of our considerations, the Linear Recovery Robust Program:

Definition 5. Let A^0 be an $m \times n$-matrix called the nominal matrix, b^0 be an m-dimensional vector called the nominal right-hand side, c be an n-dimensional vector called the nominal cost vector, $\hat{A}$ be an $m \times \hat{n}$-matrix called the recovery matrix, d be an $\hat{n}$-dimensional vector called the recovery cost vector, and D be a non-negative number called the recovery budget. Further let S be a closed set of pairs of $m \times n$-matrices and m-dimensional vectors, called the scenario set. Then the following optimization problem is called a Linear Recovery Robust Program (LRP) over S:

$$\min_{x} c'x$$

$$\text{s.t.} \qquad A^0 x \geq b^0$$

$$\forall (A, b) \in S \ \exists y \in \mathbb{R}^{\hat{n}} :$$

$$Ax + \hat{A}y \geq b$$

$$d'y \leq D$$

We refer to the A as the *planning matrix* although it is a quantified variable. The planning matrix describes how the planning x influences the feasibility in the scenario. The vectors $y \in \mathbb{R}^{\hat{n}}$ with $d'y \leq D$ are called the *admissible recovery vectors*. Note that we do not call S a scenario *space*, because primarily there is no probability distribution given for it.

We are not unnecessarily restrictive, when requiring the same number of rows for A^0 as for $\hat{A}$ and A. If this is not the case, nothing in what follows is affected, except may be readability.

If a solution x can be recovered by an admissible recovery y in a certain scenario s, we say x *covers* s.

To any LRP we can associate a linear program, which we call the scenario expansion of the LRP:

$$\min_{x, (y^s)_{s \in S}} c'x$$

$$\text{s.t.} \qquad A^0 x \geq b^0$$

$$A^s x + \hat{A}y^s \geq b^s \ \forall s \in S$$

$$d'y^s \leq D \qquad \forall s \in S$$

Note that in this formulation the set S is comprised of the scenarios s, whereas in the original formulation it contains (A^s, b^s). This ambiguity of S is convenient and should cause no confusion to the reader. Further, note that in the scenario expansion of an LRP each recovery variable y^s is indexed by its scenario. Thus the solution vector to the scenario expansion contains for each scenario a separate copy of the recovery vector. In the original formulation the recovery vector y is not indexed with a scenario, because the formulation is not a linear program but a logical expression where y is an existence quantified variable.

The scenario expansion is a first possibility to solve the LRP. But, usually, the scenario set is too big to yield a solvable scenario expansion. The scenario sets, which we will consider, are not even finite.

We will frequently use an intuitive reformulation of an LRP, that can be interpreted as a game of a *planning player* setting x, a *scenario player* choosing (A, b), and a *recovery player* deciding on the variable y. The players act one after the other:

$$\inf_{x} c'x \ \text{s.t.} \ A^0 x \geq b^0 \wedge D \geq \left\{ \sup_{(A,b) \in S} \left\{ \inf_{y} d'y \ \text{s.t.} \ Ax + \hat{A}y \geq b \right\} \right\} \qquad (20)$$

with constant vectors $c \in \mathbb{R}^n$, $b^0 \in \mathbb{R}^m$ and $d \in \mathbb{R}^{\hat{n}}$, constant matrices $A^0 \in \mathbb{R}^{m,n}$ and $\hat{A} \in \mathbb{R}^{m,\hat{n}}$, and variables $x \in \mathbb{R}^n$, $A \in \mathbb{R}^{m,n}$, $b \in \mathbb{R}^m$ and $y \in \mathbb{R}^{\hat{n}}$.

Again, when it is clear that either the extrema exist or the problem is unbounded we use the following notation:

$$\min_{x} c'x \text{ s.t. } A^0 x \geq b^0 \wedge D \geq \left\{ \max_{(A,b) \in S} \left\{ \min_{y} d'y \text{ s.t. } Ax + \hat{A}y \geq b \right\} \right\} \quad (21)$$

Observe, that an LRP, its scenario expansion and its 3-player formulation have the same feasible set of planning solutions x. Whereas, the set of recovery vectors y, that may occur as a response to some scenario (A^s, b^s) in the 3-player formulation, is only a subset of the set of feasible second stage solutions y^s in the scenario expansion. The 3-player formulation restricts the later set to those responses y, which are minimal in $d'y$. But this does not affect the feasible set for x.

The formalism of Problem (21) can also be used to express, that x and y are required to be non-negative. But it is a lot more well arranged, if we state such conditions separately:

$$\min_{x} c'x \text{ s.t. } A^0 x \geq b^0 \wedge D \geq \left\{ \max_{(A,b) \in S} \left\{ \min_{y \geq 0} d'y \text{ s.t. } Ax + \hat{A}y \geq b \right\} \right\} \quad (22)$$

and

$$\min_{x \geq 0} c'x \text{ s.t. } A^0 x \geq b^0 \wedge D \geq \left\{ \max_{(A,b) \in S} \left\{ \min_{y \geq 0} d'y \text{ s.t. } Ax + \hat{A}y \geq b \right\} \right\} \quad (23)$$

The purely deterministic condition $A^0 x \geq b^0$, which we call *nominal feasibility* condition, could also be expressed implicitly by means of S and $\hat{A}$. But, this would severely obstruct readability. In some applications the nominal feasibility plays an important role. For example, a delay resistant timetable shall be feasible for the nominal data, i.e., it must be possible to operate the published timetable unchanged at least under standard conditions. Else, trains could be scheduled in the published timetable x to depart earlier from a station than they arrive there. However, in this rather technical section the nominal feasibility plays a minor role.

Let us mention some extensions of the model. The original problem may as well be an integer or mixed integer linear program,

$$\min_{x=(\hat{x}, \bar{x}), \bar{x} \in \mathbb{Z}} c'x \text{ s.t. } A^0 x \geq b^0 \wedge D \geq \left\{ \max_{(A,b) \in S} \left\{ \min_{y} d'y \text{ s.t. } Ax + \hat{A}y \geq b \right\} \right\} \quad (24)$$

or some other optimization problem over a set of feasible solutions P and an objective function $c : \mathbb{R}^n \to \mathbb{R}$, in case the *disturbances* are confined to the right-hand side:

$$\inf_{f \in P} c(f) \text{ s.t. } D \geq \left\{ \sup_{b \in S} \left\{ \inf_{y} d'y \text{ s.t. } f + \hat{A}y \geq b \right\} \right\} \quad (25)$$

with a fixed planning matrix A.

Using the concept of planned limits to the recovery cost (cf. p. 6), the budget D can also play the role of a variable:

$$\min_{D \geq 0,\, x} c'x + D \text{ s.t. } A^0 x \geq b^0 \wedge$$

$$D \geq \left\{ \max_{(A,b) \in S} \left\{ \min_y d'y \text{ s.t. } Ax + \hat{A}y \geq b \right\} \right\} \tag{26}$$

In case of right-hand side disturbances only, we can again formulate:

$$\inf_{f \in P,\, D \geq 0} c(f) + D \text{ s.t. } D \geq \left\{ \sup_{b \in S} \left\{ \inf_y d'y \text{ s.t. } f + \hat{A}y \geq b \right\} \right\} \tag{27}$$

4.2   Solving Right-Hand Side LRPs

In this part we show that some scenario sets for the right-hand side data of an LRP yield problems that can be solved by a relatively small linear program.

Consider again the 3-player formulation of an LRP (21). Let $P := \{x \in \mathbb{R}^n : A^0 x \leq b^0\}$ be the polytope of nominally feasible solutions. If we fix the strategies of the first two players, i.e., the variables x and (A, b), we get the recovery problem of the LRP: $\min d'y$ subject to $\hat{A}y \geq b - Ax$. The dual of the latter is $\max_{\zeta \geq 0} (b - Ax)'\zeta$ s.t. $\hat{A}'\zeta \leq d$. The recovery problem is a linear program. Thus, we have strong duality, and replacing this linear program by its dual in expression (21) will not change the problem for the players optimizing x respectively (A, b).

$$\min_{x \in P} c'x \text{ s.t. } D \geq \left\{ \max_{(A,b) \in S} \left\{ \max_{\zeta \geq 0} (b - Ax)'\zeta \text{ s.t. } \hat{A}'\zeta \leq d \right\} \right\} \Leftrightarrow$$

$$\min_{x \in P} c'x \text{ s.t. } D \geq \left\{ \max_{(A,b) \in S,\, \zeta \geq 0} (b - Ax)'\zeta \text{ s.t. } \hat{A}'\zeta \leq d \right\} \tag{28}$$

Consider the maximization problem in formulation (28) for a fixed x, thus find $\max_{(A,b) \in S,\, \zeta \geq 0} (b - Ax)'\zeta$ subject to $\hat{A}'\zeta \leq d$. Assume for a moment $\|b - Ax\|_1 \leq \Delta$. In this case, for each fixed vector ζ the maximum will be attained, if we can set $\text{sign}(\zeta_i)(b - Ax)_i = \Delta$ for i with $|\zeta_i| = \|\zeta\|_\infty$ and 0 else. In other words, under the previous assumptions $(b - Ax)'\zeta$ attains its maximum when $(b - Ax) = \Delta e_i$ for some suitable $i \in [m]$. Therefore, if we have $\|b - Ax\|_1 \leq \Delta$, we can reformulate problem (28):

$$\min_{x \in P} c'x \text{ s.t. } \forall i \in [m] : D \geq \left\{ \max_{\zeta \geq 0} (\Delta e_i)'\zeta \text{ s.t. } \hat{A}'\zeta \leq d \right\} \Leftrightarrow$$

$$\min_{x \in P} c'x \text{ s.t. } \forall i \in [m] : D \geq \left\{ \min_y d'y \text{ s.t. } \hat{A}y \geq \Delta e_i \right\} \tag{29}$$

The at first sight awkward condition $\|b - Ax\|_1 \leq \Delta$ is naturally met if only the right-hand side data changes, and is limited in the set $S^1 := \{(A^s, b^s) :$

$\|b^* - b^s\|_1 \leq \Delta, A^* = A\}\}$. For an LRP over S^1 formulation (29) is equivalent to a linear program of size $\mathcal{O}(m(n + \hat{n} \cdot m))$:

$$\min_{x \in P} \quad c'x$$

$$\text{s.t.} \quad \forall i \in [m]:$$

$$Ax + \hat{A}y^i \geq b + \Delta e_i$$

$$d'y^i - D \geq 0$$

Next, consider the scenario set

$$S^k := \{(A^s, b^s) : \|b^* - b^s\|_1 \leq k \cdot \Delta, \|b^* - b^s\|_\infty \leq \Delta, A^* = A\}$$

for arbitrary $k > 1$. By the same token, the maximization over ζ in formulation (28) for fixed x and (A, b) can be achieved, by setting the maximal $\lfloor k \rfloor$ entries of the vector $(b - Ax)$ equal to 1 and the $\lceil k \rceil$-th entry equal to $k - \lfloor k \rfloor$. For example, when k is integer, we can replace the scenario set S^k by those $\binom{k}{m}$ scenarios, where exactly k entries of b deviate maximally from b^*, and the other entries equal their reference value b_i^*. So, we have:

Theorem 1. *An LRP over S^k can be solved by a linear program of size polynomial in n, $\hat{n}$, m, and $\binom{k}{m}$.*

Corollary 1. *An LRP over S^1 can be solved by a linear program of size polynomial in n, $\hat{n}$, and m.*

Corollary 2. *For fixed k an LRP over S^k can be solved by a linear program of size polynomial in n, $\hat{n}$, and m.*

Of course, in practice this approach will only work, when k is very small.

The above reasoning can give a fruitful hint to approach RROPs in general. First, try to find a small subset of the scenario set, which contains the worst-case scenarios, and then optimize over this set instead of the whole scenario set. In the above setting we can achieve this very easily, because the recovery problem fulfills strong duality. If the recovery problem is an integer program this approach fails in general. Still, one can try to find a small set of potential worst case scenarios, to replace the original scenario set. Unlike the recovery problem, the planning problem may well be an integer or mixed integer program, as we show in the following.

For the manipulations of the formulations the linearity of $P, Ax \geq b$ or c is immaterial. So we can extend the above reasoning to non-linear optimization problems. Let $c : \mathbb{R}^n \to \mathbb{R}$ be a real function, P' a set of feasible solutions and $\{g_i : \mathbb{R}^n \to \mathbb{R}\}_{i \in [m]}$ be a family of real functions, and assume that extrema in the resulting RROP are either attained, or the problem is unbounded. For the scenario set S^1 of right-hand side disturbances we have with the above notation

$$\min_{x \in P'} c(x)$$
$$\text{s.t. } D \geq \left\{ \max_{b \in S^1} \left\{ \min_y d'y \text{ s.t. } g(x) + \hat{A}y \geq b \right\} \right\}$$
$$\Leftrightarrow$$
$$\min_{x \in P'} c(x)$$
$$\text{s.t. } \forall i \in [m] : g(x) + \hat{A}y^i \geq b^* + \Delta e_i$$
$$D - d'y^i \geq 0$$

In particular we are interested in the case of an integer linear program ($\min c'x$, $Ax \geq b, x \in \mathbb{Z}^n$) with right-hand side uncertainty. We get as its recovery robust version over S^1.

$$\min_{x \in \mathbb{Z}^n} c'x$$
$$\text{s.t. } A^0 x \geq b^0$$
$$D \geq \left\{ \max_{b \in S^1} \left\{ \min_y d'y \text{ s.t. } A^* x + \hat{A}y \geq b \right\} \right\}$$
$$\Leftrightarrow$$
$$\min_{x \in \mathbb{Z}} c'x$$
$$\text{s.t. } A^0 x \geq b^0$$
$$\forall i \in [m] : A^* x + \hat{A}y^i \geq b^* + \Delta e_i$$
$$D - d'y^i \geq 0$$

Let $A^* = A^0$ and $b^* = b^0$. Defining $f := A^* x - b^*$ we can rewrite the previous program as

$$\min_{(x,f) \in \mathbb{Z}^{n+m}} \tilde{c}'(x, f)$$
$$A^* x - f = b^*$$
$$\text{s.t. } \forall i \in [m] : f + \hat{A}y^i \geq \Delta e_i \qquad (30)$$
$$d'y^i - D \geq 0$$
$$f \geq 0$$

With a suitable cost vector $\tilde{c}$. Note that the original integer linear program corresponds with the scenario part of the program only via the slack variable f. In other words, for solving the recovery robust version the solving procedures for the original, deterministic, integer linear optimization problem can be left untouched. We only have to flange a set of linear inequalities to it. The f variables function as means of communication between the original integer problem, where they correspond to the slack in each row, and the linear part, in which their effect on robustness is evaluated. In the next part we will consider this communication situation for an even more specialized type of recovery.

4.3 Robust Network Buffering

Let us use Corollary 1 for the Simple Robust Timetabling problem with right-hand side uncertainty limited in S^1. Set $g_2 = 0$ to drop the limit to the maximal delay at a node. By the corollary the Simple Robust Timetabling problem over S^1 reads as follows. Let χ_a be the indicator function of a, i.e., $\chi_a(x) = 1$ if $a = x$, and zero else.

$$\min_{\pi,f} \sum_{e=(i,j)\in E} w(e)(\pi_j - \pi_i)$$

$$\text{s.t.} \quad \pi_j - \pi_i + f_e = t(e), \quad \forall e = (i,j) \in E \tag{31}$$

$$\forall s \in E :$$

$$f_e + y_j^s - y_i^s \geq \Delta \cdot \chi_e(s), \forall e = (i,j) \in E$$
$$D - d'y^s \geq 0$$
$$f, y^s \geq 0$$

Periodic Timetabling. Many service providers operate periodic schedules. This means that—during some period of the day—equivalent events, e.g., all departures of the trains of a certain line at a certain station take place in a periodic or almost periodic manner. For example, at a subway station each departure will take place exactly, e.g., 10 minutes after the departure of the previous train. Likewise, most timetables for long-distances connections are constructed such that if a train leaves from the central station of X to central station of Y at 12:43h, then the next train to Y will leave the central station of X at (roughly) 13:43h, the next at 14:34h, and so on. This means that the long-distance connection from X to Y is operated with a *period* of one hour.

In case of periodic timetables, we do not plan the single events as in the aperiodic case, but we plan periodic events. For these we schedule a periodic time, which is understood modulo the period of the system. (There may also be differnt periods in the same system, but we restrict our consideration here to the case of a single, global period.) Assume we assign the value 5 to the variable corresponding to the periodic event that trains of line A depart from station S towards station S'. Let the period T of the system be one hour. Then—in every hour—five minutes past the hour a train of line A will depart from station S towards station S'. This leads to the Periodic Event Scheduling Problem (PESP)[2], which can be formulated as a mixed integer program of the following form. Let $G(A, V)$ be a directed graph and three functions $w, u, l : A \rightarrow \mathbb{R}$ on the arc set. Then the following problem is called a PESP.

$$\min_{k\in\mathbb{Z}^{|A|},\pi} \sum_{e=(i,j)\in A} w(e)(\pi_j - \pi_i + k_e T)$$

$$\text{s.t.} \quad u(e) \geq \pi_j - \pi_i + k_e T \geq l(e), \quad \forall e = (i,j) \in A$$

This type of problem has a broad modeling power. For a comprehensive study on periodic timetabling we refer the interested reader to [10].

To construct an RROP from an original problem, which is a PESP we have to make a choice, whether we interpret the disturbances as periodic disturbances,

---

[2] The Periodic Event Scheduling Problem was introduced in [14]. For details confer also [10].

like a construction site, that will slow down the traffic at a certain point for the whole day, or as aperiodic events, like a jammed door at a stopping event. For periodic disturbances we get the following program.

$$\min_{k\in\mathbb{Z}^{|A|},\pi,f} \sum_{e=(i,j)\in A} w(e)(\pi_j - \pi_i + k_e T)$$

$$\text{s.t.} \quad \pi_j - \pi_i + f_e + k_e T = l(e), \qquad \forall e = (i,j) \in A$$
$$\pi_j - \pi_i + \bar{f}_e + k_e T = u(e), \qquad \forall e = (i,j) \in A$$

$$\forall s \in A, \Xi \in \{0,1\}:$$

$$f_e + y_j^s - y_i^s \geq \Delta \cdot \chi_e(i) \cdot \Xi, \qquad \forall e = (i,j) \in A$$
$$\bar{f}_e + y_i^s - y_j^s \geq \Delta \cdot \chi_e(i) \cdot (1 - \Xi), \forall e = (i,j) \in A$$
$$D - d'y^s \geq 0$$
$$y^s \geq 0$$

Note that the right-hand sides are still constants, though they look like a quadratic term.

Again, the deterministic PESP instance can be flanged with a polynomial size linear program to ensure robustness. This structure can be helpful for solving such a problem, as the specialized solving techniques for the original integer program can be integrated.

As an example for this approach confer [3], where a specialized technique for an advanced platforming problem was combined with robust network buffering to get a recovery robust platforming. The method was tested on real-world data of Italian railway stations. The propagated delay through the stations was reduced by high double-digit percentages without loss in the primal objective, which is to maximizes the number of trains the station handles.

General Network Buffering. The general situation is the following: We are given an optimization problem on a network. The solution to that problem will be operated under disturbances. The disturbances propagate through the network in a way depending on the solution of the optimization problem. The solution of the original optimization problem x translates into a buffer vector f on the arcs of the network. Changing perspective, the original problem with its variables x is a cost oracle: If we fix a certain buffering f, the optimization will construct the cheapest x vector to ensure the buffering f. Let us summarize the general scheme.

Given an optimization problem P with the following features:

- A directed graph G.
- An unknown, limited, nonnegative vector of disturbances on the arcs, or on the nodes, or both.
- The disturbances cause costs on the arcs, or on the nodes, or both, which propagate through the network.
- A vector of absorbing potential on the arcs, the nodes, or both can be attributed to each solution of P.

If we further restrict the disturbance vector to lie in S^1, we get the following by the above considerations: The recovery robust version of P, in which the propagated cost must be kept below a fixed budget D, can be formulated as the original problem P plus a linear program quadratic in the size of G.

5   Platforming: A Real-World Study

In the previous section we have shown that a method to solve a linear, or convex, or linear integer optimization problem can be extended to a method for the recovery robust version of the problem with right-hand side disturbances in an efficient and simple way, provided that the recovery can be described by a linear program. Those conditions are fulfilled in particular for network buffering problems. In this case the recovery is the propagation of a disturbance along a network. The propagation in the network depends on the solution of the original optimization problem. But this original optimization problem itself need *not* be a linear program.

The advantage of this method for robustness is threefold:

- It yields an efficient algorithm, respectively it does not add to the complexity of the original problem.
- The method provides for solutions which possess a precisely defined level of robustness (in contrast to heuristics).
- The method is easy to implement. One can reuse any existing approach for the original problem and supplement it with a linear program for recoverable robustness.

To exemplify these advantages we describe a study on real-world data for the train platforming problem (cf. [3]).

The train platforming problem considers a single station and a given set of trains together with their planned departure and arrival times at the station area. The goal is a conflict-free assignment of a *pattern* to as many of these trains as possible. A pattern consists of a track in the station together with an arrival path to this track and a departure path from the track. The assignment is conflict free, if no track has a time interval during which two trains are assigned to that track, and no pair of simultaneously used paths are in spacial conflict. In the current study the spacial conflicts of paths are given explicitly in a conflict graph.

It is straight forward to formulate this problem as an integer linear program with $(0, 1)$-decision variables for each pair of a pattern and a train. Of course, there are several possibilities to phrase this problem as an integer linear program. In fact, the version used in the study is not trivial, but constructed carefully to achieve a powerful model that allows to solve large-scale instances. (For details we refer the reader to [3].) Independent of this particular study one might use a different integer programming formulation, e.g., in case the path conflicts are not given as a conflict graph, but implicitly by a digraph representing the infrastructure network. But the particular program is not relevant to the general approach on which we focus here.

The original program is reused without changes in the recovery robust program. To effect robustness we add a linear program modeling the delay propagation and a set of constraints that link the variables of the delay propagation to those of the original program. The resulting program has three sections:

1. The original train platforming program to optimize the assignment of patterns to trains (planning sub-model).
2. The linear program to model the delay propagation network (recovery sub-model).
3. The constraints linking the nominal solution to the buffer values on the delay propagation network (linking constraints).

In the delay propagation network each train has three vertices corresponding to the three events in which it will free up each of the three resources assigned to it (arrival path, platform, and departure path). Naturally, two vertices are connected by an arc whenever delay at the train-resource pair corresponding to the head-node may propagate onto the tail-node for a specic nominal solution. A delay in freeing up the platform for a train may propagate to a delay in freeing up the same platform for other trains. Something similar applies to arrival/departure paths, more precisely for paths that are in conflict. Every arc in the delay propagation network has an associated buffer value, which represents the maximum amount of delay that it is able to absorb without any propagation effect. Intuitively, a buffer corresponds to the slack among a given pair of resource occupation time intervals.

The objective function of the original problem contains three parts that are weighted (in the order given below) such that the optimal solution will also be lexicographically optimal.

1. The total number of trains that can be assigned.
2. Certain trains have a preferred set of tracks to one of which they should be assigned if possible.
3. A heuristic for robustness punishing any use of pairs of paths that are in spacial conflict during time windows that are not overlapping (i.e. the assignment is conflict-free) but close to each other.

For the recovery robust version we drop the third, heuristic objective and replace it by the exact objective to minimize the maximum delay that can occur. We use the scenario set S^1 to get a compact model for the robust platforming. It turns out that the second objective plays a role for none of the two methods in any of the considered instances, i.e., the trains that are assigned to tracks can always be assigned to their preferred tracks. Moreover, the real-world instances are such that it is not possible to assign tracks to all given trains, neither in the standard nor in the recovery robust model.

Note that by the weighting of the objective function this implies, that a conflict-free assignment of all trains is physically impossible. But, in all considered instances the recovery robust method assigns as many trains as the original method, i.e., as much as possible in general. Thus the two methods yield assignments that are equivalent in all given deterministic criteria. But they differ

Table 1. Results for Palermo Centrale

time window	# trains not platformed	D nom	CPU time nom (sec)	D RR	CPU time RR (sec)	Diff. D	Diff. D in %
A: 00:00-07:30	0	646	7	479	46	167	25.85
B: 07:30-09:00	2	729	7	579	3826	150	20.58
C: 09:00-11:00	0	487	6	356	143	131	26.90
D: 11:00-13:30	2	591	6	384	228	207	35.03
E: 13:30-15:30	1	710	9	516	2217	194	27.32
F: 15:30-18:00	1	560	7	480	18	80	14.29
G: 18:00-00:00	3	465	11	378	64	87	18.71

significantly in delay propagation. In all instances the recovery robust method yields assignments with a double-digit percentage of delay reduction. In one case the reduction is almost 50%. Averaged over all instances the reduction is roughly 1/4.

Table 1 gives the details of the study for the station Palermo Centrale. The study considers seven time windows during the day at the station Palermo Centrale. These are given in the first column of Table 1. Further, the short cut *nom* denotes values referring to the original method, whereas *RR* stands for the results of the recovery robust approach. For both the table states the CPU time required to find the solution and the maximal propagated delay. Further, we give the difference in propagated delay as absolute value (in minutes) and as percentage of delay propagation in the original method's solution. The number of non-assigned trains is given without reference to the method, because both methods achieve the same value here.

6   Conclusion

We have introduced *recoverable robustness* as an alternative concept for optimization under imperfect information. It is motivated by practical problems like delay resistant timetabling, for which classical concepts like stochastic programming and robust optimization prove inappropriate. We describe the model in full generality and demonstrate how different types of delay resistant timetabling problems can be modeled in terms of recoverable robustness. Further, we specialized the general concept of recoverable robustness to *linear recovery robust programs*. For these we provide an efficient algorithm in case of right-hand side disturbances. By means of this general method delay resistant timetabling problems can be solved efficiently. This is exemplified by a real world application of our method in a study [3] on recovery robust platforming. The platformings constructed with our method achieve maximal possible throughput at the stations, but drastically reduces the delay propagation in comparison to a state-of-the-art method.

References

1. Special issue on robust optimization. Mathematical Programming A 107(1-2) (2006)
2. Berger, A., Lorenz, U., Hoffmann, R., Stiller, S.: TOPSU-RDM: A simulation platform for railway delay management. In: Proceedings of SimuTools (2008)
3. Caprara, A., Galli, L., Stiller, S., Toth, P.: Recoverable-robust platforming by network buffering. Technical Report ARRIVAL-TR-0157, ARRIVAL Project (2008)
4. Cicerone, S., D'Angelo, G., Di Stefano, G., Frigioni, D., Navarra, A.: On the interaction between robust timetable planning and delay management. Technical Report ARRIVAL-TR-0116, ARRIVAL project (2007); Published at COCOA 2008
5. Cicerone, S., D'Angelo, G., Di Stefano, G., Frigioni, D., Navarra, A.: Robust algorithms and price of robustness in shunting problems. In: Proceedings of the 7th Workshop on Algorithmic Approaches for Transportation Modeling, Optimization, and Systems, ATOMS (2007)
6. Cicerone, S., D'Angelo, G., Di Stefano, G., Frigioni, D., Navarra, A.: Recoverable robust timetabling: Complexity results and algorithms. Technical report, ARRIVAL Project (2008)
7. Cicerone, S., D'Angelo, G., Di Stefano, G., Frigioni, D., Navarra, A., Schachtebeck, M., Schöbel, A.: Recoverable robustness in shunting and timetabling. In: Robust and Online Large-Scale Optimization, pp. 29–61. Springer, Heidelberg (2009)
8. Cicerone, S., Di Stefano, G., Schachtebeck, M., Schöbel, A.: Dynamic algorithms for recoverable robustness problems. In: Proceedings of the 8th Workshop on Algorithmic Approaches for Transportation Modeling, Optimization, and Systems (ATMOS 2008), Schloss Dagstuhl Seminar Proceedings (2008)
9. Cicerone, S., Di Stefano, G., Schachtebeck, M., Schöbel, A.: Multi-stage recoverable robustness problems. Technical report, ARRIVAL Project (2009)
10. Liebchen, C.: Periodic Timetable Optimization in Public Transport. dissertation.de – Verlag im Internet, Berlin (2006)
11. Liebchen, C., Lübbecke, M., Möhring, R.H., Stiller, S.: Recoverable robustness. Technical Report ARRIVAL-TR-0066, ARRIVAL-Project (2007)
12. Liebchen, C., Stiller, S.: Delay resistant timetabling. Preprint 024-2006, TU Berlin, Institut für Mathematik (2006)
13. Tyrrell Rockafellar, R.: Network ows and monotropic optimization. John Wiley & Sons, Inc., Chichester (1984)
14. Serafini, P., Ukovich, W.: A mathematical model for periodic scheduling problems. SIAM Journal on Discrete Mathematics 2(4), 550–581 (1989)
15. Soyster, A.L.: Convex programming with set-inclusive constraints and applications to inexact linear programming. Operations Research 21(4), 1154–1157 (1973)
16. Stiller, S.: Extending Concepts of Reliability. Network Creation Games, Real-time Scheduling, and Robust Optimization. TU Berlin, Dissertation, Berlin (2009)
17. Vromans, M., Dekker, R., Kroon, L.: Reliability and heterogeneity of railway services. European Journal of Operational Research 172, 647–665 (2006)

Recoverable Robustness in Shunting and Timetabling*

Serafino Cicerone[1], Gianlorenzo D'Angelo[1],
Gabriele Di Stefano[1], Daniele Frigioni[1], Alfredo Navarra[2],
Michael Schachtebeck[3], and Anita Schöbel[3]

[1] Dept of Electrical and Information Engineering, University of L'Aquila, Italy
{serafino.cicerone,gianlorenzo.dangelo,gabriele.distefano,
daniele.frigioni}@univaq.it
[2] Department of Mathematics and Computer Science, University of Perugia, Italy
navarra@dmi.unipg.it
[3] Institute for Numerical and Applied Mathematics, Georg-August-University
Göttingen, Germany
{schachte,schoebel}@math.uni-goettingen.de

Abstract. In practical optimization problems, disturbances to a given instance are unavoidable due to unpredictable events which can occur when the system is running. In order to face these situations, many approaches have been proposed during the last years in the area of robust optimization. The basic idea of *robustness* is to provide a solution which is able to keep feasibility even if the input instance is disturbed, at the cost of optimality. However, the notion of robustness in every day life is much broader than that pursued in the area of robust optimization so far. In fact, robustness is not always suitable unless some *recovery strategies* are introduced. Recovery strategies are some capabilities that can be used when disturbing events occur, in order to keep the feasibility of the pre-computed solution. This suggests to study robustness and recoverability in a unified framework. Recently, a first tentative of unifying the notions of robustness and recoverability into a new integrated notion of *recoverable robustness* has been done in the context of railway optimization.

In this paper, we review the recent algorithmic results achieved within the recoverable robustness model in order to evaluate the effectiveness of this model. To this aim, we concentrate our attention on two problems arising in the area of railway optimization: the *shunting* problem and the *timetabling* problem. The former problem regards the reordering of freight train cars over hump yards while the latter one consists in finding passenger train timetables in order to minimize the overall passengers traveling time. We also report on a generalization of recoverable robustness called *multi-stage recoverable robustness* which aims to extend recoverable robustness when multiple recovery phases are required.

---

* Work partially supported by the Future and Emerging Technologies Unit of EC (IST priority - 6th FP), under contract no. FP6-021235-2 (project ARRIVAL).

R.K. Ahuja et al. (Eds.): Robust and Online Large-Scale Optimization, LNCS 5868, pp. 28–60, 2009.

1   Introduction

Many real world applications are characterized by a *strategic planning* phase and an *operational* phase. The main difference between the two phases resides in the time in which they are applied. The strategic planning phase aims to plan how to optimize the use of the available resources according to some objective function before the system starts to operate. The operational phase aims to have immediate reaction to disturbing events that can occur when the system is running. In general, the objectives of strategic planning and operational phase might be in conflict with each other.

In these scenarios, it is preferable to define a strategic plan which is able to keep feasibility even if the input is disturbed instead of a plan which optimizes the available resources for the undisturbed input. It follows that disturbances have to be considered both in the *strategic planning* phase and in the *operational* phase.

To face disturbances in the operational phase, the approaches used in the literature are mainly based on the concept of *online algorithms* [5]. An online recovery strategy has to be developed when unpredictable disturbances in planned operations occur and before the entire sequence of disturbances is known. The goal is to react fast while retaining as much as possible of the quality of an optimal solution, that is, a solution that would have been achieved if the entire sequence of disturbances was known in advance.

To face disturbances in the strategic planning phase, the approaches used in the literature are mainly based on *stochastic programming* and *robust optimization*. Within stochastic programming (e.g., see [4, 27, 32]), there are two different approaches: *chance constrained programming* aims to find a solution that satisfies the constraints with high probability, while in *multi-stage stochastic programming*, an initial solution is computed in the first stage, and each time new random data is revealed, a recourse action is taken. However, stochastic programming requires detailed knowledge on the probability distributions of the disturbances which could be not available.

In robust optimization (e.g., see [1, 2, 3, 16]), the objective is purely deterministic. It aims to find a solution to an optimization problem which keeps feasibility when some disturbing events occur. For example, the notion of *strict robustness* introduced in [3] requires that a solution to an optimization problem has to be feasible for all admissible scenarios of a given set. The solution gained by this approach is fixed in the strategic planning phase and it does not need to be changed when disturbances occur. However, as the solution is fixed independently of the actual scenario, robust optimization leads to solutions that are too conservative and thus too expensive in many applications. One approach to compensate this disadvantage is the *light robustness* introduced in [17, 18]. This approach adds slacks to the constraints. A solution is considered robust if it satisfies these relaxed constraints.

Despite the increasing interest, a final answer to the question "what is *robustness* for an optimization problem?" has not yet been given. In fact, the notion of robustness in every day life is much broader than that pursued in the area of robust optimization so far. The basic idea of robustness is given by a problem and some knowledge imperfection which one has to cope with. That is, the solution provided for a given instance of the problem must hold even though some changes in such an instance occur. This kind of robustness is not always suitable unless some recovery strategies are introduced. Moreover, in many practical applications, there might be the possibility to intervene before some scheduled operations are being performed.

Usually, modifications that may occur are restricted to some specified subset of all possible ones. It is reasonable to require that, if a disturbance occurs, one would like to maintain as much as possible a pre-computed solution taking into account some "soft" recovery strategies. Recovering should be simple and fast. Moreover, there are cases where recoverability is necessary in order to still have some useful solution for a problem. A solution that undergoes slight changes is called robust even though it could require the use of some recovery capabilities. This suggests to study robustness and recoverability in a unified way.

A first tentative of unifying the notions of *robustness* and *recoverability* into a new integrated notion of *recoverable robustness* has been done in [29, 30] in the context of railway optimization. This new notion describes robustness with respect to (limited) recovery capabilities. It integrates robustness and recoverability as the solutions are required to be recoverable. The basic idea of recoverable robustness is to compute solutions that are robust against a limited set of scenarios and for a limited recovery. The quality of the robust solution is measured by its *price of robustness* that determines the trade-off between an optimal and a robust solution. Given an instance i of a problem, the price of robustness of i is the ratio between the cost of an optimal robust solution and the cost of the optimal (non robust) solution. The price of robustness of a recoverable robustness problem is then given by a worst case analysis, i.e., it is the maximum price of robustness among all the instances of the problem. Hence the price of robustness of a problem provides an upper bound to the loss that one has to pay in order to introduce recoverable robustness in an optimization problem by fixing some disturbances and recovery capabilities. In [29, 30], the aim is to provide the best robust solution, i.e., the one that minimizes the price of robustness.

In [6], algorithmic aspects of recoverable robustness have been highlighted by giving the definition of *robust algorithm* and of the corresponding *price of robustness*. A robust algorithm is an algorithm which provides a robust solution for each instance of a problem. The price of robustness of a robust algorithm A_{rob} is given by the worst case ratio, among all the possible instances of the problem, between the cost of the solution computed by A_{rob} for an instance i and the optimal (non robust) solution for i. The price of robustness of a recoverable robustness problem defined in [6] is then given by the price of the best possible robust algorithm which solves the given problem. Hence, the price of robustness of a problem here provides the loss that cannot be avoided by a robust algorithm, fixed some disturbances and recovery capabilities. If the price of robustness of an algorithm matches this minimal loss, then the algorithm is called *optimal*. If it is equal to 1, then no price has to be paid in order to achieve robustness and hence such an algorithm is called *exact*.

Notice that, given a recoverable robust problem $\mathcal{P}$, if there exists a robust algorithm that is able to find an optimal robust solution for any instance of $\mathcal{P}$, then the price of robustness of $\mathcal{P}$ defined in [29, 30] and that defined in [6] are equivalent. However, the model given in [6] is more suitable for analyzing robust algorithms than that in [29, 30] as the former concentrates on finding robust algorithms and comparing them by using their prices of robustness, while the latter concentrates on finding optimal robust solutions and recovery algorithms.

In this paper, we intend to review the algorithmic results achieved within the recoverable robustness model given in [6] in order to evaluate the effectiveness of this model in both practical and theoretical frameworks. To this aim, we focus our attention on two problems arising in the area of railway optimization: the *shunting* problem and the *timetabling* problem. The former problem regards the reordering of freight train cars over hump yards while the latter one consists in finding passenger train timetables in order to minimize the overall passengers traveling time. We also report on a generalization of the recoverable robustness model called *multi-stage recoverable robustness* model proposed in [11]. It aims to extend recoverable robustness in the case of multiple disturbances which can arise in many practical optimization problems and require a sequence of recovery phases.

This paper is organized as follows: in the next section we report the recoverable robustness model as given in [6]. In Section 3 we survey results on the shunting problem obtained in [6, 8]. In Section 4 we survey results on the timetabling problem obtained in [7, 9, 10, 11, 12]. In Section 5 we report the multi-stage recoverable robustness model given in [11] and provide an example on how to apply this model in the context of timetabling. Finally, in Section 6, we analyze the given models and propose some possible extensions, open problems and future research directions.

2   Recoverable Robustness Model

In this section, we report the recoverable robustness model given in [6] which is based on that given in [29, 30].

The recoverable robustness model aims to introduce robustness in an optimization problem. In the remainder, an optimization problem P is characterized by the following parameters.

- I, the set of instances of P;
- F, a function that associates to any instance $i \in I$ the set of all feasible solutions for i;
- $f \colon S \to \mathbb{R}^{\geq 0}$, the objective function of P, where $S = \bigcup_{i \in I} F(i)$ is the set of all feasible solutions for P.

Note that, for several optimization problems, the objective function is defined to have values in $\mathbb{R}$. However, it is possible to turn any such problem into an equivalent one having values in $\mathbb{R}^{\geq 0}$. Without loss of generality, from now on, minimization problems are considered. Additional concepts to introduce robustness requirements for a minimization problem P are needed:

- $M : I \to 2^I$ – a *modification* function for instances of P. This function models the following case. Let $i \in I$ be the considered input to the problem P, and let $s \in S$ be the planned solution for i. A *disturbance* is meant as a modification to the input i, and such a modification can be seen as a new input $j \in I$. Typically, the modification j depends on the current input i, and this fact is modeled by the constraint $j \in M(i)$. Hence, given $i \in I$, $M(i)$ represents the set of instances of P that can be obtained by applying all possible modifications to i. Of course, when a disturbance $j \in M(i)$ occurs, a new solution $s' \in F(j)$ has to be recomputed for P.
- $\mathbb{A}_{rec}$ – a class of *recovery algorithms* for P. Algorithms in $\mathbb{A}_{rec}$ represent the capability of recovering against disturbances. Since in a real-world problem the capability of recovering is limited in some way, the class $\mathbb{A}_{rec}$ can be defined in terms of some kind of *restrictions*, such as feasibility or algorithmic restrictions. An element $A_{rec} \in \mathbb{A}_{rec}$ works as follows: given a solution s for P and a modification $j \in M(i)$ of the current instance i, then $A_{rec}(s,j) = s'$ where $s' \in S$ represents the recovered solution for P.

  In what follows, some examples of recovery algorithm classes, used in the remainder of this paper, are given.

Class 1: **Strict robustness.** It models the case in which there are no recovery capabilities, that is, each algorithm $A_{rec} \in \mathbb{A}_{rec}$ fulfills the following constraint:

$$\forall i \in I, \forall s \in S, \forall j \in M(i), \ A_{rec}(s,j) = s. \tag{1}$$

Class 2: **Bounded distance from the original solution.** $\mathbb{A}_{rec}$ is defined by imposing a constraint on the solutions provided by the recovery algorithms. In particular, the new (recovered) solutions computed by a recovery algorithm must not deviate too much from the original solution s, according to a distance measure d. Formally, given a real number $\Delta \in \mathbb{R}$ and a distance function $d : S \times S \to \mathbb{R}$, each element A_{rec} in such a class fulfills the following constraint:

$$\forall i \in I, \forall s \in S, \forall j \in M(i), \ d(s, A_{rec}(s,j)) \leq \Delta. \tag{2}$$

Note that *Class 1* is contained in *Class 2*. In fact, for any distance function d, if $\Delta = 0$, then constraints (1) and (2) are equivalent.

Class 3: **Bounded computational power.** $\mathbb{A}_{rec}$ is defined by bounding the computational power of recovery algorithms. Formally, given a function $t : S \times I \to \mathbb{N}$, each element A_{rec} in such a class fulfills the following constraint:

$$\forall i \in I, \forall s \in S, \forall j \in M(i),$$

$A_{rec}(s,j)$ must be computed in $O(t(s,j))$ time.

Given an optimization problem P, it can be turned into a recoverable robustness problem $\mathcal{P}$ as described below.

Definition 1. *A* recoverable robustness problem $\mathcal{P}$ *is defined by the triple* $(P, M, \mathbb{A}_{rec})$. *All the recoverable robustness problems form the class* RRP.

Definition 2. *Let* $\mathcal{P} = (P, M, \mathbb{A}_{rec}) \in$ RRP. *Given an instance* $i \in I$ *of* P, *an element* $s \in F(i)$ *is a* feasible solution *for* i *with respect to* $\mathcal{P}$ *if and only if the following relationship holds:*

$$\exists A_{rec} \in \mathbb{A}_{rec} : \forall j \in M(i), \ A_{rec}(s, j) \in F(j).$$

In other words, $s \in F(i)$ is feasible for i with respect to $\mathcal{P}$ if it can be *recovered* by applying some algorithm $A_{rec} \in \mathbb{A}_{rec}$ for each possible disturbance $j \in M(i)$. The set of all the feasible solutions for i with respect to $\mathcal{P}$ is denoted by $F_{\mathcal{P}}(i)$. Formally:

$$F_{\mathcal{P}}(i) = \{s \in F(i) : s \text{ is a feasible solution for } i \text{ with respect to } \mathcal{P}\}.$$

In the remainder, solutions in $F_{\mathcal{P}}(i)$ are also called *robust solutions* for i with respect to the original problem P.

It is worth to mention that, if $\mathbb{A}_{rec}$ is *Class 1*, i.e., it is the class of algorithms that do not change the solution s, then the robustness problem $\mathcal{P} = (P, M, \mathbb{A}_{rec})$ represents the so-called *strict robustness problem* [3]. Note that in this case, given an instance i and a robust solution s for i, then for each possible modification $j \in M(i)$, $s \in F(j)$. This means that, since A_{rec} has no capability of recovering against possible disturbances, a robust solution has to "absorb" *any* possible disturbance.

Definition 3. *Let* $\mathcal{P} = (P, M, \mathbb{A}_{rec}) \in$ RRP. *A* robust algorithm *for* $\mathcal{P}$ *is any algorithm* A_{rob} *such that, for each* $i \in I$, $A_{rob}(i)$ *is a robust solution for* i *with respect to* $\mathcal{P}$.

A possible scenario for this situation is depicted in Figure 1. Note that, if $\bar{s}$ denotes the optimal solution for P when the input instance is i, it is possible that $\bar{s}$ is not in $F_{\mathcal{P}}(i)$; this implies that every robust solution for i may be "very far" from the optimal solution $\bar{s}$. A "good" robust algorithm should find the best solution in $F_{\mathcal{P}}(i)$ for P, for each possible input $i \in I$. The quality of a robust algorithm is measured by the so-called price of robustness. The following definitions report the concepts of the price of robustness of both a robust algorithm and a recoverable robustness problem.

Definition 4. *Let* $\mathcal{P} \in$ RRP. *The* price of robustness *of a robust algorithm* A_{rob} *for* $\mathcal{P}$ *is*

$$P_{rob}(\mathcal{P}, A_{rob}) = \max_{i \in I} \left\{ \frac{f(A_{rob}(i))}{\min\{f(x) : x \in F(i)\}} \right\}.$$

Definition 5. *Let* $\mathcal{P} \in$ RRP. *The* price of robustness *of* $\mathcal{P}$ *is*

$$P_{rob}(\mathcal{P}) = \min\{P_{rob}(\mathcal{P}, A_{rob}) : A_{rob} \text{ is a robust algorithm for } \mathcal{P}\}.$$

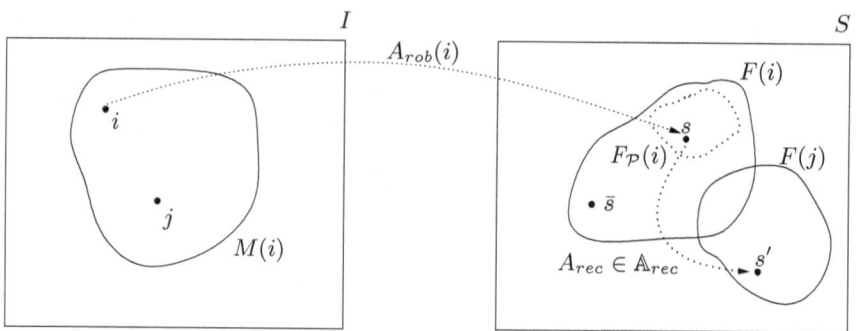

Fig. 1. A scenario for recoverable robustness problem: I, set of instances; S, set of solutions; $M(i)$, set of instances obtainable after a small modification; $F(i)$ and $F(j)$, set of feasible solutions for i and j respectively; $F_{\mathcal{P}}(i)$, set of recoverable solutions for i; $\bar{s}$, optimal non-robust solution for i; s, robust solution obtained by A_{rob}; s', recovered solution obtained by an algorithm $A_{rec} \in \mathbb{A}_{rec}$ after disturbance $j \in M(i)$

Definition 6. *Let* $\mathcal{P} \in \mathrm{RRP}$ *and let* A_{rob} *be a robust algorithm for* $\mathcal{P}$. *Then,*

- A_{rob} *is* $\mathcal{P}$-*optimal if* $P_{rob}(\mathcal{P}, A_{rob}) = P_{rob}(\mathcal{P})$;
- A_{rob} *is exact if* $P_{rob}(\mathcal{P}, A_{rob}) = 1$.

A solution provided by an optimal (exact) robust algorithm is called an *optimal* (*exact*) solution. Notice that an exact algorithm is $\mathcal{P}$-optimal.

The price of robustness of an algorithm A_{rob} represents the quality of the solutions it provides. In particular, it measures the relative worst case loss induced by the value of the solutions provided by A_{rob} compared to the value of the optimal (non robust) ones. The price of robustness of the best robust algorithm defines the price of robustness of the problem. This value represents the minimal loss due to the introduction of robustness given by some disturbances and by some recovery capabilities.

3   Recoverable Robust Shunting

This section is devoted to survey on recent results concerning robustness in shunting problems [6, 8]. First the shunting over a hump yard model provided in [23, 24, 25, 26] is described, and then, results obtained in this area in terms of recoverable robustness are reported.

3.1   Shunting over a Hump Yard

The problem is specified by an input train T_{in} composed of n cars and an output train T_{out} given by a permutation of T_{in} cars. Each car is assigned with a unique label. The considered hump yard appears as in Figure 2. The hump yard is made of an input track where trains arrive, and of a set of switches by which

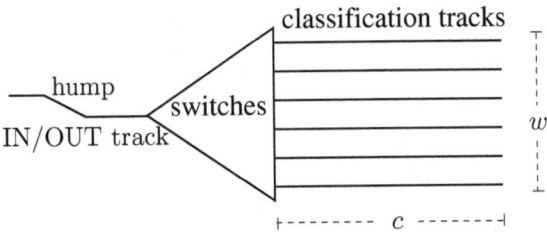

Fig. 2. Hump yard infrastructure composed of w classification tracks, each of size c

cars composing the incoming train can be shunted over the available classification tracks. A classification track is approached from a single side and works like a stack. The set of classification tracks is denoted by W, the size of W is denoted by w, and the size of each track, i.e., the number of cars that can fit into a classification track, by c. Therefore, an instance of the problem is given by a quadruple (T_{in}, T_{out}, W, c).

The hump yard supports a sorting operation by repeatedly doing the so called *track pull* operation which is made up of the following steps:

- connect the cars of one classification track into a train, called *pseudotrain*;
- pull the pseudotrain over the hump;
- disconnect the cars in the pseudotrain;
- push the pseudotrain slowly over the hump, yielding single cars that run down the hill from the hump towards the classification tracks;
- control the switches such that every single car goes to a specified track.

The goal is to reorder T_{in} according to T_{out} by repeatedly performing the track pull operation (an example of reordering by means of track pulls can be seen in Figure 3). The cost of the reordering is measured by the number of track pulls. Notice that at least one track pull must be performed as the hump yard is used only when one has to reorder or to park a train.

As in [24], three different variants of the shunting over a hump yard problem are considered by specifying constraints for parameters c and w. Namely,

Sh_1: c bounded, w unbounded;
Sh_2: c unbounded, w bounded;
Sh_3: c and w unbounded.

When convenient, Sh is used to refer to any of the above problems.

In [24] polynomial time algorithms for each of the above problems is given. In particular, a 2-approximation algorithm for Sh_1 and optimal algorithms for Sh_2 and Sh_3 are provided.

In what follows, the notation used in [6, 8, 24, 25, 26] to represent a *shunting plan* is described. A shunting plan specifies (i) a sequence S of h track pull operations given by the tracks whose cars are pulled, and (ii) for every pulled car which track it is sent to. Note that, if one track is pulled several times, then it appears in S more than once. Of course, if there is no limit on the number of tracks ($w \geq h$), then there is no need to reuse a track. Given S, the itinerary of a car can be described by the sequence of tracks it visits. For the task at hand, it is convenient to specify this sequence as a bit-string or code $b_1 \cdots b_h$ where the different bits stand for the pulled tracks and there is a 1 if and only if the car visits that track. Now, if track i is pulled, then the new destination of a car is given by the position of its next 1 in its code, i.e. the lowest index $j > i$ such that $b_j = 1$. A shunting plan must specify a track pull sequence S and it has to associate a code to each car. The length of each code is determined by the length of S and cars may share the same code.

An example is shown in Figure 3. The sequence of track pulls is given by $S = \{1, 2, 3, 4, 5\}$ from right to left among classification tracks. In the example $c = 3$ and the number of track pulls is set to 5. The set of codes of length 5 provided by a feasible solution satisfies the property that at each position at most three codes have the corresponding bit set to 1. This implements the constraint on c and implies that at most eleven different codes can be generated. Cars from 11 down to 1 are associated with codes 00000, 00001, 00010, 00011, 00100, 00110, 01000, 01100, 10000, 10001, and 11000, respectively. Figure 3 shows the sequence of configurations obtained after each track pull and reorder of the pulled cars according to their codes. The algorithm used in the example has been proposed in [24]. From now on, such an algorithm will be called A_{out}. It computes a shunting plan when c is bounded and the input train is unknown in advance. In particular, A_{out} provides n different codes, one for each car in T_{in}. Each code specifies the route that the corresponding car has to perform among the shunting yard in order to be placed in the desired position according to T_{out}. In [24] it has been shown that, when the order of cars in the incoming train is not known in advance, A_{out} is optimal with respect to the minimum number of track pulls. For the sake of simplicity, it is assumed that T_{out} is composed on a track not used for shunting operations but that can contain the full train.

Note that, when T_{in} is known in advance, two cars might be assigned with the same code. This would imply that they will have the same relative order in T_{out} as in T_{in}. Two cars that are consecutive in T_{out} can get the same code if they are in the correct order in T_{in}. A maximal set of cars in T_{out} that has this property is called a *chain*. In a shunting plan, for each code x, a *pure chain* is the set of all cars associated with x.

In practice, the number of chains in T_{in} along with the hump yard structure represent the key quantity with respect to the number of track pulls that must be performed by a shunting plan in order to obtain the desired T_{out}. The following further notation is used: $opt(k, c, w) \geq 1$ is the number of track pulls needed by an optimal shunting plan in order to manage k cars/chains over a hump yard made of w tracks, each of size c (for Sh_1 and Sh_3, $w = \infty$, while for Sh_2 and Sh_3, $c = \infty$); $apx(k, c, w)$ is the best known approximation algorithm for the corresponding shunting problem, and $apxr$ is its approximation ratio (when it is clear by the context, parameters equal to ∞ are removed from the previous notation); C denotes the set of codes assigned by an algorithm to the

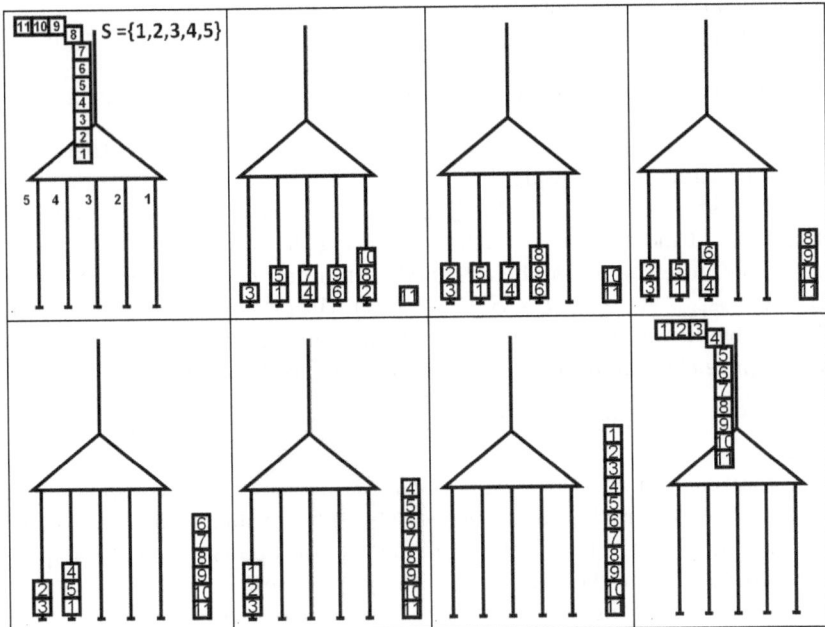

Fig. 3. Example of a shunting plan when $c = 3$ and the number of track pulls is set to 5. Cars from 11 down to 1 are associated with codes 00000, 00001, 00010, 00011, 00100, 00110, 01000, 01100, 10000, 10001, 11000 respectively. T_{out} is composed outside the hump yard and the corresponding track is not shown.

cars. Furthermore, for every instance $i = (T_{in}, T_{out}, W, c)$, r_i and n_i denote the number of chains and cars in T_{in}, respectively.

3.2   Robust Algorithms

This section presents a surveys of the results obtained in [6, 8] on robust algorithms for the shunting problems described above. For example, Sh_1 is defined by

- I: set of quadruples (T_{in}, T_{out}, W, c) where train T_{in} is defined as a sequence of cars and train T_{out} is a permutation of T_{in};
- $F(i)$: set of all feasible solutions for a given instance $i \equiv (T_{in}, T_{out}, W, c) \in I$, i.e., any sequence of track pulls combined with a set of codes (one per car) that transforms T_{in} in T_{out} when c is bounded;
- f: number of track pulls.

Regarding the modification function M, four different possibilities are considered:

M_1: one car can arrive in an unexpected incoming position;
M_2: the incoming train contains one additional unexpected car;

M_3: the incoming train contains one car less than expected;

M_4: one of the classification tracks composing the hump yard may fault.

Concerning recovery algorithms, the following three classes are considered:

$\mathbb{A}_{rec}^1$: $\forall A \in \mathbb{A}_{rec}^1$, $\forall (i, s) \in I \times S$, $\forall j \in M(i)$, $A(s, j) = s$, i.e., there are no recovery strategies to apply;

$\mathbb{A}_{rec}^2$: $\forall A \in \mathbb{A}_{rec}^2$, $\forall (i, s) \in I \times S$, $\forall j \in M(i)$, $A(s, j) = s'$ where s' may differ from s by at most one code without affecting the track pull sequence, i.e., at most one pure chain may be assigned with a new code of the same length;

$\mathbb{A}_{rec}^3$: $\forall A \in \mathbb{A}_{rec}^3$, $\forall (i, s) \in I \times S$, $\forall j \in M(i)$, $A(s, j) = s'$ where s' may differ from s by all the set of codes without affecting the track pull sequence, i.e., every pure chain may be assigned with a new code of the same length.

The class $\mathbb{A}_{rec}^1$ is equivalent to *Class 1*, while classes $\mathbb{A}_{rec}^2$ and $\mathbb{A}_{rec}^3$ belong to *Class 2*.

Note that each of the three defined classes of recovery algorithms does not affect the scheduled track pulls sequence defined by a robust shunting algorithm. This is motivated by the fact that modifying the track pulls sequence is expensive as it requires to change the switches setting or increase the number of track pulls. Recovery capabilities, instead, should be cheap operations since they cannot be planned a priori but are used during the operational phase. By definition, every upper bound to the price of robustness of each shunting problem with $\mathbb{A}_{rec}^1$ holds for the same problem with $\mathbb{A}_{rec}^2$ as well as every upper bound obtained with $\mathbb{A}_{rec}^2$ holds for $\mathbb{A}_{rec}^3$. Moreover, every lower bound obtained with $\mathbb{A}_{rec}^3$ holds for $\mathbb{A}_{rec}^2$ as well as every lower bound obtained with $\mathbb{A}_{rec}^2$ holds for $\mathbb{A}_{rec}^1$.

Given a shunting problem Sh, a modification function M and a class of recovery algorithms $\mathbb{A}_{rec}$, the corresponding recoverable robustness problem is denoted by $\mathcal{SH} = (Sh, M, \mathbb{A}_{rec})$. Tables 1, 2 and 3 summarize the obtained results for all the considered robustness problems arising from Sh_1, Sh_2 and Sh_3, respectively. In these tables, A_{rob} denotes the best robust algorithm given in [6, 8] for the specific problem at hand.

3.3   One Car in an Unexpected Incoming Position

In order to better understand the ideas behind recoverable robust algorithms, in this section, more details are provided concerning the case when the modification function M_1 is considered. Given an instance $i = (T_{in}, T_{out}, W, c)$ of a shunting optimization problem Sh, let $M_1(i)$ represent all possible instances (T_{in}', T_{out}, W, c) obtainable from i by changing the position of just one car in T_{in}. The following lemma describes which practical situation a robust plan must be able to absorb/recover with respect to a car incoming at an unexpected position.

Lemma 1 ([6, 8]). *Let v be a car arriving at the hump yard in a different position than expected. At most one additional pure chain must be managed with respect to the expected case.*

Table 1. Price of Robustness for Sh_1

Modifications		$\mathbb{A}^1_{rec}$	$\mathbb{A}^2_{rec}$	$\mathbb{A}^3_{rec}$
		Shunting Problem Sh_1		
M_1	$P_{rob}(\mathcal{SH})$	$\geq \max\limits_{i \in I} \frac{opt(n_i,c)}{opt(r_i,c)}$	≥ 2	≥ 2
	$P_{rob}(\mathcal{SH}, A_{rob})$	$\max\limits_{i \in I} \frac{opt(n_i,c)}{opt(r_i,c)}$	3	3
M_2	$P_{rob}(\mathcal{SH})$	indefinite	$\geq \max\limits_{i \in I} \frac{opt\left(\frac{n_i+1}{3},c\right)}{opt(r_i,c)}$	$\geq \max\limits_{i \in I} \frac{opt(r_i+1,c)}{opt(r_i,c)}$
	$P_{rob}(\mathcal{SH}, A_{rob})$	no solution	$\max\limits_{i \in I} \frac{opt(n_i+1,c-1)+1}{opt(r_i,c)}$	$\max\limits_{i \in I} \frac{apx(r_i+1,c)}{opt(r_i,c)}$
M_3	$P_{rob}(\mathcal{SH})$	1	1	1
	$P_{rob}(\mathcal{SH}, A_{rob})$	2	2	2
M_4	$P_{rob}(\mathcal{SH})$	indefinite	≥ 2	≥ 2
	$P_{rob}(\mathcal{SH}, A_{rob})$	no solution	$|C| + 1$	3

Table 2. Price of Robustness for Sh_2

Modifications		$\mathbb{A}^1_{rec}$	$\mathbb{A}^2_{rec}$	$\mathbb{A}^3_{rec}$
		Shunting Problem Sh_2		
M_1	$P_{rob}(\mathcal{SH})$	$\geq \max\limits_{i \in I} \frac{opt(n_i,w)}{opt(r_i,w)}$	≥ 2	≥ 2
	$P_{rob}(\mathcal{SH}, A_{rob})$	$\max\limits_{i \in I} \frac{opt(n_i,w)}{opt(r_i,w)}$	2	2
M_2	$P_{rob}(\mathcal{SH})$	indefinite	$\geq \max\limits_{i \in I} \frac{opt\left(\frac{n_i+1}{3},w\right)}{opt(r_i,w)}$	$\geq \max\limits_{i \in I} \frac{opt(r_i+1,w)}{opt(r_i,w)}$
	$P_{rob}(\mathcal{SH}, A_{rob})$	no solution	$\max\limits_{i \in I} \frac{opt(n_i+1,w)+1}{opt(r_i,w)}$	$\max\limits_{i \in I} \frac{opt(r_i+1,w)}{opt(r_i,w)}$
M_3	$P_{rob}(\mathcal{SH})$	1	1	1
	$P_{rob}(\mathcal{SH}, A_{rob})$	1	1	1
M_4	$P_{rob}(\mathcal{SH})$	indefinite	indefinite	≥ 2
	$P_{rob}(\mathcal{SH}, A_{rob})$	no solution	no solution	$\max\limits_{i \in I} \frac{opt(r_i,w-1)+1}{opt(r_i,w)}$

Table 3. Price of Robustness for Sh_3

Modifications		$\mathbb{A}^1_{rec}$	$\mathbb{A}^2_{rec}$	$\mathbb{A}^3_{rec}$
		Shunting Problem Sh_3		
M_1	$P_{rob}(\mathcal{SH})$	$\geq \max\limits_{i \in I} \frac{opt(n_i)}{opt(r_i)}$	≥ 2	≥ 2
	$P_{rob}(\mathcal{SH}, A_{rob})$	$\max\limits_{i \in I} \frac{opt(n_i)}{opt(r_i)}$	2	2
M_2	$P_{rob}(\mathcal{SH})$	indefinite	$\geq \max\limits_{i \in I} \frac{opt((n_i+1)/3)}{opt(r_i)}$	$\geq \max\limits_{i \in I} \frac{opt(r_i+1)}{opt(r_i)}$
	$P_{rob}(\mathcal{SH}, A_{rob})$	no solution	$\max\limits_{i \in I} \frac{opt(n_i+1)+1}{opt(r_i)}$	$\max\limits_{i \in I} \frac{opt(r_i+1)}{opt(r_i)}$
M_3	$P_{rob}(\mathcal{SH})$	1	1	1
	$P_{rob}(\mathcal{SH}, A_{rob})$	1	1	1
M_4	$P_{rob}(\mathcal{SH})$	indefinite	≥ 2	≥ 2
	$P_{rob}(\mathcal{SH}, A_{rob})$	no solution	$|C| + 1$	2

In a shunting plan, Lemma 1 is reflected in the need of at most one additional code. However, if the class of available recovery algorithm is $\mathbb{A}_{rec}^1$, then the following lemma indicates how an optimal robust algorithm should behave.

Lemma 2 ([6, 8]). *Let $\mathcal{SH} = (Sh, M_1, \mathbb{A}_{rec}^1)$. For every input train T_{in}, any robust shunting algorithm A_{rob} must provide a unique code to each car of T_{in}.*

Let us consider problem Sh_1. As mentioned in Section 3.1, two solutions have been proposed in [24] for this case. The first solution provides a 2-approximation of the optimum, i.e., $apxr = 2$, but it cannot be used for robustness purposes when considering $\mathbb{A}_{rec}^1$ since it does not fulfill the condition of Lemma 2. The second solution, i.e., algorithm A_{out} described before, turns out to be $\mathcal{SH}$-optimal when $\mathcal{SH} = (Sh, M_1, \mathbb{A}_{rec}^1)$.

Theorem 1 ([6, 8]). *Let $\mathcal{SH} = (Sh_1, M_1, \mathbb{A}_{rec}^1)$. There exists a $\mathcal{SH}$-optimal robust shunting algorithm A_{rob} such that $P_{rob}(\mathcal{SH}, A_{rob}) = \max\limits_{i \in I} \frac{opt(n_i, c)}{opt(r_i, c)}$.*

Even though A_{out} is $\mathcal{SH}$-optimal for $\mathbb{A}_{rec}^1$, i.e., $P_{rob}(\mathcal{SH}, A_{rob}) = P_{rob}(\mathcal{SH})$, it is not exact since it could exist an instance i such that $opt(n_i, c) > opt(r_i, c)$.

4   Recoverable Robust Timetabling

In this section, we survey results in [7, 9, 10, 11, 12] concerning recoverable robust problems in the field of timetabling. First, we describe the problems at hand and then, for each problem, we give its complexity and some solving algorithms. Furthermore, we report experimental results in real world scenarios.

The problem of *timetable planning* (in short *timetabling*) arises in the strategic planning phase for transportation systems and requires to compute a timetable for passenger trains that determines minimal passengers traveling times. However, many disturbing events can occur during the operational phase that might completely change the scheduled activities. The main effect of such disturbing events is the arising of delays, caused by malfunctioning infrastructure/devices, special events, or extreme weather conditions. The conflicting objectives of strategic against operational planning are evident in timetable optimization. In fact, a timetable that lets trains sit at stations for some time will not suffer from small delays of arriving trains because delayed passengers can still catch potential connecting trains. On the other hand, big delays can cause passengers to lose trains and hence imply extra traveling time. The problem of deciding when to guarantee connections from a delayed train to a connecting train is known in the literature as *delay management problem* [15, 19, 20, 21, 22, 33, 34] and has a twofold impact. On the one hand, if a connection is maintained, the passengers arriving late still catch their connection, but passengers in the connecting train now face a delay and may miss subsequent connections. On the other hand, if a connection is not maintained, the passengers on the departing train are on time, but those arriving late have to wait for the next train. The latter implies that the delay can propagate through the railway network. The trade-off between these two effects leads to the natural objective of minimizing the overall delay faced by the total passenger population. Despite its natural formalization, the problem turns out to be very complicated to be optimally solved. In fact, it has been shown to be NP-hard in the general case, while it is polynomial in some particular cases (see [19, 20, 33]).

In railway systems, events and dependencies among events are modeled by means of an *event activity network* (see [34]). This is a directed graph where the vertices represent events (e.g., arrivals or departures of trains) and the arcs represent activities occurring between events (e.g., waiting in a train, driving between stations or changing to another train). Event activity networks are a particular class of Directed Acyclic Graphs (DAGs). Hence, in the remainder of the section, we will survey on more general results concerning DAGs.

Let us consider a DAG $G = (V, A)$ with one specified vertex v_0 such that there exists a directed path from v_0 to each other vertex.

A solution for a timetabling problem requires to assign a time π_v to each event $v \in V$ in such a way that all the constraints provided by the set of activities are respected. In detail, the acyclic timetabling problem is then given as

$$(TT) \quad \min f(\pi) = \sum_{u \in V} w_u \pi_u \tag{3}$$

such that

$$\pi_v - \pi_u \geq L_a \quad \forall a = (u, v) \in A \tag{4}$$

$$\pi_u \in \mathbb{R}^{\geq 0} \quad \forall u \in V \tag{5}$$

where $w_u \in \mathbb{R}$ are weights representing the importance of the corresponding events and $L_a \in \mathbb{R}^{>0}$ are given lower bounds indicating the minimal duration that is needed for activity $a \in A$. An instance i of TT is specified by a triple $i = (G, w, L)$.

In the following, two subproblems of TT are considered:

Timetabling with Nonnegative Node Weights (TT^v): In this subproblem, the weights associated to the vertices are nonnegative. Formally:

$$(TT^v) \quad \min f(\pi) = \sum_{u \in V} w_u \pi_u$$

such that

$$\pi_v - \pi_u \geq L_a \quad \forall a = (u, v) \in A$$

$$\pi_u \in \mathbb{R}^{\geq 0} \quad \forall u \in V$$

with $w_u \geq 0 \; \forall \, u \in V$.

Timetabling with arc Weights (TT^a): In this subproblem, (nonnegative) weights are associated to arcs instead of vertices. Formally:

$$(TT^a) \quad \min f_{\text{arcs}}(\pi) = \sum_{a=(u,v) \in A} w_a (\pi_v - \pi_u)$$

such that

$$\pi_v - \pi_u \geq L_a \quad \forall a = (u,v) \in A$$
$$\pi_u \in \mathbb{R}^{\geq 0} \quad \forall u \in V$$

with $w_a \geq 0 \ \forall \ a \in A$.

In [11] it has been shown that TT^a is a subproblem of TT. In order to turn problems TT^v and TT^a into recoverable robustness problems, it is needed to define a modification function and a class of recovery algorithms.

The modification function is defined by admitting a single delay of at most α time. It is modelled as an increase of the minimal duration time of the delayed activity. Formally, given an instance i of TT, $M(i)$ is defined as follows:

$$M(i) = \{(G,w,L') \mid \exists \ \bar{a} \in A : L_{\bar{a}} \leq L'_{\bar{a}} \leq L_{\bar{a}} + \alpha \text{ and } L'_a = L_a \ \forall a \neq \bar{a}\}.$$

In this section, we consider recovery algorithms which are allowed to change the time of at most Δ events where $\Delta \in \mathbb{N}$. The class of these algorithms is denoted by $\mathbb{A}_\Delta$. Formally:

Limited-Events ($\mathbb{A}_\Delta$): Given an instance i of TT, a solution π for i and a modification $i' \in M(i)$, then a solution π' computed by a recovery algorithm in $\mathbb{A}_\Delta$ fulfills

$$|\{u \in V : \pi'_u \neq \pi_u\}| \leq \Delta.$$

Notice that class $\mathbb{A}_\Delta$ belongs to *Class 2*.

Throughout this section, we denote by $TT^v = (TT^v, M, \mathbb{A}_\Delta)$ and $TT^a = (TT^a, M, \mathbb{A}_\Delta)$ the robust problems derived from TT^v and TT^a by imposing the restriction described above on the recovery algorithms.

A general approach for tackling TT^v and TT^a is to add a *slack time* s_a to each activity a, i.e., to find a timetable π such that for each $a = (u,v)$

$$\pi_v - \pi_u \geq L_a + s_a.$$

It follows that TT^v and TT^a can be solved in two steps. The first step consists in finding a slack times assignment $s : L \to \mathbb{R}^{\geq 0}$ which ensures the robustness for an instance i, while the second step consists in solving the instance i' of the (non robust) problem TT obtained by adding s to the minimal duration times of activities of i. The second step is performed by algorithm Alg_s^+ defined as follows.

ALGORITHM Alg_s^+

INPUT: An instance $i = (G,w,L)$ of TT.

ALGORITHM: 1. $\bar{L}_a := L_a + s_a$ for all $a \in A$.
               2. Solve $\bar{i} = (G,w,\bar{L})$ optimally.

Instead of adding a positive slack time to the lower bounds L_a, one can also multiply them with factors $t_a \geq 1$. This approach is known as *proportional* *buffering* and has been studied and applied in the context of timetabling [31]. It leads to the following variant Alg_t^* which differs from Alg_s^+ only in Step 1:

ALGORITHM Alg_t^*

INPUT: An instance $i = (G, w, L)$ of TT.

ALGORITHM: 1. $\bar{L}_a := t_a * L_a$ for all $a \in A$.

              2. Solve $\bar{i} = (G, w, \bar{L})$ optimally.

A special case of algorithm Alg_s^+ (Alg_s^*, respectively) is $\mathsf{Alg}_{\bar{s}}^+$ ($\mathsf{Alg}_{\bar{s}}^*$) where the same slack is added (multiplied) to each activity, i.e., $\bar{s} = (s, \ldots, s)$. Given two positive real numbers a, b, when $s = (a, \ldots, a) \in \mathbb{R}^{|A|}$ and $t = (b, \ldots, b) \in \mathbb{R}^{|A|}$, we denote such algorithms by Alg_a^+ and Alg_b^*, respectively. It is intuitively clear that algorithms Alg_s^+ and Alg_t^* are robust if s is large enough and that the price of robustness increases in s. These observations are formally stated in the following lemma (formulated only for Alg_s^+, the statement for Alg_t^* is similar).

Lemma 3 ([11]). *Consider Alg_s^+ as an algorithm for solving both TT^v and TT^a. Then:*

1. *Alg_s^+ is robust for $\Delta = 0$ (strict robustness) if $s_a \geq \alpha$ for all $a \in A$.*
2. *Let Alg_s^+ be robust. Then $\mathsf{Alg}_{s'}^+$ is robust if $s_a' \geq s_a$ for all $a \in A$.*
3. *The price of robustness is monotone in s. In particular, if $\mathsf{Alg}_{s^1}^+$ and $\mathsf{Alg}_{s^2}^+$ are robust and $s_a^2 \geq s_a^1$ for all $a \in A$, then:*
   - *$P_{rob}(TT^v, \mathsf{Alg}_{s^1}^+) \leq P_{rob}(TT^v, \mathsf{Alg}_{s^2}^+)$.*
   - *$P_{rob}(TT^a, \mathsf{Alg}_{s^1}^+) \leq P_{rob}(TT^a, \mathsf{Alg}_{s^2}^+)$.*

Problem TT (as well as Step 2 of algorithms Alg_s^+ and Alg_t^*) can be solved in polynomial time by linear programming, but, if G is a tree or if it is a DAG and $w_u \geq 0$ for all $u \in V$, it can be solved by the *Critical Path Method* (*CPM*) of project planning (see e.g. [28]) which requires linear time. This method also plays an important role in the recovery algorithms in $\mathbb{A}_\Delta$. Given an instance $i = (G, w, L)$ of TT, *CPM* works as follows.

ALGORITHM *CPM*

INPUT: An instance $i = (G, w, L)$ of TT.

ALGORITHM: 1. For each $v \in V$

              2.   if $v = v_o$ then $\pi_v := 0$

              3.   else $\pi_v := \max \{\pi_u + L_a \; : \; a = (u, v) \in A\}$

The next two subsections are devoted to problems TT^v and TT^a, respectively. For each problem, we first give results for general DAGs, then we restrict the topologies of the graphs to trees or linear graphs.

4.1 Problem TT^v

Results for General DAGs. In [11], it has been shown that computing the price of robustness of problem TT^v is NP-hard by a reduction from 3SAT. Furthermore, when the value of Δ is fixed and is not in the input of the problem, then computing the price of robustness of such a problem remains NP-hard for any $\Delta \geq 3$.

As the problem is NP-hard, non-optimal algorithms with a bounded price of robustness have been provided. The focus has been posed only on the strict robustness problem, i.e. the recovery algorithms have been restricted to *Class 1* by considering $\Delta = 0$. Note that, if an algorithm is robust for such a problem, then it is robust also for the problem arising by allowing any $\Delta > 0$. However, in these cases, the price of robustness of a robust algorithm could be far from the price of robustness of the problem and hence the solution obtained could be too expensive. With the aim of finding such solutions, algorithms Alg_α^+ and Alg_γ^* have been used where $\gamma = (1 + \frac{\alpha}{L_{min}})$, α is the maximum delay allowed and L_{min} is the minimum value assigned by function L with respect to all the possible instances of TT^v.

For $\Delta = 0$, every robust algorithm for TT^v must provide solutions that assign a slack time of at least α to each activity. Then, it follows that Alg_α^+ is a robust algorithm for TT^v. To show that also Alg_γ^* is a robust algorithm for TT^v, it is sufficient to observe that for each activity $a \in A$,

$$\gamma L_a = (1 + \frac{\alpha}{L_{min}})L_a = L_a + \alpha \frac{L_a}{L_{min}} \geq L_a + \alpha.$$

The following theorem shows the price of robustness of Alg_γ^*.

Theorem 2 ([11]). $P_{rob}(TT^v, \mathsf{Alg}_\gamma^*) = 1 + \alpha/L_{min}$.

It has been shown that for each instance i of TT^v, $f(\mathsf{Alg}_\alpha^+(i)) \leq f(\mathsf{Alg}_\gamma^*(i))$, that is, Alg_α^+ is always better than Alg_γ^*. It implies the following result.

Corollary 1 ([11]). $P_{rob}(TT^v, \mathsf{Alg}_\alpha^+) \leq 1 + \alpha/L_{min}$.

Results for Trees. In [12, 13, 14], the attention is devoted to the subproblem of TT^v where the DAG G is a tree. In particular, in [12] theoretical results on this subproblem are given, while in [14] and [13] modelling issues and practical analysis have been addressed.

The motivation for focussing on trees is that, also in practice, a tree can be very useful to model dependencies among trains in the particular case where the railway system considered is a *single-line corridor*. A corridor is a sequence of stations, represented by a line, where each station is served by many trains of different types. Types of trains mostly concern the locations that each train serves and its maximal speed. For an example, see Figure 4. In these systems, it is a practical evidence that slow trains wait for faster trains in order to allow passenger to reach the small stations that are not served by fast trains. This situation is modelled with the only assumption that the changes of passengers from one train to another at a station must be guaranteed only when the second train is starting its journey from the current station.

Let us consider the real world example provided in Figure 4 where three trains serve the same line. The slowest train, the Espresso, goes from Verona to

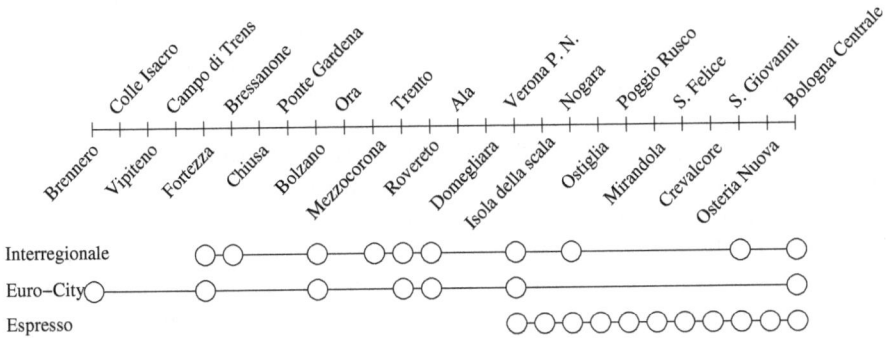

Fig. 4. Example of three trains serving a same line

Bologna, the Interregionale goes from Fortezza to Bologna, and the Euro-City goes from Brennero to Bologna. The Euro-City starts its journey before all the other trains, and it arrives at Verona station before the Interregionale. There, the Espresso is scheduled to start its journey before the Interregionale arrives. Hence, there is an arc between the Euro-City and the starting event corresponding to the Interregionale at Fortezza station, and another arc connecting the Euro-City to the starting event of the Espresso at Verona station. As described above, an arc which represents a changing activity can only connect one vertex to the head of a branch. The DAG obtained by this procedure is a tree, the tree corresponding to the scenario in Figure 4 is shown in Figure 5.

Let $T = (V, A)$ be a tree rooted in v_0. If $v \in V$, T_v denotes the subtree of T rooted in v. Given a subtree T_v, $N_o(T_v)$ denotes the set of vertices y such that $(x, y) \in A$, $x \in T_v$ and $y \notin T_v$; $deg(v)$ denotes $|N_o(T_v)|$.

In [12], it has been shown that the problem TT^v restricted to trees remains NP-hard and a pseudo-polynomial time algorithm SA_Δ for fixed Δ has been provided.

The proposed algorithm looks for solutions which assign only slack times of size α as it has been proved that for any instance, there exists a TT^v-optimal solution which fulfills this condition.

In order to characterize a solution π, the following definition and lemma are needed.

Definition 7. *Given a solution π to TT^v and a vertex $v \in V$, a ball is the maximal subtree $B_\pi(v)$ rooted in v such that for each arc $a = (x, y)$ in $B_\pi(v)$, $s_a = 0$.*

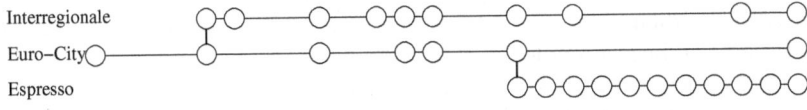

Fig. 5. A tree obtainable from the example provided by Figure 4

Lemma 4 ([12]). *For each instance of TT^v, there exists a TT^v-optimal solution such that for each $v \in V$, $B_\pi(v)$ cannot be extended by adding any vertex from $N_o(B_\pi(v))$ while keeping feasibility and, unless $\Delta = 0$, at most one of two consecutive arcs has a slack time of α.*

For any $\Delta \geq 1$, there exists a TT^v-optimal solution π with the following structure. By Lemma 4, for each arc a outgoing from the root v_0, $s_a = 0$. Then, for each $v \in N_o(v_0)$, π induces a ball $B_\pi(v)$ such that $|B_\pi(v)| \leq \Delta$. In particular, $|B_\pi(v)| < \Delta$ only if $|T_v| < \Delta$. As a consequence, $|B_\pi(v_0)| \leq 1 + \Delta \cdot deg(v_0)$. For each arc $a = (x, y)$ with $x \in B_\pi(v_0)$ and $y \notin B_\pi(v_0)$, $s_a = \alpha$. By Lemma 4, for each arc a outgoing from y, $s_a = 0$, and the same arguments used for $B_\pi(v_0)$ can be used to characterize $B_\pi(y)$.

A possible approach can be that of enumerating all the solutions with the above structure and choosing the cheapest one. Note that such an approach has a computational time which is exponential in the number of nodes. In what follows we report the recursive approach of [12] which avoids to consider a large number of solutions and thus reduces the computational time to a polynomial in n. The algorithm works as follows. It assigns $\pi(v_0) = 0$ and no slack times to arcs outgoing v_0. Then, for each $v \in N_o(v_0)$ it has to decide which subtree of T_v belongs to $B_\pi(v_0)$. To do this, it evaluates the cost, in terms of the value of the objective function, of any possible subtree B of T_v rooted in v of size at most Δ and then chooses the subtree which implies the cheapest solution.

For each already defined ball B_π, this procedure is then repeated for each vertex $v \in N_o(B_\pi)$ which does not belong to an already defined ball by using v as the root.

The cost of a subtree B rooted in v is computed as the value of the objective function when B is chosen as a ball rooted in v. That is, for each arc $a \in B$, $s_a = 0$; for each $a = (x, y) \in A$ with $x \in B$ and $y \notin B$, $s_a = \alpha$; and for each vertex in T_y, an optimal solution is chosen. Computing this cost requires to know the optimal solution of a subtree, this is done by recursively using the algorithm.

The following theorems provide the theoretical results concerning the performances of SA_Δ, $\Delta \geq 1$, on a tree with n vertices.

Theorem 3 ([12]).

- SA_Δ is TT^v-optimal;
- $Prob(TT^v, SA_\Delta) \leq 1 + \frac{\alpha}{2}$;
- $Prob(TT^v) \geq 1 + \frac{\alpha}{\Delta+1}$;
- SA_Δ requires $O(n^{\Delta+1})$ time and $O(n^2)$ space.

In [12], faster algorithms for the special cases of $\Delta \in \{0, 1, 2\}$ have been provided. In particular, when $\Delta \in \{0, 1\}$ ($\Delta = 2$, resp.), linear (quadratic, resp.) time algorithms have been provided.

In [13, 14] these algorithms have also been experimentally studied in practical, real world scenarios. Table 4 shows the data used in the experiments referring to 4 corridors provided by Trenitalia [35].

Table 4. Data used in the experiments

Corridor	Line	Stations	Trains
BrBo	Brennero–Bologna	48	68
MdMi	Modane–Milano	54	291
BzVr	Bolzano–Verona	27	65
PzBo	Piacenza–Bologna	17	25

Starting from the provided data and according to the described requirements, the authors derived event activity networks having tree topologies whose sizes are reported in Table 5.

Table 5. Sizes of the trees

Corridor	N. of nodes	Max. time of traveling	Avg activity time	Max. N. of hops
BrBo	1103	516	9	66
MdMi	4358	318	8	27
BzVr	648	197	5	37
PzBo	163	187	10	14

One of the experimental results of [14] concerning a particular real-world corridor and three values of α is reported in Figure 6. It shows the price of robustness of the solutions obtained by SA_Δ and the computational time needed as functions of Δ. It turned out that algorithm SA_Δ is very fast in practice (it only needs less than 30 milliseconds in the case reported in Figure 6 and a few seconds in the worst cases), and its price of robustness rapidly tends to 1 while Δ increases. In all the cases analyzed, the real price of robustness is even less than the theoretical lower bound which is computed on the worst case instance.

Results for Linear Graphs. In [11], it has been shown that, if the subproblem of TT^v where the DAG $G = (V, A)$ is a linear graph is considered, the price of robustness can be optimally computed in polynomial time. In detail, in [11], a linear time algorithm is given. The idea of the algorithm is to add slack times "as late as possible". Let a linear graph be defined by a set of vertices $V = \{v_1, \ldots, v_{|V|}\}$, ordered such that $A = \{a_1 = (v_1, v_2), \ldots, a_{|A|} = (v_{|V|-1}, v_{|V|})\}$. Define s^α by

$$s_{a_j}^\alpha := \begin{cases} \alpha & \text{if } (\Delta + 1)|j \\ 0 & \text{else} \end{cases} \tag{6}$$

for all arcs $a_j \in A$. Then, the algorithm $\mathsf{Alg}_{s^\alpha}^+$ adds s_a^α to L_a for each $a \in A$ and calculates an optimal robust solution for TT^v by applying CPM on the resulting instance.

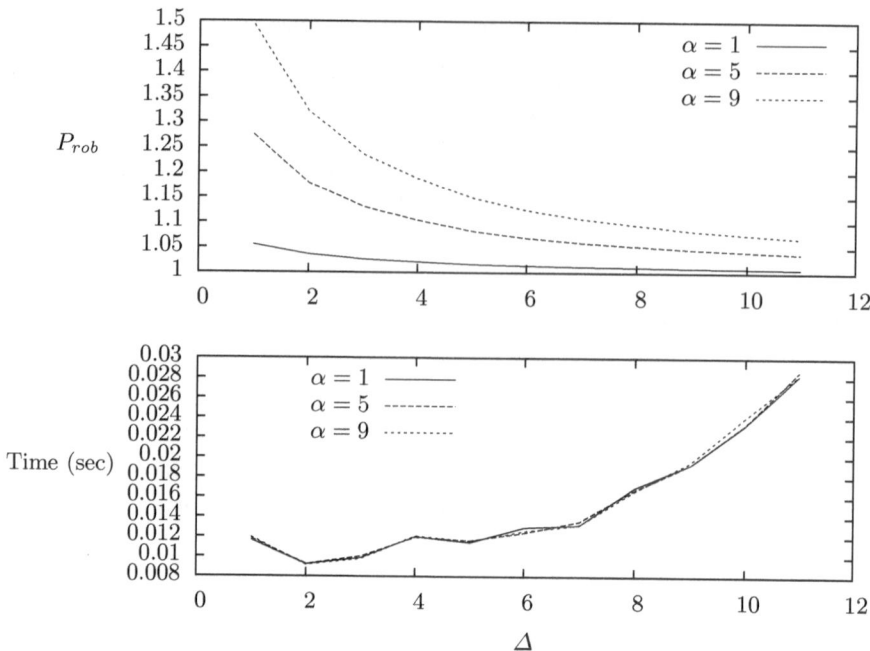

Fig. 6. Price of robustness and computational time needed by SA$_\Delta$ in a particular real-world corridor

4.2   Problem TT^a

Results for General DAGs. By exploiting similar arguments used for TT^v, in [9] it has been shown that computing the price of robustness of problem TT^a is NP-hard. Furthermore, in [10] it has been shown that computing the price of robustness TT^a remains NP-hard for any fixed $\Delta \geq 3$ and it remains NP-hard for any fixed $\Delta \geq 5$ when the considered DAG is restricted to an event activity network.

As computing the price of robustness of TT^a is NP-hard, there exist no polynomial TT^a-optimal algorithms, unless P=NP. It makes sense then to investigate restricted sub-problems which may be of practical interest.

In [9, 10], problem TT^a has been analyzed by using three different subproblems obtained by imposing constraints to the input instances I of TT^a. In detail, the following constraints to functions L and w are imposed:

I_1: L is constant;
I_2: L and w are constant;
I_3: w is constant.

Note that $I_2 \equiv I_1 \cap I_3$.

In the remainder of this section we use the following notation. The minimum and maximum values assigned by the function w (L, resp.) with respect to all the possible instances of TT^a are denoted by w_{min} and w_{max} (L_{min} and L_{max}, resp.). The maximum out degree of a DAG is denoted by deg. Moreover, given $i = (G, L, w) \in I$, $\Delta \in \mathbb{N}$ and $\alpha \in \mathbb{N}$, we denote

$$\gamma(\Delta) = 1 + \frac{\alpha}{(k+1)L_{min}} \quad \text{where } k \geq 0 \text{ is such that} \quad \sum_{p=0}^{k-1} \deg^p \leq \Delta < \sum_{p=0}^{k} \deg^p.$$

In [10] it has been proposed to use a proportional slack time of size $\gamma(\Delta)$ for each activity of a given instance, that is, algorithm $\mathsf{Alg}^*_{\gamma(\Delta)}$ has been used.

The next theorems characterize the price of robustness of problem TT^a and algorithm $\mathsf{Alg}^*_{\gamma(\Delta)}$. In particular, they provide the price of robustness of TT^a when the input instances are those of I_1; show that algorithm $\mathsf{Alg}^*_{\gamma(\Delta)}$ is robust for any instance in I; provide the price of robustness of $\mathsf{Alg}^*_{\gamma(\Delta)}$ when the input instances are in I and that $\mathsf{Alg}^*_{\gamma(\Delta)}$ is TT^a-optimal when the input instances are those of I_2.

Theorem 4 ([10]). *If L is constant, then $P_{rob}(TT^a) \geq \gamma(\Delta)\frac{w_{min}}{w_{max}}$ for any fixed Δ.*

Theorem 5 ([10]). *For any $\Delta \geq 0$, $\mathsf{Alg}^*_{\gamma(\Delta)}$ is robust for TT^a.*

Theorem 6 ([10]).

- *For any fixed $\Delta \geq 0$, $P_{rob}(TT^a, \mathsf{Alg}^*_{\gamma(\Delta)}) \leq \gamma(\Delta)\frac{w_{max}}{w_{min}}$.*
- *If w and L are constant, then $P_{rob}(TT^a, \mathsf{Alg}^*_{\gamma(\Delta)}) = \gamma(\Delta)$, and $\mathsf{Alg}^*_{\gamma(\Delta)}$ is TT^a-optimal.*

In [10] an algorithm $\mathsf{Alg}^+_{\alpha(\Delta)}$ which uses an additive constant slack time which depends on Δ instead of a proportional one has been proposed. In particular, the slack time added to each activity is:

$$\alpha(\Delta) = \frac{\alpha}{(k+1)} \quad \text{where } k \geq 0 \text{ is such that} \quad \sum_{p=0}^{k-1} \deg(\mathcal{N})^p \leq \Delta < \sum_{p=0}^{k} \deg(\mathcal{N})^p.$$

It has been shown that $\mathsf{Alg}^+_{\alpha(\Delta)}$ returns the same timetable as $\mathsf{Alg}^*_{\gamma(\Delta)}$ when the input instances are those of I_1 and that $\mathsf{Alg}^+_{\alpha(\Delta)}$ is TT^a-optimal when the input instances are those of I_3. When the input instances are those of the whole I, other than one would expect, in some cases the timetable returned by $\mathsf{Alg}^*_{\gamma(\Delta)}$ costs less than that returned by $\mathsf{Alg}^+_{\alpha(\Delta)}$. For an extended example where $\mathsf{Alg}^*_{\gamma(\Delta)}$ performs better than $\mathsf{Alg}^+_{\alpha(\Delta)}$ see [10].

Results for Linear Graphs. In [10] the attention also has been concentrated on linear graph with n nodes and no restrictions imposed on functions L and w. In this case, a dynamic programming algorithm of time complexity $O(\Delta n)$ which is TT^a-optimal has been given.

5   Multi-stage Recoverable Robustness and Application to Timetabling

As shown so far, the recoverable robustness model represents a significant improvement in the optimization area. It nevertheless has the following drawback: in many applications, one is typically not facing only one disturbing event, but several disturbances $i^1, i^2, \ldots, i^\sigma$ may occur. For example, assume that we expect at most two disturbances i^1 and i^2. In this case, a robust solution for i^1 should be also recoverable against the next disturbance i^2. This means that under all solutions which are robust for i^1, we should choose one that is again robust against the next disturbance i^2 (if it exists). This example can be extended to more than two disturbances, see Figure 7 for an illustration.

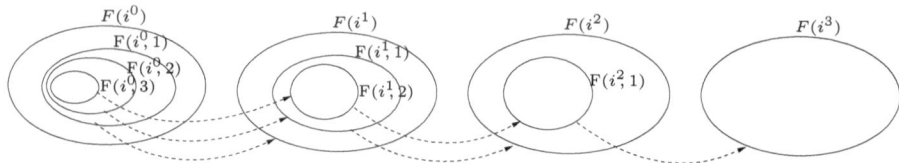

Fig. 7. The set of solutions that are recoverable against 1, 2, and 3 disturbances. Symbol $F(i, n)$ is used to denote the set of feasible solutions for a problem which has to be solved against n disturbances. Dotted arrows represent recovery algorithms.

In this section, we survey results in [7, 11] concerning an extension of the recoverable robustness model (see Section 2) to the *multi-stage recoverable robustness* model which takes into account more than one disturbance and recovery stage. Note that in [11], this model also has been called *dynamic recoverable robustness* model.

To report the new model, some notation has to be recalled. Also in the multi-stage case, the task is to introduce robustness to an arbitrary minimization problems P. Problem P is characterized by the parameters I (set of instances), $F(i)$ for each $i \in I$ (feasible solutions), and f (objective function). The goal of a multi-stage recoverable robust problem is to find a *robust solution* for some given initial instance $i \in I$ of P. It is hence assumed again that a modification function M and a class $\mathbb{A}_{rec}$ of recovery algorithms for P are given. Any recovery algorithm $A_{rec} \in \mathbb{A}_{rec}$ is defined as $A_{rec} : S \times I \to S$, where the initial feasible solution s^0 (of the initial undisturbed instance i^0) and the first modification $i^1 \in M(i^0)$ define the *minimal* amount of information necessary to recover the solution. However, A_{rec} can also require additional information. In particular, when A_{rec} is used in the k-th stage, it can use everything that has been processed in the previous stages (in particular $i^0, \ldots, i^{k-1}$ and $s^0, \ldots, s^{k-1}$). Concerning possible classes of recovery algorithms, all the classes defined in Section 2 can be also used for multi-stage recoverable robustness. Additionally, the value $\sigma \in \mathbb{N}$ is used to denote the maximum number of expected modifications.

In [11], the multi-stage recoverable robustness model has been introduced according to the following definitions.

Definition 8. *A multi-stage recoverable robustness problem is defined by $(P, M, \mathbb{A}_{rec}, \sigma)$ where $(P, M, \mathbb{A}_{rec}) \in$ RRP and $\sigma \in \mathbb{N}$. The class $\mathrm{RRP}(\sigma)$ contains all the multi-stage recoverable robustness problems, that is, the recoverable robustness problems that have to be solved against $\sigma \geq 1$ possible disturbances.*

Definition 9. *Let $\sigma \in \mathbb{N}$ and $\mathcal{P} = (P, M, \mathbb{A}_{rec}, \sigma)$ be an element of $\mathrm{RRP}(\sigma)$. Given an instance $i^0 \in I$ for P, s^0 is a feasible solution for i^0 with respect to $\mathcal{P}$ if and only if the following relationship holds:*

$$\exists A_{rec} \in \mathbb{A}_{rec} : \quad s^0 \in F(i^0) \tag{7}$$
$$s^k := A_{rec}(s^{k-1}, i^k) \in F(i^k), \ \forall i^k \in M(i^{k-1}), \forall k \in [1..\sigma]. \tag{8}$$

This definition ensures that for each stage k, for any possible modification $i^k \in M(i^{k-1})$, and for any feasible solution s^{k-1} computed in the previous stage, the output s^k of algorithm A_{rec} is a feasible solution for i^k with respect to $\mathcal{P}$. If it is clear to which problem P, M and $\mathbb{A}_{rec}$ we refer to, we also say in short that s^0 is *feasible for i with respect to σ recoveries*.

Notice that $\mathrm{RRP}(1) = \mathrm{RRP}$. Hence, each problem in $\mathrm{RRP}(1)$ is called a *single-stage problem* and each problem in $\mathrm{RRP}(\sigma)$, $\sigma > 1$, is called a *multi-stage problem*.

Using the definition of a feasible solution, the robust algorithm that is used to compute the initial solution s^0 for the initial (undisturbed) instance i^0 is defined in the following.

Definition 10. *Given $\mathcal{P} = (P, M, \mathbb{A}_{rec}, \sigma) \in \mathrm{RRP}(\sigma)$, a multi-stage robust algorithm for $\mathcal{P}$ is any algorithm $A_{rob} : I \to S$ such that for each $i \in I$, $A_{rob}(i)$ is feasible for i with respect to P, i.e., such that A_{rob} outputs a solution that can be recovered against σ disturbances.*

Notice that in the case of strict robustness, a robust algorithm A_{rob} for $\mathcal{P}$ must provide a solution s^0 for i^0 such that for each possible modification $i^k \in M(i^{k-1})$, we have $s^0 \in F_{\mathcal{P}}(i^k)$ for all $k \in [1..\sigma]$. The meaning is the following: If A_{rec} has no recovery capability, then A_{rob} has to find solutions that "absorb" *any* possible sequence of disturbances.

Analogous to the single-stage case, the price of robustness has also been defined for multi-stage recovery algorithms. The following definition differs from the corresponding definition for the single-stage case only in the problem $\mathcal{P}$ (it now belongs to the larger class $\mathrm{RRP}(\sigma)$ instead of RRP).

Definition 11. *Let $\mathcal{P} \in \mathrm{RRP}(\sigma)$, $\sigma \geq 1$, and A_{rob} be a robust algorithm for $\mathcal{P}$. Then:*

– *The* price of robustness *of A_{rob} is*

$$P_{rob}(\mathcal{P}, A_{rob}) = \max_{i \in I} \left\{ \frac{f(A_{rob}(i))}{\min\{f(x) : x \in F(i)\}} \right\}.$$

- *The* price of robustness *of* $\mathcal{P}$ *is given by*

$$Prob(\mathcal{P}) = \min\{Prob(\mathcal{P}, A_{rob}) : A_{rob} \text{ is a robust algorithm for } \mathcal{P}\}.$$

- A_{rob} *is* $\mathcal{P}$*-optimal if* $Prob(\mathcal{P}, A_{rob}) = Prob(\mathcal{P})$.
- A_{rob} *is exact if* $Prob(\mathcal{P}, A_{rob}) = 1$.

We report a first, but important observation concerning the price of robustness:

Lemma 5 ([11]). *For fixed* P, M *and* $\mathbb{A}_{rec}$, *consider a family of problems* $\mathcal{P}_\sigma = (P, M, \mathbb{A}_{rec}, \sigma)$ *for different values of* σ, *i.e., these problems vary in the expected number of recoveries only. For* $\sigma_1 < \sigma_2$, *the following holds:*

- $F_{\mathcal{P}_{\sigma_2}}(i) \subseteq F_{\mathcal{P}_{\sigma_1}}(i)$ *for all instances* $i \in I$,
- $Prob(\mathcal{P}_{\sigma_1}) \leq Prob(\mathcal{P}_{\sigma_2})$, *i.e., the price of robustness grows in the number of expected recoveries.*

An important aspect when discussing multi-stage recovery algorithms is to analyze how $Prob(\mathcal{P}_\sigma)$ increases in σ, i.e., what it costs to establish a solution which is robust with respect to σ recoveries. In the next section we present such an analysis for the application of timetabling.

5.1   An Application: Timetabling

An example of real world systems where the multi-stage model plays an important role is the timetable problem TT introduced in Section 4. In this problem, *many* unforeseen delays (caused by disturbing events such as bad weather conditions, repair work, signaling problems, or accidents) might occur during the operational phase, and they might completely change the schedule. In order to be able to deal with more than one delay, it is possible to consider the multi-stage recoverable robust version of the timetabling problem. To this end, it is necessary to formalize the modification function M and the class $\mathbb{A}_{rec}$ as follows:

- Given an instance $i^{k-1} = (G, w, L^{k-1})$ and a constant $\alpha \in \mathbb{R}^{>0}$, then $M(i^{k-1}) =$

$$= \{(G, w, L^k) : \exists\, \bar{a} \in A : L_{\bar{a}}^0 \leq L_{\bar{a}}^k \leq L_{\bar{a}}^0 + \alpha \text{ and } L_a^k = L_a^{k-1} \,\forall a \neq \bar{a}\},$$

  i.e., one additional delay (whose size is bounded by α) is allowed in every stage k.
- In general, it is interesting to use *Class 2* with the two limitations specified next. One of them already has been defined for the single-stage case in Section 4, the other one has not yet been defined before. Let π be a solution of TT and consider a recovery algorithm A_{rec}. Let π' be the recovered timetable computed by A_{rec} with input π and $i^k \in M(i^{k-1})$.
  Limited-Events ($\mathbb{A}_\Delta$)**:** For each recovery stage, it is required that the scheduled times of only a limited number of events might be changed during the recovery with respect to π. Formally, this class of recovery

algorithm is denoted by $\mathbb{A}_\Delta$ and contains the algorithm A_{rec} if and only if

$$|\{u \in V : \pi'_u \neq \pi_u\}| \leq \Delta,$$

for all $\pi' = A_{rec}(\pi, i^k)$, for any feasible disturbance $i^k \in M(i^{k-1})$ of step k with $k = 1, \ldots, \sigma$. Notice that this class has been already defined in Section 4 where it has been used for the single-stage case.

Limited-Delay $(\mathbb{A}_\delta)$: The class $\mathbb{A}_\delta$ contains all polynomial algorithms producing a solution for which the sum of all delays is less than or equal to δ. Formally, the class $\mathbb{A}_\delta$ contains the algorithm A_{rec} if and only if

$$\|\pi' - \pi\|_1 \leq \delta,$$

for all $\pi' = \mathbb{A}_{rec}(\pi, i^k)$, for any feasible disturbance $i^k \in M(i^{k-1})$ of step k with $k = 1, \ldots, \sigma$.

Analogously to Section 4, by using the multi-stage recoverable robustness model, the following problems in $\mathrm{RRP}(\sigma)$ have been investigated:

- Robust timetabling with nonnegative node weights and limited events: $TT^v_\sigma = (TT^v, M, \mathbb{A}_\Delta, \sigma)$ where TT^v denotes the timetabling problem TT with nonnegative weights w_u. Note that studying TT^v_σ corresponds to investigating the problem TT^v (see Section 4) with respect to σ disturbances instead of just one.
- Robust timetabling with arc weights and limited delay: $TT^a_\sigma = (TT^a, M, \mathbb{A}_\delta, \sigma)$ where TT^a denotes the timetabling problem defined for the special case f_{arcs} with nonnegative weights w_a (see Section 4). However, note that in contrast to Section 4, class $\mathbb{A}_\delta$ instead of class $\mathbb{A}_\Delta$ is considered here and that σ disturbances instead of just one are taken into account.

The general approach of using the algorithm Alg_s^+ (with the same slack s for all the activities) has been adopted to face both the previous problems. Of course, in this scenario, the main task is to find the smallest value for s such that Alg_s^+ is robust. Finding a bound for the price of robustness of Alg_s^+ is important as well.

In the following paragraphs, we summarize the results obtained in [7, 11] for both subproblems TT^v_σ and TT^a_σ:

Robust Timetabling with Nonnegative Node Weights and Limited Events (TT^v_σ). The following observations are easily obtained:

- π is robust if $\Delta \geq |V| - 1$.
- Let $\sigma > \Delta$. Then the following holds: π is robust $\Longleftrightarrow s_a \geq \alpha$ for all $a \in A \Longleftrightarrow \pi$ is strictly robust.
- Let $\Delta = 0$: π is robust for $\sigma = 1 \Longleftrightarrow \pi$ is robust for $\sigma > 1$.

Table 6. Computational complexity of calculating an optimal robust solution

Problem	graph	σ	Δ	Complexity
TT_σ^v	arbitrary	1	≥ 3	NP-hard
TT_σ^v	linear	any	any	linear

- The set of robust solutions w.r.t $\sigma > 1$ is strictly contained in the set of robust solutions for $\sigma = 1$.

However, the problem of calculating the price of robustness with number of events as limitation is NP-hard for all fixed $\Delta \geq 3$ even for the case $\sigma = 1$ which corresponds to the problem faced in Section 4. Table 6 summarizes the computational complexity of calculating an optimal robust solution for TT_σ^v with respect to different graph topologies.

Notice that, in the case of a sequence of $\sigma \geq 1$ disturbances, it has been proposed a solution for TT_σ^v for arbitrary σ when the underlying graph is a linear graph. As mentioned above, the solution uses the algorithm Alg_{s*}^+ assigning the same slack

$$s^* = \min\left\{\alpha, \frac{\sigma\alpha}{\Delta+1}\right\} \qquad (9)$$

to each arc. The following result has been obtained:

Theorem 7 ([11]). *Let s^* be defined as in Eq. (9), and let G be a linear graph. Then:*

- Alg_s^+ *is a robust algorithm if and only if $s \geq s^*$;*
- $Prob(TT_\sigma^v, \mathsf{Alg}_{s*}^+) \leq 1 + \frac{s^*}{L_{min}}$;
- Alg_{s*}^+ *is optimal compared to all robust algorithms that add an equal slack s to all arcs.*

Robust Timetabling with arc Weights and Limited Delay (TT_σ^a). Also for this problem, the algorithm Alg_s^+ (with the same slack s for all the activities) has been used to solve TT_σ^a. The following properties hold:

- Alg_α^+ is strictly robust for all σ;
- $\mathsf{Alg}_{\alpha-\Delta}^+$ is robust for $\sigma = 1$ and $\Delta \leq \frac{\alpha}{2}$.

If G is a tree and Alg_s^+ is robust, the price of robustness can be bounded by $Prob(TT_\sigma^a, \mathsf{Alg}_s^+) \leq 1 + \frac{s}{L_{min}}$ where the bound is obtained in the case of strict robustness. This leads to the results in Table 7.

For arbitrary σ and Δ, it has been shown that there exists some s^* such that Alg_s^+ is robust for all $s \geq s^*$. In linear graphs this value s^* can be calculated as

$$s^* = \frac{2\sigma\alpha\left(\left\lceil\frac{2\Delta}{\sigma\alpha}\right\rceil + \sigma\right) - \sigma\alpha(\sigma+1) - 2\Delta}{\left(\left\lceil\frac{2\Delta}{\sigma\alpha}\right\rceil + \sigma\right)\left(\left\lceil\frac{2\Delta}{\sigma\alpha}\right\rceil + \sigma - 1\right)},$$

leading to a price of robustness equal to $1 + s^*$.

Table 7. Upper bounds for P_{rob} in some special cases

problem	graph	σ	Δ	P_{rob}
TT_σ^v	DAG	1	any	$1 + \frac{\alpha}{L_{\min}}$
TT_σ^v	tree	1	any	$1 + \min\left\{\frac{\alpha}{L_{\min}}, \frac{\alpha}{2}\right\}$
TT_σ^v	linear	any	any	$1 + \frac{1}{L_{\min}} \min\left\{\alpha, \frac{\sigma\alpha}{\Delta+1}\right\}$
TT_σ^a	tree	any	any	$1 + \frac{\alpha}{L_{min}}$
TT_σ^a	tree	1	$\Delta \le \frac{\alpha}{2}$	$1 + \frac{\alpha-\Delta}{L_{min}}$
TT_σ^a	linear	any	any	$1 + \frac{2\sigma\alpha\left(\left\lceil\frac{2\Delta}{\sigma\alpha}\right\rceil+\sigma\right)-\sigma\alpha(\sigma+1)-2\Delta}{\left(\left\lceil\frac{2\Delta}{\sigma\alpha}\right\rceil+\sigma\right)\left(\left\lceil\frac{2\Delta}{\sigma\alpha}\right\rceil+\sigma-1\right)}$

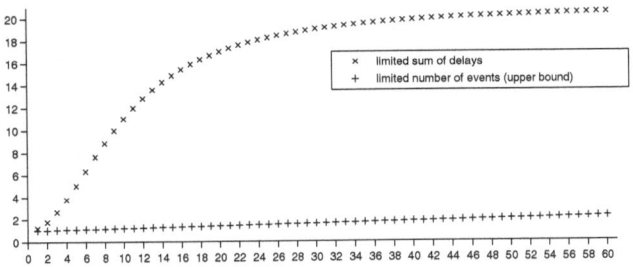

Fig. 8. The price of robustness of algorithm Alg_s^+ for $\alpha = 20$ and $\Delta = 1000$ as a function of σ

In conclusion, the price of robustness can be exactly calculated in special cases only; in many cases, an approximation is possible. It is clear that the price of robustness increases with the number of expected recoveries, but its growth is smaller than expected as can be seen in Figure 8.

6 Conclusions

The paper surveys on some recent algorithmic results achieved within the recoverable robustness model. This model provides the unification between the standard notion of robustness (a solution must remain feasible although some disturbances may occur) and the possibility to apply limited recovery strategies once the feasibility of the current solution is lost. The attention has been addressed to two main problems arising in the area of railway optimization: the shunting problem and the timetabling problem. The former problem regards the reordering of freight train cars over hump yards while the latter one consists in finding passenger train timetables in order to minimize the overall passengers traveling time. In the reviewed papers, algorithms for both problems have been investigated with respect to different possible disturbances and different recovery capabilities. Moreover, the timetabling problem has been also used as a testbed for empirical experiments and for investigating on a natural extension of the original model, i.e. the so called multi-stage recoverable robustness. This extension aims to provide recoverable robustness in the case that multiple recovery phases are allowed.

The presented results reveal an interesting and practical applicability of the model under which useful algorithms have been designed in the studied contexts. The proposed notion of recoverable robustness provides a mean to compare the performances of different robust algorithms in terms of distance from the optimality. In particular, the price of robustness for an algorithm measures such a distance and provides a practical tool to rank among algorithms. This implies the possibility to apply standard techniques developed in algorithmic theory for choosing the "best" algorithm, robust with respect to the required constraints. Moreover, the presented experimental results confirm the effectiveness for the defined price of robustness. In fact, the evaluations show how the algorithmic performances are affected by the variation on both the recovery capabilities and the modification function. As theoretically expected, the less restrictive the available recovery capabilities are, the smaller the price of robustness of an algorithm is. Viceversa, the larger the set of modification is, the larger the price of robustness of an algorithm is. Both observations can be seen in Figure 6 where the price of robustness decreases with Δ and increases with α.

The reported results represent some initial contributions to the applicability of the recoverable robustness model. It might be worth investigating the feasibility of the model on different optimization problem, varying on modification functions and the class of recovery algorithms. Concerning this last point, an important issue that has not yet been addressed concerns the investigation of robust algorithms when the allowed recovery capabilities are that of *Class 3*, i.e. when the computational power of the recovery algorithms is bounded.

Extensions to the proposed model are also of interest. In particular, it is worth to note that the proposed notion of recoverable robustness does not enforce the requirement to design efficient robust algorithms in time and space. This is a key point in introducing recoverable robustness in most of the problems studied in the context of railway optimization as well as in the more general algorithmic optimization turn out to be NP-hard. Hence, the optimal algorithm on which the price of robustness depends might be exponential due to the computational complexity of the underlying robust problem. Then, it would be interesting to study the case in which the computational power of the robust algorithms is limited and hence to define the price of robustness of the problem according to the admitted class of robust algorithms.

Another interesting direction of research for the extension of the model is about the design of robust and recovery algorithms. Consider the typical scenario in the context of recoverable robustness: it is necessary to face a problem $\mathcal{P} = (P, M, \mathbb{A}_{rec})$ and, to this aim, in the operational phase, a plan s (computed by a robust algorithm) is being used. At this time, if a disturbing event $j \in M(i)$ occurs, the recoverable robustness model guarantees that an algorithm $A_{rec} \in$

$\mathbb{A}_{rec}$ exists to *recover*, that is, to compute a new plan $s' = A_{rec}(s, j)$ which is feasible for j.

Notice that, the recoverable robustness model ensures only that A_{rec} exists, and it does not take care of defining details about such a recovery algorithm. In an operative environment, such details are needed, and hence a recovery algorithm has to be carefully designed. In this context, relevant questions are: How to measure the quality of a given recovery algorithm? When to design recovery algorithms? Concerning the first question, a method could be to define a sort of *price of recovery*. Concerning the other question, in general, a robust algorithm (planning stage) and a recovery algorithm (operational stage) can be designed independently of each other. The main motivation for this lies on the fact that $\mathbb{A}_{rec}$ can be defined by means of just some mathematical properties, and hence, the input of each recovery algorithm is completely defined, regardless of the definition of any possible robust algorithm. This leads to two possible scenarios:

- Considering the robust and recovery algorithms as belonging to different modules of the same system. If A_{rec} is designed *a priori*, i.e., independently of any information concerning robust algorithms, then the robust and recovery modules are decoupled. This guarantees that, in case, each module can be re-engineered without affecting the whole system.

  In this scenario, a possible definition for a price of recovery could be the following:

Definition 12. *Let $\mathcal{P} = (P, M, \mathbb{A}_{rec})$ be a problem in RRP, and let $A_{rec} \in \mathbb{A}_{rec}$. The price of recovery of A_{rec} is:*

$$P_{rec}(\mathcal{P}, A_{rec}) = \max_{i \in I,\, s \in F_{\mathcal{P}}(i),\, j \in M(i)} \left\{ \frac{f(A_{rec}(s, j))}{\min\{f(x) : x \in F(j)\}} \right\}.$$

This price of recovery measures "how far" the recovered solution computed by A_{rec} is away from the optimum one, independently of which robust algorithm has been used.

Unfortunately, this measure could be useless in most cases. For instance, consider the robustness problem TT^v for linear graphs defined at the end of Section 4.1 where each algorithm in the class $\mathbb{A}_{rec}$ can change the time of at most Δ events. In this case, it is easy to see that the price of recovery is unbounded since in $F_{\mathcal{P}}(i)$, for $i \in I$, there are solutions s which are robust due to slack times arbitrarily large.

- Designing a recovery algorithm A_{rec} in hindsight (i.e., after the definition of a *good* robust algorithm A_{rob}) could imply the design of a *specialized* recovery algorithm, since A_{rec} could exploit specific properties of the robust solutions computed by A_{rob}.

  In this scenario, a possible definition for the price of recovery could be the following:

Definition 13. *Let* $\mathcal{P} = (P, M, \mathbb{A}_{rec})$ *be a problem in* RRP, A_{rob} *be a robust algorithm for* $\mathcal{P}$, *and* $A_{rec} \in \mathbb{A}_{rec}$. *The* price of A_{rob}-recovery *of* A_{rec} *is:*

$$P_{rec}(\mathcal{P}, A_{rob}, A_{rec}) = \max_{i \in I, \, j \in M(i)} \left\{ \frac{f(A_{rec}(A_{rob}(i), j))}{\min\{f(x) : x \in F(j)\}} \right\}.$$

This measure could represent a good parameter for the quality of a recovery algorithm.

Additionally, notice that a recovery algorithm A_{rec} with a small price of recovery is useless when its traditional *worst case execution time* is bad. For instance, if we consider the timetabling problem TT along with the class $\mathbb{A}_{\Delta}$ (remember, any algorithm in $\mathbb{A}_{\Delta}$ can change the time of at most Δ events), then we expect that an algorithm in $\mathbb{A}_{\Delta}$ should be computationally efficient, that is, running in $O(\Delta)$ time. Hence, the quality of a recovery algorithm should be measured by means of two parameters: the price of recovery and the worst case execution time. A trade-off between these two parameters could define a *good recovery algorithm*.

References

1. Bayer, H.G., Sendhoff, B.: Robust Optimization - A Comprehensive Survey. Computer Methods in Applied Mechanics and Engineering 196(33-34), 3190–3218 (2007)
2. Ben-Tal, A., El Ghaoui, L., Nemirovski, A.: Mathematical Programming: Special Issue on Robust Optimization, vol. 107. Springer, Berlin (2006)
3. Bertsimas, D., Sim, M.: The price of robustness. Operations Research 52(1), 35–53 (2004)
4. Birge, J.R., Louveaux, F.V.: Introduction to Stochastic Programming. Springer, New York (1997)
5. Borodin, A., El-Yaniv, R. (eds.): Online Computation and Competitive Analysis. Cambridge University Press, Cambridge (1998)
6. Cicerone, S., D'Angelo, G., Di Stefano, G., Frigioni, D., Navarra, A.: Robust Algorithms and Price of Robustness in Shunting Problems. In: Proceedings of the 7th Workshop on Algorithmic Approaches for Transportation Modeling, Optimization, and Systems (ATMOS 2007),Schloss Dagstuhl Seminar proceedings, pp. 175–190 (2007)
7. Cicerone, S., Di Stefano, G., Schachtebeck, M., Schöbel, A.: Multi-stage recoverable robustness problems. Technical Report ARRIVAL-TR-0226, ARRIVAL Project (2009)
8. Cicerone, S., D'Angelo, G., Di Stefano, G., Frigioni, D., Navarra, A.: Robust algorithms and price of robustness in shunting problems. Technical Report ARRIVAL-TR-0072, ARRIVAL project (2007)
9. Cicerone, S., D'Angelo, G., Di Stefano, G., Frigioni, D., Navarra, A.: Delay management problem: Complexity results and robust algorithms. In: Yang, B., Du, D.-Z., Wang, C.A. (eds.) COCOA 2008. LNCS, vol. 5165, pp. 458–468. Springer, Heidelberg (2008)

10. Cicerone, S., D'Angelo, G., Di Stefano, G., Frigioni, D., Navarra, A.: Recoverable robust timetabling for single delay: Complexity and polynomial algorithms for special cases. Journal of Combinatorial Optimization. Tech. Rep. ARRIVAL-TR-0172, ARRIVAL project (to appear, 2009)
11. Cicerone, S., Di Stefano, G., Schachtebeck, M., Schöbel, A.: Dynamic algorithms for recoverable robustness problems. In: Proceedings of the 8th Workshop on Algorithmic Approaches for Transportation Modeling, Optimization, and Systems (ATMOS 2008). Schloss Dagstuhl Seminar proceedings, Technical Report ARRIVAL-TR-0130 (2008)
12. D'Angelo, G., Di Stefano, G., Navarra, A.: Recoverable robust timetables on trees. Technical Report ARRIVAL-TR-0163, ARRIVAL project (2008)
13. D'Angelo, G., Di Stefano, G., Navarra, A.: Evaluation of recoverable-robust timetables on tree networks. In: Proceedings of the 20th International Workshop on Combinatorial Algorithms (IWOCA2009) (to appear, 2009)
14. D'Angelo, G., Di Stefano, G., Navarra, A.: Recoverable-robust timetables for trains on single-line corridors. In: Proceedings of the 3rd International Seminar on Railway Operations Modelling and Analysis - Engineering and Optimisation Approaches, RailZurich 2009 (to appear, 2009)
15. De Giovanni, L., Heilporn, G., Labbé, M.: Optimization models for the delay management problem in public transportation. European Journal of Operational Research 189(3), 762–774 (2007)
16. Fischetti, M., Monaci, M.: Robust optimization through branch-and-price. In: Proceedings of the 37th Annual Conference of the Italian Operations Research Society, AIRO (2006)
17. Fischetti, M., Monaci, M.: Light robustness. Technical Report ARRIVAL-TR-0066, ARRIVAL project (2008)
18. Fischetti, M., Monaci, M.: Light robustness. In: Robust and Online Large-Scale Optimization, pp. 62–85. Springer, Heidelberg (2009)
19. Gatto, M., Glaus, B., Jacob, R., Peeters, L., Widmayer, P.: Railway delay management: Exploring its algorithmic complexity. In: Hagerup, T., Katajainen, J. (eds.) SWAT 2004. LNCS, vol. 3111, pp. 199–211. Springer, Heidelberg (2004)
20. Gatto, M., Jacob, R., Peeters, L., Schöbel, A.: The Computational Complexity of Delay Management. In: Kratsch, D. (ed.) WG 2005. LNCS, vol. 3787, pp. 227–238. Springer, Heidelberg (2005)
21. Gatto, M., Jacob, R., Peeters, L., Widmayer, P.: Online Delay Management on a Single Train Line. In: Geraets, F., Kroon, L.G., Schoebel, A., Wagner, D., Zaroliagis, C.D. (eds.) Railway Optimization 2004. LNCS, vol. 4359, pp. 306–320. Springer, Heidelberg (2007)
22. Ginkel, A., Schöbel, A.: The bicriteria delay management problem. Transportation Science 41(4), 527–538 (2007)
23. Hansman, R.S., Zimmermann, U.T.: Optimal sorting of rolling stock at hump yard. In: Mathematics - Key Technology for the Future: Joint Project Between Universities and Industry, pp. 189–203. Springer, Heidelberg (2008)
24. Jacob, R.: On shunting over a hump. Technical Report 576, Institute of Theoretical Computer Science, ETH Zürich (2007)
25. Jacob, R., Marton, P., Maue, J., Nunkesser, M.: Multistage methods for freight train classification. In: Proceedings of the 7th Workshop on Algorithmic Approaches for Transportation Modeling, Optimization, and Systems (ATMOS 2007), pp. 158–174. Schloss Dagstuhl Seminar proceedings (2007)

26. Jacob, R., Marton, P., Maue, J., Nunkesser, M.: Multistage methods for freight train classification. Technical Report ARRIVAL-TR-0140, ARRIVAL project (2007)
27. Kall, P., Wallace, S.W.: Stochastic Programming. Wiley, Chichester (1994)
28. Levy, F.K., Thompson, G.L., Wies, J.D.: The ABCs of the Critical Path Method. Graduate School of Business Administration. Harvard University (1963)
29. Liebchen, C., Lüebbecke, M., Möhring, R.H., Stiller, S.: Recoverable robustness. Technical Report ARRIVAL-TR-0066, ARRIVAL project (2007)
30. Liebchen, C., Lüebbecke, M., Möhring, R.H., Stiller, S.: The concept of recoverable robustness, linear programming recovery, and railway applications. In: Ahuja, R.K., Möhring, R.H., Zaroliagis, C.D. (eds.) Robust and Online Large-Scale Optimization. LNCS, vol. 5868, pp. 1–27. Springer, Heidelberg (2009)
31. Liebchen, C., Stiller, S.: Delay resistant timetabling. Public Transport, Tech. Rep. ARRIVAL-TR-0056, ARRIVAL project (2008)(To appear)
32. Ruszczynski, A., Shapiro, A. (eds.): Stochastic Programming, Handbooks in Operations Research and Management Science, vol. 10. North-Holland, Amsterdam (2003)
33. Schöbel, A.: A model for the delay management problem based on mixed integer programming. Electronic Notes in Theoretical Computer Science 50(1) (2001)
34. Schöbel, A.: Integer programming approaches for solving the delay management problem. In: Geraets, F., Kroon, L.G., Schoebel, A., Wagner, D., Zaroliagis, C.D. (eds.) Railway Optimization 2004. LNCS, vol. 4359, pp. 145–170. Springer, Heidelberg (2007)
35. Trenitalia, http://www.trenitalia.com/

Light Robustness

Matteo Fischetti and Michele Monaci

DEI, University of Padova,
Via Gradenigo 6/A, I-35131 Padova, Italy
{matteo.fischetti, michele.monaci}@unipd.it

Abstract. We consider optimization problems where the exact value of
the input data is not known in advance and can be affected by uncertainty. For these problems, one is typically required to determine a robust
solution, i.e., a possibly suboptimal solution whose feasibility and cost
is not affected heavily by the change of certain input coefficients. Two
main classes of methods have been proposed in the literature to handle
uncertainty: stochastic programming (offering great flexibility, but often
leading to models too large in size to be handled efficiently), and robust
optimization (whose models are easier to solve but sometimes lead to
very conservative solutions of little practical use). In this paper we investigate a heuristic way to model uncertainty, leading to a modelling
framework that we call *Light Robustness*. Light Robustness couples robust optimization with a simplified two-stage stochastic programming
approach, and has a number of important advantages in terms of flexibility and ease to use. In particular, experiments on both random and
real word problems show that Light Robustness is sometimes able to produce solutions whose quality is comparable with that obtained through
stochastic programming or robust models, though it requires less effort
in terms of model formulation and solution time.

Keywords: Robust optimization, Stochastic Programming, Integer Linear Programming, Multi-dimensional Knapsack, Train Timetabling.

1 Introduction

One of the basic assumptions in mathematical programming is that the exact
value of the input data is fixed and known in advance. This assumption can
however be violated in many situations arising when real world problems are
considered. This can be due to the fact that the parameters used in the model are
just estimates of real parameters, or more generally to the effect of uncertainty
affecting some parameters. When uncertainty is taken into account, an optimal
solution with respect to the nominal values of the parameters can be suboptimal
(or even infeasible) according to the actual parameters. Hence, small uncertainty
in the input data can make the nominal optimal solution completely meaningless
from a practical viewpoint.

Within the above setting, a main request when dealing with real world applications is to determine a *robust* solution, i.e., a solution that remains feasible

R.K. Ahuja et al. (Eds.): Robust and Online Large-Scale Optimization, LNCS 5868, pp. 61–84, 2009.

even if some of the input coefficients change. In other words, one is required to determine a solution that is not necessarily optimal for the nominal objective function, but such that its feasibility and cost is not affected heavily by the change of some coefficients—at least for certain meaningful realizations of the input data.

In this paper we mainly focus on a Linear Program (LP) of the form

$$\min \sum_{j \in N} c_j \, x_j \tag{1}$$

$$\sum_{j \in N} a_{ij} \, x_j \le b_i \ \ i \in M \tag{2}$$

$$x_j \ge 0 \ \ j \in N \tag{3}$$

where some coefficients of constraint matrix A can take a value, say $\tilde{a}_{ij} \in [a_{ij}, a_{ij} + \hat{a}_{ij}]$, which is different from the nominal one (namely, a_{ij}). Uncertainty on vectors b and c is not dealt with explicitly, as it can be handled in a straight-forward way by just adding suitable artificial variables and/or constraints. We denote by $n = |N|$ and $m = |M|$ the number of variables and constraints in the LP model, respectively. Our approach extends easily to the Mixed-Integer Programming (MIP) case, where certain variables can only assume integer values.

Classical approaches for dealing with uncertainty can be classified as follows:

- *Stochastic Programming* (SP): find a solution that is optimal by considering possible recourse variables y^ω implementing corrective actions to be performed after a certain scenario $\omega \in \Omega$ has taken place (see, e.g., Birge and Louveaux [8], Ruszczynski and Shapiro [16], and Linderoth, Shapiro, and Wright [13]). This approach typically does not restrict the original solution space, but penalizes the feasible solutions by taking into account the cost of the corrective actions needed to face a certain scenario. The approach is quite powerful but requires the knowledge of the probability and main features of the various scenarios, and almost invariably leads to huge LPs that require very large computing time—though clever decomposition techniques have been proposed in the literature to speed-up their resolution.
- *Robust Optimization* (RO): uncertainty is associated with hard constraints restricting the solution space, i.e., one is required to find a solution that is still feasible for worst-case parameters chosen within a certain uncertainty domain (see, e.g., Ben-Tal and Nemirovski [2] and Bertsimas and Sim [6]). This is an effective way to model uncertainty, but it can lead to overconservative solutions that are quite bad in terms of cost (actually, a feasible solution may not exist at all).

In the present paper we analyze a heuristic way to model uncertainty, leading to a modelling framework that we call *Light Robustness* (LR). Light Robustness can be viewed as a "flexible counterpart" of robust optimization, obtained through the following modelling steps. We first fix the maximum objective function deterioration that we are willing to accept in our model, by introducing a linear constraint of the type $c^T x \leq \overline{z}$. Then we define a "robustness goal" that we would like to achieve, and model it by using a classical robust optimization framework (e.g., through the Ben-Tal and Nemirovski [2] or Bertsimas and Sim [6] methods). In this way we obtain a robust model with no objective function, that however is likely to be infeasible. To cope with infeasibility, we introduce appropriate slack variables that allow for "local violations" of the robustness requirements, and define an auxiliary objective function aimed at minimizing the slacks. The LR slack variables play a role similar to second-stage recourse variables in SP models, as they penalize the corrective actions needed to restore feasibility. In this view, LR is a heuristic framework combining the flexibility of SP (due to the presence of second-stage variables) and the modelling ease of RO. The underlying assumption is that the robust model already captures uncertainty in a sufficiently detailed way, so we hopefully do not need a cumbersome second-stage set of variables and constraints—simple slack variables are enough. Whether this is a reasonable assumption for a specific application (and model) can only be verified experimentally, through simulations.

LR models are easy to formulate and to solve, and their applicability is potentially larger than robust models. However, it is not clear whether such a simple heuristic approach can deliver solutions that are comparable with those obtained through more involved stochastic programming or robust models. The computational experience reported in the present paper confirms the viability of the LR approach—at least in some practically relevant contexts.

The rest of the paper is organized as follows. In Section 2 we briefly outline the Stochastic Programming approach, whereas in Section 3 we address Robust Optimization and review the Bertsimas and Sim method [6]. Two Light Robustness variants are described in Sections 4 and 5, respectively, and are computationally tested on random instances in Section 6. It is known that random instances may produce "smooth" test cases that hide the hard situations. A real-world application is therefore addressed in Section 7 and evaluated experimentally on real data provided by Trenitalia, the main Italian railway operator for passengers. Finally, Section 8 draws some conclusions.

2   Stochastic Programming

As already mentioned, SP is a framework for modelling optimization problems that involve uncertainty in the parameter set. SP models take advantage of the fact that probability distributions governing the data are known or can be estimated. The goal here is to find some policy that is feasible for all (or almost all) the possible data instances and maximizes the expectation of some function of the decisions and the random variables. During the last four decades a vast amount of literature on stochastic programming appeared. Two comprehensive textbooks on the subject are [8] and [15]; an easily accessible introduction is given in [11].

The most widely applied and studied SP models are 2-stage linear programs. Here, the decision maker takes some action in the first stage, after which a random event occurs that affects the outcome of the first-stage decision. A recourse decision can then be made in the second stage that compensates for the bad effects that might have been experienced because of the first-stage decision. The optimal policy for the resulting model is a single first-stage policy and a collection of recourse decisions defining which second-stage action should be taken in response to each random outcome. In a 2-stage SP, the set of constraints is decomposed into *structural* constraints, which represent the deterministic part of the model, and *control* constraints which have a stochastic nature and whose coefficients depend on the particular scenario. Roughly speaking, the approach allows one to take decisions in the first stage by ignoring the stochastic part of the model, but enforces some costly *recourse* action when indeterminacy will eventually occur. More specifically, a generic 2-stage SP formulation for linear problems is given by

$$\min\{c^T x + Q(x) \mid Ax \le b, x \ge 0\},$$

where

$$Q(x) = \mathbb{E}[\min\{q^T y \mid Wy \le h - Tx, \ y \ge 0\}],$$

expresses the expected recourse cost associated with the first-stage decision x, and the triple (q, T, h) modelling the recourse is assumed to be affected by randomness.

Solution methods for SP problems typically address the so-called deterministic equivalent problem [8]. The basic assumption here is that the realizations of the random parameters are specified in the form of K (say) *scenarios* $\omega_1 = (q^1, T^1, h^1), \omega_2 = (q^2, T^2, h^2), \ldots, \omega_K = (q^K, T^K, h^K)$, each with a probability $p^1, \ldots p^K$ of occurrence. The problem can therefore be formulated as

$$\begin{aligned}
\min \ & c^T x + \sum_{k=1}^{K} p^k (q^k)^T y^k \\
& A x \qquad\qquad \le b, \ x \ge 0 \\
& T^k x + W y^k \le h^k, \ y^k \ge 0, \qquad k = 1, \ldots, K.
\end{aligned} \tag{4}$$

and solved through methods that take advantage of its structure, e.g., Benders' decomposition approaches.

3   Robust Optimization: The Bertsimas and Sim Approach

The first attempt to handle data uncertainty through mathematical models was performed by Soyster [18], who considered uncertain problems of the form

$$\min \ \{\sum_{j \in N} c_j \, x_j \mid \sum_{j \in N} A_j \, x_j \le b, \ \forall A_j \in \mathcal{K}_j, j \in N\}$$

where $\mathcal{K}_j$ are convex sets associated with "column-wise" uncertainty. This approach tends to lead to overconservative models, thus to poor solutions in term of optimality. Ben-Tal and Nemirovski [2], [3] and [4] defined less conservative models by considering ellipsoidal uncertainties. Moreover, [2] shows that the robust counterpart of an uncertain LP is equivalent to an explicit computationally tractable problem, provided that the uncertainty is itself "tractable". On the contrary, when the problem to be considered is an ILP, these nonlinear (convex) models become computationally hard problems.

Later on, Bertsimas and Sim (BS) considered a different concept of robustness (see [5] and [6]). Their approach is based on the observation that, in real situations, it is unrealistic to assume that all coefficients take, at the same time, their worst-case value. So, it makes sense to define a robust model whose optimal solution remains feasible for every change of (at most) Γ_i coefficients in each row $i \in M$, where Γ_i is an input parameter associated to the expected robustness of the solution. (For sake of simplicity, we will implicitly assume that Γ_i is integer, although this is not required in the approach proposed in [6] nor by our LR method.) The robust counterpart of (1)–(3) is therefore defined by replacing each row $i \in M$ with the new constraint:

$$\sum_{j \in N} a_{ij}\, x_j + \beta(x, \Gamma_i) \le b_i \tag{5}$$

where $\beta(x, \Gamma_i)$ is related to the level of protection with respect to uncertainty in the coefficients of row i, and is defined as

$$\beta(x, \Gamma_i) = \max_{S \subseteq N:|S| \le \Gamma_i} \sum_{j \in S} \hat{a}_{ij}\, x_j \tag{6}$$

So, $\beta(x, \Gamma_i)$ is the maximum increase in the left-hand side of the i-th constraint evaluated for x^*, when at most Γ_i coefficients in row i take their worst-case value.

As already mentioned, parameter Γ_i allows the modeler to control the solution robustness: $\Gamma_i = 0$ means that robustness is not taken into account and the nominal constraint is considered, whereas $\Gamma_i = n$ means that each coefficient in row i can take its worst-case value, and corresponds to the conservative method by Soyster [18].

By using LP duality, the robust model can be formulated through the following LP:

$$\min \sum_{j \in N} c_j\, x_j \tag{7}$$

$$\sum_{j \in N} a_{ij}\, x_j + \Gamma_i\, z_i + \sum_{j \in N} p_{ij} \le b_i \quad i \in M \tag{8}$$

$$-\hat{a}_{ij}\, x_j + z_i + p_{ij} \ge 0 \quad i \in M, j \in N \tag{9}$$

$$z_i \ge 0 \quad i \in M \tag{10}$$

$$p_{ij} \ge 0 \quad i \in M, j \in N \tag{11}$$

$$x_j \ge 0 \quad j \in N \tag{12}$$

The robust formulation above, referred to as BS in the sequel, involves a number of variables and constraints that is polynomial in the input size. Note that the approach remains valid when MIPs are considered instead of just LPs, the only requirement being that term $\beta(x, \Gamma_i)$ can be formulated as an LP whose size is polynomial in the input size.

The BS approach provides solutions that are deterministically feasible if the coefficients change under the assumptions above, and are feasible with a high probability if more than Γ_i coefficients in row i are allowed to change.

4   A First Light Robustness Heuristic

Very often, the optimal robust solution found according to the BS definition can be considerably worse (with respect to the objective function value) than the optimal nominal solution, even if few coefficients are allowed to change in each row. This fact is dramatically emphasized for those problems where most of the coefficients are "structural" and the number of uncertain coefficients in each row is very small (as, e.g., in the train timetabling problem addressed in Section 7).

As already outlined in the introduction, our definition of Light Robustness is a compromise between the robustness of the solution with respect to uncertainty of the matrix coefficients, and the quality of the solution with respect to the objective function. Indeed, in our scheme we look for the most robust solution among those which are "not too far" from optimality for the nominal problem. To be more specific, given the robust counterpart, such as the BS one, for model (1)–(3), we define the LR counterpart as:

$$\min \sum_{i \in M} w_i \gamma_i \tag{13}$$

$$\sum_{j \in N} a_{ij}\, x_j + \beta(x, \Gamma_i) - \gamma_i \le b_i \quad i \in M \tag{14}$$

$$\sum_{j \in N} a_{ij}\, x_j \le b_i \quad i \in M \tag{15}$$

$$\sum_{j \in N} c_j\, x_j \le (1 + \delta)\, z^* \tag{16}$$

$$x_j \ge 0 \quad j \in N \tag{17}$$

$$\gamma_i \ge 0 \quad i \in M \tag{18}$$

Slack variables γ_i play the role of second-stage recourse variables used to recover from a possible infeasibility, whose weighted sum is minimized by objective function (13). Each variable γ_i defines the level of robustness of the solution with respect to uncertainty of parameters in row $i \in M$: in particular, γ_i takes a strictly positive value if the corresponding robust constraint i is violated. Constraint (16) imposes a maximum worsening of the objective function value with respect to z^*, defined as the value of the optimal solution of the nominal problem. The role of the input parameter δ in (16) is to balance the quality (optimality) and the feasibility (robustness) of the solution: $\delta = 0$ corresponds to the nominal problem (i.e., robustness is only taken into account to break ties among equivalent optimal solutions), while for $\delta = \infty$ the nominal objective function is not considered at all.

Note that the presence of constraints (14) combined with nominal constraints (15) and objective function (13), implies $0 \leq \gamma_i \leq \beta(x, \Gamma_i)$ for all $i \in M$. In other words, for each row i the model gives a prize (to be maximized) that is proportional to the slack quantity $s_i := b_i - \sum_{j \in N} a_{ij} x_j \ (= \beta(x, \Gamma_i) - \gamma_i)$, but only till the target value $\beta(x, \Gamma_i)$ is reached—a larger s_i does not receive any extra prize.

Weights w_i appearing in the objective function (13) are intended to compensate for possibly different scales for the constraints and can be set, e.g., to the Euclidean norm of each left-hand side coefficient vector. In the sequel we assume implicitly that all the constraints are stated in a comparable unit, hence we set $w_i = 1$ for all i. It is worth noting that the BS approach itself is intrinsically dependent on the specific formulation of LP model at hand, in the sense that it is not invariant with respect to transformations of the constraints that leave the feasible space of the nominal problem unchanged. In other words, the practical applicability of the BS approach (and hence of its LR counterpart) implicitly assumes that the original model is stated in a form that is "suited for robustness"—taking an LP-equivalent model can lead to meaningless results.

By using LP duality as in the BS approach, the LR counterpart of (1)–(3) becomes:

$$\min \ \sum_{i \in M} \gamma_i \tag{19}$$

$$\sum_{j \in N} a_{ij} \, x_j + \Gamma_i \, z_i + \sum_{j \in N} p_{ij} - \gamma_i \leq b_i \ \ i \in M \tag{20}$$

$$-\hat{a}_{ij} \, x_j + z_i + p_{ij} \geq 0 \ \ i \in M, j \in N \tag{21}$$

$$z_i \geq 0 \ \ i \in M \tag{22}$$

$$p_{ij} \geq 0 \ \ i \in M, j \in N \tag{23}$$

$$\sum_{j \in N} a_{ij} \, x_j \leq b_i \ \ i \in M \tag{24}$$

$$\sum_{j \in N} c_j \, x_j \leq (1 + \delta) \, z^* \tag{25}$$

$$\gamma_i \geq 0 \ \ i \in M \tag{26}$$

$$x_j \geq 0 \ \ j \in N \tag{27}$$

As stated, Light Robustness is strongly dependent on the BS definition of robustness, hence it can be applied only in those cases in which uncertainty can be described by means of a linear formulation. However, different LR variants can be defined for specific problems. In fact, in our view LR is not a rigid technique, but a modelling framework where robustness is achieved by first enforcing a demanding robustness/optimality goal, and then by allowing for local violations of the constraints (absorbed by the slack variables) to deal with possible infeasibility issues. In the next section we analyze a different (and simpler) LR version whose definition does not rely on the BS model. A problem-specific LR definition will be addressed in Section 7.

5   A Second Light Robustness Heuristic

We next describe a modified LR scheme (called MLR in the sequel) that is not based on the BS approach, but deals directly with the slack variables associated with the constraints of the nominal problem. The underlying assumption here is that the degree of robustness of a solution is somehow proportional to the slack left in the uncertain rows, to be used to absorb variations of the left-hand side coefficients. Determining the exact value of the slack in each row is of course a difficult task that depends on the whole solution x^* (and not just on the constraint slacks) and has to take into account interactions among the constraints, but it can be approached heuristically as follows.

Let x^* be an optimal solution of nominal problem (1)–(3), and let

$$L_i^* = \sum_{j \in N} (a_{ij} + \hat{a}_{ij})\, x_j^* - b_i$$

denote the maximum violation of constraint i with respect to solution x^*. We define by

$$U = \{i \in M : L_i^* > 0\}$$

the set of constraints that may be affected by uncertainty with respect to x^*. In other words, U contains the rows we want to take care of in terms of uncertainty, i.e., those rows for which enough slack should be given. We can assume without loss of generality $|U| \geq 1$, since otherwise the optimal solution x^* of the nominal problem would be feasible (and hence optimal) in any realization of the data.

We first solve the following LP

$$\max \sigma \tag{28}$$

$$\sum_{j \in N} a_{ij}\, x_j + s_i = b_i \quad i \in M \tag{29}$$

$$\sigma \leq \frac{s_i}{L_i^*} \quad i \in U \tag{30}$$

$$\sum_{j \in N} c_j\, x_j \leq (1 + \delta)\, z^* \tag{31}$$

$$x_j \geq 0 \quad j \in N \tag{32}$$

$$s_i \geq 0 \quad i \in M \tag{33}$$

which maximizes the minimum slack that can be assigned to any uncertain row. In order to take into account uncertainty on each row separately, the slack variable s_i in the i-th uncertain constraint (30) is heuristically normalized by dividing it by L_i^* ($i \in U$).

The LP above typically has several equivalent optimal solutions, due to its max-min nature. Indeed, objective function (28) only considers the row corresponding to the minimum normalized slack, hence there is no incentive in giving a large slack to the remaining rows—whereas this is very important for improving robustness. Thus, a second LP is solved in order to balance the slack among uncertain rows, while keeping the total amount of slack large enough. Given an optimal solution (x^*, s^*, σ^*) of model (28)–(33), we define the average and minimum value for the normalized slack as

$$s_{\mathrm{avg}} = \frac{\sum_{i \in U} s_i^* / L_i^*}{|U|}$$

$$s_{\min} = \min\{s_i^* / L_i^* : i \in U\} \quad (= \sigma^*)$$

and solve the following LP

$$\min \sum_{i \in U} t_i \tag{34}$$

$$\sum_{j \in N} a_{ij}\, x_j + s_i = b_i \quad i \in M \tag{35}$$

$$\sum_{j \in N} c_j\, x_j \leq (1 + \delta)\, z^* \tag{36}$$

$$\frac{s_i}{L_i^*} + t_i \geq s_{\mathrm{avg}} \quad i \in U \tag{37}$$

$$x_j \geq 0 \quad j \in N \tag{38}$$

$$s_i / L_i^* \geq s_{\min} \quad i \in U \tag{39}$$

$$s_i \geq 0,\ t_i \geq 0 \quad i \in U \tag{40}$$

In this model, for each uncertain constraint $i \in U$ we introduce an auxiliary variable t_i assuming a positive value if the associated normalized slack is smaller than the average. Objective function (34) penalizes the sum of these variables, so as to balance the normalized slack among all constraints.

Although this method requires the solution of two LPs (actually, three if the nominal problem is also considered), our computational experiments reported in Section 6 show that the corresponding extra computing time is quite small in practice, due to the use of fast parametric reoptimization techniques.

6   Computational Experiments on Random Data

In order to test the two LR approaches described in the previous sections, we performed computational experiments on knapsack and portfolio instances similar to those considered by Bertsimas and Sim in [6], and on variants of these instances.

Our computational measure of robustness for a given feasible solution $\tilde{x}$ of the nominal model (1)–(3) is provided by an external tool (called the external *validation tool* in the sequel) that generates 10,000 random *scenarios*, i.e., realizations of the input data according to a uniform distribution. The validation tool receives solution $\tilde{x}$ on input, and returns the probability of infeasibility of $\tilde{x}$ (of course, violation of a single constraint in model (1)–(3) is enough to declare $\tilde{x}$ infeasible for a certain scenario).

Since our LR heuristics require an optimality threshold on input, namely $\bar{z} := (1 + \delta)z^*$, a fair comparison with respect to BS is not immediate. In our experiments we implemented the following scheme.

We first solved the BS model (7)–(12) so as to test the BS approach alone. Since a main difficulty in using the BS approach is the definition of coefficients Γ_i to be used in (6), we heuristically fixed $\Gamma_i = \Gamma$ for all $i \in M$, and solved the corresponding BS model for increasing values of Γ, until a value was found, say $\Gamma^{\max}$, for which the corresponding solution is always feasible according to our external validation tool. We will refer to this solution as the *always-feasible* solution. The gap between the value of the optimal solution of the nominal problem and the value of the always-feasible solution is then used for defining threshold values $\bar{z}$. More specifically, we considered 9 threshold values obtained by allowing for a worsening (with respect to the optimal nominal solution) of 1%, 5%, 10%, 25%, 50%, 60%, 70%, 80%, and 90% of such a gap.

Once the threshold value $\bar{z}$ is fixed, we ran all models and evaluated the robustness of the corresponding solution $\tilde{x}$ through our external validation tool. The basic LR model (13)–(18) was solved by setting all Γ_i's to a constant (quite large) value, so as to require a high level of protection against uncertainty.

Our second heuristic LR model (MLR) does not require any other parameter and was solved as described. As to the BS model (7)–(12), we embedded it into a binary search procedure that finds the maximum real value of $\Gamma \in [0, n]$ such that the optimal solution value for model (7)–(12) does not exceed $\bar{z}$. In order to limit computing time, binary search is halted as soon as the difference between the maximum and minimum Γ values is smaller than 0.1. The procedure is further speeded-up, at each binary-search iteration, by stopping the solution of model (7)–(12) as soon as a solution with value not greater than $\bar{z}$ is found. In a similar way, each iteration is halted whenever a proof is given that no such solution exists. The value of Γ produced by the binary search procedure, say Γ^*, is therefore an approximation of the best possible value for model (7)–(12) when a solution having cost at most $\bar{z}$ is required. In the following we refer to this method as BinBS.

A fair comparison of BinBS and LR computing times is not immediate, since our experiment design is biased somehow in favor of the LR approach. Indeed, one could symmetrically fix the Γ value and apply binary search to LR to find the corresponding threshold value $\bar{z}$. According to our experience, in practical cases working with an optimality threshold is more natural than providing the Γ coefficient(s). In any case, the reported computing times for BinBS and LR have to be compared with some caution.

The following tables report, separately for each problem, the results of each method for each threshold value $\bar{z}$, showing the probability that the solution found is infeasible along with the corresponding computing time. In addition, for method BinBS we report the value Γ^* found by the binary search procedure, and the average time required to perform a single binary-search iteration. All experiments described in this section have been performed on a AMD Athlon 64 Processor 3500+ using ILOG-Cplex 10.1 as LP/ILP solver, and all computing times in Tables 1–7 are expressed in CPU seconds.

6.1 Single Knapsack Problem

One of the most famous problems in Combinatorial Optimization is the *Knapsack Problem* (KP) in which one is given a set $N = \{1, \ldots, n\}$ of *items* and a knapsack of capacity W. Each item $j \in N$ has associated a positive *profit* p_j and a positive weight w_j, and the aim is to select a set of items in such a way that (i) the sum of the weights of the selected items does not exceed W, and (ii) the sum of the profits of the selected items is maximized. By introducing, for each item $j \in N$ a binary variable x_j taking value 1 iff item j is selected, the problem can be formulated as follows:

$$\max \sum_{j \in N} p_j \, x_j \tag{41}$$

$$\sum_{j \in N} w_j \, x_j \leq W \tag{42}$$

$$x_j \in \{0, 1\} \quad j \in N \tag{43}$$

This problem is NP-hard, although pseudo-polynomial solution algorithms exist. For extensive studies on approaches to the knapsack problem, as well as to its variants or extensions, the reader is referred to the books by Martello and Toth [14] and by Kellerer, Pferschy and Pisinger [12].

Following Berstimas and Sim [6], we tested our robust approach by generating a KP instance with $|N| = 200$, integer profits p_j randomly generated in $[16, 77]$, integer weights w_j randomly generated in $[20, 29]$, and $W = 4000$. Uncertainty was modelled by allowing each weight to differ by at most 10% with respect to its nominal value.

Table 1 reports the value of the optimal solution of model (7)–(12) using different values for Γ. In addition, the table gives the percentage worsening in the solution value, the required computing time, and the probability that the provided solution is infeasible. The results of Table 1 experimentally confirm the theoretical results provided in [6] for what concerns both the worsening of the solution value and the probability of infeasibility.

The optimal solution of the nominal (maximization) problem has value 8801, while the always feasible solution, provided by model (7)–(12) with $\Gamma = 22$, has value 8732. The derived threshold values and the corresponding results for each robust method are reported in Table 2. The first table row has the following meaning: fixing a lower bound of $\bar{z} = 8800$ on the solution profit, one can find a solution with infeasibility probability of 43.18% (43.10% for MLR); this solution if found in 0.06 CPU seconds by BinBS (each binary-search iteration taking 0.01

Table 1. Results on BS model (7)–(12) on a random knapsack problem

Γ	z	% wors.	% Infeas	Time
0	8801	0.0000	47.57	0.00
1	8800	0.0114	43.18	0.01
5	8786	0.1704	19.96	0.12
10	8773	0.3181	5.08	0.02
15	8754	0.5340	0.57	0.06
20	8740	0.6931	0.02	0.13
22	8732	0.7840	0.00	0.14

Table 2. Results on a random knapsack problem

$\bar{z}$		BinBS			LR$_{(\Gamma=20)}$		MLR	
	Γ^*	% Infeas	Time	Avg.Time	% Infeas	Time	% Infeas	Time
8800	0.98	43.18	0.06	0.01	43.18	0.01	43.10	0.02
8797	1.95	36.82	0.07	0.01	36.82	0.01	36.54	0.01
8794	2.93	30.61	0.06	0.01	30.61	0.01	30.61	0.02
8783	5.47	18.60	0.19	0.02	18.60	0.01	18.60	0.19
8766	11.33	3.13	0.16	0.02	3.25	0.02	3.20	0.02
8759	13.28	1.48	0.25	0.03	1.48	0.02	1.48	0.15
8752	16.60	0.31	0.13	0.01	0.31	0.02	0.31	0.01
8745	18.75	0.12	0.09	0.01	0.12	0.02	0.12	0.03
8738	20.90	0.01	0.20	0.02	0.02	0.18	0.01	0.02

seconds on average), and corresponds to the choice $\Gamma^* = 0.98$, whereas LR and MLR require 0.01 and 0.02 seconds, respectively.

According to the table, for each threshold value $\bar{z}$ the three methods deliver solutions with negligible differences in terms of robustness. This is not surprising, due to the very simple structure of the KP problem. As expected, the LR approaches are faster than BinBS as no binary search is required.

6.2 Multi-dimensional Knapsack Problem

In order to validate our methods on a problem involving several constraints, we considered a Multi-dimensional Knapsack instance with $|M| = 10$ constraints. All coefficients and the associated deviations are generated as for the KP instance of Section 6.1, i.e., in the same way used in [6]. For this instance, the optimal solution of the nominal problem has value 8316, while the always feasible solution is provided by model (7)–(12) with $\Gamma = 24$ and has value 8238.

Computational results in Table 3 show that the solutions provided by BinBS and LR are quite similar in terms of robustness. On the other hand, computing times for LR are considerably smaller (often by two orders of magnitude) than those required by BinBS. E.g., for threshold $\bar{z} = 8296$ BinBS required 27.37 seconds, whereas LR took just 0.24 seconds. At first glance, this is quite surprising since the average BS time for a single binary-search iteration is 2.74

Table 3. Results on the multi-dimensional knapsack instance

		BinBS			LR$_{(\Gamma\,=\,20)}$		MLR	
$\bar{z}$	Γ^*	% Infeas	Time	Avg.Time	% Infeas	Time	% Infeas	Time
8315	0.59	86.97	1.88	0.19	88.42	0.21	86.97	0.10
8312	2.34	81.01	0.84	0.08	82.25	0.16	81.01	0.11
8308	3.32	78.20	0.88	0.09	71.53	0.11	71.58	0.08
8296	5.47	64.70	27.37	2.74	61.42	0.24	53.64	0.09
8277	12.70	5.62	4.45	0.45	6.74	0.14	5.62	0.08
8269	14.45	4.02	37.26	3.73	4.77	0.17	4.02	0.08
8261	18.75	0.30	2.38	0.24	0.30	0.30	0.30	0.11
8253	20.90	0.08	2.75	0.28	0.09	0.28	0.08	0.06
8245	22.66	0.05	12.08	1.21	0.15	2.05	0.05	0.09

Table 4. Results on the multi-dimensional knapsack instance with larger uncertainty of the coefficients in the first row

		BinBS			LR$_{(\Gamma\,=\,20)}$		MLR	
$\bar{z}$	Γ^*	% Infeas	Time	Avg.Time	% Infeas	Time	% Infeas	Time
8313	1.56	89.68	1.24	0.12	86.44	0.27	89.68	0.08
8301	4.30	73.34	6.32	0.63	73.35	0.10	74.51	0.08
8286	7.03	41.21	0.72	0.07	28.20	0.29	40.85	0.05
8241	10.16	8.99	0.96	0.10	7.30	1.35	8.99	0.04
8167	14.84	0.77	0.69	0.07	0.84	0.08	0.84	0.04
8137	16.60	0.29	1.23	0.12	0.29	0.06	0.29	0.04
8108	18.36	0.14	3.05	0.31	0.12	0.26	0.13	0.06
8078	20.31	0.09	0.70	0.07	0.08	0.74	0.08	0.07
8048	22.07	0.02	0.55	0.06	0.04	0.06	0.01	0.06

seconds, i.e., 10 times larger than LR. A similar situation arises for $\bar{z} = 8277$ and 8269. The explanation is that, during binary search, the BS model has to deal with weird (noninteger) values for the Γ_i coefficients appearing in (8), which makes these constraints numerically nasty and the solution of the overall problem much harder. The LR models, instead, do not suffer from this problem, due to the greater flexibility granted by the presence of slack variables.

As to MLR, it provides even (slightly) better results than BinBS and LR in terms of robustness, and requires much shorter computing times. E.g., for $\bar{z} = 8296$ it provides a solution with about 10% less probability of infeasibility than BinBS, and requires about 3 orders of magnitude less computing time.

In order to analyze the performance of various robust approaches on more demanding settings, we performed additional experiments on the multi-dimensional knapsack instance described above.

We first considered the situation arising when coefficients of the first constraint have more uncertainty than those of the other constraints. In particular, the original instance is considered, but each coefficient in the first constraint is allowed to differ by at most 50% with respect to its nominal value, while uncertainty for coefficients in the remaining rows is at most 10%, as in the previous

Table 5. Results on the multi-dimensional knapsack instance when the number of uncertain coefficients in each row is not a constant

		BinBS			LR$_{(\theta = 1.0)}$		MLR	
$\bar{z}$	θ^*	% Infeas	Time	Avg.Time	% Infeas	Time	% Infeas	Time
8315	0.015	77.89	0.58	0.06	77.80	0.09	77.89	0.08
8313	0.024	65.70	0.57	0.06	62.27	0.03	65.70	0.10
8311	0.043	40.60	0.39	0.04	60.39	0.05	40.60	0.04
8304	0.073	22.40	0.31	0.03	32.08	0.05	23.39	0.06
8292	0.108	20.64	6.82	0.68	13.28	1.24	27.12	0.11
8287	0.148	0.39	0.52	0.05	6.58	4.13	0.39	0.04
8283	0.167	0.09	0.58	0.06	2.47	1.00	0.20	0.05
8278	0.186	0.04	0.53	0.05	0.31	0.34	0.04	0.08
8273	0.204	0.00	0.75	0.08	0.14	0.02	0.02	0.06

Table 6. Results on the multi-dimensional knapsack instance when the number of uncertain coefficients in each row is not a constant, and high correlation among uncertainty in different rows exists

		BinBS			LR$_{(\theta = 1.0)}$		MLR	
$\bar{z}$	θ^*	% Infeas	Time	Avg.Time	% Infeas	Time	% Infeas	Time
8315	0.015	78.27	0.52	0.05	78.27	0.08	78.27	0.09
8313	0.024	64.52	0.58	0.06	59.38	0.02	64.52	0.10
8311	0.043	40.95	0.41	0.04	77.45	0.05	40.95	0.04
8304	0.073	20.60	0.33	0.03	59.56	0.05	22.86	0.06
8292	0.109	19.93	5.74	0.57	12.93	1.70	22.00	0.14
8287	0.149	0.45	0.76	0.08	2.16	1.31	0.45	0.04
8283	0.167	0.15	0.92	0.09	0.78	0.12	0.25	0.06
8278	0.186	0.09	0.64	0.06	0.13	0.10	0.10	0.07
8273	0.204	0.01	0.94	0.09	0.03	0.02	0.06	0.06

experiment. The corresponding computational results are given in Table 4 and confirm the previous findings: all three methods provided solutions with similar robustness (the only exception being $\bar{z} = 8313$ and 8286, where LR produced significantly more robust solutions), and MLR is faster than LR, which is in turn much faster than BinBS.

Finally, we considered two cases where the number of uncertain coefficients in each row is not a constant. In particular, let $J_i \subseteq N$ denote the index set of the uncertain coefficients in each row i ($i = 1, \ldots, 10$). We considered case $|J_i| = 10 * (11 - i)$, i.e., 100 uncertain coefficients arise in the first row, 90 in the second, and 10 in the last row. The set of uncertain coefficients in each row is generated according to a uniform distribution, and each uncertain coefficient can differ by at most 10% with respect to the nominal value.

Note that, in the new setting, defining a same value for all Γ_i's does not make sense for BS. Thus, according to [6], we considered a value θ representing the normalized number of uncertain coefficients, and defined $\Gamma_i = \theta |J_i|$ for each row i. Accordingly, binary search was executed with an accuracy equal to 10^{-3} on the value of θ, while LR was executed with $\theta = 1$.

In the instance addressed in Table 5, there is no correlation among uncertain coefficients in different rows. On the contrary, in the instance of Table 6 uncertainty was generated so that a coefficient can be uncertain in row i only if the coefficient in the same column is uncertain in row $i-1$, thus inducing a certain degree of correlation among uncertain coefficients in different rows.

Results in Tables 5 and 6 confirm once again that the LR heuristic approaches, in spite of their simplicity, are able to produce solutions that turn out to be equally (or even more) robust than those produced by BinBS, in much shorter computing times. In particular, MLR qualifies as the method of choice for producing robust solutions for multi-dimensional knapsack problems.

6.3 A Simple Portfolio Problem

The two previous subsections showed the effectiveness of the LR approach in the context of knapsack problems. In fact, these problems are very well suited for the LR models, as the slack variables in the model correspond to empty space in the the the knapsacks, so encouraging large slacks has a clear impact on the robustness of the final solution. There are however other contexts where the correlation between slacks and robustness is more subtle, hence the LR approach is less likely to be effective. However, it is important to stress that the LR performance depends heavily on the *model* used (rather than on the problem itself), in the sense that different models can lead to drastically different results in terms of robustness—a property shared by other approaches to robustness, including the BS one.

To illustrate this point, we consider a simplified portfolio problem taken again from [6]. Given a set $N = \{1, \ldots, n\}$ of stocks, the i-th having an estimated return p_i, a simplified portfolio problem requires to select the fraction x_i of wealth invested in stock i so as to maximize the portfolio value equal to $\sum_{i \in N} p_i x_i$. In real applications, the return value for stock i ($i \in N$) is subject to uncertainty, i.e., it can differ by at most σ_i from the nominal value.

A linear formulation for the portfolio problem can be obtained as follows:

$$\max \ z \tag{44}$$

$$z \le \sum_{i \in N} p_i \, x_i \tag{45}$$

$$\sum_{i \in N} x_i = 1 \tag{46}$$

$$x_i \ge 0 \ \ i \in N \tag{47}$$

We generated a portfolio instance as done in [6], using $n = 150$, and generating p_i and σ_i values as follows:

$$p_i = 1.15 + i\frac{0.05}{150} \qquad \text{and} \qquad \sigma_i = \frac{0.05}{450}\sqrt{2n(n+1)i}$$

so that stocks with higher return are also more risky. Note that the only uncertain constraint in the above model is (45).

Table 7. Results on a portfolio instance

	BinBS		LR$_{(\Gamma = 15)}$	MLR	MLR*
$\bar{z}$	Γ^*	% Infeas	% Infeas	% Infeas	% Infeas
1.1994	–	–	50.09	50.09	47.96
1.1971	–	–	49.97	49.60	42.88
1.1942	0.15	46.92	48.94	49.12	36.97
1.1855	1.03	37.79	49.10	47.63	19.03
1.1710	4.83	18.38	47.06	45.24	0.00
1.1652	7.18	11.23	22.72	44.30	0.00
1.1595	10.25	5.04	7.61	43.11	0.00
1.1537	14.06	2.02	1.98	42.19	0.00
1.1479	19.34	0.26	0.46	41.20	0.00

Table 7 gives the results on this problem, providing the same information as in the previous tables; computing times are negligible for all approaches and are omitted. Column MLR* refers to MLR applied to a different model, to be described later. Note that BinBS fails in finding a feasible solution for the first two threshold values, for which the value Γ^* is so small to be below our binary search precision (of course, one could modify the binary search procedure so as to deal with case $\Gamma^* = 0$).

According to the table, LR provides results (in terms of robustness) somehow worse than BinBS, which suggests that the LR slack variables are not effective in this context. This is confirmed by the very bad performance of MLR, that returns solutions whose robustness seems to be independent of the threshold. A closer look to the portfolio model clarifies the situation. Given a threshold value $\bar{z}$, an optimal solution for the LR counterpart of model (44)–(47) is given by $x_i = x_i^*, z = \bar{z}, s = z^* - \bar{z}$, where x^* denotes the optimal solution of the nominal problem and z^* its value. Hence, MLR will always keep the same solution x^* and use the slack variable s to absorb the allowed worsening of the objective function.

The above considerations would suggest that MLR is not applicable to the the portfolio application. However this is not true, in that one can derive an alternative model where the slack variables do play a role in terms of robustness (see also Bienstock [7] for a recent paper based on a similar idea). Indeed, consider the alternative LP model

$$\max z \tag{48}$$

$$z = \sum_{i \in N} z_i \tag{49}$$

$$z_i \le p_i \, x_i \quad i \in N \tag{50}$$

$$\sum_{i \in N} x_i = 1 \tag{51}$$

$$x_i \ge 0 \quad i \in N \tag{52}$$

By applying MLR to the above model one gets the results in column MLR* of Table 7, showing that our heuristic LR approach produces much better solutions than those found by BinBS with the original formulation.

7 A Real-World Application: The Train Timetabling Problem

In order to illustrate a possible application of the LR idea in a real world context, in this section we review the approach recently proposed by Fischetti, Salvagnin and Zanette [10] for finding robust railway timetables. We only give a brief sketch of the method and of the corresponding computational results; the reader is addressed to [10] for details.

The Train Timetabling Problem (TTP) consists in finding an effective train schedule on a given railway network. The schedule needs to satisfy some operational constraints given by capacities of the network and security measures. Moreover, one is required to exploit efficiently the resources of the railway infrastructure. In practice, however, the maximization of some objective function is not enough: the solution is also required to be robust against delays/disturbances along the network. Very often, the robustness of optimal solutions of the original problem turns out to be not enough for their practical applicability, whereas easy-to-compute robust solutions tend to be too conservative and thus unnecessarily inefficient. As a result, practitioners call for a fast yet accurate method to find the most robust timetable whose efficiency is only slightly smaller than the theoretical optimal one.

Fischetti, Salvagnin and Zanette (FSZ) [10] proposed and analyzed computationally alternative methods to find robust and efficient solutions to the TTP, in its aperiodic (non cyclic) version described in [9]. Their method is based on an event-based MIP model for the nominal TTP, akin to the formulation proposed in [17] for the periodic (cyclic) case, and will be outlined briefly in the sequel.

7.1 Measuring Timetable Robustness

FSZ implemented an external simulation-based validation module that is independent from the optimization model itself, so that it can be of general applicability and allows one to compare solutions coming from different methods. The module is required to simulate the reaction of the railways system to the occurrence of delays, by introducing small adjustments to the planned timetable (received as an input parameter). The underlying assumption here is that timetabling robustness is not concerned with major disruptions (which are to be handled by the real time control system and require human intervention) but is a way to control delay propagation, i.e., a robust timetable has to favor delay compensation without heavy human action. As a consequence, at validation time no train cancellation is allowed, and event precedences are fixed with respect to the planned timetable.

The validation model analyzes a single delay scenario at a time. As all event precedences are fixed according to the input solution to be evaluated, the nominal TTP constraints simplify to linear inequalities of the form:

$$t_i - t_j \geq d_{i,j} \tag{53}$$

where t_i and t_j are time variables associated with significant events (typically, arrival and departure of a train from a certain station), and $d_{i,j}$ is a minimum trip time or minimum rest/headway time. Let $\mathcal{P}$ denote the set of ordered pairs (i, j) for which a constraint of type (53) can be written, and E denote the set of events.

The problem of adjusting the given timetable t under a certain delay scenario δ^ω can thus be rephrased as the following simple LP model with decision variables t^ω describing the best possible adjustment of the published timetable t for the considered delay scenario:

$$\min \sum_{j \in E} \left(t_j^\omega - t_j \right) \tag{54}$$

$$t_i^\omega - t_j^\omega \geq d_{i,j} + \delta_{i,j}^\omega \quad (i,j) \in \mathcal{P} \tag{55}$$

$$t_i^\omega \geq t_i \quad i \in E \tag{56}$$

Constraints (55) correspond to linear inequalities just explained, in which the nominal right-hand-side value $\delta_{i,j}$ is updated by adding the (possibly zero) extra-time $\delta_{i,j}^\omega$ from the current scenario ω.

Constraints (56) are non-anticipatory constraints stating the obvious condition that one is not allowed to anticipate any event with respect to its published value in the timetable.

The objective function is to minimize the "cumulative delay" on the whole network.

Given a feasible solution t, the validation tool keeps testing it against a large set of scenarios, one at a time, gathering statistical information on the value of the objective function and yielding a concise figure (the average cumulative delay) of the robustness of the timetable.

7.2   Finding Robust Solutions

Different techniques to enforce robustness were implemented by FSZ.

A fat stochastic model. The first attempt to solve the robust version of the TTP was to use a standard scenario-based SP formulation whose structure can informally be sketched as follows:

$$\min \frac{1}{|\Omega|} \sum_{j \in E, \omega \in \Omega} \left(t_j^\omega - t_j \right) \tag{57}$$

$$\sum_{h \in T} \rho_h \geq (1 - \delta) z^* \tag{58}$$

$$t_i^\omega - t_j^\omega \geq d_{i,j} + \delta_{i,j}^\omega \quad (i,j) \in \mathcal{P}, \omega \in \Omega \tag{59}$$

$$t_i^\omega \geq t_i \quad i \in E, \omega \in \Omega \tag{60}$$

$$t_i - t_j \geq d_{i,j} \quad (i,j) \in \mathcal{P} \tag{61}$$

$$l_i \leq t_i \leq u_i \quad i \in E \tag{62}$$

The model is similar to that used in the validation tool, but takes into account several scenarios $\omega \in \Omega$ at the same time. Moreover, the nominal timetable values t_j are now viewed as decision variables to be optimized—their optimal value will define the final timetable to be published. The model keeps a copy of the original (linear) model with a modified right hand side for each scenario, along with the original model; the original variables and the correspondent second-stage copies in each scenario are linked through non-anticipatory constraints.

The objective is to minimize the cumulative delay over all events and scenarios. The original objective function (namely, the total train profit $\sum_{h \in T} \rho_h$, to be maximized, where T is the set of trains) is taken into account through constraint (58), where $\delta \geq 0$ is the tradeoff parameter and z^* is the objective value of the reference solution. As to the single-train profit variables ρ_h that appear in (58), they are linked to the timetable variables through appropriate constraints (not shown in the model); see [10] for a complete model.

For realistic instances and number of scenarios this model becomes very time consuming (if not impossible) to solve–hence we called it "fat". On the other hand, also in view of its similarity with the validation model, the fat model plays the role of a kind of "perfect model" in terms of achieved robustness, hence it will be used for benchmark purposes.

A slim stochastic model. Given the computing time required by the full stochastic model, the following alternative SP model was designed, which is simpler yet meaningful for the TTP problem.

$$\min \sum_{(i,j) \in \mathcal{P}, \omega \in \Omega} w_{i,j}^\omega s_{i,j}^\omega \tag{63}$$

$$\sum_{h \in T} \rho_h \geq (1 - \delta) z^* \tag{64}$$

$$t_i - t_j + s_{i,j}^\omega \geq d_{i,j} + \delta_{i,j}^\omega \quad (i,j) \in \mathcal{P}, \omega \in \Omega \tag{65}$$

$$s_{i,j}^\omega \geq 0 \quad (i,j) \in \mathcal{P}, \omega \in \Omega \tag{66}$$

$$t_i - t_j \geq d_{i,j} \quad (i,j) \in \mathcal{P} \tag{67}$$

$$l_i \leq t_i \leq u_i \quad i \in E \tag{68}$$

In this model there is just one copy of the original variables, plus the recourse variables $s_{i,j}^\omega$ that account for the unabsorbed extra times $\delta_{i,j}^\omega$. It is worth noting that the above "slim" model is inherently smaller than the fat one. Moreover, one can drop all the constraints of type (65) with $\delta_{i,j}^\omega = 0$, a situation that occurs very frequently in practice since most extra-times in a given scenario are zero.

As to the objective function, it involves a weighted sum of the the recourse variables. Finding meaningful values for the weights $w^\omega_{i,j}$ turns out to be very important. Indeed, we will shown in the sequel how to define the weights so as to produce solutions whose robustness is comparable with that obtainable by solving the (much more time consuming) fat model.

Light Robustness. A LR approach was used in [10] to generate robust timetables. The resulting method is related to the *adjustable robustness* paradigm used by Ben-Tal, El Ghaoui, and Nemirovski [1] in the context of project management.

In our TTP model, a typical constraint reads

$$t_i - t_j \geq d_{i,j}$$

where $d_{i,j}$ is the coefficient affected by uncertainty, and its LR counterpart is simply defined as

$$t_i - t_j + \gamma_{i,j} \geq d_{i,j} + \Delta_{i,j} \quad \gamma_{i,j} \geq 0$$

where $\Delta_{i,j}$ is a parameter fixing the desired (overconservative) protection level, and $\gamma_{i,j}$ are the slack variables whose weighted sum has to be minimized.

7.3   Computational Results

Computational tests were performed on four single-line medium-size TTP instances provided by the Italian railway company, Trenitalia. An almost-optimal heuristic solutions for each of these instances was computed through the algorithm described in [9], and used as a reference solution to freeze the event precedences and to select the trains to schedule.

The overall framework was implemented in C++ and tested on a AMD Athlon64 X2 4200+ computer with 4GB of RAM running Linux 2.6. The MIP solver used was `ILOG-Cplex` 10.1.

As far as scenarios are concerned, for each train on the line and for each scenario FSZ generated the corresponding extra-time, 5% on average, drawn from an exponential distribution, and distributed it proportionally to its train segments.

For each reference solution, a set of experiments was performed to compare the different methods for different values of the tradeoff parameter δ giving the allowed percentage of worsening of the nominal objective function, namely 1%, 5%, 10%, 20% and 40%. In particular, we compared the following alternative methods:

- *fat*: fat stochastic model (50 scenarios only)
- *slim1*: slim stochastic model with uniform objective function–all weights equal (400 scenarios)
- *slim2*: slim stochastic model with enhanced objective function (400 scenarios), where events arising earlier in each train sequence receive a larger weight in the objective function. More specifically, if the i-th event of train h is followed by k events, its weight in the objective is set to $k+1$. The idea beyond this weighing policy is that early extra-times in a train sequence are likely to propagate to the next ones, so they are more important.

Table 8. Comparison of different methods w.r.t. computing time and robustness (cumulative delay in minutes), for different lines and tradeoff δ

δ	Line	Fat Delay	Fat Time (s)	Slim1 Delay	Slim1 Time (s)	Slim2 Delay	Slim2 Time (s)	LR Delay	LR Time (s)
0%	BZVR	16149	9667	16316	532	16294	994	16286	2.27
0%	BrBO	12156	384	12238	128	12214	173	12216	0.49
0%	MUVR	18182	377	18879	88	18240	117	18707	0.43
0%	PDBO	3141	257	3144	52	3139	63	3137	0.25
	Tot:	49628	10685	50577	800	49887	1347	50346	3.44
1%	BZVR	14399	10265	15325	549	14787	1087	14662	2.13
1%	BrBO	11423	351	11646	134	11472	156	11499	0.48
1%	MUVR	17808	391	18721	96	17903	120	18386	0.48
1%	PDBO	2907	250	3026	57	2954	60	2954	0.27
	Tot:	46537	11257	48718	836	47116	1423	47501	3.36
5%	BZVR	11345	9003	12663	601	11588	982	12220	1.99
5%	BrBO	9782	357	11000	146	9842	164	10021	0.51
5%	MUVR	16502	385	18106	86	16574	107	17003	0.45
5%	PDBO	2412	223	2610	49	2508	57	2521	0.28
	Tot:	40041	9968	44379	882	40512	1310	41765	3.23
10%	BZVR	9142	9650	10862	596	9469	979	10532	2.01
10%	BrBO	8496	387	10179	132	8552	157	8842	0.51
10%	MUVR	15153	343	17163	84	15315	114	15710	0.43
10%	PDBO	1971	229	2244	50	2062	55	2314	0.25
	Tot:	34762	10609	40448	862	35398	1305	37398	3.20
20%	BZVR	6210	9072	7986	538	6643	1019	8707	2.04
20%	BrBO	6664	375	8672	127	6763	153	7410	0.52
20%	MUVR	13004	384	15708	91	13180	116	13576	0.42
20%	PDBO	1357	230	1653	55	1486	60	1736	0.28
	Tot:	27235	10061	34019	811	28072	1348	31429	3.26
40%	BZVR	3389	10486	4707	578	3931	998	5241	2.31
40%	BrBO	4491	410	6212	130	4544	166	6221	0.53
40%	MUVR	10289	376	13613	95	10592	108	11479	0.45
40%	PDBO	676	262	879	55	776	57	1010	0.28
	Tot:	18845	11534	25411	858	19843	1329	23951	3.57

– *LR*: light robustness model, with objective function as in *slim2* and protection level parameters set to $\Delta = -\mu \ln \frac{1}{2}$, where μ is the mean of the exponential distribution. This is the protection level required to absorb a delay of such distribution with probability $\frac{1}{2}$.

The results are reported in Table 8, where for each tradeoff parameter δ and railway line we give, for each method, the level of robustness of the corresponding solution (measured by the validation tool in terms of cumulative delay, in minutes—the smaller, the better) and the required computing time (in CPU seconds). According to the table, *fat, slim2 and LR* models produce solutions of comparable robustness (at least when the tradeoff parameter δ is not unrealistically large), whereas *slim1* is clearly the worst method. As to computing times, the *fat* model is one order of magnitude slower than *slim1* and *slim2*, although it uses only 50 scenarios instead of 400. *LR* is much faster than any other method, more than two orders of magnitude w.r.t the fast stochastic models, and qualifies as the method of choice to attack even larger instances.

8   Conclusions

In this paper we have addressed optimization problems in which input data is affected by uncertainty. Although many robust and/or stochastic programming models have been proposed in the literature to handle such a situation, their applicability is sometimes not completely satisfactory. We have proposed to deal with uncertainty by means of a new heuristic framework that we called Light Robustness (LR).

LR couples robust optimization with a simplified two-stage stochastic programming approach based on the introduction of suitable slack variables. The approach can be viewed as a "flexible counterpart" of robust optimization, obtained through the following modelling steps. We first fix the maximum objective function deterioration that we are willing to accept in our model. Then we define a "robustness goal" that we would like to achieve, and model it by using a classical robust optimization framework. In this way we obtain a robust model with no objective function, that however is likely to be infeasible. To cope with infeasibility, we introduce appropriate slack variables that allow for "local violations" of the robustness requirements, and define an auxiliary objective function aimed at minimizing the slacks. The LR slack variables then play a role similar to second-stage recourse variables in stochastic programming models, as they penalize the corrective actions needed to restore feasibility. In this view, LR is a heuristic framework combining the flexibility of SP (due to the presence of second-stage variables) and the modelling ease of RO.

Because of its heuristic nature, the LR performance on a specific application can only be evaluated through experimental analysis. We have reported extensive experiments on both random and real word problems, showing that LR is often able to produce solutions whose quality is comparable with that obtained through stochastic programming or robust models, though it requires much less effort in terms of model formulation and solution time—even if this latter aspect appears to be less important in many applications.

According to our computational results, the LR approach is mostly successful when the slack variables have a direct impact on robustness. For the cases where the correlation between slacks and robustness is more subtle, the LR approach is less likely to be effective, though an appropriate reformulation of the model can be highly beneficial. We have illustrated this behavior on a simplified portfolio problem, where a simple LR scheme applied to a suitable reformulation of the initial model produces extremely good results in term of robustness.

In our view, the LR framework is not a rigid technique, but a heuristic modelling framework where robustness is achieved by first enforcing a demanding robustness/optimality goal, and then by allowing for local violations of the constraints (absorbed by the slack variables) to deal with possible infeasibility issues. As such, effective LR variants can be designed for specific problems, such as the train timetabling problem recently addressed in [10].

Future research should investigate the applicability of the LR paradigm to other real life problems, so as to highlight its pros and cons in various contexts.

Acknowledgments

This work was supported by the Future and Emerging Technologies unit of the EC (IST priority), under contract no. FP6-021235-2 (project ARRIVAL). We thank Domenico Salvagnin and Arrigo Zanette for their work on the train timetabling application. Thanks are also due to two anonymous referees for their constructive comments, leading to an improved presentation.

References

1. El Ghaoui, L., Ben-Tal, A., Nemirovski, A.: Robust Optimization. In: preparation, preliminary draft (2008), http://www2.isye.gatech.edu/~nemirovs/RBnew.pdf
2. Ben-Tal, A., Nemirovski, A.: Robust solutions to uncertain linear programs. Operations Research Letters 25, 1–13 (1999)
3. Ben-Tal, A., Nemirovski, A.: Robust solutions of linear programming problems contaminated with uncertain data. Mathematical Programming 88, 411–424 (2000)
4. Ben-Tal, A., Nemirovski, A.: Robust optimization - methodology and applications. Mathematical Programming 92, 453–480 (2002)
5. Bertsimas, D., Sim, M.: Robust discrete optimization and network flows. Mathematical Programming 98, 49–71 (2003)
6. Bertsimas, D., Sim, M.: The price of robustness. Operations Research 52, 35–53 (2004)
7. Bienstock, D.: Histogram models for robust portfolio optimization. The Journal of Computational Finance 11 (2007)
8. Birge, J.R., Louveaux, F.: Introduction to Stochastic Programming. In: Springer Series in Operations Research and Financial Engineering. Springer, Heidelberg (2000)
9. Caprara, A., Fischetti, M., Toth, P.: Modeling and solving the train timetabling problem. Operations Research 50, 851–861 (2002)
10. Fischetti, M., Salvagnin, D., Zanette, A.: Fast approaches to improve the robustness of a railway timetable. Research paper, DEI, University of Padova (2007); to apper in Transportation Science
11. Kall, P., Wallace, S.W.: Stochastic Programming. Wiley, Chichester (1994)
12. Kellerer, H., Pferschy, U., Pisinger, D.: Knapsack Problems. Springer, Berlin (2004)
13. Linderoth, J.T., Shapiro, A., Wright, S.J.: The empirical behavior of sampling methods for stochastic programming. Annals of Operations Research 142, 219–245 (2006)
14. Martello, S., Toth, P.: Knapsack Problems: Algorithms and Computer Implementations. John Wiley & Sons, Chichester (1990)
15. Prékopa, A.: Stochastic Programming. Kluwer Academic Publishers, Dordrecht (1995)

16. Ruszczynski, A., Shapiro, A.: Stochastic Programming. In: Hanbooks in Operations Research and Management Science. Elsevier, Amsterdam (2003)
17. Serafini, P., Ukovich, W.: A mathematical model for periodic scheduling problems. SIAM Journal on Discrete Mathematics 2, 550–581 (1989)
18. Soyster, A.L.: Convex programming with set-inclusive constraints and applications to inexact linear programming. Operations Research 21, 1154–1157 (1973)

Incentive-Compatible Robust Line Planning*

Apostolos Bessas[1,3], Spyros Kontogiannis[1,2], and Christos Zaroliagis[1,3]

[1] R.A. Computer Technology Institute, N. Kazantzaki Str., Patras University Campus, 26500 Patras, Greece
[2] Computer Science Department, University of Ioannina, 45110 Ioannina, Greece
[3] Department of Computer Engineering and Informatics, University of Patras, 26500 Patras, Greece

mpessas@ceid.upatras.gr, kontog@cs.uoi.gr, zaro@ceid.upatras.gr

Abstract. The problem of *robust line planning* requests for a set of origin-destination paths (lines) along with their frequencies in an underlying railway network infrastructure, which are robust to fluctuations of real-time parameters of the solution. In this work, we investigate a variant of robust line planning stemming from recent regulations in the railway sector that introduce competition and free railway markets, and set up a new application scenario: there is a (potentially large) number of *line operators* that have their lines fixed and operate as competing entities issuing frequency requests, while the management of the infrastructure itself remains the responsibility of a single entity, the *network operator*. The line operators are typically unwilling to reveal their true incentives, while the network operator strives to ensure a fair (or socially optimal) usage of the infrastructure, e.g., by maximizing the (unknown to him) aggregate incentives of the line operators.

By investigating a resource allocation mechanism (originally developed in the context of communication networks), we show that a socially optimal solution can be accomplished in certain situations via an anonymous incentive-compatible pricing scheme for the usage of the shared resources that is *robust* against the unknown incentives and the changes in the demands of the entities. This brings up a new notion of robustness, which we call *incentive-compatible robustness*, that considers as robustness of the system its tolerance to the entities' unknown incentives and elasticity of demands, aiming at an eventual stabilization to an equilibrium point that is as close as possible to the social optimum.

1 Introduction

Problem Setting. An important phase in the strategic planning process of a railway (or any public transportation) company is to establish a suitable *line plan*, i.e., to determine the routes of trains that serve the customers. In the *line planning* problem, we are given a network $G = (V, L)$ (usually referred to as

---

* This work was partially supported by the Future and Emerging Technologies Unit of EC, under contracts no. FP6-021235-2 (FP6 IST/FET Open/Project ARRIVAL), and no. ICT-215270 (FP7 ICT/FET Proactive/Project FRONTS).

R.K. Ahuja et al. (Eds.): Robust and Online Large-Scale Optimization, LNCS 5868, pp. 85–118, 2009.

the public transportation network), where the node set V represents the set of stations (including important junctions of railway tracks) and the edge set L represents the direct connections or links (of railway tracks) between elements of V. A line p is a path in G. The *frequency* of a line p is a rational number indicating how often service to customers is provided along p within the planning period considered. For an edge $\ell \in L$, the *edge frequency* f_ℓ is the sum of the frequencies of the lines containing ℓ and is upper bounded by the *capacity* c_ℓ of ℓ, i.e., a maximum edge frequency established for safety reasons (measured as the maximum number of trains per day). The goal of the line planning problem is to provide the final set of lines offered by the public transportation company, along with their frequencies (also known as the *line concept*). Typically, a *line pool* is also provided, i.e., a set of potential lines among which the final set of lines will be decided. In certain cases, there may be *multiple line pools* representing the availability of the network infrastructure at different time slots or zones. This is due to variations in customer traffic (e.g., rush-hour pool, late evening pool, night pool), maintenance (some part of the network at a specific time zone may be unavailable), dependencies between lines (e.g., the choice of a high-speed line may affect the choice of lines for other trains), etc.

The line planning problem has been mostly studied under two main approaches (see e.g., [7,10]). In the *cost-oriented* approach, the goal is to minimize the costs of the public transportation company, under the constraint that all customers can be transported. In the *customer-oriented* approach, the goal is to maximize the aggregate level of satisfaction for the customers (e.g., maximize the number of customers with direct connections, minimize the maximum number of intermediate changes of a single customer, or minimize the traveling time of the customers). A recent approach aims at minimizing the travel times over all customers including penalties for the transfers needed [18,20].

The aforementioned approaches do not take into account certain fluctuations of input parameters; for instance, due to disruptions to daily operations (e.g., delays), or due to fluctuating customer demands. This aspect introduces the so-called *robust line planning* problem: Provide a set of lines along with their frequencies, which are robust to fluctuations of input parameters. Very recently, a game theoretic approach for robust line planning was presented in [19]. In that model, the lines act as players and the strategies of the players correspond to line frequencies. Each player aims to minimize the expected delay of her own line. The delay depends on the traffic load and hence on the frequencies of all lines in the network. The objective is to provide lines *and* their frequencies, that are robust against delays. This is pursued by distributing the traffic load evenly over the network (respecting edge capacities) such that the probability of delays in the system is as small as possible.

In this work, we investigate a different perspective of robust line planning stemming from recent regulations in the railway sector (at least within Europe) that introduce competition and free railway markets, and set up a new application scenario: there is a (possibly large) number of **line operators** (LOPs in short) that should operate as commercial organizations, while the management of the network remains the responsibility of a single (typically governmental) entity; we shall refer to the latter as the **network operator** (NOP in short). Under this framework, LOPs act as competing entities for the exploitation of shared goods and are (possibly) unwilling to reveal their actual level-of-satisfaction (or utility) functions that determine their true incentives. Nevertheless, the NOP would like to ensure the maximum possible level of satisfaction of these competing entities, e.g., by maximizing the (unknown due to privacy) aggregate levels of satisfaction. This would establish a notion of a socially optimal solution, which could also be seen as a fair solution, in the sense that the average level of satisfaction is maximized. Additionally, the NOP should ensure that the operational costs of the whole system are covered by a fair cost sharing scheme announced to the competing entities. This implies that a (possibly anonymous) pricing scheme for the usage of the shared resources should be adopted, that is also *robust* against changes in the demands of the LOPs. That is, we consider as *robustness* of the system its tolerance to the entities' unknown incentives and elasticity of demand requests, and the eventual stabilization at an equilibrium point that is as close as possible to the social optimum.

Contribution. In this paper, motivated by rate allocation in communication networks [13,14,21], we explore the aforementioned rationale by considering the case where the (selfishly motivated) LOPs request frequencies over a pool of already fixed line routes (one route per LOP). In particular, we investigate the resource allocation mechanism proposed in the pioneering work of Kelly [13]. Rather than requesting end-to-end frequencies, the LOPs offer bids, which they (dynamically) update for buying frequencies. Each LOP has a utility function determining her level of satisfaction that is *private*; i.e., she is not willing to reveal it to the NOP or her competitors, due to her competitive nature. The NOP announces an (anonymous) resource pricing scheme, which indirectly implies an allocation of frequencies to the LOPs, given their own bids.

Our first contribution is to show that for the case of a single line pool an adaptation of Kelly's approach [13] provides a distributed, dynamic, (LOP) bidding and (resource) price updating scheme, whose equilibrium point is the unknown social optimum – assuming strict concavity and monotonicity of the private utility functions. All dynamic updates of bids and prices can be done at the LOP and resource level respectively, based only on *local information* that concerns the particular LOP or resource. The key assumption is that the LOPs can control only a negligible amount of frequency along a single line, compared to its total frequency.

Our second contribution is a (non-trivial) extension of the approach for a single pool to the case of multiple line pools. By assuming that the NOP can periodically exploit a whole set of (disjointly operating) line pools and that each LOP may be interested in different lines from different pools, we show that there exists a globally convergent, dynamic, (LOP) bidding and (resource) price updating scheme, whose equilibrium point is the unknown social optimum. The NOP, similarly to the single pool case, uses a mechanism (a feasible frequency allocation rule and an anonymous resource pricing scheme) aiming to maximize the aggregate level of satisfaction of LOPs. The NOP, contrary to the single pool case, decides on how to divide the whole infrastructure among the different pools so that the resource capacity constraints are preserved, aiming (again) to achieve the optimal welfare value.

Our third contribution is an experimental study on a discrete variant of the distributed, dynamic scheme developed for the single pool case on both synthetic and real-world data. We note that in both single and multiple line pool cases the proposed mechanisms assure *market clearance*, i.e., the entire network infrastructure (capacities) is eventually used by the LOPs and all the budget afforded by the LOPs is actually spent.

Our solution is robust against the imperfect knowledge imposed by the private (unknown) utility functions and the arbitrary (dynamically updated) bids, since the proposed protocol enforces convergence to an equilibrium which is the social optimum. Our approach introduces a new notion of robustness, which we call *incentive-compatible robustness*, that is complementary to the notion of *recoverable robustness* introduced in [2,16,17]. The latter appears to be more suitable in the context of railway optimization, as opposed to the classical notion of robustness within robust optimization; see [2,16,17] for a detailed discussion on the subject and for the limitations of the classical approach as suggested in [4].

Recoverable robustness is about computing solutions that are robust against a limited set of scenarios (that determine the imperfection of information) and which can be made feasible (recovered) by a limited effort. One starts from a feasible solution x of an optimization problem, which a particular scenario s, that introduces imperfect knowledge (i.e., by adding more constraints), may turn to infeasible. The goal is to have at our disposal a recovery algorithm A that takes x and turns it to a feasible solution under s (i.e., under the new set of constraints). In other words, in recoverable robustness there is uncertainty about the feasibility space: imperfect information generates infeasibility and one strives to (re-)establish feasibility.

Incentive-compatible robustness is about computing an incentive-compatible recovery scheme for achieving robustness (interpreted as convergence to optimality). By an incentive-compatible scheme, we mean that the players act (update their bids, in our application) in a selfish manner during the convergence sequence. In this context, the feasibility space is known and incomplete information refers to complete lack of information about the optimization problem, due to the unknown utility functions. The goal is to define an incentive-compatible (pricing) scheme so that the players converge (recover) to the system's optimum. In other words, in incentive-compatible robustness there is uncertainty about the objectives: feasibility is guaranteed, since imperfect knowledge does not introduce new constraints, and one strives to achieve optimality, exploiting the selfish nature of the players.

Note that incentive-compatible robustness is different from the concept of game-theoretic robustness as developed in [1]. The approach in [1] is a centralized, deterministic paradigm to uncertainty in strategic games, mainly in the flavor of the Bertsimas and Sim approach [4] to robust LP optimization. We elaborate on the differences in Section 5.

Related Work and Approaches. Related to our work is that of Borndörfer et al. [5] that considers the allocation of slots in railway networks. That work considers the improvement of existing schedules of lines and frequencies, by reconsidering the allocation of (scarce) bundles of slots (i.e., lines with given frequencies in our own terminology) that have positive synergies with each other. The remaining schedule is assumed to remain intact, so that the resulting optimization problem is solvable. Initially, the involved users (LOPs) make some bids and consequently a centralized optimization problem is solved to determine the changes in the allocation of these slots so as to maximize the welfare of the whole system. This approach is different from ours in the following points: (i) It assumes no incentive-compatibility for the involved users and the eventual allocation is determined by a centralized scheduler. In our case, there is a simple pricing policy per resource (track), which is a priori known to all the players, and the winner is determined by the players' bids. The selfish behavior of the LOPs (in our case) is not only taken into account, but also exploited by the system in order to assure convergence to the social optimum of the whole network. (ii) The approach in [5] makes some local improvements *in hope* of improving the whole system, but does not exclude being trapped at some local optimum, which may be far away from the social optimum of the system. Our proposed scheme *provably* converges towards the social optimum, even if changes in the parameters of the game (e.g., in the players' secret utilities) change in the future. (iii) In [5], it is required that a centralized optimization problem is solved (considering the data regarding the whole network) and its solution is enforced in the current schedule. In our work (at least for the single-pool case) there is no need for global knowledge of the whole network. Each player dynamically adapts her bids according to her own (secret) utility and the aggregate cost she faces along her own path.

Another way to tackle the problem we consider here would be through the celebrated Vickrey-Clarke-Groves (VCG) class of mechanisms [6,12,22]. Such a mechanism would guarantee in our application scenario the existence of a dominant strategy equilibrium [11] in which the allocation of frequencies to the LOPs indeed maximizes the sum of their utilities, by encouraging LOPs to reveal their utility functions truthfully. Unfortunately, implementing VCG mechanisms is generally a very complex task, not only due to the huge size of the centralized optimization problem to be solved, but also due to the dynamic nature of an evolving market in which the parameters of the problem (railway infrastructure, number of participating LOPs, LOP utilities, etc) change over time. For instance, each LOP may vary her own utility function over time, due to changes in her own data, or her way of thinking. It may even be the case that some LOPs are unable to fully express their utility function, simply because they do not know it (for example, determining the parameters of such a function could be a hard optimization problem to solve by itself). On the other hand, it seems more plausible for a LOP to determine whether she would like to marginally increase or decrease her budget for claiming usage of the network, given the current situation she faces in the system. Therefore, we opt to follow Kelly's approach [13] by deploying a decentralized, dynamic updating scheme for the LOP budgets and the resource prices, whose updating rules are *simple* and are based on local information as much as possible that will monotonically converge to the socially optimal solution. Of course, the price to pay is some loss of efficiency w.r.t. how fast we converge to the optimum. Nevertheless, the self-stabilizing nature of our scheme, even to also dynamically changing optima (e.g., due to changes in the system infrastructure), is a very strong characteristic that compensates this efficiency drawback, compared to the adoption of a static, centralized VCG mechanism.

Structure. The rest of this paper is organized as follows. In Section 2, we provide the set up of our modeling, and present the adaptation of Kelly's approach [13] to the case of a single line pool by showing that the social optimum can be found by a polynomial-time computable mechanism, and by providing a decentralized, dynamic scheme that globally converges to the social optimum. To adapt and cast Kelly's approach to our problem setting, we recapitulate and re-prove certain results both for the sake of completeness and for providing the road-map for the extension to the multiple pools case. In Section 3, we provide our approach for the case of multiple line pools. We show that the social optimum can be found by a polynomial-time computable mechanism, and we provide a dynamic scheme (for implementing this mechanism) that globally converges to the social optimum. In Section 4, we present an experimental evaluation of our decentralized dynamic scheme for the single pool case, using synthetic and real-world data. In Section 5, we discuss incentive-compatible robustness and its comparison to other notions of robustness. We conclude in Section 6. A preliminary version of this work appeared in [15].

2   Single Line Pool: Modeling and Solution Approach

In this section, we present the modeling and the solution approach for the robust line planning problem we consider, for the case where a single line pool is provided. The development in this section is based on an adaptation of Kelly's resource allocation mechanism [13] (originally proposed within the context of communication networks for allocating network capacity to potential users). To adapt and cast Kelly's approach to our problem setting, we recapitulate and re-prove certain results for the sake of completeness and also to make this section the road-map for the development of our approach for the case of multiple line pools in Section 3.

Suppose that a set P of LOPs behave as competing service providers, willing to offer regular (train) line routes to the end users of a railway public transportation system. The NOP provides the (aforementioned) public transportation network $G = (V, L)$. The node set V represents train stations and junctions, while the edge set L (with each edge corresponding to a railway track establishing direct

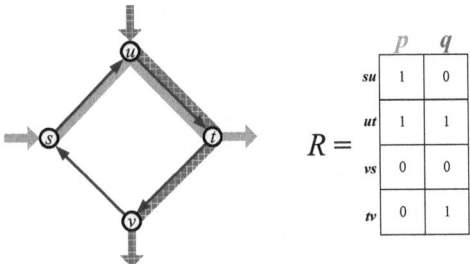

$$R = \begin{array}{c|c|c} & p & q \\ \hline su & 1 & 0 \\ \hline ut & 1 & 1 \\ \hline vs & 0 & 0 \\ \hline tv & 0 & 1 \end{array}$$

Fig. 1. A simple network with two distinct lines, and the corresponding routing matrix

connection for some pair of nodes in G) is the set of shared *resources* of the network. These resources are assumed to be subject to (fixed) capacity constraints, described by the capacity vector $\boldsymbol{c} = (c_\ell)_{\ell \in L} > 0$; for an edge $\ell \in L$, c_ℓ represents the maximum number of trains passing through ℓ over a whole time period (e.g., a day).

There is a fixed pool of line routes (i.e., origin–destination paths) that the LOPs are willing to use, and we assume that there is one line route per LOP[1]. This pool is represented by a **routing matrix** $\boldsymbol{R} \in \{0,1\}^{|L| \times |P|}$, in which each row $\boldsymbol{R}_{\ell,\star}$ corresponds to a different edge $\ell \in L$, and each column $\boldsymbol{R}_{\star,p}$ corresponds (actually, is the characteristic vector of) the line route of a distinct LOP $p \in P$. Fig. 1 demonstrates a network with two distinct lines and the corresponding routing matrix. Each LOP $p \in P$ claims a *frequency* of trains that she wishes to route over her path, $\boldsymbol{R}_{\star,p}$, given that no edge capacity constraint is violated in the network. A utility function $U_p : \mathbb{R} \mapsto \mathbb{R}$ determines the level of satisfaction of $p \in P$ for committing an end-to-end frequency $x_p > 0$ along her route $\boldsymbol{R}_{\star,p}$, for the purposes of her clients. These utility functions are assumed to be strictly increasing, strictly concave, nonnegative real functions of the end-to-end frequency x_p allocated to $p \in P$. It is also assumed that these functions are *private*: each LOP is not willing to reveal it either to the NOP, or to her competitors, due to her competitive nature.

The NOP is only interested in having a socially optimal (fair) solution. This is usually interpreted as maximizing the aggregate satisfaction of the LOPs. Therefore, the social welfare objective is considered to be the maximization of the aggregate utilities of the LOPs, subject to the capacity constraints. That is, the NOP is interested in the solution of the following convex optimization[2] problem:

$$\boxed{\text{SOCIAL (SC)}} \quad \max \left\{ \sum_{p \in P} U_p(x_p) : \boldsymbol{R}\boldsymbol{x} \leq \boldsymbol{c}; \ \boldsymbol{x} \geq \boldsymbol{0} \right\}$$

---

[1] We can always enforce this assumption by considering a LOP with more than one routes as different LOPs distinguished by the specific route.

[2] We make the tacit assumption that convex optimization refers to minimizing a convex function f, which is equivalent to maximizing the concave function $-f$.

Since all (private) utility functions are strictly concave and the feasible space is also convex, $\boxed{\text{SC}}$ has a *unique* optimal solution, which is called the **social optimum**. To solve $\boxed{\text{SC}}$ directly, the NOP, apart from the inherent difficulty in centrally solving (even convex) optimization programs of the size of a railway network, faces the additional obstacle of *not knowing* the exact shape of the objective function. Moreover, there exist some operational costs that have to be split among the LOPs who use the infrastructure, and this has to be done also in a fair way: each LOP should only be charged for the usage of the resources standing on her own route. In addition, the per–unit cost for using a line should be independent of a LOP's identity (i.e., we would like to have an *anonymous* pricing scheme for using the resources). But of course, this cost depends on the aggregate frequency induced by all the LOPs in each of these edges, due to the congestion effect. Indeed, it would be desirable for the NOP to be able to exploit the announcement of a pricing scheme not only for covering these operational costs, but also in such a way that a fair solution for all the LOPs is induced, despite the fact that there is no global knowledge of the exact utility functions of the LOPs.

In this work, we explore the possibilities of having such a frequency allocation and resource pricing mechanism. We would like this mechanism to depend only on the information affecting either a specific LOP (e.g., the amount of money she is willing to afford) or a specific resource (e.g., the aggregate frequency induced by the LOPs' demands on this resource), but as we shall see this is not always possible.

As for the LOPs (the players), each of them is interested in selfishly utilizing her own payoff, which is determined by the difference of the private utility value minus the operational cost that the NOP charges her for claiming an amount of frequency along her own route. The strategy space of a LOP is to claim (via bidding) the value of the frequency she is willing to buy, subject to the global capacity constraints (for all the players). It is mentioned here that this linear combination of the private utility and the cost share is not a real restriction, as there is no restriction for the shape of the utility function, other than the strict concavity and the monotonicity, which are quite natural assumptions.

2.1 Social Optimum – Tractability

Our first goal is to demonstrate that, despite the hidden utilities of the LOPs, it is indeed possible for the NOP to induce the social optimum, i.e., the solution of $\boxed{\text{SC}}$, as the result of the LOPs' selfish behavior. In order to study the effect of the selfish behavior in this setting, we consider the following **Frequency Game in Line Planning**:

- Each player $p \in P$ is a LOP, whose strategy is to choose a line frequency over her (already fixed) route $\boldsymbol{R}_{*,p}$ connecting her own origin–destination pair (s_p, t_p) of stations/stops.

- The strategy space for all the players is the set of feasible flows from origin to destination nodes, so that the edge capacity constraints are preserved. That is, the strategy space of the game is the set of vectors $\left\{ \boldsymbol{x} \in \mathbb{R}_{\geq 0}^{|P|} : \boldsymbol{Rx} \leq \boldsymbol{c} \right\}$.
- Each player's payoff is determined both by the value of the private utility function $U_p(x_p)$ (for having a frequency of x_p over her route) and the operational cost $C_p(\boldsymbol{x})$ she has to pay along her own route, due to the required frequency vector $\boldsymbol{x}$ induced by all the players in the network. Hence, player p's individual payoff is defined as: $IP_p(x_p, \boldsymbol{x}_{-p}) = U_p(x_p) - C_p(x_p, \boldsymbol{x}_{-p})$, where $\boldsymbol{x}_{-p}$ is the frequency vector for all the players but for player p. Therefore, the sole goal of player $p \in P$ is to choose her frequency so as to maximize her individual payoff:

$$\boxed{\text{USER}} \quad \max \left\{ IP(x_p, \boldsymbol{x}_{-p}) = U_p(x_p) - C_p(x_p, \boldsymbol{x}_{-p}) : x_p \geq 0 \right\}$$

- We consider as shared resources the capacities of the available network edges, for which the LOPs compete with each other.

As we shall explain later, we actually view this game as a mechanism–design instance, in which the NOP is the game regulator that receives the players' bids (for buying frequencies) and consequently decides both a feasible allocation of frequencies to the players and the payments that they have to provide. In this setting, the players can only affect their own eventual choice (allocation of a frequency) *indirectly* via bidding, rather than freely setting her own frequency along her route. In order to receive a (hopefully) higher frequency, a player may only offer a higher bid.

Describing the Social Optimum. Due to our assumption on the convexity of $\boxed{\text{SC}}$, we know that a frequency vector $\hat{\boldsymbol{x}}$ is the social optimum if there exists a vector of Lagrange Multipliers $\hat{\boldsymbol{\lambda}} = (\hat{\lambda}_\ell)_{\ell \in L}$ satisfying the following Karush-Kuhn-Tucker (KKT) conditions (see e.g., [3, Chap. 3]):

KKT-SOCIAL (KKT-SC)

$$U_p'(\hat{x}_p) = \hat{\boldsymbol{\lambda}}^T \cdot \boldsymbol{R}_{\star,p}, \quad \forall p \in P, \tag{1}$$

$$\hat{\lambda}_\ell \left(c_\ell - \boldsymbol{R}_{\ell,\star} \cdot \hat{\boldsymbol{x}} \right) = 0, \quad \forall \ell \in L, \tag{2}$$

$$\boldsymbol{R}_{\ell,\star} \cdot \hat{\boldsymbol{x}} \leq c_\ell, \quad \forall \ell \in L, \tag{3}$$

$$\hat{\boldsymbol{\lambda}}, \hat{\boldsymbol{x}} \geq \boldsymbol{0} \tag{4}$$

Of course, the problem with the $\boxed{\text{KKT-SC}}$ system is that the utility functions (and hence their derivatives) are unknown to the system. The question is whether there exists a way for the network designer to enforce the optimal solution of $\boxed{\text{SC}}$, also described in $\boxed{\text{KKT-SC}}$, without demanding this knowledge. The answer to this is *partially affirmative*, and this is by exploiting the selfish nature of the LOPs as we shall see shortly.

Setting the Right Pricing Scheme for the Players. In order to allow usage of his resources (the capacities of the edges in the network), the NOP has to define a pricing scheme that will (at least) pay back the operational costs of the edges. This scheme should be *anonymous*, in the sense that all the LOPs willing to use a given edge, will have to pay the same per–unit–of–frequency price for using it. But these prices may vary for different edges, depending on the popularity and the availability of each edge.

For the moment let us assume that we already know the optimal Lagrange Multipliers, $(\hat{\lambda}_\ell)_{\ell \in L}$ of $\boxed{\text{KKT-SC}}$. Interpreting these values as the per–unit–of–frequency prices of the resources, we have a pricing scheme for the frequency induced by the LOPs to their own routes. Each LOP pays exactly for the marginal cost of her own frequency at the resources she uses in her route. That is,

$$\forall p \in P, C_p(x_p, \boldsymbol{x}_{-p}) = \hat{\mu}_p \cdot x_p$$

where $\hat{\mu}_p \equiv \sum_{\ell \in L:R_{\ell,p}=1} \hat{\lambda}_\ell = \hat{\boldsymbol{\lambda}}^T \boldsymbol{R}_{\star,p}$ is the per–unit price for committing one unit of frequency along the route $\boldsymbol{R}_{\star,p}$ of player $p \in P$.

One should mention here that there is indeed an indirect effect of the other players' congestion in the marginal cost of each player, despite the fact that this seems to be only linear in her own frequency. This is because the scalar $\hat{\mu}_p$ actually depends on the optimal primal–dual pair $(\hat{\boldsymbol{x}}, \hat{\boldsymbol{\lambda}})$.

We next assume that the players are actually controlling only *negligible* amounts of frequencies compared to the aggregate ones[3]. Then, their effect in the total congestion (and therefore in the values of the marginal prices) is also negligible. This implies that the players consider the per-unit-prices they face to be constant, even if this is actually affected by the frequency vector as well. In such a case we say that the players are **price takers**, i.e., they accept the prices without anticipating to have an effect on them by their own strategy. In such a case each player solves the following optimization problem:

$$\boxed{\text{USER-I}} \ \max\{U_p(x_p) - \hat{\mu}_p x_p : x_p \geq 0\}$$

Due to the convexity of $\boxed{\text{USER-I}}$, $\tilde{x}_p \geq 0$ is an optimal solution if $U_p'(\tilde{x}_p) = \hat{\mu}_p$. That is, each player (selfishly) tries to satisfy her own part of equations (1) in $\boxed{\text{KKT-SC}}$. Of course, we still have to deal with the crucial problem that the optimal Lagrange Multipliers (that define the marginal prices for the users) cannot be directly computed, due to both the size of $\boxed{\text{SC}}$ and the lack of knowledge of the private utility functions, in the framework of railway optimization.

To tackle this situation, we transform the Frequency Game to a mechanism design instance, in order to have a more active participation of the NOP, as the game regulator. In particular, we consider the following two-level scenario for dynamically setting per–unit prices of the edges and frequencies of the selfish players. Initially each LOP $p \in P$ announces a bid $w_p \geq 0$ concerning the total amount of money she is willing to pay for buying frequency along her own route.

---

[3] For the considered application scenario, this is not unrealistic.

The exact amount of frequency that she will eventually buy, depends on the per–unit price that will be announced by the NOP, and is not yet known to her (nevertheless, it will be the case that, given the other players' bids, any LOP $p \in P$ may only increase her assigned frequency by raising, unilaterally, her own bid). Consequently, the NOP considers the following optimization problem, whose Lagrange Multipliers define the per–unit prices of the edges:

$$\boxed{\text{NETWORK (NET)}} \quad \max \left\{ \sum_{p \in P} w_p \cdot \log(x_p) : \boldsymbol{R}\boldsymbol{x} \leq \boldsymbol{c}; \ \boldsymbol{x} \geq \boldsymbol{0} \right\}$$

That is, the NOP considers that the private utility $U_p(x_p)$ is substituted by the (also strictly concave and increasing, for any given bid vector $\boldsymbol{w} \in \mathbb{R}_{\geq 0}^{|P|}$) function $w_p \log(x_p)$. The choice of this function, along with the selfishness of the LOPs, allows us to obtain a *convex program with linear inequalities*, whose KKT conditions are very similar (except for the first line) to those of $\boxed{\text{KKT-SC}}$:

$$\text{KKT-NETWORK (KKT-NET)}$$

$$\frac{w_p}{\bar{x}_p} = \bar{\boldsymbol{\lambda}}^T \cdot \boldsymbol{R}_{\star,p}, \quad \forall p \in P, \tag{5}$$

$$\bar{\lambda}_\ell \left(c_\ell - \boldsymbol{R}_{\ell,\star} \cdot \bar{\boldsymbol{x}} \right) = 0, \quad \forall \ell \in L, \tag{6}$$

$$\boldsymbol{R}_{\ell,\star} \cdot \bar{\boldsymbol{x}} \leq c_\ell, \quad \forall \ell \in L, \tag{7}$$

$$\bar{\boldsymbol{\lambda}}, \bar{\boldsymbol{x}} \geq \boldsymbol{0} \tag{8}$$

By $(\bar{\boldsymbol{x}}, \bar{\boldsymbol{\lambda}})$, we denote the optimal primal-dual pair of $\boxed{\text{KKT-NET}}$. Observe that the only difference between $\boxed{\text{KKT-NET}}$ and $\boxed{\text{KKT-SC}}$ concerns the (left-hand side of) equations (5) and (1), respectively. But we shall demonstrate now that the selfish (and price taking) behavior of the LOPs is enough to make this difference vanish. Returning to the LOPs, we initially assumed that they announce some *fixed* bids, and consequently the NOP sets the per–unit prices of the resources. Given the bid vector and the resource prices, it is then easy to determine each LOP's assigned frequency. But the truth is that, since the pricing scheme changes over time, it is in the interest of each LOP to actually vary her own bid over time. Indeed, if the players are assumed to be price takers and act myopically (i.e., without anticipating to affect the prices via their own pricing policy), then they will try to solve the following system, which is parameterized by the instantaneous set of per–unit prices $\boldsymbol{\mu}(t) = (\mu_p(t))_{p \in P}$ (now seen by the LOPs as *constants*) they are charged at time $t \geq 0$:

$$\boxed{\text{USER-II}} \quad \max \left\{ U_p \left(\frac{w_p(t)}{\mu_p(t)} \right) - w_p(t) : w_p(t) \geq 0 \right\}$$

Due to convexity, the optimal solution $\tilde{w}_p(t)$ of the unconstrained optimization program $\boxed{\text{USER-II}}$, will be the bid chosen by player $p \in P$ at time $t \geq 0$, and is be given by:

$$\frac{1}{\mu_p(t)} \cdot U'_p\left(\frac{\tilde{w}_p(t)}{\mu_p(t)}\right) = 1 \Leftrightarrow$$

$$U'_p\left(\tilde{x}_p(t)\right) = U'_p\left(\frac{\tilde{w}_p(t)}{\mu_p(t)}\right) = \mu_p(t) \Leftrightarrow$$

$$\tilde{x}_p(t)U'_p\left(\tilde{x}_p(t)\right) = \mu_p(t) \cdot \tilde{x}_p(t) = \tilde{w}_p(t)$$

That is, the price taking, myopic players have an incentive to set their bids properly so that $\forall t \geq 0, \forall p \in P, w_p(t) = x_p(t)U'_p(x_p(t))$. This will also hold at the optimal solution of $\boxed{\text{NET}}$, i.e., $\forall p \in P, \bar{w}_p = \bar{x}_p U'_p(\bar{x}_p)$. But when this is true, it also holds that $\boxed{\text{KKT-NET}}$ and $\boxed{\text{KKT-SC}}$ coincide. That is, the selfish–bidding behavior of the myopic, price taking players, under the pricing scheme $\bar{\boldsymbol{\lambda}}$ determined by the Lagrange Multipliers of $\boxed{\text{KKT-NET}}$, leads to the optimal solution $(\bar{\boldsymbol{x}}, \bar{\boldsymbol{\lambda}}) = (\hat{\boldsymbol{x}}, \hat{\boldsymbol{\lambda}})$ of $\boxed{\text{KKT-SC}}$.

The discussion within this section establishes the following result.

Theorem 1. *Consider a transportation network $G = (V, L)$ and a set P of (selfish, price taking) LOPs with hidden utilities, whose lines are determined by a routing matrix $\boldsymbol{R} \in \{0, 1\}^{|L| \times |P|}$. There exists a polynomial–time computable mechanism (i.e., a pair of a frequency allocation rule and resource pricing scheme) which induces the optimal solution of $\boxed{\text{SC}}$ as a result of the LOPs' selfish behavior.*

2.2   Social Optimum – Dynamic and Decentralized Computation

At this point, one could argue that, in order to solve the (partially determined) convex program $\boxed{\text{SC}}$, it suffices to determine the proper resource prices by the optimal solution of the (completely determined, and computationally tractable) convex program $\boxed{\text{NET}}$. The latter can be directly solved and provide the proper Lagrange Multipliers of $\boxed{\text{SC}}$. However, the huge scale of a railway network optimization instance makes this rationale rather unappealing.

Motivated by the pioneering work of Kelly et al. [13,14] and its excellent simplification and elaboration in [21], we shall try to compute an optimal solution of $\boxed{\text{NET}}$ as the stable point of a system of differential equations that determines the updates of the resource prices, and (consequently) the LOPs' bids. The crucial observation at this point is that it suffices to enforce the resource prices to gradually converge to the optimal price vector $\bar{\boldsymbol{\lambda}}$ provided by $\boxed{\text{NET}}$, and the "right bids" will follow.

We consider the following dynamic system of differential equations that actually constitutes our decentralized, dynamic algorithm for computing the social optimum.

1. Each resource (edge in the transportation network) is equipped with a dynamically updated charging mechanism, which is the same (per–unit) price for all the LOPs using it. This charging mechanism is updated according to the following system of differential equations:

$$\forall \ell \in L, \quad \dot{\lambda}_\ell(t) = \max\{y_\ell(t) - c_\ell, 0\} \cdot \mathbb{I}_{\{\lambda_\ell(t)=0\}} + (y_\ell(t) - c_\ell) \cdot \mathbb{I}_{\{\lambda_\ell(t)>0\}} \quad (9)$$

where $y_\ell(t) \equiv \sum_{p \in R: R_{\ell,p}=1} x_p(t) = \boldsymbol{R}_{\ell,\star} \cdot \boldsymbol{x}(t)$ is the aggregate frequency committed at edge $\ell \in L$ at time $t \geq 0$, and $\mathbb{I}_{\{\mathcal{E}\}}$ is the indicator variable of the truth of a logical expression $\mathcal{E}$.

2. Each LOP $p \in P$, at any time $t \geq 0$, is charged an instantaneous per-unit price $\mu_p(t) \equiv \sum_{\ell \in L: R_{\ell,p}=1} \lambda_\ell(t) = \boldsymbol{\lambda}(t)^T \cdot \boldsymbol{R}_{\star,p}$. It solves $\boxed{\text{USER-II}}$ to determine $w_p(t)$, and consequently is allocated a frequency $x_p(t) = \frac{w_p(t)}{\mu_p(t)}$.

The system of differential equations (9) is obtained from the well-known approach (see e.g., [13,21]) that considers the Lagrange Multipliers of an optimization problem as the (per unit) prices of the resources corresponding to the constraints represented by each Lagrange Multiplier. Therefore, the above system has the following intuitive interpretation. For each resource ℓ that currently has a zero price, the tendency is to increase the price only if this resource is over-used (i.e., the aggregate frequency exceeds the capacity of the resource). When a resource has positive price, then the tendency is either to increase or reduce this price, depending on whether its current frequency exceeds or is below the capacity of the resource, respectively. Thus, the only stable situation is when a resource is either under-used and has zero price (since there is no interest in using the residual capacity), or its frequency has already reached its capacity. Observe that the equilibrium of this system of differential equations has $\forall \ell \in L, \bar{y}_\ell \equiv \boldsymbol{R}_{\ell,\star} \cdot \bar{\boldsymbol{x}} = c_\ell \ \vee \ \bar{\lambda}_\ell = 0$. That is, the complementarity conditions of both $\boxed{\text{KKT-SC}}$ and $\boxed{\text{KKT-NET}}$ (equations (2) and (6)) are satisfied.

Step 2 above implies that at equilibrium player p, given its commitment on spending w_p for buying frequency, is allocated a frequency of $\bar{x}_p = \frac{w_p}{\bar{\mu}_p}$. From this we deduce that at equilibrium also the equations (5) of $\boxed{\text{KKT-NET}}$ are satisfied.

We are now ready to prove the following.

Theorem 2. *The above defined dynamic system of resource-pricing and LOP-bid-updating differential equations ensures monotonic convergence to the social optimum of* $\boxed{\text{NET}}$ *from any initial point of resource prices and LOP bids.*

Proof. The above system of differential equations is a distributed algorithm, in which each LOP reacts to signals she gets about the aggregate frequency along her route. These signals are the per-unit prices $\mu_p(t)$ that the LOP gets from the NOP at any time.

The question is whether the above system converges at all. This is indeed true, if we assume that the routing matrix $\boldsymbol{R}$ has full rank. This assures that given a set $\boldsymbol{\lambda}(t) = (\lambda_\ell(t))_{\ell \in L}$ of instantaneous per-unit prices at the resources, the set $\boldsymbol{\mu}(t) = (\mu_p(t))_{p \in P}$ of per-unit prices for the LOPs, that is computed as the solution of the system $\boldsymbol{\mu}(t) = R^T \cdot \boldsymbol{\lambda}(t)$, is *unique*. Using a proper Lyapunov function argument, it can be shown (cf. [21, Chapter 3]) that this dynamic (and distributively implemented) pricing scheme, for *fixed* player bids $(w_p)_{p \in P}$, is stable and converges to the optimal solution $(\bar{x}, \bar{\lambda})$ of $\boxed{\text{NET}}$.

In particular, consider the Lyapunov function $V(\lambda(t)) = \frac{1}{2}(\lambda(t) - \bar{\lambda})^T(\lambda(t) - \bar{\lambda})$. To show stability of our scheme, it suffices to show that $dV(\lambda(t))/dt \leq 0$. Then we have:

$$
\begin{aligned}
\frac{dV(\lambda(t))}{dt} \\
= \sum_{\ell \in L} (\lambda_\ell(t) - \bar{\lambda}_\ell) \cdot \dot{\lambda}(t) \\
= \sum_{\ell \in L} (\lambda_\ell(t) - \bar{\lambda}_\ell) \cdot [\max\{y_\ell(t) - c_\ell, 0\} \cdot \mathbb{I}_{\{\lambda_\ell(t)=0\}} + (y_\ell(t) - c_\ell) \cdot \mathbb{I}_{\{\lambda_\ell(t)>0\}}] \\
\leq \sum_{\ell \in L} (\lambda_\ell(t) - \bar{\lambda}_\ell) \cdot (y_\ell(t) - c_\ell) \\
= \sum_{\ell \in L} (\lambda_\ell(t) - \bar{\lambda}_\ell) \cdot [(y_\ell(t) - \bar{y}_\ell) + (\bar{y}_\ell - c_\ell)] \\
\leq \sum_{\ell \in L} (\lambda_\ell(t) - \bar{\lambda}_\ell) \cdot (y_\ell(t) - \bar{y}_\ell) \\
= \sum_{\ell \in L} (\lambda_\ell(t) - \bar{\lambda}_\ell) \cdot R_{\ell,\star} \cdot (x(t) - \bar{x}) \\
= \sum_{p \in P} (\mu_p(t) - \bar{\mu}_p) \cdot (x_p(t) - \bar{x}_p) \\
= \sum_{p \in P} \left(\frac{w_p}{x_p(t)} - \frac{w_p}{\bar{x}_p} \right) \cdot (x_p(t) - \bar{x}_p) = \sum_{p \in P} w_p \cdot \left(2 - \frac{x_p(t)}{\bar{x}_p} - \frac{\bar{x}_p}{x_p(t)} \right) \\
\leq 0
\end{aligned}
$$

The first inequality holds because: $\forall \ell \in L$, (i) if $\lambda_\ell(t) > 0$ then $\dot{\lambda}_\ell(t) = y_\ell - c_\ell$; (ii) if $\lambda_\ell(t) = 0$ then $\max\{y_\ell - c_\ell, 0\} \geq 0$ and $\lambda_\ell(t) - \bar{\lambda}_\ell = -\bar{\lambda}_\ell \leq 0$. Therefore, for $\lambda_\ell(t) = 0$ it holds that $(\lambda_\ell(t) - \bar{\lambda}_\ell)\max\{y_\ell(t) - c_\ell, 0\} = -\bar{\lambda}_\ell \max\{y_\ell(t) - c_\ell, 0\} \leq 0$. But so long as $\lambda(t) = 0$, it holds that the total frequency $y_\ell(t)$ is at most as large as the capacity c_ℓ (otherwise the price for this resource would have raised earlier). That is, $0 \leq -\bar{\lambda}_\ell(y_\ell(t) - c_\ell)$. The second inequality holds because at equilibrium no aggregate frequency $\bar{y}_\ell$ can exceed the capacity c_ℓ of the resource, and $\bar{\lambda}_\ell(\bar{y}_\ell - c_\ell) = 0$. The third inequality holds because $\forall z > 0, z + \frac{1}{z} \geq 2 \Rightarrow 2 - z - \frac{1}{z} \leq 0$. We have also exploited the facts that $\forall t \geq 0, y(t) = R \cdot x(t)$ and $\mu(t) = \lambda(t)^T \cdot R$. $\qquad \square$

3   Multiple Line Pools: Modeling and Solution Approach

In this section we extend the freedom of both the NOP and the LOPs. For the NOP we assume that he can now periodically exploit a whole set K of (disjointly operating) line pools, rather than a single line pool, to serve the LOPs' connection requests. It is up to the NOP how to split a whole operational period of the railway infrastructure among the different pools, so that (in overall, for the whole period) the resource capacity constraints are not violated. A first assumption that we make at this point, is that the NOP divides the usage of the whole infrastructure (rather than each resource separately) among the pools. This is because we envision the line pools to be implemented, not concurrently, but in disjoint time intervals (e.g., via some sort of time division multiplexing), and also to concern different characteristics of the involved lines (e.g., high-speed pool, regular-speed pool, local-trains pool, rush-hour pool, night-shift pool, etc.). The capacity of each resource (as in the single pool case) refers to its usage (number of trains) over the whole time period we consider (e.g., a day), and if a particular pool consumes (say) 50% of the whole infrastructure, then this implies that for all the lines in this pool, each resource may exploit at most half of its capacity.

As for the LOPs, they can now even claim different lines from different pools. In accordance with the single pool case, each LOP may express interest in at most one line per pool. For simplicity we assume that each LOP is interested for *exactly* one line per pool, adding dummy origin-destination pairs connected with an edge of zero capacity, for every LOP that has no interest in some pool. Technically, our analysis would allow even the case where a LOP expresses interest for lines with different origin–destination pair (in different pools). Nevertheless, we assume that each LOP $p \in P$ has a *single* (strictly concave, as before) utility function $U_p : \mathbb{R}_{\geq 0} \mapsto \mathbb{R}_{\geq 0}$, which depends on the *aggregate frequency* x_p that she gets from all the pools in which she is involved. In order to be in compliance with this assumption, we consider the case where each LOP expresses interest for different lines (at most one per pool) over the same origin-destination pair. Of course, in reality, different ways of dividing the same aggregate frequency x_p among the various lines, could make a huge difference for the particular LOP, but we do not account for this effect in this paper.

In analogy with the single pool case, each pool $k \in K$ is represented by its own routing matrix $\boldsymbol{R}(k) \in \{0,1\}^{|L| \times |P|}$. The frequency (number of trains over one time period) granted to the LOP $p \in P$ within the pool $k \in K$, is indicated by a nonnegative real variable $x_{p,k}$. The aggregate frequency that p gets is then $x_p = \sum_{k \in K} x_{p,k}$. The LOPs still try to have (indirect, via bidding) control over the aggregate end-to-end frequency x_p they get by the NOP along all their lines, from all the possible pools that may be of use by the NOP. It is up to the NOP to decide how to divide the whole railway infrastructure among the different pools, so that the resource capacity constraints are preserved, the goal being to achieve the optimal social welfare value. That is, the NOP now directly participates in the optimization problem via the variables $f_k : k \in K$ indicating the proportion of capacity that each pool consumes from every resource, over the whole time period we study. We will say that the NOP or the vector $\boldsymbol{f}$ *completely divides* the infrastructure, if $\sum_{k \in K} f_k = 1$.

The NOP is now interested in solving the following optimization problem:

MULTI-SOCIAL (MSC)

$$\text{maximize} \sum_{p \in P} U_p(x_p) = \sum_{p \in P} U_p \left(\sum_{k \in K} x_{p,k} \right)$$

$$\text{s.t. } \forall (\ell, k) \in L \times K, \ \sum_{p \in P} R_{\ell,p}(k) \cdot x_{p,k} \leq c_\ell \cdot f_k$$

$$\sum_{k \in K} f_k \leq 1$$

$$x, f \geq 0$$

Once more, this is a strictly convex optimization problem (due to the strict concavity of the LOP utility functions, and the linearity of the feasible space), whose objective function is unknown to the NOP. We shall explain in this section how we can tackle this issue. The overall idea is that we can handle the multiple pools case as an expanded single pool case. We have $|K|$ replicas of the same railway infrastructure, and $|K|$ replicas $p_1, \ldots, p_{|K|}$ of the same LOP $p \in P$, each being interested only in a single line (the one of interest to LOP p in the corresponding pool). Each LOP offers her bid w_p for buying aggregate frequency x_p. The NOP determines the proportions of railway infrastructure that are committed per pool. Exactly the same proportions are used (by the NOP) for splitting the LOPs' bids among the various pools.

3.1 Multi Social Optimum – Tractability

We start by an observation that exploits the economic interpretation of the Lagrange Multipliers of MSC.

Lemma 1. *Assuming that all the players adopt strictly increasing, concave utility functions, if the resource prices are determined by the vector $\hat{\Lambda}$ of optimal Lagrange Multipliers of the resource constraints in* MSC*, then the following are true: (i) Each LOP is indifferent of the way her aggregate frequency is split among different pools. (ii) All the pools have the same (weighted) aggregate cost. (iii) The NOP completely divides the whole railway infrastructure among the different pools. Facts (i) and (iii) also hold even when the NOP fixes a priori the vector of proportions, for the corresponding optimal solution.*

Proof. Let Λ be the vector of Lagrange Multipliers for the resource capacity constraints, and ζ the Lagrange Multiplier concerning the constraint for the capacity proportions per pool. The Lagrangian function is the following:

$$L(x, f, \Lambda, \zeta)$$

$$= \sum_{p \in P} U_p(x_p) - \sum_{\ell \in L} \sum_{k \in K} \Lambda_{\ell,k} \cdot \left[\sum_{p \in P} R_{\ell,p}(k) \cdot x_{p,k} - c_\ell \cdot f_k \right] - \zeta \left[\sum_{k \in K} f_k - 1 \right]$$

$$= \sum_{p \in P} \left[U_p(x_p) - \sum_{k \in K} x_{p,k} \left(\sum_{\ell \in L} \Lambda_{\ell,k} \cdot R_{\ell,p}(k) \right) \right] + \sum_{k \in K} f_k \cdot \left[c^T \Lambda_{\star,k} - \zeta \right] + \zeta$$

$$= \sum_{p \in P} \left[U_p(x_p) - \sum_{k \in K} x_{p,k} \cdot \mu_{p,k}(\Lambda) \right] + \sum_{k \in K} f_k \cdot \left[c^T \Lambda_{\star,k} - \zeta \right] + \zeta$$

where we set $\mu_{p,k}(\Lambda) \equiv \sum_{\ell \in L} \Lambda_{\ell,k} \cdot R_{\ell,p}(k)$. If we consider $\Lambda_{\ell,k}$ as the per-unit-of-frequency price of resource ℓ with respect to the pool k, then $\mu_{p,k}(\Lambda)$ is again the end-to-end per-unit cost that p has to pay in pool k. The strict concavity of the utility functions, along with the linearity of the feasible space, assure that we indeed have to solve a strictly convex optimization problem, which has a *unique* optimal solution, $(\hat{x}, \hat{f})$. The system of KKT conditions of $\boxed{\text{MSC}}$ describing this solution, is the following:

KKT-MULTI-SOCIAL (KKT-MSC)

$$U_p'(\hat{x}_p) = \hat{\mu}_{p,k} \equiv \mu_{p,k}(\hat{\Lambda}), \ \forall (p,k) \in P \times K \quad (10)$$

$$c^T \cdot \hat{\Lambda}_{\star,k} \equiv \sum_{\ell \in L} \hat{\Lambda}_{\ell,k} \cdot c_\ell = \hat{\zeta}, \ \forall k \in K \quad (11)$$

$$\hat{\Lambda}_{\ell,k} \left[\sum_{p \in P} R_{\ell,p}(k)\hat{x}_{p,k} - c_\ell \hat{f}_k \right] = 0, \ \forall (\ell,k) \in L \times K \quad (12)$$

$$\hat{\zeta} \cdot \left(\sum_{k \in K} \hat{f}_k - 1 \right) = 0 \quad (13)$$

$$\sum_{p \in P} R_{\ell,p}(k) \cdot \hat{x}_{p,k} \leq c_\ell \cdot \hat{f}_k, \ \forall (\ell,k) \in L \times K \quad (14)$$

$$\sum_{k \in K} \hat{f}_k \leq 1 \quad (15)$$

$$\hat{x} \geq 0, \ \hat{f} \geq 0, \hat{\Lambda} \geq 0, \hat{\zeta} \geq 0 \quad (16)$$

Observe that from $\boxed{\text{KKT-MSC}}$ we can easily deduce the following facts with respect to the optimal solution:

(i) By equation (10), each LOP faces exactly the same end-to-end per-unit-of-frequency cost $\hat{\mu}_p = \hat{\mu}_{p,k} = U_p'(\hat{x}_p)$, $\forall k \in K$, along any line of interest to her. This justifies the fact that p is not really concerned about how the aggregate frequency $\hat{x}_p = \sum_{k \in K} \hat{x}_{p,k}$ is distributed among the different lines of interest to her. The pricing scheme induced by $\hat{\Lambda}$ makes all these lines look of equal importance.

(ii) By equation (11), in the optimal solution all the pools have the same (weighted) aggregate per-unit-of-frequency cost, equal to $\hat{\zeta}$, if we interpret the resource capacities as their weights.

(iii) Due to equation (13), unless this optimal (identical for all pools) aggregate per-unit-of-frequency cost is zero, it holds that the resource capacities are totally distributed among the distinct pools: **if** $\hat{\zeta} > 0$ **then** $\sum_{k \in K} \hat{f}_k = 1$. But if we consider the non-trivial case in which the network has positive resource capacities, then clearly (due to strict concavity of the utilities) some of the resource prices will have to be positive. This directly implies the positivity of $\hat{\zeta}$.

Observe finally that facts (i) and (iii) still hold for the unique optimal primal-dual solution $(\bar{x}, \bar{\Lambda})$, in the case that the NOP fixes a particular vector of proportions $\bar{f}$ (that completely divides the infrastructure among the pools), which is then considered to be constant both in $\boxed{\text{MSC}}$ and in $\boxed{\text{KKT-MSC}}$. Of course, this time we cannot assure the same aggregate cost per pool. □

To tackle the issue of limited information, we consider again (as in the single pool case) a mechanism in which the LOPs are initially required to propose their own bids for buying frequencies, and consequently the NOP somehow determines the resource prices and the frequencies granted to the LOPs (per pool) according to this pricing scheme and their bids. In particular, we construct a new (strictly convex) program, by substituting the (unknown) LOP utility functions with the pseudo-utilities $w_p \log(x_p)$, where $w_p \geq 0$ is the (fixed) amount of money that $p \in P$ is willing to spend for buying frequency (across all pools). This program is the following:

MULTI-NETWORK (MNET)

$$\text{maximize} \quad \sum_{p \in P} w_p \log(x_p) = \sum_{p \in P} w_p \log \left(\sum_{k \in K} x_{p,k} \right)$$

$$\text{s.t. } \forall(\ell, k) \in L \times K, \quad \sum_{p \in P} R_{\ell,p}(k) \cdot x_{p,k} \leq c_\ell \cdot f_k$$

$$\sum_{k \in K} f_k \leq 1$$

$$x, f \geq 0$$

Therefore, for any (fixed) vector $w = (w_p)_{p \in P}$ of LOP bids, the NOP computes (in polynomial time) the optimal solution of $\boxed{\text{MNET}}$, considering as resource prices the optimal Lagrange Multipliers of the resource constraints in $\boxed{\text{KKT-MNET}}$, which is almost identical to $\boxed{\text{KKT-MSC}}$, except for the equations (10), which are substituted by the following:

$$\frac{w_p}{\bar{x}_p} = \bar{\mu}_{p,k} \equiv \mu_{p,k}(\bar{\Lambda}), \quad \forall(p, k) \in P \times K \qquad (17)$$

The properties of Lemma 1 for the optimal solution of $\boxed{\text{MSC}}$ also hold for the optimal solution (for any fixed bid vector) of $\boxed{\text{MNET}}$, even when the NOP decides to fix a particular vector of proportions $\bar{f}$. In particular, for $(\bar{x}, \bar{f}, \bar{\Lambda}, \bar{\zeta})$ it holds that each LOP faces exactly the same cost $\bar{\mu}_p = \bar{\mu}_{p,k}$ in every pool $k \in K$, this time equal to $\frac{w_p}{\bar{x}_p}$ rather than $U_p'(\bar{x}_p)$. Moreover, if the capacity proportions are also variables (rather than constants), then all the pools have the same (weighted) aggregate cost $\bar{\zeta}$.

Consequently, the NOP announces all the optimal frequencies $\bar{x}_{p,k}$ for each LOP $p \in P$ and each pool $k \in K$, for which we know that $\bar{x}_p = \sum_{k \in K} x_{p,k} = \frac{w_p}{\bar{\mu}_p}$. Based once more on our assumption that the LOPs are price taking selfish entities, as in the single pool case, we exploit the fact that each LOP will choose her own bid $\bar{w}_p$ as the optimal solution of $\boxed{\text{USER-II}}$, which assures then that $U_p'(\bar{x}_p) = \bar{\mu}_p = \bar{\mu}_{p,k}, \forall k \in K$, exactly as required in $\boxed{\text{KKT-MSC}}$. This holds for any vector of resource prices that assures for every LOP exactly the same per-unit cost in all the pools, and in particular, for the optimal Lagrange Multipliers vector $\bar{\Lambda}$ of $\boxed{\text{KKT-MNET}}$. Therefore, we again conclude that at equilibrium the LOPs will choose their bids in such a way that the optimal solution of $\boxed{\text{KKT-MSC}}$ coincides with the optimal solution of $\boxed{\text{KKT-MNET}}$, for any fixed vector of capacity proportions, $\bar{f}$. The above discussion thus leads to the following conclusion.

Theorem 3. *Consider a transportation network $G = (V, L)$ and a set P of (selfish, price taking) LOPs with private utility functions of the aggregate frequency assigned to them. Each LOP expresses interest for at most one line in each pool from a set K of pools. There exists a polynomial–time computable mechanism (i.e., a pair of a frequency allocation rule and resource pricing scheme) which induces (as the only equilibrium point) the optimal solution with respect to the aggregate utility value, as a result of the LOPs' selfish behavior.*

Once more, this tractable mechanism, which is based on the solvability of $\boxed{\text{MNET}}$, is totally centralized and rather inconvenient for a dynamically changing (over time), large-scale railway system. Therefore, in the next subsection we shall devise an almost-localized analogue to the single pool case that is based on a system of updating rules for the resource prices (determined by each resource), the LOP bids, and the vector of proportions (determined by the NOP), which converges to this optimal solution of $\boxed{\text{MSC}}$.

3.2 Multi Social Optimum – Dynamic and Decentralized Computation

Our first argument has to do with the independence of the adopted pricing scheme from the way that the NOP chooses to split the railway infrastructure among the different pools. In particular, as we shall shortly explain (Lemma 2), for any *fixed* vector of capacity proportions $\bar{f}$ that the NOP chooses, the optimal value of the corresponding dual program of $\boxed{\text{MSC}}$ exclusively depends on the choice of the vector Λ of resource prices. The dynamic updating system that we shall later propose will exploit exactly this fact and let (in a continuous fashion) the resource prices gradually converge to the optimal price vector (which then forces the LOP bids and the corresponding frequencies to get the right values), for the currently adopted vector of capacity proportions. This vector of proportions will be updated *periodically* by the NOP, only after the system has stabilized to that optimal point (of optimal prices and bids).

Lemma 2. *For any (fixed) vector $\boldsymbol{f}$ of capacity proportions that completely divides the network infrastructure among the pools, the optimal value of* $\boxed{\text{MSC}}$ *exclusively depends on the optimal vector $\bar{\boldsymbol{\Lambda}}$ of per-unit prices for the resources.*

Proof. Using the Lagrange function previously defined, the dual problem of $\boxed{\text{MSC}}$ is the following:

$$\boxed{\text{DUAL-MSC}} \quad \max\left\{D(\boldsymbol{\Lambda}, \zeta) : \forall \ell \in L, \forall k \in K, \Lambda_{\ell,k} \geq 0; \zeta \geq 0\right\}$$

where:

$$D(\boldsymbol{\Lambda}, \zeta) = \max\left\{L(\boldsymbol{x}, \boldsymbol{f}, \boldsymbol{\Lambda}, \zeta) : \boldsymbol{x}, \boldsymbol{f} \geq 0\right\}$$

$$= \max_{\boldsymbol{x}, \boldsymbol{f} \geq 0} \left\{ \sum_{p \in P} \left[U_p(x_p) - \sum_{k \in K} x_{p,k} \sum_{\ell \in L} \Lambda_{\ell,k} R_{\ell,p}(k) \right] + \sum_{k \in K} f_k \left[\sum_{\ell \in L} \Lambda_{\ell,k} c_\ell - \zeta \right] + \zeta \right\}$$

$$= \max_{\boldsymbol{x} \geq 0} \left\{ \sum_{p \in P} \left[U_p(x_p) - \sum_{k \in K} x_{p,k} \mu_{p,k}(\boldsymbol{\Lambda}) \right] \right\} + \max_{\boldsymbol{f} \geq 0} \left\{ \sum_{k \in K} f_k \left[\sum_{\ell \in L} \Lambda_{\ell,k} c_\ell - \zeta \right] \right\} + \zeta$$

Observe that the dual objective $D(\boldsymbol{\Lambda}, \zeta)$ can be split in two parts. The first part:

$$F(\boldsymbol{\Lambda}) = \max_{\boldsymbol{x} \geq 0} \left\{ \sum_{p \in P} \left[U_p(x_p) - \sum_{k \in K} x_{p,k} \cdot \mu_{p,k}(\boldsymbol{\Lambda}) \right] \right\}$$

is a maximization problem similar to the one already dealt with in the single pool case (i.e., for $|K| = 1$) of the previous single-pool case. Its value is a function of the resource prices, and the vector of proportions has no involvement at this point. The only difference from the single pool case, is that we now have distinct LOP frequencies, as well as LOP end-to-end costs, per pool. But this technical issue can be tackled by a proper choice of the dynamic updating system, as we shall see later. The second part of $D(\boldsymbol{\Lambda}, \zeta)$ is the following:

$$G(\boldsymbol{\Lambda}, \zeta) = \max_{\boldsymbol{f} \geq 0} \left\{ \sum_{k \in K} f_k \cdot \left[\sum_{\ell \in L} \Lambda_{\ell,k} \cdot c_\ell - \zeta \right] \right\} + \zeta$$

$$= \max_{\boldsymbol{f} \geq 0} \left\{ \sum_{k \in K} f_k \cdot (\boldsymbol{c}^T \boldsymbol{\Lambda}_{\star,k}) + \zeta \cdot \left(1 - \sum_{k \in K} f_k\right) \right\}$$

Recall that at global optimality (when we consider the capacity proportions as variables), the term $\zeta \cdot \left(1 - \sum_{k \in K} f_k\right)$ has zero contribution in $G(\boldsymbol{\Lambda}, \zeta)$. But this also holds for any solution in which the NOP chooses some vector $\boldsymbol{f}$ of capacity proportions that sums up to 1. Additionally, we have already seen that this is indeed the case for the optimal vector of capacity proportions as well, as was explained in Lemma 1, fact (iii). Therefore, the optimal choice $\hat{\boldsymbol{f}}$ of capacity proportions can be seen as a *probability distribution* that assigns positive mass only to pools of maximum aggregate price (according to $\boldsymbol{\Lambda}$). We demand this restriction explicitly from $G(\boldsymbol{\Lambda}, \zeta)$:

$$G(\boldsymbol{\Lambda}, \zeta) = \max_{\mathbf{1}^T \boldsymbol{f}=1;\ \boldsymbol{f} \geq 0} \left\{ \sum_{k \in K} f_k \cdot \left(\boldsymbol{c}^T \cdot \boldsymbol{\Lambda}_{\star,k} \right) \right\} = \max_{k \in K} \left\{ \boldsymbol{c}^T \cdot \boldsymbol{\Lambda}_{\star,k} \right\}$$
$$= \min \left\{ z : z \cdot \mathbf{1}^T \geq \boldsymbol{c}^T \cdot \boldsymbol{\Lambda} \right\}$$

That is, $G(\boldsymbol{\Lambda}, \zeta)$ simply calculates the maximum (rather than the average, indicated by ζ) aggregate (per-unit) cost among the pools, which only depends on the given resource pricing vector $\boldsymbol{\Lambda}$.   □

Lemma 2 is crucial in deriving a dynamic algorithm that computes the social optimum, in analogy with the one derived for the single pool case. In particular, Lemma 2 and the framework of the single pool case suggest the following dynamic algorithm, whose high-level description is as follows.

1. Each resource continuously updates its own (anonymous) per-unit-of-frequency price.
2. Each LOP updates her offer (bid) for claiming frequency, only when the resource prices (and thus her own per-unit costs in the pools) have stabilized.
3. The NOP updates *periodically* the vector of capacity proportions of the railway infrastructure given to the different line pools, only when both the resource prices and the LOP bids have stabilized.

In particular, assume that at some time $t \geq 0$ we have the following situation:

- $\boldsymbol{\Lambda}(t)$ is the vector of current resource prices. $\forall p \in P, \forall k \in K$, $\mu_{p,k}(t) = \sum_{\ell \in L} R_{\ell,p}(k) \cdot \Lambda_{\ell,k}$ is the per-unit cost of player p at pool k, while $\mu_p(t) = \frac{1}{|K|} \sum_{k \in K} \mu_{p,k}(t)$ is the *average* per-unit cost of p over all the pools.
- $\boldsymbol{w}(t) = (w_p(t))_{p \in P}$ is the vector of the LOPs' current bids.
- $\boldsymbol{f}(t) = (f_k(t))_{k \in K}$ is the current vector of proportions of resource capacities of the railway infrastructure to each of the pools (as determined by NOP). We always assure the **invariant** that the entire railway infrastructure is provided to the pools: $\mathbf{1}^T \cdot \boldsymbol{f}(t) = 1$.
- We calculate the frequencies that each LOP gets per pool as follows. We split each LOP's bid $w_p(t)$ among the different pools according to the vector of capacity proportions. Then each LOP buys the corresponding frequency, given her bid and the per-unit cost for this LOP at each particular pool: $\forall (p, k) \in P \times K$, $x_{p,k}(t) = \frac{f_k(t) \cdot w_p(t)}{\mu_{p,k}(t)}$. The aggregate frequency of the LOP $p \in P$ is obviously $x_p(t) = \sum_{k \in K} x_{p,k}(t)$.

- The resource frequencies at time t are then calculated as follows: $\forall (\ell, k) \in L \times K$, $y_{\ell,k}(t) = \sum_{p \in P} R_{\ell,p}(k) \cdot x_{p,k}(t)$ and $y_\ell(t) = \sum_{k \in K} y_{\ell,k}(t)$.

We assume that the resource price updating scheme operates continuously, the LOP bidding updating scheme applies only when the resource prices have stabilized, and finally the updating of the capacity proportions (conducted by the NOP) is carried out only when both the resource prices and the LOP bids have stabilized. This is explained as follows. Each resource continuously updates its price as a function of the aggregate frequency over it (in each pool), and this is instantly known local information to the resource. As for the LOPs, they would like to update their bids only when there is a clear picture of what should be paid in each pool. This can only happen when the resource prices have stabilized. Additionally, each LOP has to gather the pricing information along the lines she uses, which is somehow local information (only refers to resources actually used by the LOP) but not instantly available. Finally the NOP wishes to: (i) let the whole situation with the resource prices and LOP bids stabilize before it intervenes to determine the new capacity proportions of infrastructure, and (ii) avoid too frequent changes in the capacity proportions, since this updating scheme does not depend only on local information (either on each LOP, or on each resource) but on the aggregate costs of all the pools, as we shall see shortly. Therefore, the NOP prefers this update to happen only occasionally, in order to be able to amortize its heavy cost over a large period of time.

Let's now see the exact shape of the dynamic protocol at time $t \geq 0$.

Resource Price Updating. $\forall t \geq 0$, the resource per-unit prices are updated according to the following differential equation: $\forall \ell \in L, \forall k \in K$,

$$\dot{\Lambda}_{\ell,k}(t) = \max\left\{0, y_{\ell,k}(t) - c_\ell f_k\right\} \cdot \mathbb{I}_{\{\Lambda_{\ell,k}(t)=0\}} + (y_{\ell,k}(t) - c_\ell f_k) \cdot \mathbb{I}_{\{\Lambda_{\ell,k}(t)>0\}}$$

LOP Bid Updating. Assuming now that the LOPs are selfish entities, their (instantaneous) bids are chosen as the solutions of the analogue of $\boxed{\text{USER-II}}$ (per LOP), which is the following:

$$\boxed{\text{MUSER-II}} \quad \text{maximize} \left\{ U_p \left(\sum_{k \in K} \frac{f_k w_p}{\bar{\mu}_{p,k}} \right) - w_p : w_p \geq 0 \right\}$$

where, $\forall k \in K$, $\bar{\mu}_{p,k} = \bar{\mu}_p$ is the common per-unit cost that the LOP $p \in P$ faces in each pool, as soon as the resource prices stabilize. The optimality condition of $\boxed{\text{MUSER-II}}$ is now that

$$U'_p \left(\sum_{k \in K} \frac{f_k w_p}{\bar{\mu}_{p,k}} \right) \cdot \sum_{k \in K} \frac{f_k}{\bar{\mu}_{p,k}} = 1$$

$$\Leftrightarrow U'_p(x_p) = U'_p \left(\sum_{k \in K} \frac{f_k w_p}{\bar{\mu}_{p,k}} \right) = \left(\sum_{k \in K} \frac{f_k}{\bar{\mu}_{p,k}} \right)^{-1} = \bar{\mu}_p$$

Capacity Proportions Updating. After the LOPs have stabilized the bids $(w_p(t))_{p\in P}$ and the resources have updated their per-unit prices in each pool $(\Lambda_{\ell,k}(t))_{\ell\in L, k\in K}$, the NOP sets $\zeta(t)$ to the *average* price of a pool:

$$\zeta(t) = \frac{1}{|K|}\sum_{k\in K} c^T \cdot \Lambda_{\star,k}(t) \qquad (18)$$

Then the NOP updates the proportions of the railway infrastructure granted to each of the pools, so that pools exceeding the current average cost $\zeta(t)$ increase their proportion (in hope of decreasing their weighted cost), while pools that are cheaper than the average price slightly decrease their proportion (recall that in the optimal solution all the pools have exactly the same weighted aggregate cost). The proportions are updated according to the following system of differential equations:

$$\forall k \in K, \; \dot{f}_k(t) = \max\left\{0, c^T \cdot \Lambda_{\star,k}(t) - \zeta(t)\right\} \qquad (19)$$

It should be noted here that, in order for the vector $f(t+1)$ of capacity proportions to sum up to 1, we must divide the resulting vector of new proportions by a proper scaling factor $\phi(t) > 1$ (since the expensive pools increased their proportions, while the cheap pools kept their old proportion, according to the proposed derivative in equation (19)).

The resource updating in this differential system assures the validity of equations (12) at equilibrium, for any fixed vector f of capacity proportions provided by the NOP, and any fixed bid vector w provided by the LOPs. Moreover, if we assume that the LOPs are price taking and myopic entities, the LOP bid updating again leads us to the validity of equations (10). We shall now prove the convergence to the optimal resource prices, with respect to any given vector of capacity proportions, and any given vector of LOP bids.

Lemma 3. *For any choice of* fixed *bid vector* $\bar{w} = (w_p)_{p\in P}$ *offered by the LOPs, and any* fixed *vector of proportions* $\bar{f} = (f_k)_{k\in K}$ *determined by the NOP, the resource price updating scheme makes the resource prices converge to the corresponding optimal vector* $\bar{\Lambda}$ *(for these particular given bids and proportions).*

Proof. We use again the Lyapunov function $V(\Lambda(t)) = \frac{1}{2}\cdot(\Lambda(t)-\bar{\Lambda})^T\cdot(\Lambda(t)-\bar{\Lambda})$, we can once more prove convergence to the optimal resource prices, $\bar{\Lambda}$, for any fixed vector of LOP bids, $\bar{w}$ and any vector of pool proportions, $\bar{f}$ (determined by the NOP):

$$\frac{dV(\Lambda(t))}{dt} = \sum_{\ell\in L}\sum_{k\in K}(\Lambda_{\ell,k}(t) - \bar{\Lambda}_{\ell,k})\cdot \dot{\Lambda}(t)$$

$$= \sum_{\ell\in L}\sum_{k\in K}(\Lambda_{\ell,k}(t) - \bar{\Lambda}_{\ell,k})\cdot \left[\max\left\{0, y_{\ell,k}(t) - c_\ell\bar{f}_k\right\}\cdot \mathbb{I}_{\{\Lambda_{\ell,k}(t)=0\}}\right.$$

$$\left. + \left(y_{\ell,k}(t) - c_\ell\bar{f}_k\right)\cdot \mathbb{I}_{\{\Lambda_{\ell,k}(t)>0\}}\right]$$

$$\leq \sum_{\ell \in L} \sum_{k \in K} (\Lambda_{\ell,k}(t) - \bar{\Lambda}_{\ell,k}) \cdot [y_{\ell,k}(t) - c_\ell \bar{f}_k]$$

$$= \sum_{\ell \in L} \sum_{k \in K} (\Lambda_{\ell,k}(t) - \bar{\Lambda}_{\ell,k}) \cdot [y_{\ell,k}(t) - \bar{y}_{\ell,k} + \bar{y}_{\ell,k} - c_\ell \bar{f}_k]$$

$$\leq \sum_{\ell \in L} \sum_{k \in K} (\Lambda_{\ell,k}(t) - \bar{\Lambda}_{\ell,k}) \cdot [y_{\ell,k}(t) - \bar{y}_{\ell,k}]$$

$$= \sum_{\ell \in L} \sum_{k \in K} (\Lambda_{\ell,k}(t) - \bar{\Lambda}_{\ell,k}) \cdot \sum_{p \in P} R_{\ell,p}(k) \cdot [x_{p,k}(t) - \bar{x}_{p,k}]$$

$$= \sum_{p \in P} \sum_{k \in K} [x_{p,k}(t) - \bar{x}_{p,k}] \cdot (\mu_{p,k}(t) - \bar{\mu}_{p,k})$$

$$= \sum_{p \in P} \sum_{k \in K} [x_{p,k}(t) - \bar{x}_{p,k}] \cdot \left(\frac{\bar{f}_k \bar{w}_p}{x_{p,k}(t)} - \frac{\bar{f}_k \bar{w}_p}{\bar{x}_{p,k}} \right)$$

$$= \sum_{p \in P} \sum_{k \in K} \bar{f}_k \bar{w}_p \cdot \left[2 - \frac{x_{p,k}(t)}{\bar{x}_{p,k}} - \frac{\bar{x}_{p,k}}{x_{p,k}(t)} \right]$$

$$\leq 0$$

The first inequality holds trivially for each $(\ell, k) : \Lambda_{\ell,k}(t) > 0$, but also holds when for $(\ell, k) : \Lambda_{\ell,k}(t) = 0$, because then either $y_{\ell,k}(t) - c_\ell \bar{f}_k \geq 0$ and

$$(\Lambda_{\ell,k}(t) - \bar{\Lambda}_{\ell,k}) \cdot \max\{0, y_{\ell,k}(t) - c_\ell \bar{f}_k\} = -\bar{\Lambda}_{\ell,k} \cdot [y_{\ell,k}(t) - c_\ell \bar{f}_k]$$

or $y_{\ell,k}(t) - c_\ell \bar{f}_k < 0$ and then:

$$(\Lambda_{\ell,k}(t) - \bar{\Lambda}_{\ell,k}) \cdot \max\{0, y_{\ell,k}(t) - c_\ell \bar{f}_k\} = -\bar{\Lambda}_{\ell,k} \cdot 0 < -\bar{\Lambda}_{\ell,k} \cdot [y_{\ell,k}(t) - c_\ell \bar{f}_k]$$

The second inequality holds because for the optimal vector $\bar{\Lambda}$ (for the given vectors $\bar{w}$ and $\bar{f}$) it holds that $\sum_{\ell \in L} \sum_{k \in K} \bar{\Lambda}_{\ell,k}(\bar{y}_{\ell,k} - c_\ell \bar{f}_k) = 0$ (cf. equation (12), which is also a KKT condition for $\boxed{\text{MNET}}$). The third inequality holds again because $\forall z > 0, 2 - z - \frac{1}{z} \leq 0$. □

Of course, when the resource prices and LOP bids have stabilized, we still cannot be sure that we have reached the optimal solution of $\boxed{\text{MNET}}$, because we cannot guarantee for the time being that all the pools have the same (weighted) aggregate cost, as required by equation (11). Due to the strict concavity of $\boxed{\text{MSC}}$, we know that the current optimal value (for the given proportions) of its dual is strictly less than the globally optimal value (with respect to the optimal proportions). This is because we obviously have not chosen yet the optimal vector of proportions. But, as it was shown in Lemma 2, the optimal value of $\boxed{\text{DUAL-MSC}}$ exclusively depends on the vector of resource prices. Therefore we also know that we do not have the optimal resource prices as well. At this point exactly, the NOP intervenes with the capacity proportions updating procedure, which increases (in a continuous fashion) the proportions of pools that are more expensive than the current value of the average cost $\zeta(t)$. That is, the NOP chooses to increase the pool-capacity proportions to already expensive pools (therefore allowing, at the next optimal point, lower aggregate costs for them) and decreases the proportion of infrastructure for cheap pools (which can afford slightly larger aggregate costs). This way we get closer to the optimal vector of capacity proportions in $\boxed{\text{MSC}}$, since the vector of aggregate pool costs will now become smoother. Consequently, the new optimal value of $\boxed{\text{DUAL-MSC}}$ (with respect to the new capacity proportions) will strictly increase due to the intervention of the NOP, because the dominant parameter for it is the vector of resource prices.

Eventually, by Lemmata 2 and 3, we shall converge to an equilibrium point of the whole system in which equation (10) is guaranteed by the selfish, price taking behavior of the LOPs, equation (11) is assured by the NOP, equation (12) is assured by the resource price updating scheme, and equation (13) is assured by our invariant on the vector of capacity proportions. This is exactly the optimal solution of both $\boxed{\text{KKT-MSC}}$ and $\boxed{\text{MNET}}$, as required. The following theorem summarizes the previous discussion.

Theorem 4. *The aforementioned dynamic system of resource-pricing, LOP-bid-updating and capacity-proportions-updating differential equations ensures monotonic convergence to the social optimum of* $\boxed{\text{MNET}}$ *from any initial point of resource prices, LOP bids and proportions of capacities for the pools.*

4   Implementation and Experimental Evaluation

In this section, we present the implementation of a discrete version of our decentralized algorithm for the single pool case and its experimental evaluation on synthetic and real-world data.

4.1   The Algorithm

We have implemented a discrete version of our distributed algorithm given in Section 2.2, and which is provided below. Parameter b determines the desired accuracy at equilibrium, B_ℓ is an upper bound on the value of $\lambda_\ell(t)$, ε_ℓ represents the interval upon which $\lambda_\ell(t)$ is defined and which gradually reduces via the UP-DATE routine, and the boolean variable S_ℓ is used to determine the termination condition of the repeat-until loop. The algorithm is as follows.

1. INITIALIZATION $(t = 0)$.
    (a) For all $p \in P$: { $w_p(0) = 1$; $x_p(0) = \min_{\ell \in p}\{c_\ell\}$; }
    (b) For all $\ell \in L$: { $\lambda_\ell(0) = 0$; $\varepsilon_\ell = 1$; $\delta_\ell = 10^{-b}$; $B_\ell = +\infty$; $S_\ell = $ FALSE; }
2. REPEAT FOR $t > 0$
    (a) For all $\ell \in L$:
        i. $y_\ell(t) = \sum_{\ell \in p} x_p(t-1)$;
        ii. $\alpha_\ell(t) = y_\ell(t) - c_\ell$;
        iii. $\dot{\lambda}_\ell(t) = \max\{0, \alpha_\ell(t)\} - \min\{\frac{\lambda_\ell(t)}{2}, \max\{0, -\alpha_\ell(t)\}\}$;
        iv. **if** $\lambda_\ell(t-1) = 0 \wedge \max\{0, \alpha_\ell(t)\} < \delta_\ell$ **then** $S_\ell = $ TRUE;

     v. **if** $\lambda_\ell(t-1) > 0 \wedge |\alpha_\ell(t)| < \delta_\ell$ **then** $S_\ell = $ TRUE;
    vi. UPDATE(ε_ℓ);
   vii. $\lambda_\ell(t) = \lambda_\ell(t-1) + \varepsilon_\ell \dot{\lambda}_\ell(t)$;
 (b) **if** $\bigcap_{\ell \in L} S_\ell = $ TRUE **then** BREAK;
 (c) For all $p \in P$:
    i. $\mu_p(t) = \sum_{\ell \in p} \lambda_\ell(t)$;
   ii. Solve $\boxed{\text{USER-II}}$ to determine $w_p(t)$;
  iii. $x_p(t) = \frac{w_p(t)}{\mu_p(t)}$;
 (d) $t = t + 1$;
UNTIL TRUE

The routine for updating ε_ℓ is as follows.

UPDATE(ε_ℓ)

1. **if** $\dot{\lambda}_\ell(t) < 0 \wedge \dot{\lambda}_\ell(t-1) > 0$ **then**
        if $\lambda_\ell(t-1) < B_\ell$ **then** $B_\ell = \lambda_\ell(t-1)$ **else** $\varepsilon_\ell = \varepsilon_\ell/2$;
2. RETURN ε_ℓ;

4.2   Experimental Setup

The algorithm was implemented in C++ using the GNU g++ compiler (version 4.3.2) with -O2 optimization level, and the LEDA C++ library (version 6.2). Our experiments were performed on a computer having an Intel Core 2 Duo Processor clocked at 2.00GHz (T7300 model) and a total of 2GB RAM.

Synthetic data consisted of grid graphs having a vertical dimension of 3 and a horizontal dimension ranging from 120 to 36000 (i.e., the graph sizes range from 120×3 to 36000×3). The edge capacity was set to 10. In these graphs, we define three paths in a way that they have a fair amount of edges in common. By considering the graph nodes as points in the plane, we define three directions:

UP: The next edge of the path is headed upwards.
RIGHT: The next edge of the path is headed to the right.
DOWN: The next edge of the path is headed downwards.

We consider two families of three paths. The first family is defined as follows; see Figure 2. All paths start at node $(0, 1)$. The first path follows the RIGHT direction until it can no longer proceed. We will call this the *middle path* (yellow-grey path in Fig. 2. The second path (red path in Fig. 2) first goes RIGHT, then UP, then RIGHT, then DOWN, and then continues the same pattern until it can no longer proceed. The third path (green path in Fig. 2) first goes RIGHT, then DOWN, then RIGHT, then UP, and then continues the same pattern until it can no longer proceed. Observe that all paths share the odd edges of the middle path (i.e., the edges $(2i, 2i+1)$, $i = 0, 1, 2, \ldots$).

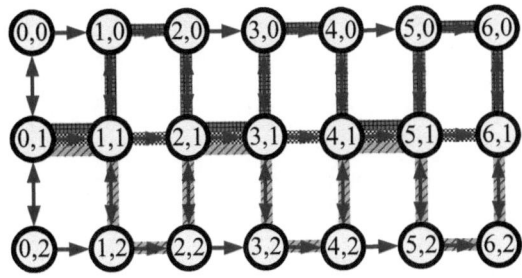

Fig. 2. Deterministic paths on grids

The second family of paths is defined as follows. All paths start at node $(0, 1)$ and go RIGHT. Then, at node $(2i + 1, 1)$, $i = 0, 1, 2, \ldots$, each path makes a (uniformly) random decision on whether it will go UP, RIGHT, or DOWN. If the random choice is to go UP, then it follows the pattern RIGHT, DOWN, RIGHT, reaching the next node where it will make a new random decision. If the random choice is to go DOWN, then it follows the pattern RIGHT, UP, RIGHT, reaching the next node where it will make a new random decision. If the random choice is to go RIGHT, then it follows the pattern RIGHT, RIGHT, reaching the next node where it will make a new random decision. These choices ensure that all three paths share the edges $(2i, 2i + 1)$, $i = 0, 1, 2, \ldots$.

Real-world data concern parts of the German railway network (concerning mainly intercity train connections). We have considered three instances with 280 (354), 296 (393), and 319 (452) nodes (edges), and a single line pool of varying size. The capacity of the edges varied from 8 up to 16.

For both synthetic and real-world data, we used the function $U_p(x) = a\sqrt{x}$ as utility function of all LOPs $p \in P$, where a is a constant ($a \geq 10^4$).

4.3   Experimental Results

We start with the experimental results on our synthetic data. Figure 3 shows the number of iterations required by our distributed algorithm to converge to the social optimum in the grid graphs of sizes 12000×3 to 36000×3. The top diagram does this for the first family of paths (deterministically defined paths), while the bottom diagram does it for the second family of paths (that include random choices at certain nodes). We observe that despite the graph size, the algorithm converges quite fast to the optimal solution. It is worth mentioning that the real execution time never exceeded 1.5 minutes.

We now turn to the real-world graphs. Figures 4 and 5 show the number of iterations required by our distributed algorithm to converge to the social optimum with respect to the size of the line pool, which varies from 100 to 2000 lines (in Fig. 5). We observe again that the algorithm converges fast to the optimal solution; the maximum execution time never exceeded 2 minutes.

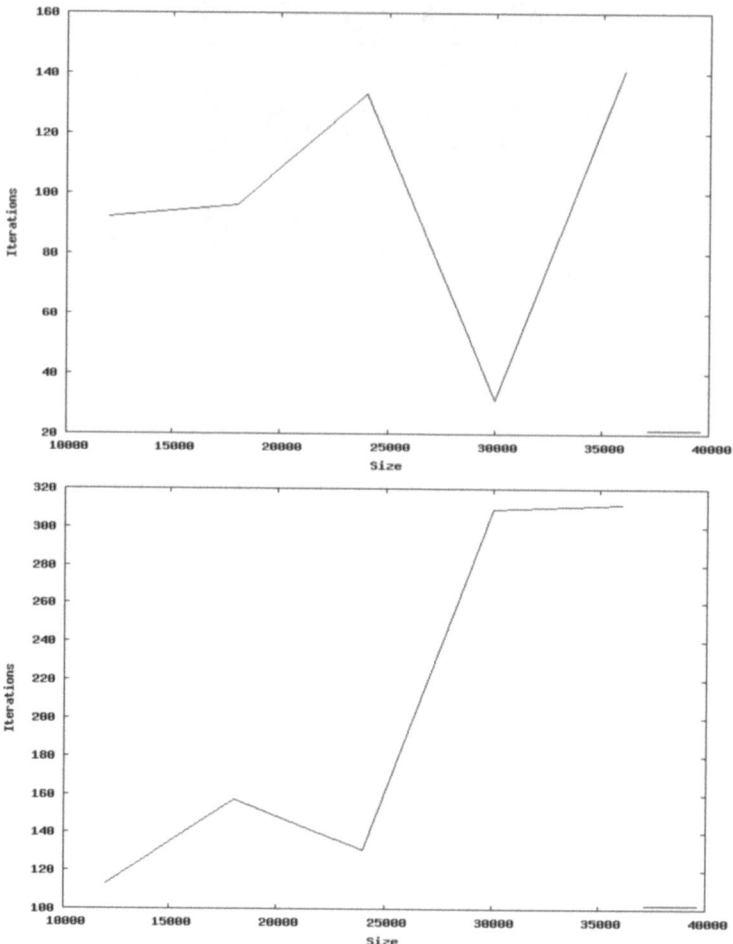

Fig. 3. Grid graphs. Top: deterministic paths. Bottom: random paths.

In both synthetic and real-world experiments, we observe that the convergence rate, determined by the number of iterations required to reach the optimum, varies not only between graph classes but also within the same graph class. This can be explained as follows.

It is clear from the description of the algorithm that the number of iterations depends on how fast $\lambda_\ell(t)$ reach their optimal values, which also depends on $\dot{\lambda}_\ell(t)$. The latter depends on $\alpha_\ell(t)$, which in turn depends on $y_\ell(t)$.

The quantity $y_\ell(t)$ depends on the number of paths that use edge ℓ. Fast convergence implies an as small as possible value for $\alpha_\ell(t)$, which implies a value for $y_\ell(t)$ that is as close as possible to c_ℓ. At a first place observe that initially $y_\ell(t)$ can be much larger than c_ℓ, especially in the case where many paths use edge ℓ. At a second place observe that the initial value of $w_p(t)$ can be quite

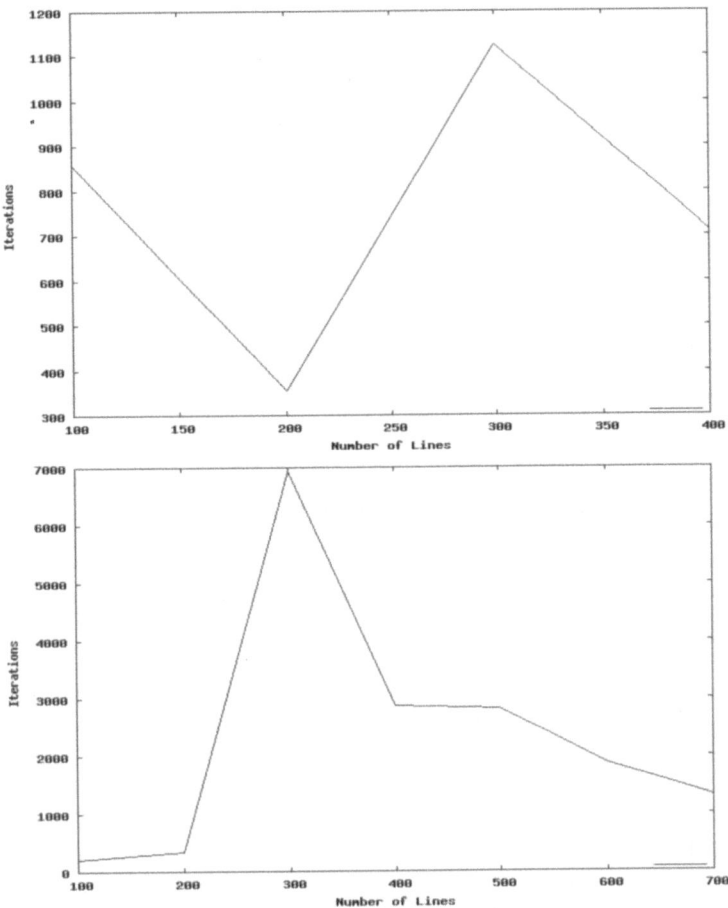

Fig. 4. Real-world graphs. Top: 280 nodes and 354 edges. Bottom: 296 nodes and 393edges.

large since its value depends on the utility function. This in turn implies that the initial frequency value $x_p(t)$ can be very large, leading to large values of $y_\ell(t)$ in subsequent iterations. For these two reasons, the value of $\alpha_\ell(t)$ (that depends on $y_\ell(t)$) can start, for a specific input instance, from a rather high value and therefore it may take more iterations to reach its proper value, demonstrating a slower rate of convergence.

In conclusion, the convergence rate depends on a combination of input-specific factors that can vary considerably even between instances of the same graph class. These factors include the edge capacity values, the specific form of the utility function, the number of edges in a line route, and the number of line routes that share edges.

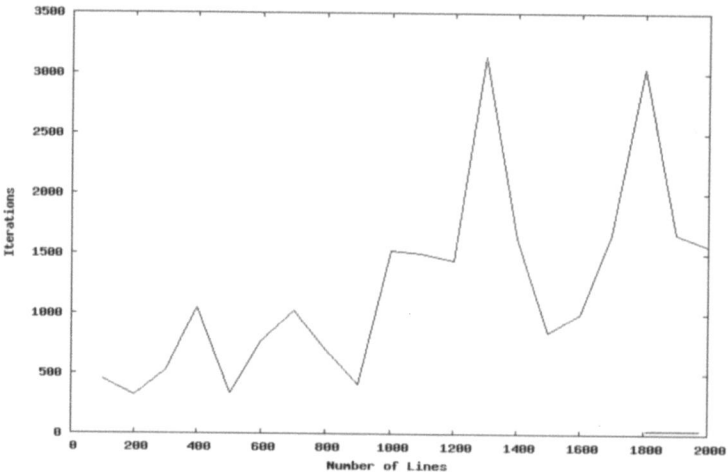

Fig. 5. Real-world graph with 319 nodes and 452 edges

5 Incentive-Compatible Robustness and Railway Optimization

The approach pursued in the preceding sections can be generalized to deal with robustness issues in the broader context of railway optimization (i.e., not only within line planning). In this section, we will argue on this matter and also compare incentive-compatible robustness with other known notions of robustness.

Railway optimization deals with large-scale planning and scheduling problems over several time horizons. Due to their complexity and sheer size, quite often such problems are provided with incomplete or uncertain data. For instance, some data may in advance be completely unknown or of low precision, while other data are subject to changes during the operational phase (e.g., due to delays). As there are several different types of imperfect information, there are also several different concepts for optimizing with respect to imperfect information. Each of the known concepts, such as multi-stage stochastic programming, chance constraint programming, robust optimization or online optimization, has its own strengths and weaknesses, making the various concepts the models of choice for different practical problems.

Two main approaches have been pursued in the literature to handle uncertainty: (i) Stochastic programming models, offering great flexibility but often too large in size to be handled efficiently, and also requiring that probability distributions are given a priori and can be handled in the solution procedure. (ii) Robust optimization models, which are easier to solve but sometimes leading to very conservative solutions of little practical use. Under this classical concept, feasibility is guaranteed if the number of constraints affected by data changes is bounded; see, for instance, the seminal work of Betsimas and Sim [4] on this subject.

A third way to model uncertainty, leading to a modeling framework called *light robustness*, was recently proposed in [8,9]. It couples robust optimization with a simplified two-stage stochastic programming approach, and constitutes a flexible counterpart of (classical) robust models that turned out to be quite promising within railway optimization. It is also worth mentioning that a variant of the Bertsimas and Sim method [4] for robustness was also applied to a game-theoretic scenario, in which competing entities act selfishly so as to optimize their own utility functions. In this case, uncertainty can involve both the rules of the game and the players' utility functions. It is showed that the robust counterpart of a game (under bounded, polyhedral uncertainty) is approximately as hard as the nominal game, at least in certain cases of finite games [1].

The classical robustness framework neglects the realistic possibility of a recovery phase; i.e., that small infeasibilities may be corrected. If such a phase is excluded, then solutions should be either unacceptably expensive or too conservative in order to be feasible in all scenarios that determine the imperfection of information.

Recoverable robustness is a new notion introduced in [2,16,17] that appears more suitable to deal with data uncertainty and recovery in the context of railway optimization. Recoverable robustness is about computing solutions that are robust against a limited set of scenarios and which can be made feasible (recovered) by a limited effort. One starts from a feasible solution x of an optimization problem which a particular scenario s, that introduces imperfect knowledge (i.e., by adding more constraints), may turn to infeasible. The goal is to have handy a recovery algorithm A that takes x and turns it to a feasible solution under s (i.e., under the new set of constraints). In other words, in recoverable robustness there is uncertainty about the feasibility space: imperfect information generates infeasibility and one strives to (re-)achieve feasibility.

The aforementioned approaches provide a quite powerful set of methods to deal with some kind of *predictable* and statically described level of uncertainty mainly in the constraints[4]. But what happens when the exact shape of the global objective function is unknown to the system? This may happen in application scenarios where many entities compete for common resources and each one acts selfishly. For example, in the line planning setting which we considered in the previous sections, where commercial line operators (competing entities) with fixed choices of lines compete for the utilization of these lines (common resources) via frequency negotiation with the (possibly public-sector or governmental) network operator. In such settings, for obvious reasons, each competing entity is not willing to reveal her own (private) utility function; that is, to reveal her level of satisfaction for acquiring a specific usage of resources. Nevertheless, the goal of the dynamic market designer (network operator in the line planning setting) – corresponding to the socially optimal solution – is to guarantee a *fair and feasible* solution, that is, a certain usage of subsets of resources to the competing entities in such a way that constraints regarding the usage of the resources

---

[4] Uncertainty can also be transferred to the objective function as well, by incorporating it into the constraints.

are not violated and at the same time the average satisfaction of the competing entities (players) is maximized.

All aforementioned approaches seem to be inadequate to deal with such an application scenario, because the nature of the uncertainty itself is not quantified in any sense, and indeed may vary with time. Additionally, this situation should not be dealt with as a static problem to be centrally solved, but rather as a dynamic decentralized scheme, that continuously adapts the usage of resources to the players, in order to always keep them as close as possible to the socially optimal solution, as the utility functions of the players may also evolve with time.

In this work, we propose a new notion of robustness along with a corresponding solution framework, which we call *incentive-compatible robustness* and which is complementary to the notion of recoverable robustness. It is concerned with the computation of an *incentive-compatible recovery scheme* that achieves robustness by enforcing the system to converge to its optimal solution. By an incentive-compatible recovery scheme we mean a decentralized price-updating and resource-usage allocation method, that exploits the selfish nature of the competing entities, in order to lead them back to the socially optimal solution, even if the social optimum itself varies with time. Each resource gets a dynamic pricing scheme, and each competing entity is allowed to continuously change her bidding for getting (in the near future) usage of resources. In this context, the feasibility space is known and incomplete information refers to lack of information about the optimization problem, due to the unknown utility functions. In incentive-compatible robustness, there is uncertainty about the objectives: feasibility is guaranteed, since imperfect knowledge does not introduce new constraints, and one strives to achieve optimality, exploiting the selfish nature of the players.

Note that incentive-compatible robustness is different from the concept of game-theoretic robustness as developed in [1]. The approach in [1] is a centralized, deterministic paradigm to uncertainty in strategic games. Our approach differs from that in the following: (i) It is decentralized to a large extent, based only on local information that the participating entities (line operators and resources) have at any time; (ii) we impose no restriction on the kind of the utility functions of the players other than their strict concavity, whereas the approach in [1] has to somehow quantify the "magnitude" of uncertainty of the constraints and/or the payoffs, in order to keep the solvability of the problem comparable to that of the nominal counterpart; (iii) the solvability of the robust counterpart in [1] is largely based on the solvability of the nominal counterpart (which is strongly questionable for the general game-theoretic framework).

To summarize, incentive-compatible robust optimization suggests a generic approach to deal with robustness issues in railway optimization applications that require setting up a dynamic market for negotiating usage of resources, over subsets of resources, by selfish entities that do not reveal their incentives and having non-fixed (elastic) demands.

6   Conclusions and Open Issues

We investigated a new application scenario in line planning that achieves *incentive-compatible robust* solutions by exploiting a resource allocation mechanism introduced by Kelly [13] in the context of communication networks. For the case of a single line pool, an adaptation of Kelly's approach can provide (under certain assumptions) a decentralized algorithm that provably converges to the socially optimal solution. For the case of multiple line pools, an extension and further elaboration of Kelly's approach is required in order to derive such an algorithm. We also conducted experiments on a discrete variant of the pricing scheme for the single-pool case over synthetic and real-world data. Our algorithms allow LOPs to negotiate line frequencies over fixed lines in a dynamic fashion. In a broader context, our approach comprises a generic technique to set up a dynamic market for (re-)negotiating usage of resources over subsets of resources. Consequently, it could be applied to set up a dynamic frequency market over other transportation settings (e.g., in the airline domain).

A crucial question would be to devise protocols that demonstrate faster convergence to the equilibrium point, even approximately. Additionally, it would be interesting to find ways to tackle the assumption on price taking and myopic behavior of the users. It would be nice to do this even at the cost of suboptimal equilibrium points. It is noted that when the LOPs are not price takers and myopic (called *price anticipators* in the congestion control jargon), then the above scheme does not lead to socially optimal solutions, even for the case where there is only a single resource to share. Nevertheless, it would be quite interesting to know how far one can be from the social optimum, given that a decentralized updating scheme is adopted for the user requests and the prices of the resources.

Further open issues concern: (i) a theoretical analysis of the discrete variants of our algorithms; (ii) an extension of our approach to introduce proportions per resource (rather than per line pool); (iii) the investigation of other types of LOP's utility functions, or the case for a LOP to pursue a different utility function per line pool; (iv) looking for other parameters of robustness and recoverability (e.g., introduction of delays).

Acknowledgements. We would like to thank the referees for their valuable comments and suggestions that improved the presentation. We are also indebted to Rolf Möhring, Christian Liebchen, and Sebastian Stiller for their comments on an earlier draft of this work, and to Kostas Tsihlas for many fruitful discussions.

References

1. Aghassi, M., Bertsimas, D.: Robust Game Theory. Mathematical Programming Ser. B 107, 231–273 (2006)
2. ARRIVAL Deliverable D1.2, New Theoretical Notion of the Prices of Robustness and Recoverability, ARRIVAL Project, Version 2 (July 2007)
3. Bertsekas, D.: Nonlinear Programming, 2nd edn. Athena Scientific (1999)

4. Bertsimas, D., Sim, M.: The Price of Robustness. Operations Research 52(1), 35–53 (2004)
5. Borndörfer, R., Grötschel, M., Lukac, S., Mitusch, M., Schlechte, T., Schultz, S., Tanner, A.: An Auctioning Approach to Railway Slot Allocation. Competition and Regulation in Network Industries 1(2), 163–196 (2006)
6. Clarke, E.H.: Multipart pricing of public goods. Public Choice 11, 19–33 (1971)
7. Dienst, H.: Linienplanung im spurgeführten Personenverkehr mit Hilfe eines heuristischen Verfahrens, PhD thesis, Technische Universität Braunschweig (1978)
8. Fischetti, M., Monaci, M.: Light Robustness. Technical Report ARRIVAL-TR-0119, ARRIVAL Project (January 2008)
9. Fischetti, M., Monaci, M.: Light Robustness. In: Ahuja, R.K., Möhring, R.H., Zaroliagis, C.D. (eds.) Robust and Online Large-Scale Optimization. LNCS, vol. 5868, pp. 61–84. Springer, Heidelberg (2009)
10. Goossens, J., van Hoesel, C., Kroon, L.: A branch and cut approach for solving line planning problems. Transportation Science 38, 379–393 (2004)
11. Green, J.R., Laffont, J.J.: Incentives in Public Decision-Making. North-Holland Publishing Company, Amsterdam (1979)
12. Groves, T.: Incentives in teams. Econometrica 41(4), 617–631 (1973)
13. Kelly, F.: Charging and rate control for elastic traffic. European Transactions on Telecommunications 8, 33–37 (1997)
14. Kelly, F., Maulloo, A., Tan, D.: Rate control in communication networks: shadow prices, proportional fairness and stability. Journal of the Operational Research Society 49, 237–252 (1998)
15. Kontogiannis, S., Zaroliagis, C.: Robust Line Planning under Unknown Incentives and Elasticity of Frequencies. In: Proc. 8th Workshop on Algorithmic Approaches for Transportation Modeling, Optimization, and Systems – ATMOS 2008 (2008)
16. Liebchen, C., Lübbecke, M., Möhring, R., Stiller, S.: Recoverable Robustness. Technical Report ARRIVAL-TR-0066, ARRIVAL Project (August 2007)
17. Liebchen, C., Lübbecke, M., Möhring, R., Stiller, S.: The Concept of Recoverable Robustness, Linear Programming Recovery, and Railway Applications. In: Ahuja, R.K., Möhring, R.H., Zaroliagis, C.D. (eds.) Robust and Online Large-Scale Optimization. LNCS, vol. 5868, pp. 1–27. Springer, Heidelberg (2009)
18. Schöbel, A., Scholl, S.: Line Planning with Minimal Traveling Time. In: Proc. 5th Workshop on Algorithmic Methods and Models for Optimization of Railways – ATMOS (2005)
19. Schöbel, A., Schwarze, S.: A Game-Theoretic Approach to Line Planning. In: Proc. 6th Workshop on Algorithmic Methods and Models for Optimization of Railways – ATMOS (2006)
20. Scholl, S.: Customer-oriented line planning. PhD thesis, Technische Universität Kaiserslautern (2005)
21. Srikant, R.: The Mathematics of Internet Congestion Control. Birkauser (2004)
22. Vickrey, W.: Counterspeculation, auctions, and competitive sealed tender. Journal of Finance 16(1), 8–37 (1961)

A Bicriteria Approach for Robust Timetabling*

Anita Schöbel and Albrecht Kratz

Institut für Numerische und Angewandte Mathematik
Georg-August Universität Göttingen, Germany
schoebel@math.uni-goettingen.de, Albrecht.Kratz@gmx.de

Abstract. Finding robust solutions of an optimization problem is an important issue in practice. Various concepts on how to define the robustness of an algorithm or of a solution have been suggested. However, there is always a trade-off between the best possible solution and a robust solution, called the price of robustness. In this paper, we analyze this trade-off using the following bicriteria approach. We treat an optimization problem as a bicriteria problem adding the robustness of its solution as an additional objective function. We demonstrate this approach at the aperiodic timetabling problem in which a timetable which is robust under delays is sought. We are able to derive necessary conditions for the resulting Pareto-optimal timetables. For the case in which the robustness is defined as the largest delay for which all connections are maintained we show the bicriteria problem can be solved with the same time complexity as the original single-criteria problem.

1 Introduction

In many applications optimization tools can nowadays be used to calculate good (or even optimal) solutions. Unfortunately, there is one major drawback that prevents many solutions being established in real-world applications: nearly always there will be some kind of disturbance, e.g. input data changes, disruptions, delays or any other unforeseen event. To overcome such difficulties and make solutions applicable for real-world problems, researchers are working on various concepts of *robustness*. The goal of these concepts is to find not the best solution to the *nominal* (undisturbed) problem but to calculate a *robust* solution which is still good or at least recoverable in case of a disturbance. The ratio between the optimal solution and the robust solution is called its *price of robustness*. It is intuitively clear that the two objectives, to optimize the solution for the undisturbed scenario and to maximize its robustness are conflicting in most cases.

In this paper we hence suggest to treat these two objectives, the *nominal* objective of the undisturbed, original problem and the *robustness* of the solution as two objective functions of a bicriteria optimization problem. The corresponding

---

* This work was partially supported by the Future and Emerging Technologies Unit of EC (IST priority - 6th FP), under contract no. FP6-021235-2 (project ARRIVAL).

R.K. Ahuja et al. (Eds.): Robust and Online Large-Scale Optimization, LNCS 5868, pp. 119–144, 2009.

Pareto solutions hence yield decisions that can not be improved in both objectives simultaneously: Whenever the robustness is increased the nominal objective will get worse (and vice versa).

Let us first mention some related literature. Dealing with expected disturbances already in the strategic planning phase has been done by using *stochastic programming*, or within the area of *robust optimization*. Within stochastic programming (e.g., see [4,12,21]), there are two different approaches: *chance constrained programming* aims to find a solution that satisfies the constraints in most scenarios (i.e. with a high probability) instead of satisfying them for all possible realizations of the random variables, while in *multi-stage stochastic programming*, an initial solution is computed in the first stage, and each time when some new random data is revealed, a recourse action is taken. However, stochastic programming requires detailed knowledge on the probability distributions of the random variables.

In robust optimization (e.g., see [1,2,3,8]), the objective – in contrast to stochastic programming – is purely deterministic. In the concept of *strict robustness*, the solution has to be feasible for all likely scenarios. The solution gained by this approach can then be fixed since by construction it needs not be changed when disruptions occur. However, as the solution is fixed independently of the actual scenario, robust optimization leads to solutions that are too conservative in many applications. One approach to compensate this disadvantage is the idea of *light robustness* introduced in [9,10]. This approach relaxes the constraints by adding slack to them. A solution is considered as robust if it satisfies the relaxed constraints.

Recently, [15,16] suggested the concept of the *recoverable robustness*. They start from the practical point of view that a solution is robust if it can be recovered easily in case of a disruption. This means the solution has no longer to be feasible for all possible scenarios, but a recovery phase is allowed in which a recovery algorithm is applied to turn an infeasible solution into a feasible one. To obtain a good solution, some limitations on the recovery phase have to be taken into account. For example, the recovery should be quick enough and the quality of the recovered solution should not be too bad. The initial model of recovery robustness has been extended in [5], a multi-stage approach that can handle not only one, but a sequence of disturbances has been developed in [6].

Finding timetables in public transportation is used as an example of robust optimization in many of the studies mentioned above. Timetabling arises in the strategic planning phase for transportation systems. The problem is well known and well researched (see [14], [19], [17] and references therein for approaches on solving the NP hard problem of periodic timetabling). In our case we restrict ourselves to the tractable variant of aperiodic timetabling, also known as feasible differential problem (FDS, see [20]). It can be solved by linear programming or by shortest path techniques. In timetabling, disturbances during the operational phase often occur. In public transportation systems such disturbances can be caused e.g. by bad weather conditions, repair work, signalling problems, or accidents. Hence timetabling is a prominent and important example on which robustness concepts can be tested and should be implemented.

Most of the robustness approaches use a given level of robustness that has to be determined beforehand. In this paper we suggest to treat the timetabling problem as a bicriteria problem (see [7] for an introduction to multicriteria optimization) with the two conflicting objectives of minimizing the waiting time of the passengers and maximizing the robustness of the timetable. Theoretically any of the robustness concepts mentioned above can be used as robustness function; in our case we define the robustness of a timetable as the largest possible delay such that all transfers are maintained under some given strategy.

In Section 2 we summarize properties of aperiodic timetables and describe our robustness model in Section 3. The model is illustrated in Section 4 where we analyze a basic example. Other structures are investigated in Section 5. In Section 6 we derive properties of Pareto solutions. In Section 7 we present an efficient approach for solving the bicriteria problem.

2   Finding Timetables in Public Transportation

In order to describe a timetable, we use an event-activity network $\mathcal{N} = (\mathcal{E}, \mathcal{A})$ in which the set $\mathcal{E}$ represents the *events* and its directed edges are denoted as *activities* $\mathcal{A} \subseteq \mathcal{E} \times \mathcal{E}$. We assume $\mathcal{N}$ to be connected. A *timetable* $\Pi \in \mathbb{R}^{|\mathcal{E}|}$ assigns a time $\Pi_i \in \mathbb{R}$ to each event $i \in \mathcal{E}$. For each activity $a = (i, j)$ linking two events i and j in $\mathcal{E}$ we are given a lower bound l_a and an upper bound u_a that have to be respected, i.e. if event i takes place at some specified time Π_i then event j cannot take place earlier than $\Pi_i + l_a$ and must not be scheduled later than $\Pi_i + u_a$. Consequently, a timetable Π is *feasible* if $\Pi_j - \Pi_i \in [l_a, u_a]$ for all $a = (i, j) \in \mathcal{A}$. We further define $m_a := u_a - l_a \geq 0$ for all $a \in \mathcal{A}$.

Additionally, we need the following notation:

A *path* P in $\mathcal{N}$ will be given as a sequence of events $P = (i_1, \ldots, i_k)$ such that either a *forward activity* $(i_l, i_{l+1}) \in \mathcal{A}$ or a *backward activity* $(i_{l+1}, i_l) \in \mathcal{A}$ exists for each pair of consecutive events $l = 1, \ldots, k-1$. The forward activities of P are denoted as P^+ and the backward activities as P^-. If all activities are forward activities, the path is called *directed path*. A cycle is a path with $i_1 = i_k$. Note that no feasible timetable exists if $\mathcal{N}$ contains a directed cycle with at least one activity a with $l_a > 0$.

Given a timetable, one can calculate its *slack times* $s_a := \Pi_j - \Pi_i - l_a$. The slack time s_a represents the additional time available for activity a and can be used to reduce delays. It is well known that the problem of finding a feasible timetable Π can equivalently be formulated with respect to the variables s_a, see e.g. [20,18]. This is needed later and hence specified in the next two lemmas.

Lemma 1. *Let Π be a feasible timetable for $\mathcal{N} = (\mathcal{E}, \mathcal{A})$. Let $s_a = \Pi_j - \Pi_i - l_a$ for all $a \in A$. Then*

(i) $0 \leq s_a \leq m_a$ for all $a \in \mathcal{A}$.
(ii) For any circle C of $\mathcal{N}$:

$$\sum_{a\in C^+} s_a - \sum_{a\in C^-} s_a = -\sum_{a\in C^+} l_a + \sum_{a\in C^-} l_a. \qquad (1)$$

On the other hand, if some s_a are given for each activity $a \in \mathcal{A}$ that satisfy the conditions (i) and (ii) one can construct a feasible timetable with slack times s_a:

Lemma 2. *Let $s \in \mathbb{R}^{|\mathcal{A}|}$ be given such that (i) and (ii) of Lemma 1 are satisfied. Let $i \in \mathcal{E}$ be an (arbitrary) event and let P_{ij} be any path from i to j. Then Π given as*

$$\Pi_i = 0$$
$$\Pi_j = \sum_{a\in P_{ij}^+} (l_a + s_a) - \sum_{a\in P_{ij}^-} (l_a + s_a) \ \text{for all } j \neq i \qquad (2)$$

is a feasible timetable with slack times s_a.

Note that (1) ensures that the above definition of Π is independent of the specific paths P_{ij} used.

Before introducing the *nominal* objective function and the *robustness* of a timetable, let us describe the application which motivates our research in more detail. It stems from the following railway timetabling problem: given a public transportation network with a set of trains, the goal is to find arrival and departure times for each train at each station. Hence, $\mathcal{E}$ consists of all arrival and departure events of trains at stations. The activities are divided into driving activities $\mathcal{A}^{drive}$, stopping activities $\mathcal{A}^{stop}$ (modeling the time a train stands at the platform allowing passengers to board or un-board), and transfer activities $\mathcal{A}^{trans}$. The transfer activities model that passengers can transfer from an incoming train at a station into an outgoing train at the same station. A small example of an event-activity network in timetabling is depicted in Figure 1. Although the following results can be transfered directly to bus schedules, and can be adapted also to other applications of project planning, we will refer to *trains* and *stations* in the following.

The usual goal when designing a timetable Π is to minimize the overall traveling time of the passengers. If w_a passengers are traveling along activity a we obtain

$$\tilde{F}(\Pi) := \sum_{a=(i,j)\in\mathcal{A}} w_a(\Pi_j - \Pi_i) = \sum_{a=(i,j)\in\mathcal{A}} w_a(\Pi_j - \Pi_i - l_a) + \sum_{a\in A} w_a l_a$$
$$= \sum_{a\in\mathcal{A}} w_a s_a + \sum_{a\in A} w_a l_a =: F(s) + \sum_{a\in A} w_a l_a,$$

i.e. the objective can be formulated either using the timetable Π or its slack times $s_a = \Pi_j - \Pi_i - l_a$. For the latter we use $F(s)$ omitting the constant part $\sum_{a\in A} w_a l_a$.

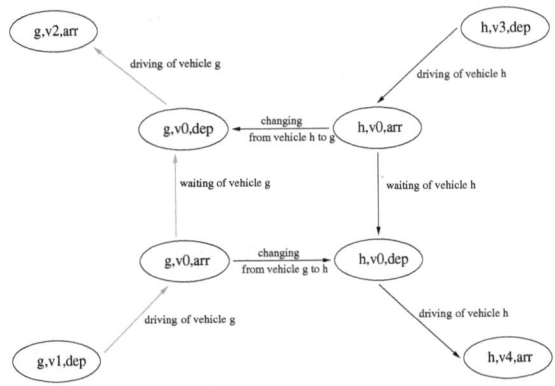

Fig. 1. An event-activity network with two trains g and h meeting at the station v_0. The events are labelled with triples $(train, v, arr/dep)$ indicate that a train arrives/departs at station v.

Note that most authors assume that the weights w_a are given. In our work we will however go a step back and start with an origin-destination (OD) matrix $(w_{ij})_{i,j \in \mathcal{E}}$. For a departure event i and an arrival event j the value w_{ij} represents the number of passengers traveling from i to j. For each OD-pair we define a directed path P_{ij} that passengers are likely to use and consequently obtain the weight

$$w_a = \sum_{\substack{i,j \in \mathcal{E}: \\ a \in P_{ij}}} w_{ij} \tag{3}$$

for activity $a \in \mathcal{A}$. We will utilize this fact in Theorem 4.

Summarizing, there are the following two equivalent formulations for the timetabling problem:

$(TT - \Pi)$ (Timetabling using variables $\Pi_i, i \in \mathcal{E}$)

$$\min \; \tilde{F}(\Pi) = \sum_{a=(i,j) \in \mathcal{A}} w_a (\Pi_j - \Pi_i)$$
$$\text{s.t.} \; l_a \leq \Pi_j - \Pi_i \leq u_a \; \text{for all } a = (i,j) \in \mathcal{A}$$
$$\Pi_i \in \mathbb{N}.$$

$(TT - s)$ (Timetabling using variables $s_a, a \in \mathcal{A}$)

$$\min \; F(s) = \sum_{a \in \mathcal{A}} w_a s_a$$
$$\text{s.t.} \quad 0 \leq s_a \leq m_a$$
$$\sum_{a \in C^+} s_a - \sum_{a \in C^-} s_a = - \sum_{a \in C^+} l_a + \sum_{a \in C^-} l_a.$$

Lemma 3. $(TT - \Pi)$ and $(TT - s)$ are equivalent. In particular,

- Let Π be an optimal solution of $(TT-\Pi)$. Then its slack times s are optimal for $(TT - s)$.
- Vice versa, if s is optimal for $(TT - s)$, any timetable Π obtained by (2) is optimal for $(TT - \Pi)$

3   The Robustness of a Timetable

Roughly speaking, the robustness of a timetable evaluates its sensitivity to unforeseen delays. Before presenting a definition of robustness we have to specify how a timetable is updated in case a delay occurs.

Let a source delay V be given at some event i and consider a stopping or driving activity $a = (i,j) \in \mathcal{A}^{stop} \cup \mathcal{A}^{drive}$ starting at i. Given the delay of V at i, activity a will start at $\Pi_i + V$. In order to keep up with the delay we allow that the activity is performed faster than planned, but still respecting its lower bound l_a. Hence we obtain $\max\{\Pi_j, \Pi_i+V+l_a\}$ as new end time of a. The delay y_j of event j is $[V - s_a]^+$, i.e. exactly the slack time s_a is used to decrease the delay. Note that the maximum is needed since no train is allowed to be ahead of its schedule.

In case of a transfer activity $a = (i,j)$ one has two possibilities:

- Either the outgoing train waits for the delayed incoming train to allow passengers to transfer. In this case, the transfer activity is treated analogously to a driving or stopping activity. The new departure time at event j must be at least $\max\{\Pi_j, \Pi_i + V + l_a\}$, its delay can be reduced to $[V - s_a]^+$ (if there is no other delayed transfer activity ending at j).
- Or, the train departs on time, i.e. without waiting for transferring passengers. In this case, no delay is transferred.

The problem of determining which transfers should be maintained and which need not be respected is known as *delay management problem*, see [23], [24], [11], [22]. Its objective is to minimize the overall delay of the passengers. Here we assume that the delay management problem has already been solved and that some strategy given as a *waiting time rule* is at hand. We consider three such waiting time rules (WTR) which will be described next.

Let i be an arrival event of train 1 and let $a = (i,j)$ be a transfer activity to train 2. Furthermore, let $\tilde{a} = (j,k)$ be the next driving activity of train 2, see Figure 2 for an illustration. Assume that train 1 arrives at i with a delay of y_i. The following three rules determine if train 2 should wait for train 1 or depart on time.

WTR1: Train 2 is not allowed to have a delay at its *next* station. Hence the maximal allowed waiting time at event j is given by the slack time $s_{\tilde{a}}$ of its next driving activity $\tilde{a} = (j,k)$. The transfer is maintained if and only if $y_i \leq s_a + s_{\tilde{a}}$.

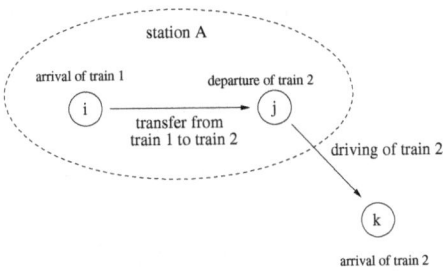

Fig. 2. Train 1 arrives at station A with a delay. Should train 2 wait or depart on time?

WTR2: The maximal allowed waiting time at event j is n minutes where n is fixed beforehand. The transfer is maintained if and only if $y_i \leq s_a + n$.

WTR3: Train 2 is not allowed to have a delay of more than m (minutes) at its *next* station. Hence the maximal allowed waiting time at event j is given by m plus the slack time $s_{\tilde{a}}$ of its next driving activity. The transfer is maintained if and only if $y_i \leq s_a + s_{\tilde{a}} + m$.

Note that each transfer activity $a = (i, j)$ is followed by exactly one driving activity starting at event j which we call $d(a)$ in the following.

Given a transfer activity $a = (i, j)$ together with a timetable Π_i, Π_j and an actual delay $y_i \geq 0$ at i as well as the slack time s of the next driving activity $d(a)$, WTR(y_i, s) gives back the decision "drive on time" or "wait and maintain transfer". All other activities have to be performed as fast as possible in order to reduce the delay. More precisely, given a timetable Π, a set of source-delayed event $\mathcal{E}^{del} \subseteq \mathcal{E}$ with delays V_i for $i \in \mathcal{E}^{del}$ and a waiting time rule WTR, we can iteratively calculate the delays y similar to the critical path method of project planning as follows:

Algorithm 1 to calculate the delayed timetable when a waiting time rule is fixed

1. Order the events in $\mathcal{E}$ according to the timetable Π, i.e. such that $\Pi_1 \leq \Pi_2 \leq \ldots \Pi_{|\mathcal{E}|}$.
2. For $i = 1, \ldots |\mathcal{E}|$ do:
   (a) $\mathcal{A}^{fix} := \{$ driving activities $a = (j, i)\} \cup \{$ waiting activities $a = (j, i)\}$
   (b) For each transfer activity $a = (j, i) \in \mathcal{A}$ its delay y_j is already known. Determine if the transfer is maintained or not according to WTR. If the transfer is maintained add a to $\mathcal{A}^{fix}$.
   (c) If $i \notin \mathcal{E}^{del}$ set $y_i := \max_{a=(j,i) \in \mathcal{A}^{fix}} [y_j - s_a]^+$,
       if $i \in \mathcal{E}^{del}$ set $y_i := \max\{V_i, \max_{a=(j,i) \in \mathcal{A}^{fix}} y_j - s_a\}$.

(Note that the waiting time rule together with Algorithm 1 is a *recovery strategy* as defined within the concept of recovery robustness in [15,16].) Throughout this paper we assume that only one waiting time rule is used within a public transport system. Combinations of different waiting time rules are investigated in [13].

We are finally in the position of defining three robustness functions. The first definition of robustness calculates the maximal size of the source delay for which no passenger misses a transfer.

Definition 1. *Let a fixed waiting time rule (according to WTR 1,2, or 3 above) be given as well as a set of source-delayed events $\mathcal{E}_{del} \subseteq \mathcal{E}$. A timetable (given by its slack values $s \in \mathbb{R}^{|\mathcal{A}|}$) has the robustness $R(s)$ if all its transfers are maintained whenever all source delays are smaller than or equal to R.*

The other two definitions of robustness evaluate how badly the passengers are affected by the source delays.

Definition 2. *Let a fixed waiting time rule (according to WTR 1,2, or 3 above) be given. Furthermore, let $s \in \mathbb{R}^{|\mathcal{A}|}$ be a timetable and consider a set of source-delayed events $\mathcal{E}^{del}$ with delays $V_i \leq V$ for all $i \in \mathcal{E}^{del}$.*

- *$R_{no}(s, V)$ is defined as the maximal number of passengers who miss a transfer if all source delays are smaller than V.*
- *$R_{del}(s, V)$ is the maximal sum of all passengers' delays if all source delays are smaller than V. The delay of a passenger missing a transfer is approximated as T assuming that the timetable is repeated after T minutes and that the passenger can then use the transfer of the next period.*

We hence obtain three problems, each of them bicriteria, and each of them can be discussed for all three waiting time rules:

$$(P) \quad \begin{pmatrix} \min F(s) \\ \max R(s) \end{pmatrix} \text{ s.t. } s \text{ is a feasible timetable}$$

$$(P_{no}) \quad \text{Given } V, \begin{pmatrix} \min F(s) \\ \min R_{no}(s, V) \end{pmatrix} \text{ s.t. } s \text{ is a feasible timetable}$$

$$(P_{del}) \quad \text{Given } V, \begin{pmatrix} \min F(s) \\ \min R_{del}(s, V) \end{pmatrix} \text{ s.t. } s \text{ is a feasible timetable}$$

The goal is to find *Pareto solutions*, i.e. timetables s such that there does not exist another timetable s' which is not worse in one of the two objectives and strictly better in the other one. A timetable is called *weak Pareto* if there does not exist another timetable which is strictly better in both objectives. Note that a Pareto timetable is always weak Pareto, but the reverse does in general not hold. The tuples $(F(s), R(s))$ (or $(F(s), R_{no}(s))$ in the case of (P_{no}), or $(F(s), R_{del}(s))$ in the case of (P_{del}), respectively) belonging to (weak) Pareto solutions are called *(weakly) non-dominated* in multicriteria optimization.

In the next section we first illustrate these definitions at the basic example of one single transfer. We show how to calculate the robustness R as well as R_{no}, and R_{del} w.r.t the three waiting time rules, set up the bicriteria models and show their Pareto solutions. In the subsequent sections we study properties of the models and present solution approaches.

4   The Basic Example: A Single Transfer

We start with the simplest case, namely two trains, train 1 running from station A to station B, and train 2 running from station B to station C. There is a transfer activity possible at station B, see Figure 3. We assume that a delay of size V occurs at station A.

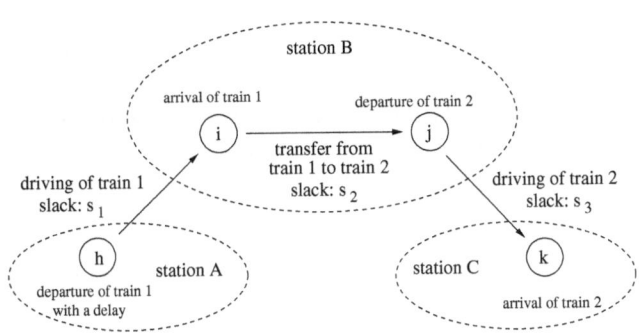

Fig. 3. The basic example: One single transfer

We first determine how long train 2 will wait for train 1 according to the three waiting time rules WTR1, WTR2, and WTR3.

- With a delay of V at station A, train 1 will reach station B with a delay of $[V - s_1]^+$. Passengers transferring to train B will hence arrive at train B with a delay of $[V - s_1 - s_2]^+$. According to WTR1, train 2 will wait at most s_3 minutes, hence the transfer is maintained if and only if $V - s_1 - s_2 \leq s_3$, or equivalently, if $V \leq s_1 + s_2 + s_3$.
- According to WTR2, train 2 waits up to n minutes, i.e. the transfer is maintained if and only if $V - s_1 - s_2 \leq n$, i.e. if and only if $V \leq s_1 + s_2 + n$.
- Analogously, we obtain for WTR3 that the transfer is maintained if and only if $V \leq s_1 + s_2 + s_3 + m$.

With the help of these observations, we obtain

$$R(s_1, s_2, s_3) = s_1 + s_2 + s_3 \text{ in case of WTR1}$$
$$R(s_1, s_2, s_3) = s_1 + s_2 + n \text{ in case of WTR2} \qquad (4)$$
$$R(s_1, s_2, s_3) = s_1 + s_2 + s_3 + m \text{ in case of WTR3}$$

For the calculation of R_{del} and R_{no} let w_{AB}, w_{AC}, and w_{BC} be given. The customers traveling from A to C are the only ones who can miss their transfer (in case $V > R(s)$). We hence obtain

$$R_{no}(s, V) = \begin{cases} w_{AC} & \text{if } V > R(s_1, s_2, s_3) \\ 0 & \text{if } V \leq R(s_1, s_2, s_3) \end{cases}$$

covering all three waiting time rules. Finally, for the calculation of R_{del} we have to consider all three OD-pairs: The passengers from A to B gain a delay of $[V - s_1]^+$, the passengers from A to C get a delay of T if they miss their transfer, otherwise they get the same delay as the passengers from B to C, namely $[V - s_1 - s_2 - s_3]^+$. Note that the delay of the latter is zero if train 2 does not wait for train 1. Summarizing, we obtain

$$R_{del}(s, V) = \begin{cases} w_{AB}[V - s_1]^+ + Tw_{AC} & \text{if } V > R(s_1, s_2, s_3) \\ w_{AB}[V - s_1]^+ + (w_{AC} + w_{BC})[V - s_1 - s_2 - s_3]^+ & \text{if } V \leq R(s_1, s_2, s_3) \end{cases}$$

We remark that in case of WTR1 train 2 will never be delayed since $R(s) = s_1 + s_2 + s_3$ (or, since train 2 never waits longer than its slack time). The formula hence simplifies to

$$R_{del}(s, V) = \begin{cases} w_{AB}[V - s_1]^+ + Tw_{AC} & \text{if } V > s_1 + s_2 + s_3 \\ w_{AB}[V - s_1]^+ & \text{if } V \leq s_1 + s_2 + s_3 \end{cases} \quad \text{in the case of WTR1.}$$

The Bicriteria Problems for the Basic Example

In order to formulate the robustness problem as a bicriteria decision problem we use the slack times s_a as variables (as in $[(TT - s)]$ on page 123). We further assume that $w_{AC} > 0$ otherwise no transfer needs to be maintained.

The first objective function in all thee bicriteria problems is the minimization of the traveling time, given as

$$F(s_1, s_2, s_3) = (w_{AB} + w_{AC})s_1 + w_{AC}s_2 + (w_{AC} + w_{BC})s_3.$$

This function is linear in the three variables s_1, s_2, s_3 and is independent of the waiting time rule and the particular robustness definition used.

Solving (P): Maximizing the Robustness R. Using (4) we obtain the linear bicriteria model

$$\begin{aligned} \min \ & F(s_1, s_2, s_3) \\ \max \ & R(s_1, s_2, s_3) \\ \text{s.t. } & 0 \leq s_i \leq m_i \text{ for } i = 1, 2, 3 \end{aligned} \quad (5)$$

for all three waiting time rules WTR1, WTR2, and WTR3. The set of Pareto solutions can be precisely described for all three waiting time rules. To this end, we use the following lemma.

Lemma 4. *Consider the following bicriteria problem*

$$\begin{pmatrix} \min \sum_{a=1}^{m} w_a s_a \\ \max \sum_{a=1}^{m} s_a \end{pmatrix} \quad s.t. \ 0 \leq s_a \leq m_a, a = 1, \ldots, m$$

where $0 \leq w_1 \leq \ldots \leq w_m$ and $m_a \geq 0, a = 1, \ldots, m$. Then the following holds:
$s = (s_1, \ldots, s_m)$ is weakly Pareto if and only if there exists $k \in \{1, \ldots, m\}$
such that $s_1 = m_1, \ldots, s_{k-1} = m_{k-1}$ and $s_{k+1} = 0, \ldots, s_m = 0$. If all $w_a > 0$
the condition yields a Pareto solution.

Proof. s is a Pareto solution if and only if s is an optimal solution to the problem

$$\min \ \sum_{a=1}^{m} w_a s_a$$
$$\text{s.t.} \ \ 0 \leq s_a \leq m_a, a = 1, \ldots, m$$
$$\sum_{a=1}^{m} s_a \geq R$$

This is a continuous knapsack problem with unit weights and hence has the solution as claimed. If all weights are strictly positive, any improvement in R leads to an increase of the objective, hence the corresponding solution is Pareto. □

Lemma 4 allows to derive the following characterization of Pareto solutions for the basic example.

Theorem 1. Let $0 \leq s_i \leq m_i$ for $i = 1, 2, 3$. (s_1, s_2, s_3) is a Pareto solution of (5) if and only if the following conditions hold:

for WTR1:
  – in case that $w_{BC} < w_{AB}$:
    $s_3 = s_1 = 0$ or $(s_1 = 0$ and $s_2 = m_2)$ or $(s_2 = m_2$ and $s_3 = m_3)$
  – in case that $w_{AB} < w_{BC}$:
    $s_3 = s_1 = 0$ or $(s_3 = 0$ and $s_2 = m_2)$ or $(s_2 = m_2$ and $s_1 = m_1)$

for WTR2: $s_1 = 0$ or $s_2 = m_2$,
for WTR3: the same condition that applies for WTR1.

Proof. First, note that $w_2 = w_{AC} < w_{AC} + w_{AB} = w_1$ and $w_2 = w_{AC} < w_{AC} + w_{BC} = w_3$ and that $w_i > 0, i = 1, 2, 3$.
For WTR1 (and WTR3) we hence have to consider two cases: either $w_2 \leq w_1 \leq w_3$ which is the case if and only if $w_{AB} \leq w_{BC}$, or we have $w_2 \leq w_3 \leq w_1$ which occurs if and only if $w_{BC} \leq w_{AB}$. Applying Lemma 4 yields the results.
For WTR2 it is even simpler since we have to consider

$$\begin{pmatrix} \min w_1 s_1 + w_2 s_2 + w_3 s_3 \\ \max s_1 + s_2 + n \end{pmatrix} \ \text{s.t.} \ 0 \leq s_a \leq m_a, a = 1, 2, 3$$

where $0 < w_2 < w_1$. Since $w_3 > 0$ any solution with $s_3 > 0$ is dominated by the solution obtained by setting $w_3 = 0$. We are hence left with a problem of Lemma 4 with $m = 2$ and the result follows. □

The interpretation of the basic example is that slack should be put on the transfer and not on the driving activities.

Solving (P_{no}): Minimizing R_{no}. In order to formulate the minimization of R_{no} we introduce a binary variable z with value zero if train 2 waits for train 1 and value one otherwise. This leads to the following formulation.

$$
\begin{aligned}
\min \quad & F(s_1, s_2, s_3) \\
\min \quad & z\, w_{AC} \\
\text{s.t.} \quad & Vz + R(s_1, s_2, s_3) \geq V \\
& 0 \leq s_i \leq m_i \quad \text{for } i = 1, 2, 3 \\
& z \in \{0, 1\}.
\end{aligned}
\tag{6}
$$

Note that the resulting problem is again a bicriteria linear optimization program which is correct due to the following observation:

Lemma 5. *Let (s_1, s_2, s_3, z) be a Pareto solution of (6). Then $z = 0$ if no passenger misses a transfer, and $z = 1$ if the passengers traveling from A to C miss their transfer.*

Proof. First, let $z = 0$. This means that $R(s_1, s_2, s_3) \geq V$, hence no passenger misses his or her transfer. On the other hand, if $z = 1$ in an optimal solution, we know that $z = 0$ is not feasible for the same values s_1, s_2, s_3 (due to $w_{AC} > 0$). This yields $R(s_1, s_2, s_3) < V$, i.e. the transfer from train 1 to train 2 is missed. □

Theorem 2. *Let $V > n$ in case of WTR2 and $V > m$ in case of WTR3. Then (P_{no}) has exactly two non-dominated points, one corresponding to the Pareto solution $s = 0$ and the other being $(F(s), 0)$, where s can be found as an optimal solution to*

$$
\begin{aligned}
\min \quad & \sum_{a=1}^{m} w_a s_a \\
\text{s.t.} \quad & 0 \leq s_a \leq m_a, a = 1, \dots, m \\
& R(s) \geq V
\end{aligned}
$$

If $V \leq n$ (for WTR2) or $V \leq m$ (for WTR3) there exists exactly one non-dominated solution, namely $(0, 0)$, corresponding to the Pareto solution $s = 0$.

Proof. For any s we have that $R_{no}(s) \in \{w_{AC}, 0\}$, i.e. the second objective can only obtain two different values, hence there exist at most two non-dominated solutions.

- The first case, $R_{no} = 0$ minimizes R_{no} and hence is non-dominated since it is lexicographically optimal. Due to $w_{AC} > 0$ it yields $z = 0$, hence the corresponding Pareto solution can be found by solving the optimization problem in which z is fixed to zero.
- The second case, $z = 1$ equals the lexicographic solution if we first minimize $F(s)$: Since $w_{AC} > 0$ we obtain $s = 0$ as unique optimal solution. If $V > n$ (for WTR2) or $V > m$ (for WTR3) the transfer is missed and we obtain the non-dominated solution $(0, w_{AC})$. If $V \leq n$ (for WTR2) or $V \leq m$ (for WTR3) then the transfer is maintained even in the case of no slack time and the two lexicographic solutions coincide. □

The result can be interpreted as follows: either distribute no slack at all or take the minimal amount of slack such that the transfer is maintained.

Solving (P_{del}): Minimizing R_{del}. For the calculation of R_{del} we use the same variable z but need furthermore to take the delay of the passengers from A to B and from B to C into account. This delay also depends on z.

Lemma 6. *Let s_1, s_2, s_3 be a Pareto solution of the bicriteria problem* $\min F(s_1, s_2, s_3)$ *and* $\min R_{del}$ *(min R_{no}, respectively). Then $s_1 + s_2 + s_3 \leq V$.*

Proof. Assume the contrary, that $s_1 + s_2 + s_3 > V$. Define a feasible timetable t through $t_1 = \min\{V, s_1\}$, $t_2 = \min\{V - t_1, s_2\}$, and, $t_3 = \min\{V - t_1 - t_2, s_3\}$. It can easily be shown that $F(t) < F(s)$ and $R(t) = R(s)$, hence t dominates s.   □

Note that Lemma 6 is a special case of Lemma 10, see also there for a detailed proof. Here we use Lemma 6 to obtain the following model for (P_{del}) in the basic example.

$$
\begin{aligned}
&\min F(s_1, s_2, s_3) \\
&\min (V - s_1)w_{AB} + zTw_{AC} + (1 - z)(w_{AC} + w_{BC})(V - s_1 - s_2 - s_3) \\
&\text{s.t. } Vz + R(s_1, s_2, s_3) \geq V \\
&\qquad s_1 + s_2 + s_3 \leq V \\
&\qquad 0 \leq s_i \leq m_i \text{ for } i = 1, 2, 3 \\
&\qquad z \in \{0, 1\}.
\end{aligned}
\tag{7}
$$

This model is quadratic. In the following we present linear formulation for the three waiting time rules.

Linear formulation for WTR1. Note that $z = 0$ yields $R(s) \geq V$, i.e. for WTR1 we obtain $s_1 + s_2 + s_3 \geq V$ such that $[V - s_1 - s_2 - s_3]^+ = 0$ and the model simplifies to the following linear formulation:

$$
\begin{aligned}
&\min && F(s_1, s_2, s_3) \\
&\min (V - s_1)w_{AB} + zTw_{AC} \\
&\text{s.t.} && Vz + s_1 + s_2 + s_3 \geq V \\
&&& s_1 + s_2 + s_3 \leq V \\
&&& 0 \leq s_i \leq m_i \text{ for } i = 1, 2, 3 \\
&&& z \in \{0, 1\}.
\end{aligned}
$$

Linear formulations for WTR2 and WTR3. For the other two rules we obtain the following linearization of (7).

$$
\begin{aligned}
&\min && F(s_1, s_2, s_3) \\
&\min (V - s_1)w_{AB} + zTw_{AC} + (w_{AC} + w_{BC})q \\
&\text{s.t.} && q + Vz + s_1 + s_2 + s_3 \geq V \\
&&& Vz + R(s_1, s_2, s_3) \geq V \\
&&& s_1 + s_2 + s_3 \leq V \\
&&& 0 \leq s_i \leq m_i \text{ for } i = 1, 2, 3 \\
&&& z \in \{0, 1\}. \\
&&& q \geq 0
\end{aligned}
\tag{8}
$$

Lemma 7. *(7) and (8) are equivalent.*

Proof. We consider the two cases $z = 0$ and $z = 1$ separately.

$z = 1$: If $(s_1, s_2, s_3, 1)$ is feasible for (7), then $(s_1, s_2, s_3, 1)$ and $q = 0$ is feasible for (8) with the same objective values. On the other hand, if a solution $s, z = 1, q$ of (8) is given, then $s, z = 1$ is a feasible solution of (7) and due to $q \geq 0$ we obtain that it has the same or a better objective value in both objective functions.

$z = 0$: If $(s_1, s_2, s_3, 0)$ is feasible for (7), then $(s_1, s_2, s_3, 0)$ and $q = V - s_1 - s_2 - s_3$ is feasible for (8) with the same objective values. Again, if a solution $s, z = 0, q$ of (8) is given, then $s, z = 0$ is a feasible solution of (7). Its objective values are the same or better than the objective values of (8) since $q \geq V - s_1 - s_2 - s_3$. $\qquad\square$

Instead of deriving general results about the Pareto solution we illustrate the Pareto space for WTR1 and WTR2 in the following example: Let $w_{AB} = w_{AC} = w_{BC} = 1$ be unique passengers' weights and define $m_1 = m_3 = n$, $m_2 = 2n$ and $T = 12n$ (a realistic value for n can be 5). In both Figure 4 and Figure 5 the x-axis contains the objective values of R_{del} and the y-axis contains F. Figure 4 uses WTR1 and shows the set of weakly non-dominated solutions (also called the *efficient frontier*) for two source delays: $V = \frac{3}{2}n$ (lower line) and $\tilde{V} = 3n$ (upper line). As expected, both sets of weakly non-dominated solutions are piecewise linear curves. In Figure 5 the weakly non-dominated solutions for WTR2 are

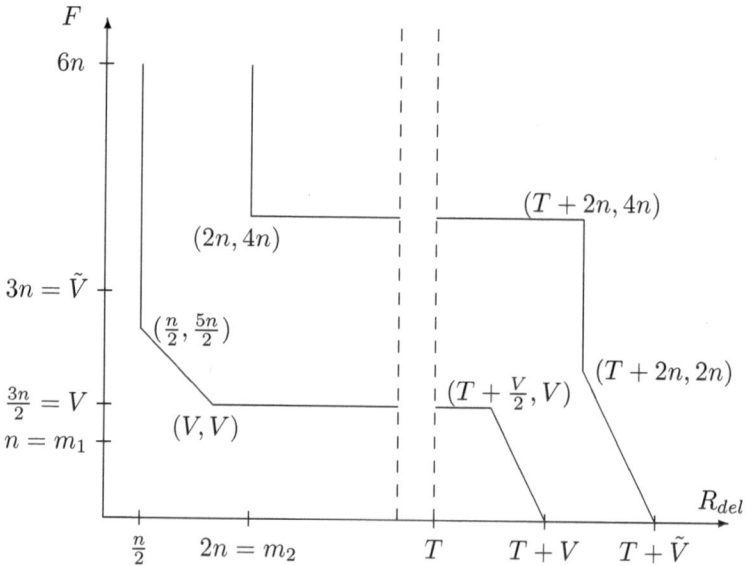

Fig. 4. Weakly non-dominated solutions for (P_{del}) in case of WTR1. Two different examples of source delays are depicted: $V = \frac{3}{2}n$ (lower line) and $\tilde{V} = 3n$ (upper line).

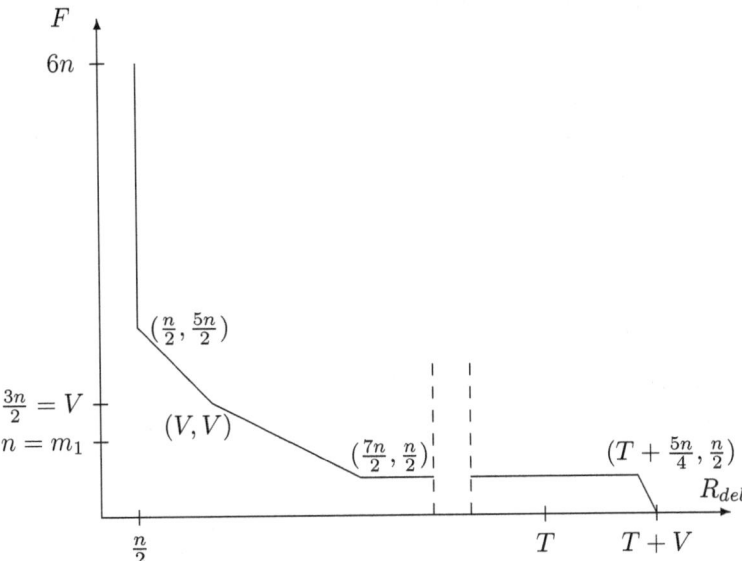

Fig. 5. Weakly non-dominated solutions for (P_{del}) in case of WTR2. The source delay of the example is $V = \frac{3}{2}n$.

depicted for a source delay of $V = \frac{3}{2}n$. Also in this case the efficient frontier is piecewise linear.

5   Subsequent Transfers

In this section we consider another basic substructure, namely a network $\mathcal{N}$ which is a path. We assume $\mathcal{E} = \{i_1, i_2, \dots, i_K\}$ and $\mathcal{A} = \{a_j = (i_j, i_{j+1}) : j = 1, \dots, K-1\}$ are ordered along the path. As usual, $\mathcal{A}$ may consist of driving, waiting and transfer activities. Let us call the transfer activity with smallest index the *first* transfer activity.

Furthermore, assume that i_1 is the only source-delayed event. The following result will be useful in the next section.

Lemma 8. *Let $\mathcal{N}$ be a path and let its first node i_1 be delayed. Let s be a feasible timetable and $R \in \mathbb{R}$. For all three waiting time rules we have:*
$R(s) \leq R$ *if and only if the first transfer activity is maintained for a delay of* $V_{i_1} = R$.

Proof. Let $a = (i, j)$ be the first transfer activity and $d(a) = (j, l)$ be the next driving activity. First note that a is maintained for all delays $V \leq R$ if and only if it is maintained for $V = R$. We hence have to show that also all other transfers are maintained if a is maintained. Let us consider the waiting time rules separately.

In case of WTR1, a is maintained if $y_i - s_a \leq s_{d(a)}$. This yields $y_l = 0$, hence there is no delay at all subsequent events, such that all subsequent transfers will be maintained.

For WTR2, a is maintained if $y_i - s_a \leq n$, hence $y_l \leq n$ and also $y_k \leq n$ for all events k following l. Hence, according to WTR2, the next transfer (and hence all others) will be maintained.

In case of WTR3, we obtain that no event after l has a delay larger than m and hence all subsequent transfers are again maintained. $\square$

The following corollaries use this observation to describe Pareto solutions for the problems (P) and (P_{no}).

Corollary 1. *Let $\mathcal{N}$ be a path and let its first node i_1 be delayed. Let $w_a > 0$ for all $a \in \mathcal{A}$. Let a_k be the first transfer activity. If s is a Pareto timetable for (P) or (P_{no}), we have $s_{a_l} = 0$ for all $l > k + 1$ in case of WTR1 and WTR3, $s_{a_l} = 0$ for all $l > k$ in case of WTR2.*

Proof. Assume s is a timetable with $s_{a_l} > 0$ for some $l > k + 1$ ($l > k$ in case of WTR2). Then the timetable t with $t_{a_l} := 0$, $t_a = s_a$ for all $a \neq a_l$ is feasible and has a better objective $F(t) < F(s)$ if $w_{a_l} > 0$. It remains to show that its robustness does not increase.

- Due to Lemma 8 we know that s is robust with $R(s) = V$ if and only if a_k is maintained for all $V_{i_1} \leq V$. Further, the first transfer a_k is maintained for the timetable s if and only if it is maintained for the timetable t, hence also t is robust with $R(t) = V$.
- For (P_{no}), let $V_{i_1} = V$ be given.
  - If a_k is missed for s it is also missed for t. In this case everything is on time for all events after a_K, hence all subsequent transfers are maintained and we obtain $R_{no}(t) = w_{a_k} = R_{no}(s)$.
  - On the other hand, if a_k is maintained for s it is also maintained for t. From Lemma 8 we know that in this case also all other transfers are maintained. Hence $R_{no}(s) = R_{no}(t) = 0$.

Summarizing, the robustness of both timetables is the same. $\square$

Note that the corollary needs not hold for (P_{del}), a counterexample is provided in Section 6 (page 139).

Corollary 2. *Let $\mathcal{N}$ be a path and let its first node i_1 be delayed. There exist at most two non-dominated timetables for problem (P_{no}).*

Proof. The result follows since $R_{no}(s) \in \{0, w_{a_k}\}$ only can obtain two different values. $\square$

Further results about two subsequent transfers in special cases can be found in [13].

6   Pareto Solutions in General Networks

Intuitively one might think that we only have to look at the worst delays that can occur when determining the robustness of a timetable. This is true in linear graphs and has already been used in the proof of Lemma 8. However, if s is robust with delays $V_i = V$ for all $i \in \mathcal{E}^{del}$ then s need **not** be robust with respect to delays $V_i \leq V$ for all $i \in \mathcal{E}^{del}$. This is due to the fact that transfers might be maintained "accidentally" as the following counterexample (see Figure 6) shows: Here we have two trains, both with source delays, namely V_1 at station A and V_2 at stations B. Both trains meet at station C where passengers can transfer from train 1 to train 2. We assume that all slack times are 2. Let e.g. $V = 20$ and $n = m = 5$. Although the waiting time rules would not maintain the transfer from train 1 to train 2, it is maintained accidentally if both trains have the (same) maximum delay $V_1 = V_2 = V = 20$. On the other hand, if only train 1 has a delay of $V_1 = 20$, but the delay of train 2 is smaller than 12, then the transfer is not maintained for all three waiting time rules. Hence the timetable s is robust for $V_1 = V_2 = V = 20$, but it is not robust for $V_1 = V = 20$ and $V_2 \leq 12$.

The example shows that each train must be able to reduce its delay by itself before a transfer activity is reached. This will be formalized next.

We first analyze how a delay spreads out. Let us call a path P in $\mathcal{N}$ a *maintained path* if all its transfers are maintained. For $i^{del} \in \mathcal{E}^{del}$ and $j \in \mathcal{E}$ let $mpath(i^{del}, j)$ denote the set of all maintained paths starting in i^{del} and ending at j. Finally, for a path P let $s(P) = \sum_{a \in P} s_a$ be the sum of slacks gathered along P.

The next lemma follows from Algorithm 1.

Lemma 9. *Let $\mathcal{E}^{del}$ be a set of source-delayed events with delays $V_{i^{del}}$ for $i^{del} \in \mathcal{E}^{del}$. Then*

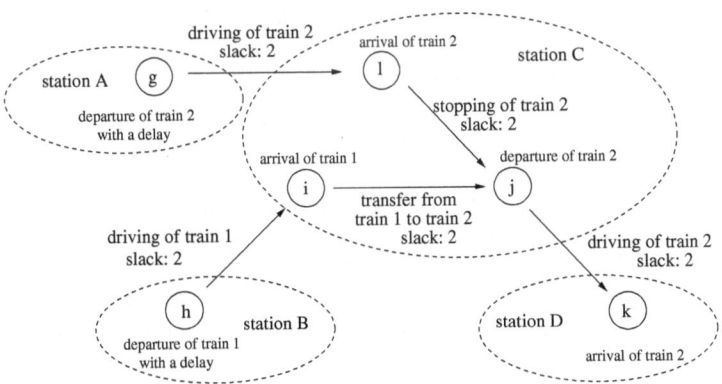

Fig. 6. Counterexample

$$y_i = \max_{i^{del} \in \mathcal{E}^{del}, P \in mpath(i^{del}, i)} [V_{i^{del}} - s(P)]^+.$$

Let us consider the subnetwork $\bar{\mathcal{N}} = (\mathcal{E}, \mathcal{A}^{wait} \cup \mathcal{A}^{drive})$ that is obtained by deleting all transfer activities. It consists of a set of paths, one for each train. Let $\bar{P}_{ij}$ denote the unique path from i to j in $\bar{\mathcal{N}}$ (if it exists).

First, we reduce the set of events that need to be considered. It is obvious that only events need to be considered that can get a delay, i.e. events which can be reached by a path from a source-delayed event. In our case we can restrict the set of relevant events even further, namely to such events which can be reached by a path in $\bar{\mathcal{N}}$:

$$\mathcal{E}^{rel} := \{j \in \mathcal{E} : \text{ there exists a directed path in } \bar{\mathcal{N}} \text{ from } i^{del} \in \mathcal{E}^{del} \text{ to } j\}.$$

For all $j \in \mathcal{E}^{rel}$ we further define

$$\bar{s}_j := \min\{s(P) : P \text{ is a directed path in } \bar{\mathcal{N}} \text{ from } i^{del} \in \mathcal{E}^{del} \text{ to } j\}. \quad (9)$$

Note that the path P for which the minimum is obtained, is unique.

For the next lemma, recall that $d(a)$ denotes the unique driving activity that directly follows the transfer a.

Theorem 3. *Let $\mathcal{E}^{del}$ be the set of potential source-delayed events and let s be a feasible timetable. $R(s) = R$ if and only if for all transfers $a = (i, j) \in \mathcal{A}^{trans}$ with $i \in \mathcal{E}^{rel}$ we have:*

$$R - \bar{s}_i \le s_{d(a)} + s_a \text{ in the case of WTR1,} \quad (10)$$
$$R - \bar{s}_i \le n + s_a \text{ in the case of WTR2,} \quad (11)$$
$$R - \bar{s}_i \le m + s_{d(a)} + s_a \text{ in the case of WTR3.} \quad (12)$$

Proof.

- Let the conditions (10), (11), or (12), respectively, be satisfied. Let a set of source-delayed events $\mathcal{E}^{del}$ be given with source delays $V_i \le R$ for all $i \in \mathcal{E}^{del}$. We want to show that all transfers are maintained. Let $a = (i, j) \in \mathcal{A}^{trans}$. According to Lemma 9 we know that

$$y_i = \max_{i^{del} \in \mathcal{E}^{del}, P \in mpath(i^{del}, i)} [V_{i^{del}} - s(P)]^+.$$

  Let P be the path from i^{del} to i for which the maximum is obtained. The following cases can occur:
  - $y_i = 0$. Then the transfer is maintained.
  - P is a path in $\bar{\mathcal{N}}$. Then $i \in \mathcal{E}^{rel}$ such that we obtain $y_i = V_{i^{del}} - s(P) \le R - \bar{s}_i$ and the transfer is maintained due to (10) (or (11),(12), respectively).
  - P is a path which is not in $\bar{\mathcal{N}}$. It hence contains at least one transfer activity $\tilde{a}$ which is maintained. Lemma 8 yields that all transfers that follow $\tilde{a}$ on a path are also maintained, in particular we conclude that a is maintained.

– Conversely, assume that s is robust with level $R(s) = R$. Assume that (10), (11), or (12), respectively, are not satisfied for some transfer activity $a = (i,j) \in \mathcal{A}^{trans}$ with $i \in \mathcal{E}^{rel}$. Then there exists $l \in \mathcal{E}^{del}$ such that

- $s(\bar{P}_{li}) = \bar{s}_i < R - s_a - s_{d(a)}$ for WTR1
- $s(\bar{P}_{li}) = \bar{s}_i < R - s_a - n$ for WTR2,
- $s(\bar{P}_{li}) = \bar{s}_i < R - s_a - s_{d(a)} - m$ for WTR3.

Set $V_l = R$ and $V_k = 0$ for all $k \in \mathcal{E}^{del} \backslash \{k\}$. Then $a = (i,j)$ is not maintained, hence $R(s) > R$.   □

The lemma will be the basis for the algorithms provided in Section 7. For the other two problems (P_{no}) and (P_{del}) we obtain the following results. Again we assume that $w_a > 0$ for all $a \in \mathcal{A}$, i.e. there is at least one passenger traveling along all of the activities.

Lemma 10. *Let a maximum delay V be given and let s be a Pareto solution w.r.t R_{no} or R_{del}, respectively. Let $j \in \mathcal{E}$. Let P_j be a shortest path w.r.t the weights s_a from a source delayed event $i^{del} \in \mathcal{E}^{del}$ to j. Then we have $s(P_j) \leq V$.*

Proof. Assume $s(P_j) > V$ for some shortest path P_j from a source-delayed event $i^{del} \in \mathcal{E}^{del}$ to another event $j \in \mathcal{E}$. Let $\tilde{a} = (\tilde{i}, \tilde{j})$ be the last activity of P_j with $s_a > 0$. Without loss of generality we can assume that $\tilde{a} = (i,j)$, otherwise we continue with $P_{\tilde{j}}$. Define a new timetable t through $t_{\tilde{a}} := [s_{\tilde{a}} - (s(P_j) - V)]^+$, $t_a := s_a$ for all $a \in \mathcal{A} \setminus \{\tilde{a}\}$. We want to show that t dominates s.

Note that due to $s(P_j) - V \geq 0$ we have $0 \leq t_a \leq s_a$ for all $a \in \mathcal{A}$, hence t is a feasible timetable and $F(t) \leq F(s)$. It remains to show that $R_{no}(t) \leq R_{no}(s)$ and that $R_{del}(t) \leq R_{del}(s)$. To this end, we calculate the delays y_j for both the timetables s and t. According to Lemma 9 we get for the timetable s that

$$y_j = \max_{i^{del} \in \mathcal{E}^{del}, P \in mpath(i^{del}, j)} [V_{i^{del}} - s(P)]^+$$
$$\leq [V - \min_{i^{del} \in \mathcal{E}^{del}, P \in mpath(i^{del}, j)} s(P)]^+$$
$$= [V - s(P_j)]^+ = 0.$$

For t we analogously obtain

$$y_j \leq [V - \min_{i^{del} \in \mathcal{E}^{del}, P \in mpath(i^{del}, j)} t(P)]^+$$
$$= [V - t(P_j)]^+ \quad \text{since } P_j \text{ is also a shortest path w.r.t } t$$
$$= 0,$$

where the last step holds because $t(P_j) = s(P_j) + t_{\tilde{a}} - s_{\tilde{a}} \geq s(P_j) - (s(P_j) - V) = V$.

Changing the slack on activity $\tilde{a}$ hence has no effect on the delay when reaching j and hence also no effect on any of the other delays y_i. Consequently, none of the robustness functions changes, i.e. we obtain $R_{no}(t) = R_{no}(s)$ and $R_{del}(t) = R_{del}(s)$ and the result follows.   □

Note that calculating $s(P_j)$ can be done easily by starting at j and backwards determining the shortest distance to all $i \in \mathcal{E}^{del}$. The lemma has two direct consequences, the second of them showing Lemma 6.

Corollary 3. *Let s be a Pareto solution w.r.t R_{no}, or, R_{del}, respectively. Then:*

1. *$s_a \leq V$ for all $a \in A$*
2. *Let $\mathcal{N}$ be a linear graph. Then the sum of all slack variables is smaller than V.*

We conclude this section with the following result about the distribution of the slack times in a Pareto solution. It says that for each driving activity either its slack is zero or one of its preceding activities uses the maximal allowed slack.

Theorem 4. *Let $\tilde{a} \in \mathcal{A}^{drive}$. Let $a^0 \in \mathcal{A}^{stop}$ its preceding stopping activity and*

$$prec(\tilde{a}) = \{a \in \mathcal{A}^{trans} : d(a) = \tilde{a}\} \cup \{a^0\}$$

be the set of all its preceding activities.

1. *Let s be a Pareto solution for (P) or (P_{no}). Then for all three waiting time rules s satisfies:*
   $s_{\tilde{a}} = 0$ or $s_a = m_a$ for some $a \in prec(\tilde{a})$.
2. *If s is a Pareto solution for (P_{del}), the result holds for WTR1.*

Proof. Let $\tilde{a} = (\tilde{i}, \tilde{j}) \in \mathcal{A}^{drive}$ and assume that $s_{\tilde{a}} > 0$ and $s_a < m_a$ for all $a \in prec(\tilde{a}) \cup \{a^0\}$. This means,

$$\epsilon := \min\{s_{\tilde{a}}, \min_{a \in prec(\tilde{a})} m_a - s_a\} > 0.$$

We define the following timetable t:

$$t_a := s_a + \epsilon \quad \text{for all } a \in prec(\tilde{a})$$
$$t_{\tilde{a}} := s_{\tilde{a}} - \epsilon$$
$$t_a := s_a \quad \text{otherwise.}$$

We show that t dominates s. Due to the choice of ϵ, t is feasible. Note that the passengers are provided as OD-pairs between some departure event and some arrival event; hence no passenger leaves the system at the departure event $\tilde{i}$. (Each path between two stations that contains a transfer or a stopping activity also contains the subsequent driving activity). For the objective F we hence first calculate

$$w_{\tilde{a}} = \sum_{P:\tilde{a} \in P} w_P \geq \sum_{a \in prec(\tilde{a})} \sum_{P:a \in P} w_P = \sum_{a \in prec(\tilde{a})} w_a$$

and hence conclude that

$$F(t) = F(s) + \sum_{a \in prec(\tilde{a})} (t_a - s_a)w_a + (t_{\tilde{a}} - s_{\tilde{a}})w_{\tilde{a}}$$

$$= F(s) + \underbrace{\epsilon}_{>0} \underbrace{\left(\sum_{a \in prec(\tilde{a})} w_a - w_{\tilde{a}} \right)}_{\leq 0}$$

$$< F(s).$$

We finally have to show that the robustness of t is not worse than the robustness of s. Given a delay V, we use Theorem 3 to show that all transfers $a = (i, j)$ with $i \in \mathcal{E}^{rel}$ that are maintained in s are also maintained in t. This shows that the robustness functions R and R_{no} do not increase.

To this end, let $a = (i, j)$ be a transfer that is maintained in s, i.e.

$$R - \bar{s}_i \leq s_{d(a)} + s_a \text{ in the case of WTR1,}$$
$$R - \bar{s}_i \leq n + s_a \text{ in the case of WTR2,}$$
$$R - \bar{s}_i \leq m + s_{d(a)} + s_a \text{ in the case of WTR3.}$$

Since $s_{a^0} + s_{\tilde{a}} = t_{a^0} + t_{\tilde{a}}$ we know that $\bar{s}_i = \bar{t}_i$ for all $i \neq \tilde{i}$. We further obtain $\bar{s}_{\tilde{i}} < \bar{t}_{\tilde{i}}$. Moreover, $s_{a'} + s_{\tilde{a}'} = t_{a'} + t_{\tilde{a}'}$ for all a' with $d(a') = \tilde{a}'$ and $t_a > s_a$ for all transfer activities a. Together we obtain that

$$R - \bar{t}_i \leq R - \bar{s}_i \leq s_{d(a)} + s_a = t_{d(a)} + t_a \text{ in the case of WTR1,}$$
$$R - \bar{t}_i \leq R - \bar{s}_i \leq n + s_a \leq n + t_a \text{ in the case of WTR2,}$$
$$R - \bar{t}_i \leq R - \bar{s}_i \leq m + s_{d(a)} + s_a \leq m + t_{d(a)} + t_a \text{ in the case of WTR3.}$$

For R_{del} the theorem only holds in the case of WTR1: Given some source delays, let y^t be the delays corresponding to t and y^s be the delays corresponding to s. For WTR1 we obtain that $y^s \geq y^t$. Since all transfers that are maintained in s are also maintained in t we conclude that $R_{del}(t) \leq R_{del}(s)$ in this case. $\square$

We remark that the idea of the proof cannot be used for WTR2 (or WTR3) in the case of (P_{del}) which can be seen at the following example: Let $\tilde{a} = (\tilde{i}, \tilde{j})$ be a driving activity that is used by 100 passengers all boarding at $\tilde{i}$ and getting off at $\tilde{j}$. Let $a' = (j', \tilde{i})$ be the only transfer activity with $d(a') = \tilde{a}$ and let only one passenger use this transfer, also getting off at $\tilde{j}$. The slack of a' should not be bounded from above; here we choose $m_{a'} \geq 4$. Let the period be $T = 60$ minutes. Finally, let one source delay $V = 4$ be given at the start event of a' and consider WTR2 with $n = 2$.

Now consider a timetable $s_{a'} = 0, s_{\tilde{a}} = 2$. In the proof of Theorem 4 we improve F and the robustness by shifting the slack to the preceding activities which in our example yields a timetable t with $t_{a'} = 2, t_{\tilde{a}} = 0$. We obtain $F(t) < F(s)$ but the robustness of t increases:

- For s the transfer is not maintained and $y_{\tilde{i}}^s = y_{\tilde{j}}^s = 0$. This yields $R_{del}(s) = Tw_{a'} + y_{\tilde{j}}^s w_{\tilde{a}} = 60$.

– On the other hand, we obtain for t: $y_i^t = 2$, hence the transfer is maintained and $y_{\bar{j}} = 2$. This yields $R_{del}(t) = y_{\bar{j}} w_{\bar{a}} = 202$.

Hence, $R_{del}(t) > R_{del}(s)$. This example also shows that in the case of R_{del} it may be better not to maintain all transfers.

7   Finding Pareto Solutions for (P)

In this section we discuss solution approaches for problem

$$(P) \quad \begin{pmatrix} \min F(s) \\ \max R(s) \end{pmatrix} \quad \text{s.t. } s \text{ is a feasible timetable}$$

for the three waiting time rules.

In order to determine Pareto solutions we use the ϵ-constraint method. We hence fix a robustness $R(s) := R$ and consider the resulting single criteria problem (SiP) of finding slack variables s such that

a) s is a feasible timetable,
b) $R(s) \leq R$, and
c) $F(s)$ is minimal.

It is well known that every optimal solution of (SiP) is a weak Pareto solution of (P).

We start with the first waiting time rule WTR1. We know from Theorem 3 that the timetable is robust with robustness $R \in \mathbb{R}$ if and only if the slack variables satisfy for any transfer activity $a = (i, j)$ that the accumulated slack time $\bar{s}_i$ between the closest event to i is at least $R - s_{d(a)} - s_a$, i.e. if

$$s_a + s_{d(a)} + \bar{s}_i \geq R \text{ for all } a = (i, j) \in \mathcal{A}^{trans} \text{ with } i \in \mathcal{E}^{rel}. \quad (13)$$

To simplify the notation let us define $\mathcal{A}^{rel} := \{a = (i, j) \in \mathcal{A}^{trans} : i \in \mathcal{E}^{rel}\}$ as the set of transfers that have to be considered.

Now consider one fixed activity $a = (i, j) \in \mathcal{A}^{rel}$. Using (9) we know that $\bar{s}_i = s(\bar{P}_{li})$ for some event $l \in \mathcal{E}^{del}$ of the same train. We define the set of edges used in (13), i.e.

$$P_a := \bar{P}_{li} \cup \{a, d(a)\}.$$

Note that P_a is a path. It can be used to reformulate (13) as

$$\sum_{a' \in P_a} s_{a'} \geq R \text{ for all } a = (i, j) \in \mathcal{A}^{rel}$$

and we are left with a multi covering problem if we neglect the feasibility condition (1) for the slack variables. Its covering matrix A is an $|\mathcal{A}^{rel}| \times |\mathcal{A}|$ matrix with coefficients

$$A_{a,a'} = \begin{cases} 1 & \text{if } a' \in P_a \\ 0 & \text{otherwise} \end{cases}, \quad (14)$$

hence we obtain $\min\{w^t s : As \geq R\}$. Since the constraints in (1) only appear if the network contains cycles, they can be neglected if $\mathcal{N}$ is a tree. The following algorithm uses this fact:

Algorithm 2 to calculate a Pareto solution for (P) with $R(s) = R$ in the case of a tree

1. Determine P_a for all $a \in \mathcal{A}^{rel}$. Define the covering matrix A according to (14).
2. Solve the multi covering problem $\min\{c^t s : As \geq R\}$, let s^* be an optimal solution.
3. Set $s^*_{\bar{a}} := 0$ for all $\bar{a} \notin \cup_{a \in \mathcal{A}^{rel}} P_a$. Output: s^* is a Pareto solution.

If $s_a \in \mathbb{R}$ are allowed, step 2 can be solved by linear programming. If integer values for s_a are required the multi-covering problem in general is NP hard. Furthermore, if $\mathcal{N}$ is not a tree, the solution of the covering problem will usually not be a feasible timetable since it does not respect (1). Surprisingly, adding the condition simplifies the problem; it even can be solved efficiently if integer values are required. This will be shown next.

The idea is the following. We use the timetabling model $(TT - \Pi)$ based on the variables Π, but add an additional activity for each constraint of type (13) as follows: Let $a = (i, j) \in \mathcal{A}^{rel}$ and let

$$\sum_{a' \in P_a} s_{a'} \geq R$$

its corresponding constraint. Let $start_a$ and end_a be the first and the last node of the path P_a, i.e. $start_a$ is the source-delayed event closest to a and end_a is the arrival event of the driving activity which follows a. We then additionally require that $\Pi_{end_a} - \Pi_{start_a} \geq \sum_{a' \in P_a} l_{a'} + R$ in the formulation of the timetabling model.

$$\begin{aligned}
\min \quad & \sum_{a=(i,j) \in \mathcal{A}} w_a(\Pi_j - \Pi_i) \\
\text{s. t.} \quad & l_a \leq \Pi_j - \Pi_i \leq u_a \text{ for all } a = (i, j) \in \mathcal{A} \qquad (15) \\
& \Pi_{end_a} - \Pi_{start_a} \geq \sum_{a' \in P_a} l_{a'} + R \text{ for all } a = (i, j) \in \mathcal{A}^{rel} \\
& \Pi_i \in \mathbb{N}.
\end{aligned}$$

Before we show the equivalence of (15) to (SiP) in Theorem 5 let us remark that the additional constraints

$$\Pi_{end_a} - \Pi_{start_a} \geq \sum_{a' \in P_a} l_{a'} + R$$

in (15) can be interpreted as lower and upper bounds on a new set of (virtual) activities $\mathcal{A}^{new} := \{(start_a, end_a) : a \in \mathcal{A}^{rel}\}$ such that problem (15) turns out to be an aperiodic timetabling problem on the network $\mathcal{N}^{new} = (\mathcal{E}, \mathcal{A} \cup \mathcal{A}^{new})$ and hence can be solved efficiently by shortest path techniques.

Theorem 5. *Let Π^* be optimal solution of (15). Then its slack times $s_a^* :=$ $\Pi_j^* - \Pi_i^* - l_a$ are optimal for (SiP) and hence are a Pareto solution of (P) for WTR1.*

Proof. Let Π^* be an optimal solution of (15) and s^* its slack times. We want to show that s^* satisfies properties a), b), and c) of (SiP). Since Π is a feasible timetable, according to Lemma 1 we have that s^* satisfies property a). Further, $R(s^*) \leq R$ if and only if $R - \bar{s}_i \leq s_{d(a)} + s_a$ for all transfers $a \in \mathcal{A}^{rel}$ (Theorem 3).

Consider a transfer $a \in \mathcal{A}^{rel}$. We know that $\Pi_{end_a} - \Pi_{start_a} \geq \sum_{a' \in P_a} l_{a'} + R$. Note that for any path P from i to j we have that

$$\Pi_i - \Pi_j = \sum_{a \in P} l_a + \sum_{a \in P} s_a. \tag{16}$$

Using (16) for the path P_a we obtain

$$\sum_{a' \in P_a} l_{a'} + \sum_{a' \in P_a} s_{a'} = \Pi_{end_a} - \Pi_{start_a} \geq \sum_{a' \in P_a} l_{a'} + R,$$

and hence $\sum_{a' \in P_a} s_{a'} \geq R$ and b) holds.

It remains to show that s^* is minimal. Assume that s' is a feasible timetable for (SiP) with strictly smaller objective $F(s') < F(s)$. Determine a feasible timetable Π' for s' according to Lemma 2. From Lemma 3 we then know that $\tilde{F}(\Pi') < \tilde{F}(\Pi^*)$. Since $R(s') \leq R$ we obtain

$$R \leq \sum_{a' \in P_a} s_{a'} = \Pi_{end_a} - \Pi_{start_a} - \sum_{a' \in P_a} l_{a'},$$

hence Π' is a feasible solution to (15) with a strictly better objective value, a contradiction. $\qquad\square$

The waiting time rules WTR2 and WTR3 can be treated analogously to the above approach. For WTR3 we just replace R be $R' = R - m$ and can use (15) as for WTR1. For WTR2, constraints (13) have to be changed to

$$s_a + \bar{s}_i \geq R - n \text{ for all } a = (i, j) \in \mathcal{A}^{rel} \tag{17}$$

For a fixed activity $a = (i, j) \in \mathcal{A}^{rel}$, we again use (9) to obtain that $\bar{s}_i = s(\bar{P}_{li})$ for some event $l \in \mathcal{E}^{del}$ of the same train. The set of edges used in (17) then has to be changed to

$$P_a := \bar{P}_{li} \cup \{a\}$$

since the slack of $d(a)$ is not relevant for WTR2. Again, P_a is a path that can be used to reformulate (17) as

$$\sum_{a' \in P_a} s_{a'} \geq R - n \text{ for all } a = (i, j) \in \mathcal{A}^{rel}$$

and the equivalence to the modified problem (15) still applies. Note that in case of WTR2 $\mathcal{A}^{rel}$ can be further reduced since we only have to consider the "earliest" transfer activities for each source-delayed train, i.e. for WTR2 it is sufficient to use the set

$$\mathcal{E}^{rel} := \{j \in \mathcal{E} : \text{ there exists a directed path in } \bar{\mathcal{N}} \text{ from } i^{del} \in \mathcal{E}^{del} \text{ to } j$$
$$\text{not containing any other path from } j^{del} \in \mathcal{E}^{del} \text{ to } j \text{ as subpath}\}.$$

8 Conclusion

In the paper we suggest to model robust optimization problems as bicriteria problems with one objective describing the original (undisturbed) objective function and the other objective modelling the robustness of the problem. We illustrate the approach using a timetabling problem. It turned out that determining Pareto solutions can be reduced to a timetabling problem with modified data and hence be solved efficiently for any given level of robustness.

More properties of Pareto solutions for the specific problem considered here are currently under research. The goal is to collect essential properties of such solutions such that in practice one can quickly judge if the solution is non-dominated or not. Moreover, a combination of different waiting time rules and an extension to semi-robustness (taking into account that missing a transfer is not too bad if the next possible connection is provided e.g. in only half of the period length) have been sketched in [13].

The bicriteria model presented in this paper can be applied to any robust optimization problem. Extensions to other problems than timetabling are interesting. A general investigation on how to model and solve robust optimization problems as bicriteria problems are under research.

References

1. Bayer, H.G., Sendhoff, B.: Robust Optimization - A Comprehensive Survey. Computer Methods in Applied Mechanics and Engineering 196(33-34), 3190–3218 (2007)
2. Ben-Tal, A., El Ghaoui, L., Nemirovski, A.: Mathematical Programming: Special Issue on Robust Optimization, vol. 107. Springer, Berlin (2006)
3. Bertsimas, D., Sim, M.: The price of robustness. Operations Research 52(1), 35–53 (2004)
4. Birge, J.R., Louveaux, F.V.: Introduction to Stochastic Programming. Springer, New York (1997)
5. Cicerone, S., D'Angelo, G., Di Stefano, G., Frigioni, D., Navarra, A.: Robust Algorithms and Price of Robustness in Shunting Problems. In: Proc. of the 7th Workshop on Algorithmic Approaches for Transportation Modeling, Optimization, and Systems, ATMOS 2007 (2007)
6. Cicerone, S., Di Stefano, G., Schachtebeck, M., Schöbel, A.: Dynamic algorithms for recoverable robustness problems. In: Fischetti, M., Widmayer, P. (eds.) ATMOS 2008 - 8th Workshop on Algorithmic Approaches for Transportation Modeling, Optimization, and Systems, Dagstuhl Seminar proceedings (2008)
7. Ehrgott, M.: Multiple Criteria Optimization. Lecture Notes in Economics and Mathematical Systems, vol. 491. Springer, Berlin (2005)

8. Fischetti, M., Monaci, M.: Robust optimization through branch-and-price. In: Proceedings of the 37th Annual Conference of the Italian Operations Research Society, AIRO (2006)
9. Fischetti, M., Monaci, M.: Light robustness. Research Paper ARRIVAL-TR-0119, ARRIVAL project (2008)
10. Fischetti, M., Monaci, M.: Light robustness. In: Ahuja, R.K., Möhring, R.H., Zaroliagis, C.D. (eds.) Robust and Online Large-Scale Optimization. LNCS, vol. 5868, pp. 61–84. Springer, Heidelberg (2009)
11. Gatto, M.: On the Impact of Uncertainty on Some Optimization Problems: Combinatorial Aspects of Delay Management and Robust Online Scheduling. PhD thesis, ETH Zürich (2007)
12. Kall, P., Wallace, S.W.: Stochastic Programming. Wiley, Chichester (1994)
13. Kratz, A.: Robuste Fahrplangestaltung als bikriterielles Optimierungsproblem. Master's thesis, Georg-August Universitt Gttingen (2009) (in German)
14. Liebchen, C.: Periodic Timetable Optimization in Public Transport. dissertation.de – Verlag im Internet, Berlin (2006)
15. Liebchen, C., Lüebbecke, M., Möhring, R.H., Stiller, S.: Recoverable robustness. Technical Report ARRIVAL-TR-0066, ARRIVAL Project (2007)
16. Liebchen, C., Lüebbecke, M., Möhring, R.H., Stiller, S.: The concept of recoverable robustness, linear programming recovery, and railway applications. In: Ahuja, R.K., Möhring, R.H., Zaroliagis, C.D. (eds.) Robust and Online Large-Scale Optimization. LNCS, vol. 5868, pp. 1–27. Springer, Heidelberg (2009)
17. Nachtigall, K.: Exact solution methods for periodic programs. Technical Report 14/93, Hildesheimer Informatik-Berichte (1993)
18. Nachtigall, K.: Optimization of the integrated fixed interval timetable. Technical Report 20/95, Hildesheimer Informatik-Berichte (1995)
19. Peeters, L.: Cyclic Railway Timetabling Optimization. PhD thesis, ERIM, Rotterdam School of Management (2002)
20. Tyrrell Rockafellar, R.: Network flows and monotropic optimization. John Wiley & Sons, Inc., Chichester (1984)
21. Ruszczynski, A., Shapiro, A. (eds.): Stochastic Programming, Handbooks in Operations Research and Management Science, vol. 10. North-Holland, Amsterdam (2003)
22. Schachtebeck, M., Schöbel, A.: IP-based techniques for delay management with priority decisions. In: Fischetti, M., Widmayer, P. (eds.) ATMOS 2008 - 8th Workshop on Algorithmic Approaches for Transportation Modeling, Optimization, and Systems, Dagstuhl Seminar proceedings (2008)
23. Schöbel, A.: Customer-oriented optimization in public transportation. In: Optimization and Its Applications. Springer, New York (2006)
24. Schöbel, A.: Integer programming approaches for solving the delay management problem. In: Geraets, F., Kroon, L.G., Schoebel, A., Wagner, D., Zaroliagis, C.D. (eds.) Railway Optimization 2004. LNCS, vol. 4359, pp. 145–170. Springer, Heidelberg (2007)

Meta-heuristic and Constraint-Based Approaches for Single-Line Railway Timetabling

Federico Barber, Laura Ingolotti, Antonio Lova, Pilar Tormos, and Miguel A. Salido

Instituto de Automática e Informática Industrial
Universidad Politécnica de Valencia
{fbarber,lingolotti,msalido}@dsic.upv.es, {allova,ptormos}@eio.upv.es

1 Introduction

This chapter is devoted to recent advances in heuristic and metaheuristic procedures, arising from the areas of Computer Science and Artificial Intelligence, which are able to cope with large scale problems as those in single-line railway timetable optimization. Timetable design is a central problem in railway planning. In the basic timetabling problem, we are given a line plan as well as demand and infrastructure information. The goal is to compute timetables for passengers and cargo trains that satisfy infrastructure capacity and achieve multicriteria objectives: minimal passenger waiting time (both at changeovers and onboard), efficient use of trains, etc. Due to its central role in the planning process of railway scheduling, timetable design has many interfaces with other classical problems: line planning, vehicle scheduling, and delay management.

In this chapter, we focus our attention on the development of metaheuristic and constraint-based techniques for solving the single-line Train Timetabling Problem (TTP). This problem is related to obtaining and optimizing timetables of periodic and non-periodic heterogeneous trains that share a railway line with single and double track sections. Thus, in section 2, we describe the problem and present the notation used to describe the problem (parameters, constraints, objective functions, etc).

Section 3 is focused on solving the Train Timetabling Problem from the perspective of Constraint Satisfaction Problems. This approach has been applied to solve the Train Timetabling Problem (TTP) using different partitioning techniques. Furthermore, the performance of distributed approaches with respect to centralized ones has been tested, showing that distributed approaches outperform the centralized ones. These new algorithms are especially useful for solving distributable, large-scale problems such as those arising in railway optimization.

In section 4, we focus on solving the Train Timetabling Problem using a metaheuristic approach based on variable ordering. A constructive approach is detailed under the assumption that the TTP is a special type of job-shop problem.

In section 5, we focus on solving the Train Timetabling Problem using a metaheuristic approach based on Genetic Algorithms (GAs) that have been designed

R.K. Ahuja et al. (Eds.): Robust and Online Large-Scale Optimization, LNCS 5868, pp. 145–181, 2009.

and successfully applied to solve real world Train Timetabling. The schedule for the new trains is obtained using a Genetic Algorithm that includes a guided process to build the initial population. This algorithm has been used to solve real-world instances, and its performance has been compared against constructive approaches. This approach is an appropriate method for exploring the search space of this complex problem. Thus, further research in the design of efficient GA is justified.

We present an alternative approach to solve the periodic railway timetabling problem in section 6. This approach is based on the topological properties of the line. Finally, we present conclusions for the above sections.

2   The Single-Line Train Timetabling Problem (TTP)

Given a railway single-line that may have single as well as double-track sections, the Train Timetabling Problem (TTP) consists of computing timetables that satisfy existing constraints and that optimize a multicriteria objective function for both, passenger and cargo trains. The railway line may be occupied by other trains whose priority is higher than that of the new ones, and the new trains to be added may belong to different train operators. The locations to be visited by each train may also be different from each other. The timetable given to each new train must be feasible, that is, it must satisfy a given set of constraints. Of the constraints that may arise in this problem, is the requirement for periodicity of the timetables. Periodicity leads to the classification of TTP as (*i*) Periodic (or cyclic) Train Timetabling and (*ii*) Non-Periodic Train Timetabling.

In **Periodic Timetabling**, each trip is operated in a periodic way. That is, each period of the timetable is the same. One advantage of a periodic railway system is the timetable is easy for the passengers to remember. One drawback is that the system is expensive to operate from the point of view of resources such as crews and rolling stock. The mathematical model called Periodic Event Scheduling Problem (PESP) by Serafini and Ukovich [30] is the most widely used in the literature. In PESP, a set of repetitive events is scheduled under periodic time window constraints. Hence, the events are scheduled for one cycle in such a way that the cycle can be repeated. The PESP model has been used by Nachtigall and Voget [21], Odijk [23], Kroon and Peeters [18], Liebchen [20], etc.

Non-Periodic Train Timetabling is especially relevant on heavy-traffic, long-distance corridors where the capacity of the infrastructure is limited due to great traffic densities. This timetabling allows the Infrastructure Manager to optimally allocate the train paths requested by the Train Operators and proceed with the overall timetable design process, possibly with final local refinements and minor adjustments made by the planner. Many references consider Mixed Integer Problem formulations in which the arrival and departures times are represented by continuous variables and there are binary variables expressing the order of the train departures from each station. The non-periodic train timetabling problem has been considered by several authors: Szpigel [32], Javanovic and Harker [15], Cai and Goh [8], Carey and Lockwood [10],

Higgins et al. [13], Silva de Oliveira [31], Kwan and Mistry [19], Caprara et al. [9], Ingolotti et al. [14].

One of the main problems that railway managers face is the allocation of the paths requested by transport operators and the process of designing the overall timetable. These timetables are generally non-periodic and have to meet a wide set of constraints as well as achieve a multicriteria objective function. A detailed formal description of both the constraints and the objective function is given in the following subsections. First, we introduce the notation that will be used hereafter.

2.1   Notation

In this subsection, the TTP is formally described. The notation used to describe the problem is the following.
Parameters

- T: finite set of trains t considered in the problem. $T = \{t_1, t_2, ..., t_k\}$
- $T_C \subset T$: subset of trains that are in circulation and whose timetables cannot be modified (T_C can be empty).
- $T_{new} \subseteq T$: subset of non-scheduled trains that do not yet have a timetable and that must be added to the railway line with a feasible timetable. Thus, $T = T_C \cup T_{new}$ and $T_C \cap T_{new} = \emptyset$
- l_i: location (station, halt, junction). The types of locations considered are described as follows:

  - Station: Place for trains to park, stop or pass through. Each station is associated with a unique station identifier. There are two or more tracks in a station where crossings or overtaking can be performed.
  - Halt: Place for trains to stop, pass through, but not park. Each halt is associated with a unique halt identifier.
  - Junction: Place where two different tracks fork. There is no stop time.

- N_i: number of tracks in location l_i.
- NP_i: number of tracks with platform (necessary for commercial stops) in location l_i.
- $L = \{l_0, l_1, ..., l_m\}$: railway line that is composed by an ordered sequence of locations that may be visited by trains $t \in T$. The contiguous locations l_i and l_{i+1} are linked by a single or double track section.
- $J_t = \{l_0^t, l_1^t, ..., l_{n_t}^t\}$: journey of train t. It is described by an ordered sequence of locations to be visited by a train t such that $\forall t \in T, \exists J_t : J_t \subseteq L$. The journey J_t shows the order that is used by train t to visit a given set of locations. Thus, l_i^t and $l_{n_t}^t$ represent the i_{th} and $last$ location visited by train t, respectively.
- T_D: set of trains travelling in the $down$ direction.
  $t \in T_D \leftrightarrow (\forall l_i^t : 0 \le i < n_t, \exists l_j \in \{L \setminus \{l_m\}\} : l_i^t = l_j \wedge l_{i+1}^t = l_{j+1})$.
- T_U: set of trains travelling in the up direction.
  $t \in T_U \leftrightarrow (\forall l_i^t : 0 \le i < n_t, \exists l_j \in \{L \setminus \{l_0\}\} : l_i^t = l_j \wedge l_{i+1}^t = l_{j-1})$. Thus $T = T_D \cup T_U$ and $T_D \cap T_U = \emptyset$

- C_i^t minimum time required for train t to perform commercial operations (such as boarding or leaving passengers) at station i (commercial stop).
- $\Delta_{i \rightarrow (i+1)}^t$: running time for train t from location l_i^t to l_{i+1}^t.
- $\lfloor I_L^t, I_U^t \rfloor$: interval for departure time of train $t \in T_{new}$ from the initial station of its journey.
- $\lfloor F_L^t, F_U^t \rfloor$: interval for arrival time of train $t \in T_{new}$ to the final station of its journey.

Variables

- dep_i^t departure time of train $t \in T$ from the location i, where $i \in J_t \setminus \{l_{n_t}^t\}$.
- arr_i^t arrival time of train $t \in T$ to the location i, where $i \in J_t \setminus \{l_0^t\}$.

Planners usually use running maps as graphic tools to help them in the planning process. A running map is a time-space diagram like the one shown in Figure 1 where several train crossings can be observed. The names of the stations are presented on the left side, and the vertical line represents the number of tracks between stations (one-way or two-way). Horizontal dotted lines represent halts or junctions, while solid lines represent stations. On a railway network, the planner needs to schedule the paths of n_k trains going in one direction and m_k trains going in the opposite direction for trains of a given type. The trains to schedule may or may not require a given frequency.

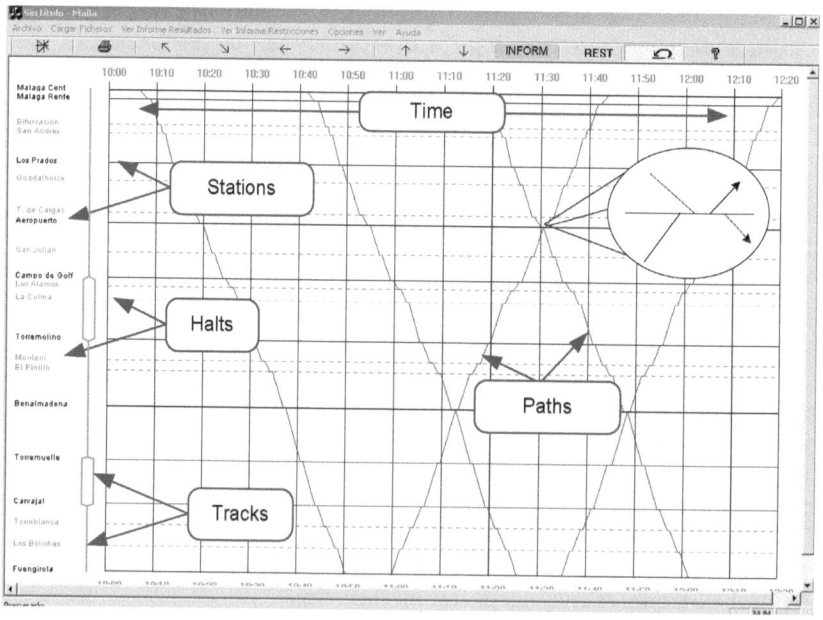

Fig. 1. Running Map

2.2   Feasibility of a Solution - Set of Constraints

In order to be feasible, a timetable has to fulfill a set of constraints that can be classified in three main groups, depending on whether they are concerned with: (i) user requirements (parameters of trains to be scheduled), (ii) traffic rules, (iii) railway infrastructure topology. The constraints described in this work have been defined together with the Manager of Railway Infrastructure of Spain (ADIF) in such a way that the resulting timetable is feasible and practicable.

User Requirements:

− *Interval for the Initial Departure*: Each train $t \in T_{\text{new}}$ should leave its initial station l_0^t at a time dep_0^t such that,

$$I_L^t \leq dep_0^t \leq I_U^t \ . \tag{1}$$

− *Interval for the Arrival Time*: Each train $t \in T_{\text{new}}$ should arrive to its final station $l_{n_t}^t$ at a time $arr_{n_t}^t$ such that,

$$F_L^t \leq arr_{n_t}^t \leq F_U^t \ . \tag{2}$$

− *Maximum Delay*: A maximum delay Λ_t and a minimum running time M_t are specified for each train $t \in T_{\text{new}}$; thus, the upper bound for the running time of $t \in T_{\text{new}}$ is given by the following expression:

$$\frac{(arr_{n_t}^t - dep_0^t - M_t)}{M_t} \leq \Lambda_t \ . \tag{3}$$

Traffic constraints:

− *Running Time*: For each train and each track section, a running time is given by $\Delta_{i \to (i+1)}^t$, which represents the time the train t should employ to go from location l_i^t to location l_{i+1}^t. Therefore, the following expression must be fulfilled

$$arr_{i+1}^t = dep_i^t + \Delta_{i \to (i+1)}^t \ . \tag{4}$$

− *Crossing*: According to the following expression, a single-track section ($i \to i+1$, *down* direction) cannot be occupied by two trains going in opposite directions ($t \in T_D$ and $t' \in T_U$).

$$dep_{i+1}^{t'} > arr_{i+1}^t \vee dep_i^t > arr_i^{t'} \ . \tag{5}$$

− *Commercial Stop*: Each train $t \in T_{\text{new}}$ is required to remain in a station l_i^t at least C_i^t time units

$$dep_i^t \geq arr_i^t + C_i^t \ . \tag{6}$$

− *Overtaking on the track section*: Overtaking must be avoided between any two heterogeneous (different speeds) trains, $\{t, t'\} \subseteq T_{\text{new}}$, going in the same direction on any track section, $k \to (k + 1)$, of their journeys:

$$(arr_{k+1}^t > arr_{k+1}^{t'}) \leftrightarrow (dep_k^t > dep_k^{t'}) \ . \tag{7}$$

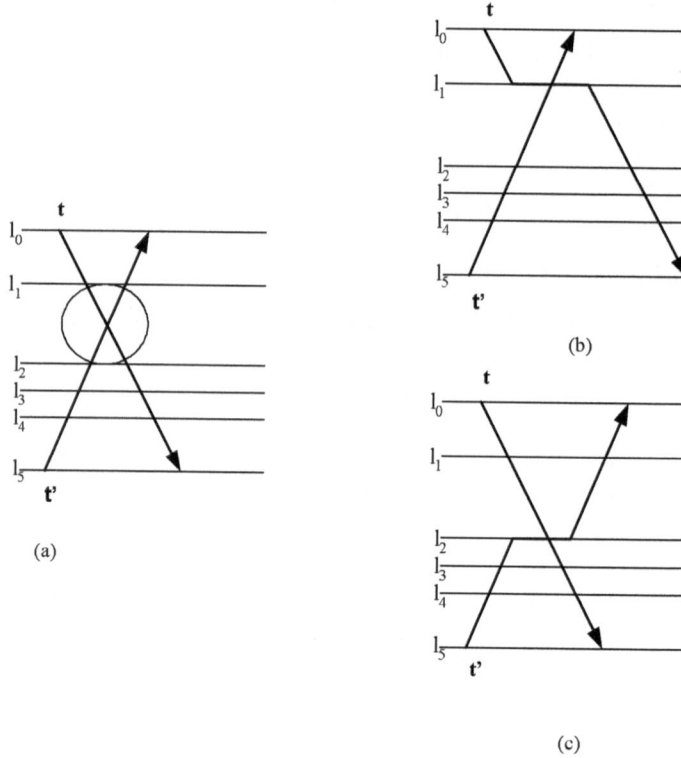

Fig. 2. (a)Crossing conflict. (b) Train in Down direction waits. (c) Train in Up direction waits.

- *Delay for unexpected stop*: When a train t stops in a station j to avoid conflicts with other trains (overtaking/crossing) and no commercial stop was planned ($C_j^t = 0$) in this station, the running time of train t that corresponds to the previous ($l_{j-1}^t \to l_j^t$) and next ($l_j^t \to l_{j+1}^t$) track sections of j must be increased by Γ_t time units. This increase represents the speed reduction of the train due to the braking and speeding up in the station.

$$dep_j^t - arr_j^t > 0 \wedge C_j^t = 0 \to \Delta_{j-1\to j} = \Delta_{j-1\to j} + \Gamma_t \wedge \Delta_{j\to j+1} = \Delta_{j\to j+1} + \Gamma_t . \tag{8}$$

In crossing and overtaking operations, the reception and expedition times are required for trains which are detoured from the main track.

- *Reception Time*: The difference between the arrival times of any two trains $\{t, t'\} \subseteq T_{new}$ in the same station l is defined by the expression below, where R_t is the reception time specified for the train that arrives to l first.

$$arr_l^{t'} \geq arr_l^t \to arr_l^{t'} - arr_l^t \geq R_t . \tag{9}$$

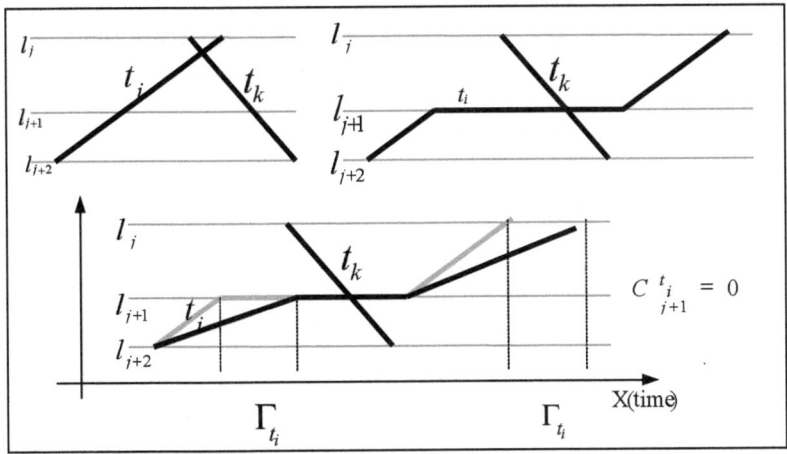

Fig. 3. Unexpected Stop

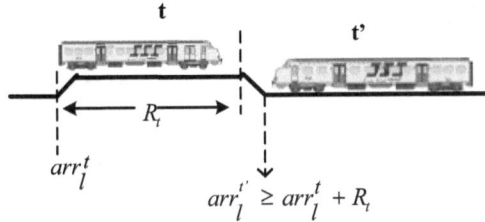

Fig. 4. Reception time between train t and train t'

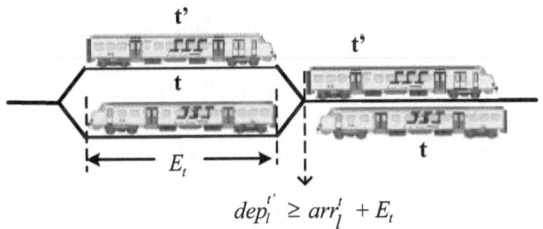

Fig. 5. Expedition time between train t and train t'

- *Expedition Time*: The difference between the departure and arrival times of
  any two trains $\{t, t'\} \subseteq T_{\text{new}}$ in the same station l is defined by the expression
  below, where E_t is the expedition time specified for t.

$$|dep_l^{t'} - arr_l^t| \geq E_t \ . \tag{10}$$

- *Simultaneous Departure*: When two trains going in opposite directions stop in the same station, the difference between their departure times from that station must be at least S. This constraint is formulated as:

$$\forall t, t' \in \mathrm{T}_{\text{new}} : dep_i^t - arr_i^t > 0 \wedge dep_i^{t'} - arr_i^{t'} > 0 \rightarrow |dep_i^t - dep_i^{t'}| \geq S \ . \quad (11)$$

Infrastructure constraints:

- *Finite Capacity of Stations*: A train $t \in \mathrm{T}_{\text{new}}$ could arrive to a location l_i^t if and only if it has at least one available track (with platform, if $C_i^t > 0$). In order to formulate this constraint, consider:

$$\forall x \in \mathrm{T}_{\text{new}} : T_x = \{t \in \mathrm{T} : t \neq x, J_t \cap J_x \neq \emptyset\} \text{ and}$$

$$\mathrm{Meet}(x, t, l) = \begin{cases} 1 \text{ if } [arr_l^x, dep_l^x] \cap [arr_l^t, dep_l^t] \neq \emptyset \wedge C_l^t = 0 \\ 0 \text{ else} \end{cases}$$

$$\mathrm{Meet_P}(x, t, l) = \begin{cases} 1 \text{ if } [arr_l^x, dep_l^x] \cap [arr_l^t, dep_l^t] \neq \emptyset \wedge C_l^t > 0 \\ 0 \text{ else} \end{cases}$$

Hence, the constraint of finite capacity of stations is formulated as follows:

$$\forall x \in \mathrm{T}_{\text{new}}, \forall l \in J_x : ((\sum_{t \in T_x} \mathrm{Meet}(x, t, l) + \sum_{t \in T_x} \mathrm{Meet_P}(x, t, l) < N_l) \wedge$$

$$(C_l^x > 0 \rightarrow \sum_{t \in T_x} \mathrm{Meet_P}(x, t, l) < N_{P_l})) \ . \quad (12)$$

- *Closing Time*: Let $[H_l^1, H_l^2]$ be the closing time interval for maintenance operations of station l. The closing time imposes constraints over regular operations -trains can pass but cannot stop in the station (see the next expression). Closing time can even forbid regular operations; for example, trains can neither pass nor stop (i.e., the number of tracks in the station is decreased to one (see (12)).

$$dep_l^t < H_l^1 \vee arr_l^t > H_l^2 \ . \quad (13)$$

- *Headway Time*: If two trains, $\{t, t'\} \subseteq \mathrm{T}_{\text{new}}$, travelling in the same direction leave the same location l_k towards the location l_{k+1}, they are required to have a difference in departure times of at least φ_k^d and a difference in their arrival times of at least φ_k^a. When the blocking type in the track section is *Automatic*, then $\varphi_k^a = \varphi_k^d$. Consider the following expression

$$|dep_k^t - dep_k^{t'}| \geq \varphi_k^d \ . \quad (14)$$

$$|arr_{k+1}^t - arr_{k+1}^{t'}| \geq \varphi_k^a \ . \quad (15)$$

According to the company requirements, the method proposed should obtain the best available solution so that all the above constraints are satisfied. As we pointed out above, the line could be previously occupied by other trains whose timetable have not been changed. That is to say $\forall t \in T_C$, the variables arr_i^t and dep_i^t have been previously instantiated with given values. This means $\forall t \in T_C, \forall i \in J_t, arr_i^t \in CONSTANT, dep_i^t \in CONSTANT$. It also means that the process generates the constraints so that the arrival and departure time of trains in circulation are constants. It does not generate constraints that only involve variables corresponding to trains in circulation. Next, the process verifies that each new train satisfies each constraint taking into account the remaining new trains as well as all the trains already in circulation. In other words, if a constraint is violated and it relates new trains with trains in circulation, the only timetables that should be modified are those corresponding to new trains.

Note that this set of constraints corresponds specifically to the Spanish railway company requirements and does not match exactly with other published works.

Several criteria can exist to assess the quality of the solution, for example: minimize travel time, minimize the passenger waiting time in the case of changeovers, balance the delay of trains in both directions, etc. The objective function considered in this work is described in the next section.

2.3   Optimality of a Solution - Objective Function

In order to assess the quality of each solution, we assess the *optimal solution* (optimal travel time) for each specific train $t \in T_{new}$. The optimal solution of train t is computed by the scheduling of the new train t (verifying all problem constraints) on the line being occupied only by trains in circulation (T_C). This optimal solution for train t (Γ_{opt}^t) is the lowest time required by t to complete its journey. The other trains to be scheduled in T_{new} are ignored.

Once the optimal time for each new train to be scheduled has been computed, the criterion to measure the quality of each solution will be the average delay of new trains with respect to their optimum (δ). That is:

$$\delta_t = \frac{(arr_{n_t}^t - dep_0^t) - \Gamma_{opt}^t}{\Gamma_{opt}^t}; \; \delta_U = \frac{\sum_{t \in T_U \cap T_{new}} \delta_t}{|T_U \cap T_{new}|}; \; \delta_D = \frac{\sum_{t \in T_D \cap T_{new}} \delta_t}{|T_D \cap T_{new}|}; \; \delta = \frac{\delta_U + \delta_D}{|T_{new}|}$$

Finally, assuming that TTABLE is a solution for the TTP problem (therefore, the timetable for all new trains), the objective function of this problem is formulated as:

$$f(TTABLE) = MIN(\delta) \tag{16}$$

If there are no trains in circulation in L ($T = T_{new}$), the optimal time of a new train t would be:

$$M_t = \Gamma_{opt}^t = \sum_{i=0}^{n_t-1} \Delta_{i \to (i+1)}^t + \sum_{i=0}^{n_t-1} C_i^t \tag{17}$$

3   The TTP from the Constraint Satisfaction Problem Perspective

The research in the field of constraint satisfaction problems (CSP) has incorporated new ways of dealing with optimization problems, including scheduling problems, by providing flexibility and robustness through constraint propagation [22]. In practical applications, there are many more parameters and side constraints than the ones considered in theoretical studies, and that can be handled as CSP in a flexible way. Another advantage of CSPs is that they can be represented and solved in a distributed way (distributed CSP).

A CSP consists of a set of variables $X = \{x_1, x_2, ..., x_n\}$; each variable $x_i \in X$ has a set D_i of possible values (its domain) and a finite collection of constraints $C = \{c_1, c_2, ..., c_p\}$ restricting the values that the variables can simultaneously take. A solution to a CSP is an assignment of values to all the variables so that all constraints are satisfied; a problem is satisfiable or consistent when it has at least one solution. A partition of a set C is a set of disjoint subsets of C whose union is C. The subsets are called the blocks of the partition.

A distributed CSP (DCSP) is a CSP in which the variables and constraints are distributed among automated agents. Each agent has one or more variables and attempts to determine their values. However, there are interagent constraints, and the value assignment must satisfy these constraints. In our model, there are k agents $1, 2, ..., k$. Each agent knows a set of constraints and the domains of variables involved in these constraints.

Distributed-CSP is a promising area for coping with large-scale problems such those arising in a real-life railway instance. Specifically, new models for distributing CSPs, both as general-oriented and problem-oriented approaches, can be developed for solving railway scheduling problems [26].

With regard to robustness, there are several previous works that analyze the stability of solutions in dynamic-CSPs, in which constraints, variables or domains change over time [34], [2]. With the aim of obtaining stable solutions, CSP techniques have been developed to avoid re-computing new solutions when the constraints changes. These techniques are based on preference management at the level of individual constraints [6], at the level of variables [7], or at the solution level [24]. There are also techniques based on the management of distances between the ideal and acceptable solution. Finally, it is important to point out the techniques based on the fuzzy-CSP [11], probabilistic-CSP Fargier, 1993), weighted-CSP [5], and other techniques that take into account the trade-off between optimality and stability [34]. Most of the latter techniques are founded on the soft-computing paradigm. This paradigm focuses on stochastic, vague, empirical and associative situations, which are typical features of industrial and manufacturing environments. For that very reason, in this kind of dynamic environments, the available data is often numerical and the desired solutions should be obtained in real time and should be reliable; therefore robust and stable solutions are required.

With regard to the distributed paradigm, distributed constraint satisfaction problems arise when it is important to separate by means of communicating agents [35] information that is related to a variable and/or constraints of the problem. Distributing provides a promising setting for dealing with a wide variety of real problems that are inherently distributed, including large-scale problems.

The new challenges posed by the scientific community working on distributed CSPs include constraint management on the resources, exploiting cooperation opportunities, as well as the design of strategies to solve complex problems.

3.1   The Railway Scheduling Problem as a Distributed CSP

The train scheduling problem can be modeled as a distributed CSP by considering the scheduling process as a collection of interacting, autonomous and flexible components (distributed CSPs), which are aimed at representing the problem among several distributed processes, solving sub-problems incrementally and aggregating sub-problem solutions in order to achieve a global solution [26]. Therefore, constraints should be distributed among several CSPs. Different approaches can be applied to distribute the problem.

Approach 1. The first way to distribute the problem can be carried out following a general-oriented approach. This approach is based on distributing the CSP by means of trees so that the resulting sub-CSPs are efficiently solved without backtracking [27]. The main difficulty of this approach is how to obtaining the selected trees. To carry out this partition, METIS, a well-known graph partitioning tool [16] can be used. METIS provides two programs, *metis* and *kmetis*, for partitioning an unstructured graph into k equal size parts. However, this software does not take into account additional information about the railway infrastructure or the type of trains to guide the partition; therefore, the generated clusters may not be the most appropriate and the results may not be appropriate.

Figure 6 shows the distribution carried out by METIS. The red agent is committed to assigning variables to train 0 and train 1 at the beginning and end of their respective journeys. The green agent studies these two trains in disjoint parts of each train. METIS carries out a partition of the constraint network generated by the corresponding CSP. However, it can be observed visually that the best partition generated by a well-known software is not the most appropriate for this problem. To improve the partition procedure, we extract additional information from the railway topology to obtain better partitions as shown in the following approaches.

Approach 2. Problem-oriented approaches provide the opportunity to refine the general distributed technique into a domain-dependent distributed model. One way of distributing the problem following a problem-oriented approach is to distribute the original railway problem by means of train type. Each agent is committed to assigning values to variables with respect to a train or trains in order to minimize the running time. Depending on the number of partitions selected, each agent will manage one or more trains. Figure 7 left, shows a running map with 20 partition, where each agent manages one train. This partition model has two important advantages:

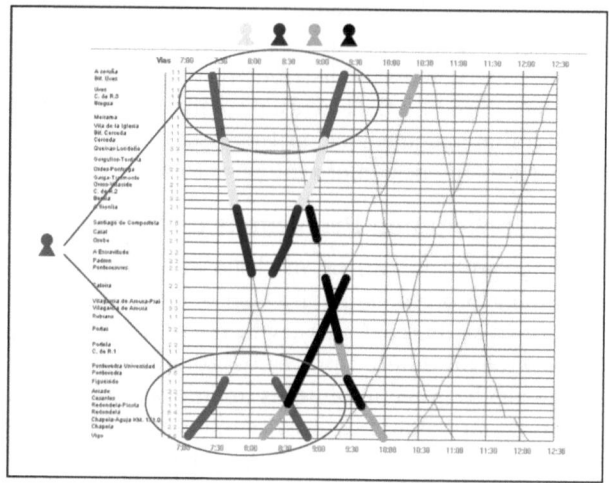

Fig. 6. Railway Scheduling Problem distributed by METIS (Approach 1)

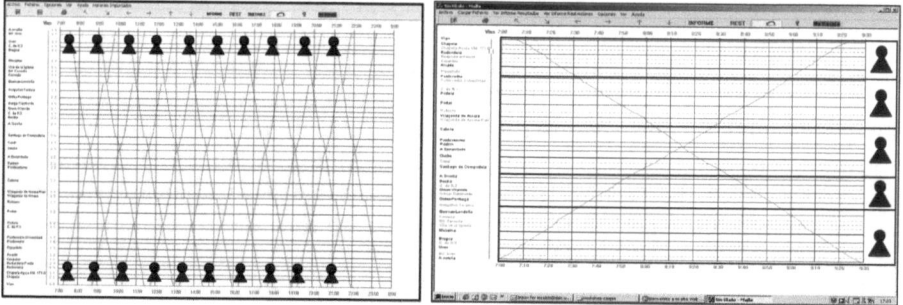

Fig. 7. Railway Scheduling Problem distributed by trains (left) and by contiguous stations (right)

- Firstly, this model allows us to improve privacy. Currently, due to the policy of deregulation in the European railways, trains from different operators work in the same railway infrastructure. Thus, the partition model gives us the possibility of partitioning the problem so that each agent is committed to an operator. Thus, different operators can maintain privacy about strategic data.
- Secondly, this model allow us to efficiently manage priorities between different types of trains (regional trains, high speed trains, freight trains). Thus, agents committed to priority trains (high speed trains) will first carry out value assignment to variables in order to achieve better running times [1].

Approach 3. Another model is based on distributing the original railway problem by means of contiguous stations. Therefore, a logical partition of the railway network can be carried out by means of regions (contiguous stations). To carry out this type of partition, it is important to analyze the railway infrastructure and detect restricted regions (bottlenecks). An agent can manage many stations if they are not restricted stations, whereas an agent can manage only a few stations if they are bottlenecks. Furthermore, the agents with bottleneck have preferences in assigning values to variables since their domains are reduced (variable ordering).To balance the problem, each agent handles a different number of stations. The set of stations will be partitioned into sets of contiguous stations, and a set of agents will be coordinated in order to achieve a global solution. Thus, it is possible to obtain very useful results such as railway capacity, consistent timetable, etc.

The performance of the different distributing models should be analyzed using real-life instances. The main conclusions indicate that general graph partitioning applications work well in general graphs. However, in the railway scheduling problem, these softwares do not obtain good results. In the partitions generated by METIS, the journey of a train is partitioned into several clusters, and each cluster is composed by tracks for trains going in opposite directions. This cluster is easy to solve, but it is very difficult to propagate to other agents. Furthermore, the following partition proposals do the opposite, that is, they never join tracks of trains going in opposite directions.

3.2   Evaluation

In this section, we present an evaluation between our distributed model and a centralized model. We also evaluate the behavior of a distributed model generated by a general software program called METIS and two proposed partition models. To this end, we have used a well-known CSP solver called Forward Checking (FC)[1].

This empirical evaluation was carried out for two different types of problems: random problems and benchmark problems.

Random Problems. In our evaluation, each set of random CSPs was defined by the 3-tuple $< n, a, p >$, where n was the number of variables, a the arity of binary constraints, and p the number of partitions. The problems were randomly generated by modifying these parameters.

Table 1 compares the execution time of the model distributed by METIS with the centralized model. In Table 1 left, the arity of binary constraints and the size of the partition were fixed, and the number of variables was increased from 100 to 500. The table shows that the execution time for small problems was worse for the distributed model than for the centralized model. However, when the number of variables increased, the behavior of the distributed problem improved. In Table 1 right, the number of variables and the arity of binary constraints were fixed, and the size of the partition was increased from 3 to 20. The table shows that the size of the partition is important for the distributed

---

[1] Forward Checking was obtained from CON'FLEX. It can be found at: http://www-bia.inra.fr/T/conflex/ Logiciels/adressesConflex.html

Table 1. Random instances $< n, a, p >$, n: variables, a: arity and p: partition size

Problem	Distributed Model (sc.)	Centralized Model (sc.)	Problem	Distributed Model (sc.)	Centralized Model (sc.)
$< 100, 25, 10 >$	12	14	$< 200, 25, 3 >$	26	75
$< 200, 25, 10 >$	16	75	$< 200, 25, 5 >$	19	75
$< 300, 25, 10 >$	19	140	$< 200, 25, 7 >$	14	75
$< 400, 25, 10 >$	30	327	$< 200, 25, 9 >$	16	75
$< 500, 25, 10 >$	42	532	$< 200, 25, 20 >$	22	75

model. For small problems, the number of partitions must be low. However, for large CSPs (railway Scheduling Problems), the size of the partition must be higher. In this case, the appropriate number of partitions was 7.

As can be observed, the distributed model using METIS works well for random instances. However, in real scheduling problems, domain-dependent distributed models are necessary to optimize execution times.

Benchmark Real-Word Railway Problems. This empirical evaluation was carried out over a real railway infrastructure that joins two important Spanish cities (La Coruña and Vigo). The journey between these two cities is currently divided by 40 stations. In our empirical evaluation, each set of random instances was defined by the 3-tuple $< n, s, f >$, where n was the number of periodic trains in each direction, s the number of stations, and f the frequency. The problems were randomly generated by modifying these parameters.

General graph partitioning software programs work well in general graphs. However, in the railway scheduling problem, we did not obtain good results using these programs. We evaluated partition approach 1 by using METIS in several instances $< n, 20, 120 >$. Figure 9 (left) shows that the obtained results were even worse in the distributed model than in the centralized model. For a low number of trains, the behavior was better than the complete model. However, with more than 8 trains, the distributed model was unable to solve the problem in 1,000,000 seconds, so the program was aborted. We studied the partitions generated by METIS and we observed that the journey of a train was partitioned in several clusters, and that each cluster was composed by tracks of trains going in opposite directions. This cluster was easy to solve, but it was very difficult to propagate to other agents. In contrast, partition approaches 2 and 3 never join tracks of trains going in opposite directions.

Figure 8 shows the behavior of partition approach 2, where the number of partition/agents was equal to the number of trains. In both graphs, it can be observed that the execution time increased when the number of trains increased (Figure 8 right) and when the number of stations increased (Figure 8 left). However, in both graphs, the distributed model maintained better behavior than the centralized model.

Partition approach 2 (distributed by trains) was similar to partition approach 3 (distributed by stations) but had better behavior, mainly when the number of

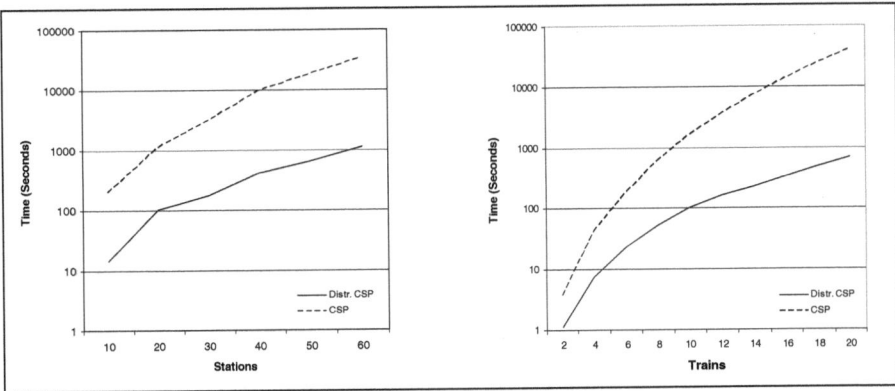

Fig. 8. Execution Time when the number of trains and stations increased (Approach 2)

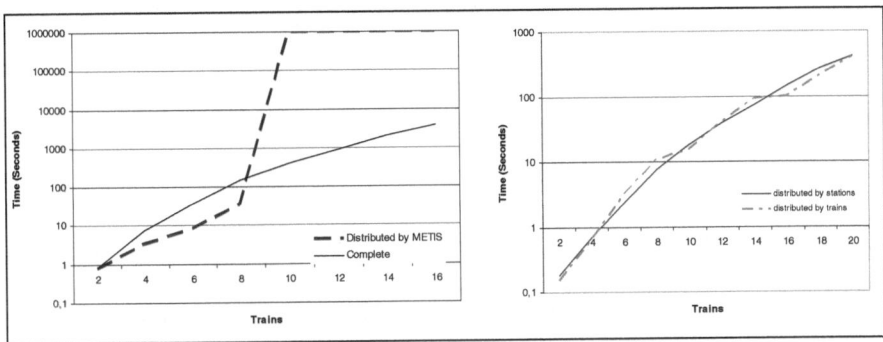

Fig. 9. Execution Time using Metis and Complete (left). Comparison of Approaches 2 and 3 (right).

trains increased (see Figure 9 right). However, partition proposal 3 maintained a uniform behavior.

4   The TTP from a Metaheuristic Approach Based on Variable Ordering

The problem that was described in Section 2 is NP-hard. Therefore, uninformed (i.e., non-heuristically guided) search is not appropriate due to the huge search space and the exponential temporal complexity. Thus, we use an irrevocable heuristic-driven search. The intermediate states are discarded because keeping them would be unfeasible in terms of spatial cost.

In this problem, a solution is different from other solutions when the same conflict is solved in different ways. A conflict exists when a constraint that references the timetable of two trains is violated. This conflict usually appears when

```
Function Get_Train_Timetabling() As Timetabling
begin
  Set_Occupation_Fixed_Trains (T_C)
  While Not (end_cond) do
  begin
      T_open = T_new; T_closed = ∅; prune = FALSE
      While( T_open ≠ ∅  AND prune = FALSE) do
      begin
          (t_i, s_j) = Select_Node()
          TTABLE = TTABLE + Set_Timetable(t_i, s_j)
          δ_est = Estimate_Cost(TTABLE)
          if(δ_est ≥ δ_best)
          then prune = TRUE
          elseif(Last_St(t_i,s_j) then T_open = T_open \ {t_i}
      end
      if(prune = FALSE) then
      begin
          δ = Evaluate_Timetabling(TTABLE)
          if δ < δ_best then TTABLE_best = TTABLE; δ_best = δ
      end
  end
  return TTABLE_best
end
```

Fig. 10. An outline of SOBM

two trains simultaneously require the same track. We solve each conflict by delaying one of these trains until the conflict disappears by means of crossing or overtaking operations. Therefore, the problem consists of deciding which of the two trains must be delayed.

Using a Search Tree to Model the Train Timetabling Problem

We consider the problem as a search tree whose root node represents the empty timetable (*initial* in Figure 11a). For each node where no successor is possible, there is an artificial terminal node (*final* in Figure 11a). Each intermediate node is composed by a pair (t_i, s_j), which indicates that a feasible timetable must be found for the train $t_i \in T_{open}$ in its track section s_j. When the timetable of a train t_i is completed, (i.e., procedure Last_St(t_i, s_j) in Figure 10 is TRUE), this train is eliminated from the set T_{open}. Each level of the search tree indicates which part of the timetable of each train can be generated (see Figure 11a). Given a node $n = (t_i, s_j)$ in a given level l; its brothers in l and the node (t_i, s_{j+1}) (if s_j is not the last track section of t_i) are the successors of the node n in the next level $l + 1$.

The problem consists of finding a path in the search tree, from the *initial* node to a *final* node, so that the order of priorities established by this path produces the minimum average delay. Figure 10 shows the algorithm followed by the method to produce feasible solutions. If we take Figure 11a into account,

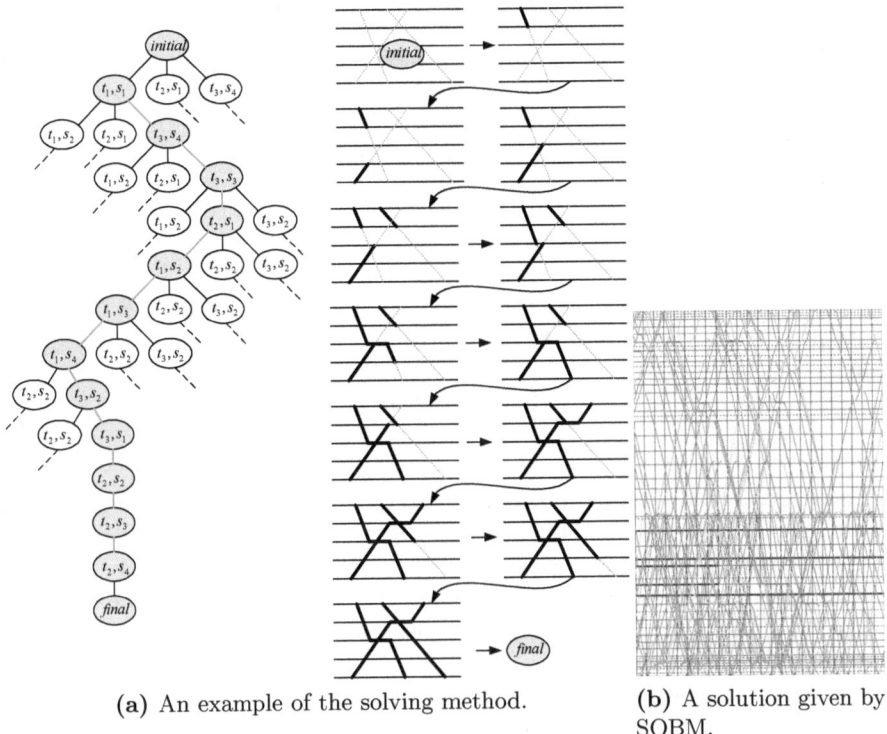

(a) An example of the solving method.   (b) A solution given by SOBM.

Fig. 11. A model for the problem and a solution obtained

the procedure `Select_Node()` decides which node of each level will form part of the final path. Each iteration of the inner loop in Figure 10 corresponds to one level in the search tree of Figure 11a. The heuristic that is used by the procedure `Select_Node()` is given in the next subsection.

Each time a new node (t_i, s_j) is chosen, the procedure `Set_Timetable`(t_i, s_j) assigns a feasible timetable to train t_i in the track section s_j. Then, this pair (t_i, s_j) with its corresponding timetable is added to the set *TTABLE*. Once *TTABLE* has been updated with the new pair, an estimated cost is computed (δ_{est}) for the current path. If δ_{est} is greater or equal to the best solution found at the time (δ_{best} in Figure 10), then the current path is pruned (*prune*=TRUE in Figure 10), and a new path is started from the *initial* node. Given that a feasible timetable was assigned to t from its initial station l_0^t until the station $l_{i_t}^t$, and $M_{i \to n_t}^t$ is the minimum running time possible for the train t to travel from $l_{i_t}^t$ until $l_{n_t}^t$, the procedure `Estimate_Cost`(*TTABLE*) in Figure 10 computes the value for δ_{est} according to the following expression:

$$\delta_{est} = \frac{\sum_{t \in \mathrm{T}_{new}} \delta_{est}^t}{|\mathrm{T}_{new}|} \quad \text{where} \quad \delta_{est}^t = \frac{arr_{i_t}^t - dep_0^t + M_{i_t \to n_t}^t - \Gamma_{opt}^t}{\Gamma_{opt}^t}$$

Heuristic for Choosing a Node for Each Level of the Tree

Here, we detail how the method determines the node that must be chosen at each level of the tree. Given that a feasible timetable was assigned to t from its initial station l_0^t until the station $l_{i_t}^t$, for each level, we measure the partial delay of each train $t \in T_{\text{open}}$ according to the following expression:

$$\delta_{\text{partial}}^t = \frac{arr_{i_t}^t - dep_0^t - \Gamma_{\text{opt}}^t}{\Gamma_{\text{opt}}^t} \, . \tag{18}$$

Given that the minimum partial delay is $\delta_{\min} = \min_{t \in T_{\text{open}}}(\delta_{\text{partial}}^t)$, the probability ρ_t of train t being selected is computed according to the following expression:

$$\rho_t = \frac{(\delta_{\text{partial}}^t - \delta_{\min} + \varepsilon)^\alpha}{\sum_{t \in T_{\text{open}}}(\delta_{\text{partial}}^t - \delta_{\min} + \varepsilon)^\alpha} \, . \tag{19}$$

A train is chosen according to the parameterized Regret-Based Biased Random Sampling (RBRS), [28] and [33], in such a way that the train with higher priority is not necessarily the train chosen, due to the random component of the RBRS method. The parameter α determines the deterministic degree of the heuristic, while the parameter ε determines the selection probability for the train with minimum delay.

An example of the method is provided in Table 2 where function Select_Node (Figure 10) has been applied. In this case, 4 trains are the candidates to be selected (column *Trains t*). The delay with respect to their minimum running time is given in column *Partial delay*, and the probability of selection in the following columns for $\alpha = 2$, $\alpha = 1$, and $\alpha = 0$, respectively.

Once the probability of selection has been computed, the next step consists of obtaining a random value between 0 and 1. For each train, its probability of selection is added until the sum is greater than the random value. When this condition is satisfied, the last train whose selection probability has been added is the selected train.

As can be observed in this table, the higher the α value, the more deterministic the Select_node function. If we consider the example in Table 2, the partial delay for the trains is the same for all the cases of α; however the gap between their probability of selection increases as the α value increases. Therefore, it is

Table 2. Example of calculation of selection probabilities

Trains (t)	Partial delay ($\delta_{\text{partial}}^t$)	ρ_t ($\alpha = 2$)	ρ_t ($\alpha = 1$)	ρ_t ($\alpha = 0$)
1	10	$\frac{10.15^2}{10.15^2+1.15^2+0.15^2+5.15^2} = 0.78$	0.61	0.25
2	1	0.01	0.067	0.25
3	0	0.0001	0.009	0.25
4	5	0.2	0.31	0.25

more probable that train t_1 be selected when $\alpha = 2$ than when $\alpha = 1$. The other parameter, ε, influences on the selection probability of the train with the minimum delay. The higher the ε value, the higher the probability of selection for this train.

4.1   Evaluation

In this subsection: (i) the SOBM and RANDOM approaches are compared with respect to their objective function cost in a given execution time; (ii) different combinations for the parameters (α and ε) of SOMB heuristic are evaluated; and (iii) an example of anytime behavior of SOBM is shown. The Manager of Railway Infrastructure of Spain (ADIF) provides us with real instances in order to obtain a realistic evaluation of the proposed heuristic. The algorithm has been implemented using C++, and the tests have been evaluated on a Pentium IV 3,6 Ghz.

The difference between the SOBM and RANDOM approaches consists in the way that instantiation of variables is ordered. In the first case, we use the heuristic described in subsection 4; in the second case, the order is determined in a random way.

In Figure 12.a (columns 2 to 9), we defined a set of test cases by means of the following: the length of the railway line; number of single/double track sections; number of stations; number of trains and track sections (TS) corresponding to trains already in circulation and new trains, respectively. The results corresponding to these test cases are shown in Figure 12.b. This Table presents the best value of the objective function and the number of feasible solutions that were obtained for each problem (columns 2 and 3 for the RANDOM approach; columns 4 and 5 for the SOBM approach). The execution time was of 300" for all the problems, and the parameters of the RBRS were set to $\alpha=1$ and $\varepsilon=0.05$.

Tables 3 and 4 show another group of tests to indicate how the values of parameters α and ε influence the selections made by the SOBM heuristic. Table 3 shows the objective function cost for the case when $\alpha = \{0, 1, 2\}$ and

Problems	Infrastructure Despcription				In Circulation		New Trains		Problems	RANDOM		SOBM	
	Km	1-Way	2-Way	Stat	Trains	TS	Trains	TS		#of Solutions	Obj%	#of Solutions	Obj%
1	209,1	25	11	22	40	472	53	543	1	169	8,6	168	5,9
2	129,4	21	0	15	27	302	30	296	2	611	10,1	608	10
3	177,8	37	4	25	11	103	11	146	3	2185	21,1	3101	18
4	225,8	33	0	23	113	1083	11	152	4	311	13,2	445	5,5
5	256,1	38	0	28	80	1049	15	235	5	396	19,3	452	17,5
6	256,1	38	0	28	81	1169	16	159	6	424	14,7	521	14,1
7	96,7	16	0	13	47	1397	16	180	7	267	18	263	15,4
8	96,7	16	0	13	22	661	40	462	8	67	50,9	85	45,5
9	298,2	46	0	24	26	330	11	173	9	1112	11,5	1129	8,7
10	401,4	37	1	24	0	0	35	499	10	405	19,2	397	17,9

(a)   (b)

Fig. 12. Results obtained with the SOBM method

Table 3. Objective Function cost according to the pair $< \varepsilon, \alpha >$

Cases	$\alpha = 0$	$\alpha = 1$			$\alpha = 2$		
		$\varepsilon = 0.05$	$\varepsilon = 0.15$	$\varepsilon = 0.30$	$\varepsilon = 0.05$	$\varepsilon = 0.15$	$\varepsilon = 0.30$
1	79,15	68,10	71,50	69,50	67,80	71,00	75,50
2	78,69	75,90	67,90	75,90	74,30	70,20	71,50
3	6,2	6,50	6,70	6,50	6,60	6,50	6,80
4	7,76	7,76	7,76	7,76	7,76	7,76	7,76
5	65,6	54,80	57,20	58,00	58,00	51,90	56,00
6	12,6	12,10	13,00	13,00	12,40	12,60	12,80
7	38,8	35,70	35,70	34,40	35,70	34,40	34,40
8	19,1	18,70	19,00	19,10	18,90	19,20	18,70
9	3,6	3,60	3,60	3,60	4,00	4,10	3,60
10	14,5	12,20	13,70	13,20	14,70	13,20	13,80
11	18,9	19,40	18,90	18,80	18,20	18,80	18,20
12	19,1	18,90	17,20	19,60	18,10	18,40	18,20
13	30,7	27,30	25,70	29,50	28,50	28,60	28,00
Prom.	30,37	27,77	27,53	28,37	28,07	27,44	28,10

Table 4. Objective Function cost according to the pair $< \varepsilon, \alpha >$

Cases	$\alpha = 3$			$\alpha = 4$		
	$\varepsilon = 0.05$	$\varepsilon = 0.15$	$\varepsilon = 0.30$	$\varepsilon = 0.05$	$\varepsilon = 0.15$	$\varepsilon = 0.30$
1	65,21	71,60	64,80	71,60	64,20	64,17
2	59,88	71,90	73,10	64,30	64,50	72,91
3	6,55	6,50	6,40	6,70	6,50	6,66
4	7,76	7,76	7,76	7,76	7,76	7,76
5	54,95	52,00	53,60	55,00	52,20	56,34
6	11,84	12,30	11,90	12,80	12,60	12,31
7	34,84	34,50	36,00	36,40	32,90	31,25
8	18,24	19,10	18,00	18,70	19,20	18,27
9	3,62	3,60	4,10	3,60	3,90	4,04
10	13,67	13,60	12,70	14,50	14,40	13,75
11	16,68	19,40	19,40	18,10	17,40	18,96
12	17,53	18,00	19,60	19,50	17,20	17,99
13	25,51	27,30	28,10	27,90	24,50	28,49
Prom.	25,87	27,50	27,34	27,45	25,94	27,15

$\varepsilon = \{0.05, 0.15, 0.3\}$, and Table 4 shows the objective function cost for the cases when $\alpha = \{3, 4\}$ and $\varepsilon = \{0.05, 0.15, 0.3\}$. Note that when $\alpha = 0$, the order instantiation is completely random.

The cost of each test case is determined by the duration of the technical stop corresponding to each train in T_{new}.

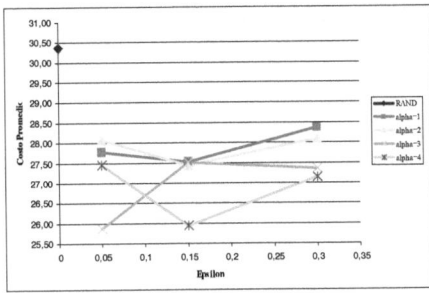

Fig. 13. Average Cost according to α for different values of ε

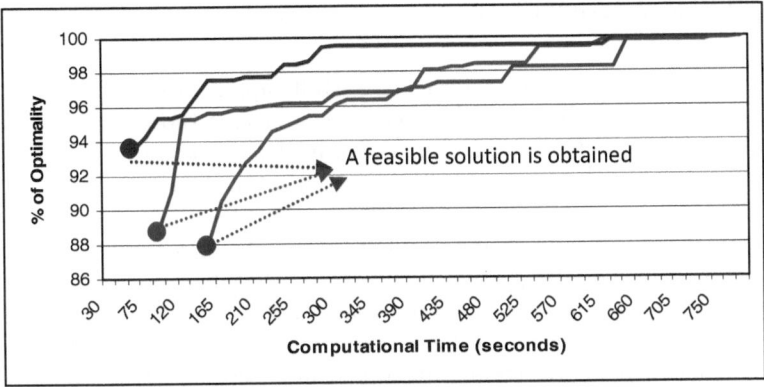

Fig. 14. Objective function cost in function of execution time

The last row of Tables 3 and 4 show the average cost, considering all the test cases for each combination of $< \alpha, \varepsilon >$.

Each point in Figure 13 represents the average cost obtained with a determined combination of $< \alpha, \varepsilon >$. All the points in a same line share the same value of α. If we compare the *SOBM* approach with the RANDOM approach, we can conclude that it is better to use the criteria based on RBRS than to use a completely random criteria, especially when the execution time is limited and results that tend to the optimality as soon as possible are needed. The combination $\alpha = 3$ and $\varepsilon = 0.05$ was the best combination for the test cases.

One of the main properties of SOBM is its anytime behavior. The computational effort to obtain a satisfactory optimized solution depends on the number of trains, the railway infrastructure and its capacity (tracks in stations, single/double-way tracks), the required traffic operations due to the load of the network, etc. Typically, the computational time required for very complex and real problems varies between a few seconds and 2-5 minutes. For instance, a railway timetabling problem that implies the scheduling of 95 new trains, with 37 trains already in circulation (with fixed timetables), on a line of 271.1 Kms, with 51 single-way track sections can be obtained in less than 60 seconds. This scheduling problem implies the solution of 136 crossing conflicts. In this case, the search space is composed of 8.7 E+40 possible solutions. The maximum optimization level obtains higher quality timetables when the computational time is longer. Figure 14 shows the anytime property of the proposed method: the longer the execution time, the more optimal the solution. Moreover, it is possible to interrupt the execution at any time, and a solution will be obtained. In addition, in the tests performed, it was observed that high quality solutions were obtained even with short execution times when compared with solutions obtained with longer execution times.

5   The TTP from a Metaheuristic Approach Based on Genetic Algorithms

Genetic Algorithms (GAs) have been successfully applied to combinatorial problems and are able to handle huge search spaces such as those arising in real-life scheduling problems. GAs perform a multidirectional stochastic search on the complete search space, and this search is intensified in the most promising areas.

The Train Timetabling Problem (TTP) is a difficult and time-consuming task in the case of real networks. The huge search space to explore when solving real-world instances of the TTP problem makes GAs a suitable approach to efficiently solve it. A feasible train timetable should specify the departure and arrival times for each train to each location of the network in such a way that the line capacity and other operational constraints are taken into account. Traditionally, plans were generated manually and adjusted so that all constraints were met. However, the new railway framework of strong competition, privatization, and deregulation along with the increase in computer speed are reasons that justify the need for automatic tools that are able to efficiently generate feasible and optimized timetables.

Assuming TTP to be a very complex problem and GA a suitable procedure to cope with it, we have designed a GA for this train scheduling problem. Once the problem was formally described, a Genetic Algorithm was designed and validated through its application on a set of real-world problem instances provided by the Manager of Railway Infrastructure of Spain (ADIF). In addition, the heuristic technique described in this work has been embedded in a computer-aided tool that is being successfully used by ADIF.

Figure 15 shows the general scheme of a generic genetic algorithm. First, the initial population (P in Figure 15), whose size is POP_SIZE, is generated and evaluated following a scheduling scheme that is described in the subsection *Initial Population*. The following steps are repeated until the terminating condition *end_cond* (execution time, number of feasible solutions or number of generations) is reached. Some individuals that compose the population P in Figure 15 are modified by applying the procedures `Selection()`, `Crossover()`, and `Mutation()`. Thus, a new population P is obtained in each generation. Each iteration of Figure 15 corresponds to a new generation of individuals.

```
Function Genetic_Algorithm(POP_SIZE, end_cond) As Timetabling
```

```
begin
    P=Generate_Initial_Population(POP_SIZE)
    While NOT (end_cond) do
    begin
        P=Selection(P)
        P=Crossover(P)
        P=Mutation(P)
        BEST_L=Evaluate_Population(P)
    end
    return BEST_L
end
```

Fig. 15. General Genetic Algorithm

5.1   Definition of Individuals: Solution Encoding

In order to apply a GA to a particular problem, an internal representation for
the solution space is needed. The choice of this component is one of the critical
aspects for the success/failure of the GA for the problem under study. In the
literature, we have found different types of representations for the solution of
different scheduling problems. In this work, we have used an activity list as the
representation of a solution. This solution representation has been widely used in
project scheduling. The solution is encoded as a precedence feasible list of pairs
(t, s_i^t), that is, if (t, s_x^t) and (t, s_y^t) are the j_{th} and k_{th} gene of a chromosome of
the same individual and $x < y$, then $j < k$.

The corresponding train schedule is generated by applying a modified version
of the Serial Schedule Generation Scheme used in Project Scheduling [17]. In the
list of pairs, all trains are merged in order to obtain a feasible solution. In our
implementation, the new trains are scheduled following the order established by
the list. Each individual in the population is represented by an array that has
as many positions as there are pairs (t, s_i^t) in the railway scheduling problem
considered. Figure 16 shows the activity list representation for a problem with
N pairs (t, s_i^t).

According to this list, (t, s_k^t) is the i_{th} pair to be scheduled. Considering
that $s_k^t = l_k^t \rightarrow l_{k+1}^t$, the departure time of train t from l_k^t will be the earliest
feasible time from $arr_k^t + C_k^t$. Note that when applying the de-codification process

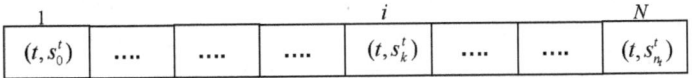

Fig. 16. Activity List Representation

described in subsection 5.1, one and only one schedule can be deduced from a given sequence, but different sequences could be transformed into the same schedule.

Fitness Computation

When applying the GA, we need to define an evaluation function that determines the probability of survival of an individual to the next generation. In this chapter, the average deviation with respect to the optimal solution for the train is returned as the fitness value. In other words, an individual χ will have the fitness value obtained from the objective function defined in section 2.3.

Initial Population

The GA starts with the generation of an initial population, that is, a set of POP_SIZE feasible solutions. This set of feasible solutions can be obtained with different scheduling techniques. For the design of a GA, the initial population should include a variety of medium to good feasible solutions in order to increase the quality of the best solution in the evolutionary process. In this work, the value of the POP_SIZE is 50. The initial population has been obtained with an iterative heuristic that is based on random sampling methods and that is repeated POP_SIZE times (Figure 17). N is the total number of track sections for all trains $t \in T_{NS}$. It is also shown how the activity list is created by selecting a train (t) and a set of track sections $s \in J_t$ for each iteration until all the (N) track sections have been scheduled for each new train. A solution is obtained once N iterations have been completed. Each solution gives a scheduling order, $L = (t_x, s_0^{t_x}), ..., (t_z, s_j^{t_z}), ..., (t_y, s_{n_{t_y}}^{t_y})$, which represents a new individual of the initial population.

The scheduling method developed implies the search of a path in a tree as the one shown in Figure 11b. At each decision point, a train with unscheduled track-sections is selected. This process obtains a feasible timetable with a value of the fitness function.

The main decision in the procedure `Select_Train()` is to determine which train has to be scheduled at each decision point. Even though a random decision (RANDOM) could be taken by selecting the train to schedule randomly, we have applied a *Regret-Based Biased Random Sampling* (RBRS) procedure that makes the selection of a train dependent on its deviation with respect to the optimal solution (optimal running time of the train calculated as indicated in Section 2.3). This approach guides the scheduling process in order to obtain better solutions.

The *Parameterized Regret-Based Biased Random Sampling* (RBRS) selects trains of T_{NS} through a random device. The use of a random device can be considered as a mapping $\psi : i \in T_{NS} \to [0..1]$ where a probability $\psi(i)$ of being selected (with $\sum_{i \in T_{NS}} \psi(i) = 1$) is assigned to each $t \in T_{NS}$. The regret value (ρ_i)

```
Function Generate_Initial_Population(POP_SIZE) As Population
```

```
begin
     ref=Get_Low_Bound_Opt_Sol()
     i=0
     P=""
     While (i<POP_SIZE)
     begin
          L= "" //L is a new list of chromosomes
          j=0
          While (j<N)
          begin
               t=Select_Train() //using the RBRS method
               s=Select_Track_Section(t)
               d=Get_Departure(t,s)
               Set_Timetable((t,s),d)
               L=L+(t,s) //(t,s) is inserted in L
               j=j+1
          end
          Set_Fitness_To_Individual(L, ref)
          P=P+L
          i=i+1
     end
     return P
end
```

Fig. 17. Procedure to obtain the Initial Population

for each train $i \subset T_{NS}$ compares the priority value of train $i -\nu(i)-$ with the worst priority value $\nu(j)$ of the trains of T_{NS} and is calculated as follows:

$$\rho_i : \max_{j \in T_{NS}} (\nu_j) - (\nu_i) \tag{20}$$

Therefore, the parameterized probability mapping $\psi(i)$ is calculated as:

$$\psi(i) : \frac{(\rho_j + \varepsilon)^\alpha}{\sum\limits_{j \in T_{NS}} (\rho_j + \varepsilon)^\alpha} \tag{21}$$

This parameterized Regret-Based Biased Random Sampling has been widely and successfully used in project scheduling (Schirmer and Riesenberg [28], Tormos and Lova [33]). The priority value of each train is calculated according to its current delay with respect to the scheduled timetable. Trains with higher delays have more probabilities of being selected. We have implemented the procedure Generate_Initial_Population() using the RBRS method to select the next train to be scheduled, with $\alpha = 1$ and $\varepsilon = 0.5$.

```
Function Random_Crossover_Point()
```

```
begin
      //Draw a random integer k, with 1<= k<= N
      //k is the random crossover-point
      //Generation of the daughter
      for i=1 to k do
            Di = Mi
      for i=k+1 to N do
      begin
            I = lowest index   1<= I<= N and Fi not in {D1, ..., D(i-1)}
            Di = Fi
      end
      //Generation of the son
      . . . . . . . .
end
```

Fig. 18. Crossover Procedure

Crossover

One of the unique and important aspects of the techniques involving Genetic Algorithms is the important role that recombination (traditionally, in the form of crossover operator) plays. Crossover combines the features of two parent chromosomes to form two offspring that inherit their characteristics. The individuals of the population are mated randomly and each pair undergoes the crossover operation with a probability of P_{cross}, producing two children by crossover. The parent population is replaced by the offspring population. The crossover is one of the most important genetic operators and must be correctly designed. Crossover must combine solutions to produce new ones. Crossover must preserve and combine "good building blocks" to build better individuals [12]. Given two individuals selected for crossover, a mother **M**, a father **F**, two offspring (a daughter **D** and a son **S**) are produced.

We have implemented the well-known one point crossover with $P_{cross} = 0.8$. First we draw a random crossover-point k, with k between 1 and N (number of Train-Track Sections in the problem). The first k positions in **D** are directly taken from **M**, in the same order. The rest of the activities in **D** are taken with their relative order in the father's sequence. In this way, the solution generated, the daughter, is a precedence feasible solution. Obviously, the generation of **S** is similar to the daughter's but **S** inherits the first positions directly from **F**, and the rest of the Train-Track Sections from **M**. The pseudocode for this crossover technique is shown in Figure 18.

Mutation

Once the crossover operator has been applied and the offspring population has replaced the parent population, the mutation operator is applied to the offspring population. Mutation alters one or more genes (positions) of a selected chromosome (solution) to reintroduce lost genetic material and introduce some extra variability into the population.

The mutation operator that we have implemented works as follows: for each pair (t, s_i^t) in the sequence, a new position is randomly chosen. In order to generate only precedence feasible solutions, this new position must be higher than its predecessor and lower than its successor. The chromosome is inserted in the new position with a probability P_{mut}. In our implementation, $P_{mut} = 0.05$.

Selection

Selection is an artificial version of the natural phenomenon called the survival of the fittest. In nature, competition among individuals for scarce resources and for mates results in the fittest individuals dominating over weaker ones. Based on their relative quality or rank, individuals receive a number of copies. A fitter individual receives a higher number of offspring and, therefore, has a higher probability of surviving in the subsequent generation. There are several ways of implementing the selection mechanism.

We have implemented *2-tournament selection*. This selection mechanism implies that two individuals are randomly chosen from the population and compete for survival. The best one (the one with the best fitness value) will appear in the subsequent population. This procedure is repeated POP_SIZE times until POP_SIZE individuals are selected to appear in the next population.

Decodification Process

In this subsection, we detail the procedure Evaluate_Population() of Figure 15. This procedure receives a population P from which it should obtain POP_SIZE solutions. Each solution will be evaluated according to the objective function that is defined in Section 2.3 (Set_Fitness_Individual() in Figure 19).

For each pair $p = (t, s_i^t) \in L$, a departure time is computed by means of the function Get_Departure(), which returns $d = arr_i^t + C_i^t$ if $i > 0$, otherwise $d = m$ such that m is the initial departure time given by the user.

Considering that s_i^t starts at station l_i^t and ends at station l_{i+1}^t, the procedure Set_Timetable() assigns a possible departure and arrival time to train t in each location between l_i^t and l_{i+1}^t, according to the running time $(\Delta_{i \to (i+1)}^t)$ defined for this train from l_i^t to l_{i+1}^t. The next step consists of verifying whether all the constraints defined in 2.2 are satisfied by the timetable given for t in s_i^t. If any constraint is not satisfied, the departure time in l_i^t is increased until that constraint is satisfied. This increment in the departure time in l_i^t causes a technical stop of the train t at this station. A backtracking may occur if the

---

Procedure Evaluate_Population(P)

```
begin
    i=0
    While(i<|P|) do
    begin
        L=Get_Individual(i,P)
        k=0
        while(k<|L|)
        begin
            p=Get_Chromosome(L,k)
            d=Get_Departure(p)
            Set_Timetable(p,d)
            k=k+1
        end
        i=i+1
        Set_Fitness_Individual(L)
    end
end
```

Fig. 19. Decodification Process

station is closed for technical operations or if the station does not have enough tracks.

Once a feasible timetable has been found in this track section for train t, the same procedure is repeated with the next chromosome $(t', s_k^{t'})$.

The priority of the trains in each track section, that is, which train should be delayed if a conflict appears, is determined by the order in which each gene is numbered in the activity list. When a conflict occurs between two trains in the same track section, the priority is for the train whose timetable in this track section was assigned first. Since different individuals define different priorities among the trains, different solutions may be obtained.

The parameter setting of the proposed GA results from previous computational experiments.

5.2   Evaluation

The performance of the developed GA has been tested using a set of real-world problems provided by the Manager of Railway Infrastructure of Spain (ADIF). The description of the instances is given in Table 1 (columns 2 to 10) by means of the following: length of the railway line, number of single/double track sections, number of locations and stations, number of trains and track sections (T-ts) corresponding to trains in circulation and new trains, respectively.

Each problem has been solved by using the two constructive methods used to generate the Initial Population that differs in the criterion to select the trains:

Table 5. Real railway problem instances provided by ADIF

Problems	Infrastructure Description					Trains in Circulation		New Trains	
	Km	1-Way	2-Way	Loc	Stat	Trains	T-ts	Trains	T-ts
1	96	16	0	13	13	47	1397	16	180
2	129	21	0	22	15	27	302	30	296
3	256	38	0	39	28	81	1169	16	159
4	401	37	1	39	24	0	0	35	499

Table 6. Results of the RANDOM, RBRS and the GA scheduling methods

Problems	RANDOM		RBRS		GA	
	# of Solutions	ADOS	# of Solutions	ADOS	# of Solutions	ADOS
1	267	18	263	15.4	255	15.1
2	611	10.1	608	10.0	313	9.6
3	424	14.7	521	14.1	382	12.4
4	405	19.2	397	17.9	285	16.0

- Random selection of each train to be scheduled in each iteration (RANDOM).
- Selection of each train using the Parameterized Regret Biased Based Random Sampling method (RBRS).
- The results obtained by means of these constructive methods are compared against those achieved by the GA with the same computational time.

Table 6 summarizes the results for each solving method with respect to the number of solutions generated and the Average Deviation with respect to the Optimal Solution (ADOS). The tests have been carried out on a Pentium IV 3.6 Ghz processor, and the execution time was of 300 seconds for all the problems. $\alpha = 1$ and $\varepsilon = 0.5$. The different number of solutions generated (depending on the method used) is mainly due to the fact that when the RANDOM and RBRS approaches are used, a prune procedure is applied. That is, when a partial schedule produces a value of the objective function that is worse than the best value obtained at the time, the current iteration is interrupted and the construction of a new one starts. However with the GA approach, the prune is not possible because each iteration must be completed to obtain a fitness value for the solution. This fitness value is necessary to obtain the next generation of individuals.

The results shown in Table 6 indicate that the GA proposed outperforms both the RANDOM and RBRS methods for all the problem instances considered. These results demonstrate the efficiency of the GA to solve Railway Scheduling problems over other constructive algorithms and also support the idea of developing more sophisticated and powerful GAs to solve complex problems such as the Train Timetabling Problem.

6   A Topological Heuristic Approach for Periodic Timetabling

In the previous sections, we have presented general purpose metaheuristics and constraint-based techniques for solving railway timetabling. In this section, we focus our attention on a simplified problem: the periodic timetabling on single track lines, where each trip is operated in a periodic way.

As was pointed out above, the majority of authors use models that are based on the Periodic Event Scheduling Problem (PESP) introduced by Serafini and Ukovich ([29]). The PESP considers the problem of scheduling as a set of periodically recurring events under periodic time-window constraints. The model generates disjunctive constraints that may cause the exponential growth of the computational complexity of the problem depending on its size.

An alternative method was presented in [25]. Salido et al. propose a topological constraint optimization technique for solving periodic train scheduling. This technique has been inserted in the system [4] and is committed to solving the cyclic timetabling problem in order to obtain timetables that are as good and feasible as possible.

In this method, the railway timetabling problem is formulated as a Constraint Optimization Problem (COP). The variables are the frequencies, arrival and departure times of trains at stations. Constraints are composed by user requirements and intrinsical constraints (railway infrastructures, rules for traffic coordination, etc.). The objective function is to minimize the running time of all trains. The problem formulation is (traditionally) translated into a formal mathematical model to be solved for optimality by means of mixed integer programming techniques. In this framework, the formal mathematical model is partitioned in two different subproblems: an integer programming problem composed by the constraints with integer variables, and a linearized problem in which there are now continuous variables remaining to be assigned. The most restricted constraints are considered to be composed of integer variables. In this way, the system studies the integer programming problem first, and then it solves the linearized problem. The integer programming problem will be partitioned into a set of subproblems so that the solution of each subproblem will generate a traffic pattern. Each block of the partition is composed by contiguous stations, so each traffic pattern represents the running map corresponding to each block of constraints.

The objective is to solve this problem by previously assigning values to integer variables so that the mixed-integer programming problem is transformed into a linear programming problem. Then, the linearized problem is easily solved. Thus, the topological constraint optimization technique is committed to assigning values to integer variables.

The topological constraint optimization technique generates the traffic patterns based on several features such as identification of bottlenecks, periodicity of running maps, number of stations, distance among stations, possible wide-paths for trains, etc.

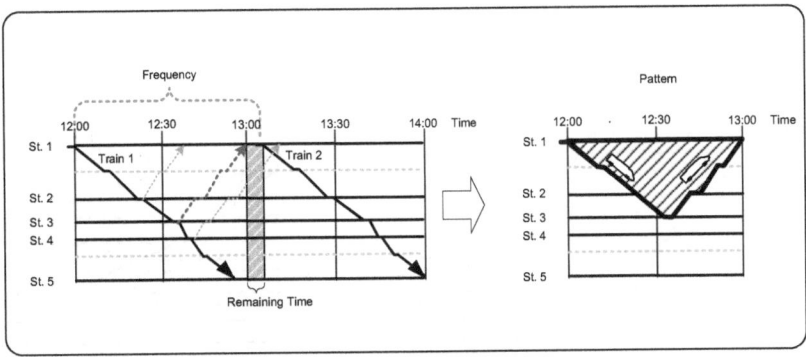

Fig. 20. First traffic pattern generation

The main idea of this technique is to generate a traffic pattern for each set of stations so that the union of these contiguous traffic patterns determines the journey of each train. Figure 20 shows a possible set of stations (block).

The block of stations will be selected taking into account the speed of the trains, the distance between stations and the frequency inserted into the problem. Each traffic pattern covers the block of stations necessary for a train to go from the first station of the block to the last station of the block and return from the last station to the first one (round trip). This round trip must arrive to the first station (St.1) as close as possible but always before the following train's departure (Train 2). Thus, our objective is to minimize the time remaining between the frequency and the round trips. Each possible round trip will involve a different set of constraints. The round trip that minimizes the remaining time will be selected as the *pattern*. This traffic pattern will be composed by a higher number of stations than the rest of the possible round trips.

Once the first traffic pattern has been generated, we study the following pattern with the remaining stations. Figure 21 shows the generation of the second pattern using the same strategy.

Therefore, when the second traffic pattern is generated, the topological technique studies the next traffic pattern until there is no station left. Figure 22 shows an example of a running map with three complete traffic patterns and some stations without traffic patterns (it is common for some stations to not be involved in any traffic pattern). These stations are not involved in any traffic pattern. We must take into account that the best traffic pattern in a block of stations implies starting the following block of stations in the last station of the previous block. We must check all traffic patterns together in order to obtain the journey. Moreover, the first combination of traffic patterns may not be the best solution due to the existence of certain combinations of traffic patterns. This first combination depends on the number of stations that are not involved in a traffic pattern. Thus, we explore all possible combinations in order to obtain the best set of traffic patterns.

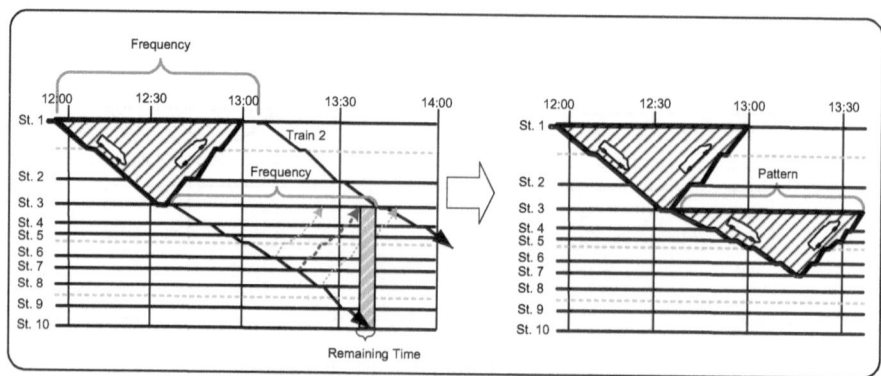

Fig. 21. Second traffic pattern generation

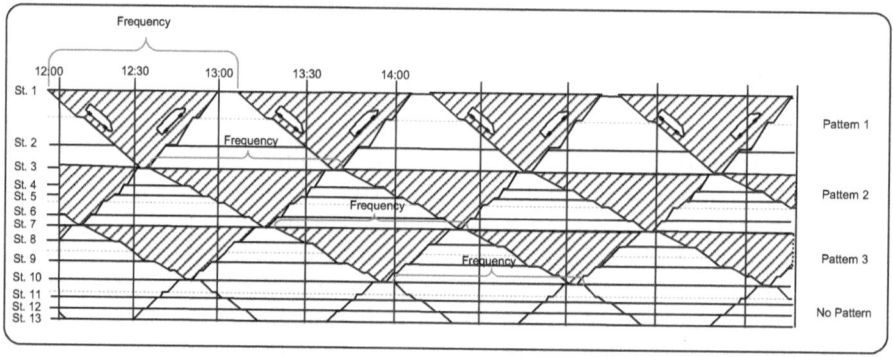

Fig. 22. Periodic Pattern generation

Figure 22 shows an example in which three stations are not involved in any traffic pattern. Therefore, some combinations are possible, and they are restricted to the set of stations involved in the first traffic pattern. These three stations can be sorted between the first and the last traffic pattern. The first traffic pattern may start at the second or third station and the last traffic pattern may finish in the penultimate or the third to the last station. However, due to efficient use of resources, or depending on the importance of the station, it is more appropriate for the first traffic pattern (last traffic pattern) to start (finishes) at the first (last) station.

This heuristic technique has been inserted in the system used by the Manager of Railway Infrastructure of Spain [4]. The topological method just described has been tested on different problem sets. The random generated instances are defined over a real railway infrastructure with single track sections. Each set of random instances was defined by the 3-tuple $< n, s, f >$, where n was the number of trains in each direction, s the number of stations/halts and f the frequency. The number of trains was increased from 5 to 50, the number of stations was increased from 10 to 60 and the frequency was increased from 60 to 140 minutes. In conclusion, the execution times were lower using the topological technique than other standard optimization tools such as CPLEX. Furthermore, the topological technique was independent of the number of trains due to the fact that it first generates the corresponding traffic patterns and then replicates these patterns according to the number of trains.

6.1   Evaluation

The application and performance of this system depends on several factors: Railway topology (locations, distances, tracks, etc.), number and type of trains (speeds, starting and stopping times, etc.), frequency ranges, initial departure interval times, etc.

In this section, we compare the performance of our topological technique with some well-known tools: LINGO, which is an Operational Research tool, and ILOG Concert Technology (CPLEX 8.0), which combines techniques of constraint programming and mathematical programming. Both are appropriate tools for solving these types of problems. However, in order to significantly reduce the size of these problems, the system carried out important preprocessing heuristics [3] before executing these well-known tools. Therefore, CPLEX and LINGO were combined with some heuristics, and they obtained the optimal solutions for their relaxed problems.

This empirical evaluation was carried out integrating two different types of problems: benchmark (real) problems and random problems. The computer used in our tests was a Pentium IV 2.8Mz with 512 Mb. of memory. Thus, we defined random instances over a real railway infrastructure that joins two important Spanish cities (La Coruña and Vigo). The journey between these two cities consists of 40 locations (23 stations and 17 halts).

In our empirical evaluation, each set of random instances was defined by the 3-tuple $< n, s, f >$, where n was the number of trains in each direction, s the number of stations/halts and f the frequency. The problems were randomly generated by modifying these parameters. Thus, each of the tables shown sets two of the parameters and varies the other one in order to evaluate the performance of the algorithm when this parameter increases.

In Table 7, we present the execution time in seconds and the running time for problems where the number of trains was increased from 5 to 50 and the number of stations/halts and the frequency were set at 40 and 90, respectively ($< n, 40, 90 >$). The results show that CPLEX obtained a better execution time and a better running time than LINGO. However, it can be observed that the execution time is lower using the topological technique than the other two COP tools. Furthermore, our technique always obtained the same running time (lower than CPLEX and LINGO) due to the fact that it generates the corresponding traffic patterns and is independent of the number of trains.

Table 8 shows the execution time in seconds and the running time in problems where the number of stations was increased from 10 to 60 and the number of

Table 7. Execution time (sec.) and running time in problems with different trains

< n, 40, 90 >	CPLEX+heuristics		LINGO+heuristics		TOPOLOGICAL	
Trains	runtime	running time	runtime	running time	runtime	running time
5	5"	2:29:33	8"	2:30:54	3"	2:22:08
10	8"	2:26:04	17"	2:31:37	4"	2:22:08
15	13"	2:26:18	24"	2:31:51	5"	2:22:08
20	16"	2:26:25	35"	2:31:58	5"	2:22:08
50	55"	2:31:09	1302"	2:32:11	10"	2:22:08

Table 8. Execution time (sec.) and running time in problems with different numbers of stations

< 10, s, 90 >	CPLEX+heuristics		LINGO+heuristics		TOPOLOGICAL	
Stations	runtime	running time	runtime	running time	runtime	running time
10	2"	0:58:36	4"	0:58:06	1"	0:57:36
20	3"	1:04:11	20"	1:04:11	2"	1:04:11
30	15"	1:45:08	42"	1:45:38	4"	1:45:08
40	56"	2:23:16	28"	2:24:36	7"	2:20:22
60	340"	3:44:28	326"	3:44:22	40"	3:32:15

Table 9. Execution time (sec.) and running time in problems with different frequencies

< 20, 40, f >	CPLEX+heuristics		LINGO+heuristics		TOPOLOGICAL	
Frequency	runtime	running time	runtime	running time	runtime	running time
60	> 43200"	-	> 43200"	-	36"	2:32:11
90	17"	2:26:25	32"	2:31:58	5"	2:22:08
100	18"	2:23:10	34"	2:22:55	3"	2:19:09
120	16"	2:16:17	27"	2:18:47	4"	2:16:00
140	17"	2:20:18	27"	2:16:19	4"	2:17:03

trains and the frequency were set at 10 and 90, respectively (< 10, s, 90 >). In this case, only stations were included to analyze the behavior of the techniques. It can be observed that the execution time was lower using our technique in all instances. The running time was also improved using our topological technique. It is important to note the difference between the instance < 10, 40, 90 > of Table 7 and the instance < 10, 40, 90 > of Table 8. These tuples represent the same instance, but in Table 8 we only used stations (no halts), so the number of possible crossing between trains was much more larger. This item reduced the running time from 2:22:08 to 2:20:22, but the number of combinations increased the execution time from 4" to 7". Furthermore, CPLEX and LINGO maintained similar behaviors.

In Table 9, we present the execution time in seconds and the running time in problems where the frequency was increased from 60 to 140 and the number of trains and stations were set at 20 and 40, respectively (< 20, 40, f >). It can be observed that the topological technique improved the running time when the frequency increased. As in previous results, the execution time of the topological technique was lower than CPLEX and LINGO.

7   Conclusions

Optimizing a train schedule on a single line track is known to be NP-Hard. This makes it difficult to determine optimum solutions to real-life problems in reasonable time and raises the need for good heuristic techniques. The Train Timetabling Problem considered in this work implies the optimization of new heterogeneous trains on a railway single-line that may or may not be occupied (or not) by other trains with fixed timetables.

In this chapter, we have presented several approaches to manage this problem. The first one is based on distributed constraint satisfaction techniques. We model the TTP as a distributed CSP. Several approaches have been developed to distribute the problem into a set of sub-problems that are as independent as possible. The second approach uses meta-heuristic models based on variable ordering. It is a constructive grasp-based approach that obtains optimized solutions with very low computational times. The third approach is based on the application of Genetic Algorithms for TTP. Finally, a topological constraint optimization technique is presented. This technique is committed to solving the cyclic timetabling problem in order to obtain timetables that are as good and feasible as possible.

All of these heuristics methods are embedded in a computer-aided tool called MOM (http://www.dsic.upv.es/grupos/gps/MOM/) which is currently being successfully used by the Manager of Railway Infrastructure of Spain (ADIF). The results obtained by these methods for solving real-world timetabling problems allow us to justify their utility and application in solving real-world instances.

Acknowledgements

This work has been partially supported by the research projects TIN2007-67943-C02-01 (Min. de Educacion y Ciencia, Spain-FEDER) , P19/08 (Min. de Fomento, Spain-FEDER), TIN2004-06354-C02-01 (Min. de Educacion y Ciencia, Spain-FEDER), the Technical University of Valencia and by the Future and Emerging Technologies Unit of EC (IST priority - 6th FP), under contract no. FP6-021235-2 (project ARRIVAL).

References

1. Abril, M., Salido, M.A., Barber, F., Ingolotti, L., Lova, A., Tormos, P.: Distributed models in railway industry. In: Proc. of the 1st Workshop on Industrial Applications of Distributed Intelligent Systems, INADIS 2006 (2006)
2. Angles-Domínguez, M.I., Terashima-Marín, H.: Stability analysis for dynamic constraint satisfaction problems. In: Monroy, R., Arroyo-Figueroa, G., Sucar, L.E., Sossa, H. (eds.) MICAI 2004. LNCS (LNAI), vol. 2972, pp. 169–178. Springer, Heidelberg (2004)
3. Barber, F., Salido, M.A., Ingolotti, L., Abril, M., Lova, A., Tormos, P.: An interactive train scheduling tool for solving and plotting running maps. In: Session of Technology Transfer on Artificial Intelligence, TTIA (2003)

4. Barber, F., Tormos, P., Lova, A., Ingolotti, L., Salido, M.A., Abril, M.: A Decision Support System for railway Timetabling (MOM): the Spanish case. In: Computers in Railways X: Computer System Design and Operation in the Railway and Other Transit Systems. Computer in Railways, vol. 10, pp. 235–244. WIT Press (2006)
5. Bistarelli, S., Fargier, H., Montanary, U., Rossi, F., Schiech, T., Verfailillie, G.: Semiring-based csps and valued csps: Frameworks, properties, and comparison. Constraints 4(3) (1999)
6. Bistarelli, S., Montanari, U., Rossi, F.: Semiring-based constraint satisfaction and optimization. Journal of ACM 44(2), 201–236 (1997)
7. Boutilier, C., Brafman, R.I., Domshlak, C., Hoos, H.H., Poole, D.: Cp-nets: A tool for representing and reasoning with conditional ceteris paribus preference statements. JAIR (21), 135–191 (2004)
8. Cai, X., Goh, C.J.: A fast heuristic for the train scheduling problem. Computers and Operation Research 21(5), 499–510 (1994)
9. Caprara, A., Monaci, M., Toth, P., Guida, P.: A lagrangian heuristic algorithm for a real -world train timetabling problem. Discrete Applied Mathematics 154, 738–753 (2006)
10. Carey, M., Lockwood, D.: A model, algorithms and strategy for train pathing. The Journal of the Operational Research Society 46(8), 988–1005 (1995)
11. Dobuis, D., Fargier, H., Prade, H.: The calculus of fuzzy restrictions as a basis for flexible constraint satisfaction. In: Proc. of FUZZ-IEEE 1993, pp. 1131–1136 (1993)
12. Goldberg, D.E.: Genetic Algorithms in Search, Optimization and Machine Learning. Addison-Wesley, Reading (1989)
13. Higgins, A., Kozan, E., Ferreira, L.: Heuristic techniques for single line train scheduling. Journal of Heuristics 3(1), 43–62 (1997)
14. Ingolotti, L., Lova, A., Barber, F., Tormos, P., Salido, M.A., Abril, M.: New heuristics to solve the csop railway timetabling problem. In: Ali, M., Dapoigny, R. (eds.) IEA/AIE 2006. LNCS (LNAI), vol. 4031, pp. 400–409. Springer, Heidelberg (2006)
15. Jovanovic, D., Harker, P.T.: Tactical scheduling of rail operations: The scan i system. Transportation Science 25(1), 46–64 (1991)
16. Karypis, G., Kumar, V.: A parallel algorithm for multilevel graph partitioning and sparse matrix ordering. Journal of Parallel and Distributed Computing 48, 71–85 (1998)
17. Kelley, J.: The critical-path method: Resources planning and scheduling. In: Industrial Scheduling (1963)
18. Kroon, L., Peeters, L.: A variable time model for cycling railway timetabling. Transportation Science 37(2), 198–212 (2003)
19. Kwan, R.K.S., Mistry, P.: A co-evolutionary algorithm for train timetabling. In: IEEE Press (ed.) Congress on Evolutionary Computation, pp. 2142–2148 (2003)
20. Liebchen, C.: Periodic Timetable Optimization in Public Transport. dissertation.de - Verlag im Internet GmbH 2006 (2006)
21. Nachtigall, K., Voget, S.: A genetic algorithm approach to periodic railway synchronization. Computers and Operations Research 23, 453–463 (1996)
22. Neagu, N., Dorer, K., Calisti, M.: Solving distributed delivery problems with agent-based technologies and constraint satisfaction techniques. In: Proceedings of AAAI Spring Symposium (2006)
23. Odijk, M.: A constraint generation algorithm for the construction of periodic railway timetables. Transportation Research Part B 30(6), 455–464 (1996)
24. Rossi, F., Sperduti, A.: Acquiring both constraint and solution preferences in interactive constraint systems. Constraints 9(4), 311–332 (2004)

25. Salido, M.A., Abril, M., Barber, F., Ingolotti, L., Tormos, P., Lova, A.: Topological Constraint in Periodic Train Scheduling. In: Planning, Scheduling and Constraint Satisfaction: from Theory to Practice. Frontiers in Artificial Intelligence and Applications, vol. 117, pp. 11–20. IOS Press, Amsterdam (2004)
26. Salido, M.A., Abril, M., Barber, F., Ingolotti, L., Tormos, P., Lova, A.: Domain dependent distributed models for railway scheduling. International Journal Knowledge Based Systems 20(2), 186–194 (2007)
27. Salido, M.A., Barber, F.: Distributed csps by graph partitioning. Applied Mathematics and Computation (Elsevier) 183, 491–498 (2006)
28. Schirmer, A., Riesenberg, S.: Parameterized heuristics for project scheduling- biased random sampling methods. Technical report, Institute fr Betriebswirtschaftslehre der UNIVERSITT KIEL (1997)
29. Serafini, P., Ukovich, W.: A mathematical for periodic scheduling problems. SIAM J. Discret. Math. 2(4), 550–581 (1989)
30. Serafini, P., Ukovich, W.: A mathematical model for periodic scheduling problems. SIAM J. on Discrete Mathematics 2, 550–581 (1989)
31. Silva de Oliveira, E.: Solving Single-Track Railway Scheduling Problem Using Constraint Programming. PhD thesis, The University of Leeds, School of Computing (September 2001)
32. Szpigel, B.: Optimal train scheduling on a single track railway. In: Roos, M. (ed.) Proceedings of IFORS Conference on Operational Research 1972, pp. 343–352 (1973)
33. Tormos, P., Lova, A.: A competitive heuristic solution technique for resource-constrained project scheduling. Annals Of Operations Research 102, 65–81 (2001)
34. Wallace, R.J., Freuder, E.C.: Stable solutions for dynamic constraint satisfaction problems. In: Maher, M.J., Puget, J.-F. (eds.) CP 1998. LNCS, vol. 1520, pp. 447–456. Springer, Heidelberg (1998)
35. Yokoo, M.: Preface. In: Proc. of AAMAS 2006. 7th Intl. Workshop on Distributed Constraint Reasoning (2006)

Engineering Time-Expanded Graphs for Faster Timetable Information*

Daniel Delling, Thomas Pajor, and Dorothea Wagner

Department of Computer Science, University of Karlsruhe, P.O. Box 6980, 76128
Karlsruhe, Germany
{delling,pajor,wagner}@informatik.uni-karlsruhe.de

Abstract. We present an extension of the well-known time-expanded approach for timetable information. By remodeling unimportant stations, we are able to obtain faster query times with less space consumption than the original model. Moreover, we show that our extensions harmonize well with speed-up techniques whose adaption to timetable networks is more challenging than one might expect.

1 Introduction

During the last years, many speed-up techniques for computing a shortest path between a given source s and target t have been developed. The main motivation is that computing shortest paths in graphs is used in many real-world applications like route planning in road networks or timetable information for railways. Although DIJKSTRA's algorithm [6] can solve this problem, it is far too slow to be used on huge datasets. Thus, several speed-up techniques have been developed (see [5] for an overview) yielding faster query times for typical instances. However, recent research focused on developing speed-up techniques for road networks, while only few work has been done on adapting techniques to graphs deriving from timetable information systems. In general, two approaches exist for modeling timetable information: The time-dependent and time-expanded approach. While the former yields smaller inputs (and hence, smaller query times), the latter allows a more flexible modeling of additional constraints. It turns out that adaption of speed-up techniques to each of these models is more challenging than one might expect.

In this work, we use a different approach for obtaining faster query times. Instead of applying a routing algorithm, e.g., plain DIJKSTRA, on the original model, we improve the *time-expanded* model itself in such a way that a routing algorithm does not exploit parts of the graph not necessary for solving the earliest arrival problem (EAP). Interestingly, it turns out that those optimizations are included in the time-dependent approach implicitly. By introducing those techniques to the time-expanded approach, query times for the time-expanded approach are comparable to the time-dependent approach.

---

* Partially supported by the Future and Emerging Technologies Unit of EC (IST priority – 6th FP), under contract no. FP6-021235-2 (project ARRIVAL).

R.K. Ahuja et al. (Eds.): Robust and Online Large-Scale Optimization, LNCS 5868, pp. 182–206, 2009.

1.1   Related Work

The simple, i.e., without realistic transfers, time-expanded model has been introduced in [22]. The model has been generalized in [19] in order to deal with realistic transfers. Since then, this realistic model has been used for many experimental studies, e.g., [2, 15, 20]; most of them focusing on faster speed-up techniques or multi-criteria optimization for timetable information. However, [22] enriched the simple time-expanded graph by shortcuts and [20] introduced minor changes to the time-expanded model itself by removing unnecessary nodes with outgoing degree 1.

1.2   Our Contributions

This paper is organized as follows. Section 2 includes formal definitions and a review of the time-expanded model for timetable information. Our main contribution is Section 3. We show how the main ingredient for high-performance speed-up techniques in road networks, i.e., *contraction*, can be adapted to time-expanded graphs. Unfortunately, it turned out that this contraction yields a tremendous growth in number of edges (unlike in road networks). However, by changing the modeling of unimportant stations, a DIJKSTRA does not exploit unnecessary parts of the network. The key observation is the following. Assume T is a station with only one line stopping. A passenger traveling via T only leaves the train if T is her target station, otherwise it never pays off to leave the train. Moreover, we are able to generalize this approach to stations with more lines stopping at that station. In Section 4 we introduce a new speed-up technique tailored to time-expanded graphs based on blocking certain connections. Furthermore, we show how existing techniques have to be adapted to timetable graphs. It turns out that certain pitfalls exist that one might not expect. However, those adapted techniques harmonize well with our new approaches, which we confirm by an experimental evaluation in Section 5. We conclude our work in Section 6 with a summary and future work.

A preliminary version of this paper has been published in [4]. Besides some minor improvements, we here provide detailed proofs of correctness.

2   Preliminaries

Throughout the whole work, we restrict ourselves to the earliest arrival problem (EAP), i.e., find a connection in a timetable network with lowest travel time. In the following we often call this single-criteria search in contrast to multi-criteria search that also minimizes number of transfers and further criteria [15, 20]. Moreover, we restrict ourselves to periodic timetables with a time-period of 1440 minutes (one day).

Moreover, we restrict ourselves to simple, directed graphs $G = (V, E, \text{length})$ with positive length function $\text{length} : E \to \mathbb{R}^+$. The reverse graph $\overline{G} = (V, \overline{E})$ is the graph obtained from G by substituting each $(u, v) \in E$ by (v, u). A *partition* of V is a family $\mathcal{P} = \{P_0, P_1, \ldots, P_k\}$ of sets $P_i \subseteq V$ such that each node $v \in V$

is contained in exactly one set P_i. An element of a partition is called a *cell*. The *boundary nodes* B_P of a cell P are all nodes $u \in P$ for which at least one node $v \in V \setminus P$ exists such that $(v, u) \in E$ or $(u, v) \in E$.

The Condensed Model is a basic representation of the network structure. Here, a node is introduced for each station and an edge is inserted iff a direct connection between two stations exists. The edge weight is set to be the minimum travel time over all possible connections between these two stations. Unfortunately, several drawbacks exist regarding timetable information. First of all, this model does not incorporate the actual departure time from a given station. Even worse, travel times highly depend on the time of the day and the time needed for changing trains is also not covered by this approach. As a result, the calculated travel time between two arbitrary stations in such a graph is only a *lower bound* of the real travel time. However, in Section 4 we show that the condensed model is helpful for certain speed-up techniques.

The (Realistic) Time-Expanded Model. Throughout this work, we use the realistic time-expanded model allowing realistic queries. Therefore, three types of nodes are used to represent certain events in the timetable. *Departure* and *arrival nodes* are used to model elementary connections in the timetable. Thus, for each elementary connection $c \in C$ one arrival and departure node is created and an edge is inserted between them. To model transfers, *transfer nodes* are introduced. For each departure event one transfer node is created which connects to the respective departure node having weight 0. To ensure a minimum transfer time TRANSFER(S) at a specific station S, an edge from each arrival node u is inserted to the smallest (considering time) transfer node v where $\Delta(\text{TIME}(u), \text{TIME}(v)) \geq \text{TRANSFER}(S)$. Here $\Delta(\cdot, \cdot)$ denotes the time difference between two points in time and TIME $: V \to \mathcal{T}$ maps each node to its timestamp with respect to the timetable. Due to the periodic nature of our timetables Δ is defined by

$$\Delta(t_1, t_2) := \begin{cases} t_2 - t_1 & \text{if } t_2 \geq t_1, \\ t_2 + 1440 - t_1 & \text{otherwise.} \end{cases}$$

To ensure the possibility to stay in the same train when passing through a station, an additional edge is created which connects the arrival node with the appropriate departure node belonging to this same train. Further to allow transfers to an arbitrary train, transfer nodes are ordered non-decreasingly. Two adjacent nodes (w.r.t. the order) are connected by an edge from the smaller to the bigger node. Furthermore, to allow transfers over midnight, an overnight-edge from the biggest to the smallest node is created. For further details, see [20].

For each edge $e = (u, v)$ in the expanded graph the weight $w(e)$ is defined as the time difference $\Delta(\text{TIME}(u), \text{TIME}(v))$ of the nodes the edge connects. Hence, we call the graph consistent in time, meaning for each path from u to v in the graph, the sum of the edge weights along the paths is equal to the time difference $\Delta(\text{TIME}(u), \text{TIME}(v))$.

For future considerations the following notation will be helpful. Let $\prec \subseteq V \times V$ be a relation which compares two events in time. Since in the expanded model nodes correspond to events with a certain timestamp, our relation is defined on the set of nodes of the graph. We say for two nodes $u, v \in V$ that $u \prec v$ if the event of u is happening *before* the event of v. Please note that it cannot be determined for u and v whether $u \prec v$ just by comparing $\text{TIME}(u)$ and $\text{TIME}(v)$ due to the periodic nature of the timetable and the fact that times are always expressed in minutes after midnight. If for example $\text{TIME}(u) = 400$ and $\text{TIME}(v) = 600$ there are two possibilities. Either $u \prec v$ with $\Delta(u, v) = 200$ or $v \prec u$ with $\Delta(v, u) = 1640$. As a consequence, the Δ function applied to a tuple (u, v) is *only* valid if $u \prec v$.

3   Engineering the Time-Expanded Model

In this section, we present approaches how to enhance the classic time-expanded model. Our first attempt applies a technique deriving from road networks, i.e., contraction, to railway graphs. However, it turns out that this approach yields a too high number of edges. Hence, we also introduce the *Route-Model* which changes the modeling of "unimportant" stations.

3.1   Basic Contraction

All speed-up techniques developed during the last years have one thing in common. During preprocessing they apply a contraction routine, i.e., a process that removes unimportant nodes from the graph and adds shortcuts to the graph to keep the distances between the remaining nodes correct. Interestingly, the fastest hierarchical technique for routing in road networks, Contraction Hierarchies [7], relies *only* on such a routine. The key observation is that in road networks, the average degree of remaining nodes does *not* explode.

At a glance, one could be optimistic that contraction also works well in railway networks. Like in road networks, some nodes in time-expanded graphs are more important than others. However, contraction does not exploit the special structure of time-expanded timetable graphs. For example, departure nodes have an outgoing degree of 1. Thus, we can safely remove such nodes and add a shortcut between the corresponding transfer and arrival node. More precisely, we propose a new contraction routine consisting of three steps. In the following we explain each step separately.

Omitting Departure Nodes. The first step of our contraction routing bypasses *all* departure nodes. In [20], the authors state that departure nodes can be omitted in time-expanded graphs which can be interpreted as bypassing those nodes.

Omitting Arrival Nodes. In a second step, we bypass *all* arrival nodes within the network. As a consequence, the degree of transfer nodes highly increases. By these two steps we reduce the number of nodes by approximately a factor of 3. However, the graph still contains all original transfer nodes of which some are more important than others.

Bypass Transfer Nodes. The final step of our contraction bypasses nodes according to their degree. We bypass nodes with low degree first yielding changes in the degree of its neighbors. Our contraction ends if all transfer nodes have a total degree at least of δ, which is a tuning parameter. We suggest to use a min-heap to determine the next node to be bypassed. The key of a node x shall be $\deg_{in}(x) + \deg_{out}(x)$.

Note that we need not apply all three steps. While the first step reduces both number of nodes and edges, the following two steps yield higher edge counts. In the following, we call a time-expanded model with shortcut departure nodes, the *phase 1* model. The *phase 2* model has neither arrival nor departure nodes. If we also remove (some) transfer nodes, we call the resulting graph a *phase 3* graph. For an experimental evaluation of this contraction routine, see Section 5.

3.2 Route-Model

In our experimental studies, it turned out that our contraction routine from the last section suffers from a dramatic growth in number of edges. Already our phase 2 model has up to 3.6 times more edges than the original graph (cf. Section 5). Hence, we here introduce a different approach, called the *route model*. In contrast to contraction, we exploit certain semantic properties of the time expanded graph regarding transferring which eventually leads to a reduction of the number of shortest paths. The classic time-expanded model allows transfers at a station from each arriving train to *all* subsequent departing trains. However, when planning an itinerary by hand, we would probably do the following intuitive pruning: During the way from the source to the target station assume we find a route which leads to some station S on the way, arriving there at time t_S. Then, we would not need to examine paths toward station S with an arrival time $t'_S > t_S$, since computing these paths is redundant as we already arrived at S earlier, and we could achieve the same result by taking the earlier computed path arriving at S at t_S and then waiting at S until t'_S. This observation is the basic idea behind the route model.

Remodeling of Stations. The modifications to the (original realistic) time-expanded graph are done locally and independently for each station S, and involve the following three steps:

1. Remove all outgoing edges from all arrival nodes. This includes edges to transfer nodes as well as edges to the departure node of the same train.
2. Insert a minimal number of new transfer-edges directly from the arrival nodes to departure nodes. This allows us to model transfers more specifically without losing any optimal shortest paths in comparison to the original time expanded model.
3. Keep the transfer nodes and their interconnecting edges as well as departure-edges from transfer to departure nodes.

   Although, there are no more edges in the graph to get from an arrival node to a transfer node, the transfer nodes are still used as source nodes for the actual DIJKSTRA query. A possible alternative approach would be to use a set of departure nodes as source nodes. However, in this case these sets have to be either precomputed (which again consumes space) or computed during the query (which yields a penalty regarding query time). Thus, we decide on using the existing transfer nodes as source nodes.

The only non-trivial modification is the second one, where for each arrival node we need to find a minimal set of departure nodes which shall become reachable from the particular arrival node. For that reason let S be the currently considered station and $\mathcal{N}_S$ all *neighbors* of S. A station $T \in \mathcal{N}_S$ is called a neighbor of S if at least one elementary connection from S to T exists. Thus, we can speak of *routes* between S and each neighbor from $\mathcal{N}_S$. We now use the following notation. u denotes an arbitrary but fixed arrival node of S from which outgoing edges are inserted. v denotes the departure node toward which the edges (u, v) are inserted. Furthermore, w denotes the arrival node corresponding to the elementary connection to which the departure node v belongs. The basic idea is to insert (at least) one edge per route toward a departure node belonging to the particular route. So, let us consider some fixed station $T \in \mathcal{N}_S$ with $T \neq R$ where R is the station where we just came from through u. Of all departure nodes v belonging to an elementary connection (v, w) from S to T we insert an edge (u, v) in S according to the following criteria.

1. The node w is the smallest (regarding time) possible (meaning it is not in violation with the second criterion) arrival node at T that is after u, i.e. $w \succ u$.
2. The node v respects the transfer time criterion at S. For that reason it has to hold that $v \succ u + \text{TRANSFER}(S)$ if u and v belong to different trains, or $v \succ u$ if they share the same train.

Obviously, by this strategy we select the edge (u, v) according to the earliest possible arrival event at the *target station* T. This yields a transfer to a train which arrives at T by the earliest possible time. Note that if we instead would have chosen v according to the earliest possible departure node at S, we could have missed a different train that departs at S later, but arrives at T earlier. Such a scenario is called overtaking of trains. Also note, that if the train belonging to u utilizes the route toward station T, it does not necessarily have to be the case, that the inserted edge (u, v) corresponds to the departure event of that specific train. It simply corresponds to the train arriving at T first, which may well be a different train.

Transfer Times at Neighboring Stations. While we did respect the transfer time criterion of S, we also have to respect the transfer time criterion at T. Figure 1 shows why this is important.

On the left side the train Z_2 is required for the optimal path. However, Z_2 is arriving at T just slightly after Z_1 and it can not be transferred to, because at S only an edge toward Z_1 is inserted and at T the transfer time is too big to reach Z_2 from Z_1. On the right picture the scenario is even worse. While the train

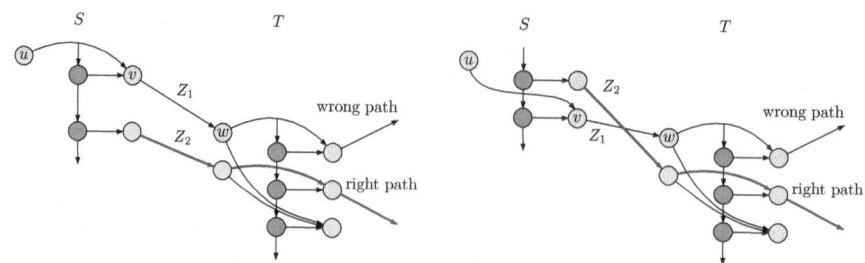

Fig. 1. Two problems concerning the transfer time criterion at station T

Z_1 is the earliest train regarding the arrival time at T, the optimal route again contains Z_2 which departs at S earlier than Z_2, but it is not reachable because it arrives at T slightly after Z_1. Again the transfer time at T is too big to enter Z_2 at T. In both cases we have to ensure that Z_2 can be entered somewhere. Since our modifications should remain local in the sense that modifications at S should not involve modifications at some other stations, we ensure that Z_2 can be reached at S.

By adding some more edges to the graph, we are able to allow those connections as well. Let w_{earl} denote the earliest arrival node at T as computed before. Then, we insert edges (u, v) (belonging to connections (v, w)) satisfying the following properties.

1. Consider all trains arriving after w_{earl} but no later than the transfer time at T, meaning $w \succ w_{\text{earl}}$ and $w \prec w_{\text{earl}} + \text{TRANSFER}(T)$.
2. Still respect the transfer time criterion at S, i.e. $v \succ u + \text{TRANSFER}(S)$ if u and v belong to different trains and $v \succ u$ otherwise.

This routine ensures that (a) it is possible to arrive at T as early as possible and (b) all trains that go through T within the margin between the earliest arrival time and the transfer time at T can be reached by entering them at S.

Uncommon Routes. Despite these modifications, we additionally have to deal with another phenomenom in railway networks. In very few cases, it might pay off to use an itinerary with a sequence of stations $R \to S \to T \to S \to R'$ instead of $R \to S \to R'$. This odd situation may arise if T and S are close to each other, a train runs from R to T, another from T to R', and $\text{TRANSFER}(S) < \text{TRANSFER}(T)$ holds. While in railway networks this case is extremely rare, it occurs more frequently in bus networks, since the average travel time between stations is significantly smaller in such networks. Moreover, we use randomly generated transfer times in our experiments (cf. Section 5) which might turn out too high for critical stations. Using real world transfer times might reduce or eliminate such paths. Figure 2 gives an example.

Our Route-Model does not allow such connections by the definitions up to now. However, we may overcome this problem by introducing edges at arrival nodes u of T toward departure nodes leading back to S if and only if the following inequation holds:

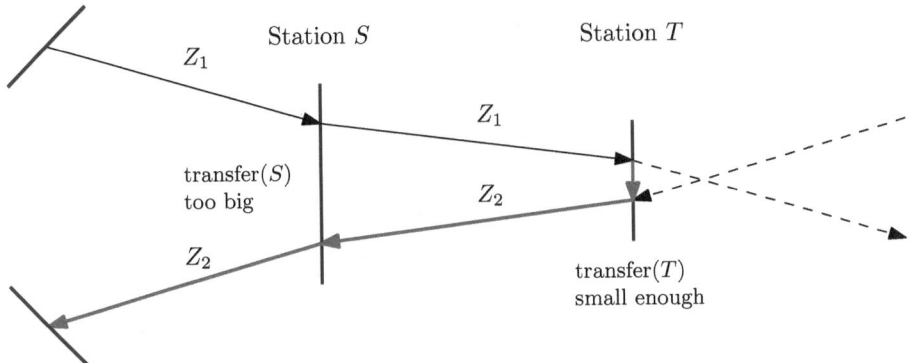

Fig. 2. Situation where it is necessary to go forth and back along the same route in order to transfer to train Z_2

$$\kappa_{S,T} + \kappa_{T,S} + \text{TRANSFER}(T) < \text{TRANSFER}(S).$$

Here $\kappa_{S,T}$ denotes the best lower bound regarding travel time from S to T. By this we ensure that no shortest paths get lost while in most cases we still get the advantage of prohibiting cycles along the same route. Please note, that we can not rule out cycles such as $\cdots \rightarrow R \rightarrow S \rightarrow T \rightarrow R \rightarrow \cdots$, however cycles of this type occur less often in general timetable networks. Generally, if we demand simple paths (with respect to visited stations), we may omit the additional edges introduced here at all.

Leaving Big Stations Untouched. It turns out that remodeling of stations with many neighbors, e.g., major train hubs, lead to a disproportionately high increase in additional edges, since for each neighbor (route) at least one edge must be inserted for each arriving train. In the original time expanded model, however, at most two edges existed for each arrival node (arrival-transfer and arrival-departure). Since our modifications are only local we can choose for each station individually whether we want to convert it to the Route-Model or not. For that reason we introduce a tuning parameter γ indicating that stations with more neighbors than γ should be left untouched. Hence, changing γ yields a trade-off between a speed-up regarding the number of touched nodes against an increasing size of the edge set of the graph.

A problem that arises when mixing Route-Model stations with classic stations is that the main advantage of the Route-Model—subsequent connections on the same route are not visited during the DIJKSTRA search—may fade. Analyzing the example in Figure 3, we observe a big station which has not been converted followed by a route containing a few small stations. While at the small stations no connections exist between connections of the same route, they are nevertheless visited, because they are all accessible through the big station. Hence, we developed *Node-Blocking* which adopts the idea behind the Route-Model as a speed-up technique, and blocks redundant connections of the same route, so they are not visited. This technique is explained in Section 4.

Big Station   Small Station   Small Station

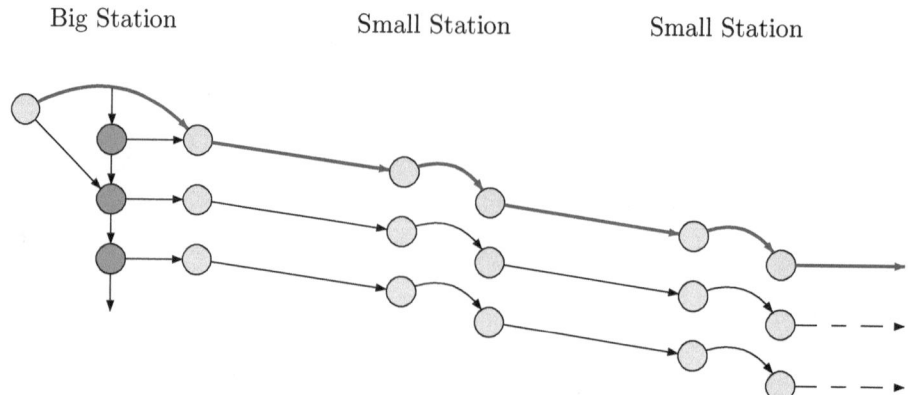

Fig. 3. When a big station which is not converted is visited during a DIJKSTRA query, all subsequent connections are visited as well, while only the red path should be relevant. Unimportant nodes are omitted in the figure.

3.3 Correctness of the Route-Model

In this section we provide an extensive correctness proof of our Route-Model, i.e. we show that applying DIJKSTRA on the Route-Model still yields correct solutions to the earliest arrival problem.

In order to conduct our proof we need to introduce some notions first. Let Π be a path in a time-expanded railway graph. Then Π covers a *sequence of stations* $\mathbf{S} = S_1 \to S_2 \to \cdots \to S_n$. A sequence of the form $S_1 \to S_2 \to \cdots \to S_{k-1} \to S_k \to S_{k-1} \to \cdots S_2 \to S_1$ is called a *cycle*. Note, that there might be more complicated "cycles" like for example $S_1 \to S_2 \to S_3 \to S_1$, but we restrict ourselves to the simple cycles as defined above. A sequence $\mathbf{S}'$ is said to be *contained* in a sequence $\mathbf{S}$, if $\mathbf{S}'$ is part of $\mathbf{S}$, i.e. the sequence $\mathbf{S}'$ occurs at some place in $\mathbf{S}$. A cycle $\mathbf{S}$ is called *dispensable* if it holds that

$$\sum_{i=1}^{k-1} \kappa_{S_i,S_{i+1}} + \text{TRANSFER}(S_k) + \sum_{i=1}^{k-1} \kappa_{S_{i+1},S_i} \geq \text{TRANSFER}(S_1).$$

Here $\kappa_{R,S}$ for two stations R and S, again, denotes the minimal travel time from R to S. Now, a sequence $\mathbf{S}$ is called *minimal*, if it does not contain any dispensable cycles. A minimal sequence can be constructed from any (non-minimal) sequence by removing every dispensable cycle from it. Think of it as continuing the journey at S_1 (of the cycle) directly instead of going through the cycle first. Since the minimal travel time of the cycle is longer than the transfer time at S_1, this is always possible.

First, we now prove the following lemma, which is essential to the proof of correctness.

Lemma 1. *Each minimal sequence of stations in the realistic time expanded graph is also contained in the Route-Model graph.*

Proof. Let **S** be a minimal sequence in the time expanded graph. If **S** does not contain any cycles, then it is trivially contained in the Route-Model by its construction rules: At each station S_i for each neighbor edges are introduced to connect to them, just as well toward S_{i+1}.

If there is a non-dispensable cycle in the sequence **S**, then the only place where no edges might be in the Route-Model graph is at the turning point S_k of the cycle. The sub-cycle $S_{k-1} \rightarrow S_k \rightarrow S_{k-1}$ must be non-dispensable itself, otherwise it would not be contained in **S**. For that reason, it must hold that

$$\kappa_{S_{k-1},S_k} + \text{TRANSFER}(S_k) + \kappa_{S_k,S_{k-1}} \quad < \quad \text{TRANSFER}(S_{k-1}).$$

But this is exactly the criterion for which edges back to S_{k-1} are inserted in the Route-Model. Hence, the path **S** is also contained in the Route-Model. □

We can now deduce the main correctness theorem.

Theorem 1. *Applying* DIJKSTRA *on the Route-Model yields correct solutions to the earliest arrival problem.*

Proof. We prove the theorem in two steps. First, we show that each shortest path in the Route-Model is also contained in the original time expanded model and second, we show the reverse, that for each shortest path in the expanded model there is an equivalently long shortest path in the Route-Model.

Route Model → *Classic Model.* Let Π be an arbitrary (shortest) path in the Route-Model covering a sequence **S** of stations. The first construction step, namely the removal of edges does not create any new paths in the Route-Model, so by that argument Π is also contained in the classic expanded graph. For the second construction step (the appropriate insertion of new outgoing edges from the arrival nodes) does not lead to any new shortest paths either. Since an edge $e = (u, v)$ at some station S_i is only inserted if it does not violate the transfer time criterion, it always corresponds to a valid path in the classic time expanded graph. If no trains are changed through e, then e is exactly the *train-edge* from the arrival to the departure node of that train. If trains are changed through e, then by the construction rules it holds that $u + \text{TRANSFER}(S_i) \prec v$. But, in this case there is also a path (through some transfer nodes) from u to v in the classic graph. By that reason, there are no shorter paths in the Route-Model than in the classic model.

Classic Model → *Route Model.* We now show that no shortest paths get lost by the Route-Model. We prove this by contradiction. Let Π be a shortest path of length λ retrieved by some query from S_1 to S_n at departure time $t_d(S_1)$. Assume that the shortest path Π' computed in the Route-Model for the same query has length $\lambda' > \lambda$. Then there are two possibilities.

1. The sequence of stations covered by Π and Π' are identical.
   Then it must hold that at some station S_i in the classic model we entered a train Z_f that arrives at S_n earlier. We assume without the loss of generality

that S_i is the latest possible station (meaning the nearest from the target station) where we entered the faster train Z_f. Because there is no possibility to enter Z_f at a later point, for all subsequent stations $S_{i+1}, \ldots, S_n$ it must hold that either Z_f arrives there before the slower train Z_s (computed by the Route-Model), or that it arrives after Z_s but within the margin of the transfer time at the particular station (otherwise S_i would not be the latest possible station to switch to Z_f).

Let w_f be the arrival node of Z_f and w_s be the arrival node of Z_s at the next station S_{i+1}. In the first case if $w_f \prec w_s$ there must have been an edge inserted in the Route-Model to board Z_f at S_i, because there is always an edge inserted to the train arriving at S_{i+1} earliest. In the second case if $w_s + \text{TRANSFER}(S_{i+1}) \succ w_f$, there is also an edge inserted at S_i to board Z_f, because edges to all trains along the route are inserted that arrive at S_{i+1} within the margin of $\text{TRANSFER}(S_{i+1})$. Hence it is possible to board Z_f at S_i in the Route-Model which is the desired contradiction.

2. The sequence of stations covered by Π and Π' are not identical.
   Let us call the sequences $\mathbf{S}$ and $\mathbf{S}'$. Because of Lemma 1 there must also exist a (potentially longer) path along $\mathbf{S}$ in the Route-Model. If we substitute Π' for that path, this case can be reduced to the first one leading to the desired contradiction.

Thus, the two models are equivalent in the sense that (a) no shorter itineraries can be computed in the Route-Model and (b) for each (shortest) itinerary computed in the classic model there is an equally short itinerary computable in the Route-Model. □

4   Speedup Techniques

In principle, we could use DIJKSTRA's algorithm for solving the EAP. However, plain DIJKSTRA visits unnecessary parts of the graph, even if we use our Route-Model. Hence, we introduce two approaches for obtaining faster query times. We adapt existing techniques—developed for road networks—to timetable graphs and introduce a new speed-up technique following the ideas from our Route-Model.

4.1   Tailored Speed-Up Techniques

Node-Blocking is a speed-up technique tailored to time-expanded networks. It basically incorporates the ideas behind the Route-Model as described in Section 3.2: if we can reach a station S at some time t_S we try to prune paths reaching S at a later time $t'_S > t_S$. Recall that the Route-Model prunes the search by removing certain edges from the graph. Node-Blocking, on the contrary, achieves a similar result by dynamically blocking departure nodes during the DIJKSTRA query. The idea is as follows. If we visit a departure node v belonging to an elementary connection targeting some station T, we can *prune* all future departure nodes b targeting T.

Preprocessing. Formally, each departure node v of an elementary connection between two stations S and T induces a set B_v of blocked nodes. A node b is contained in B_v if and only if the following conditions hold.

1. b is a departure node at S belonging to an elementary connection targeting the same station T as v.
2. $b \succ v$ holds.
3. If w and c are the arrival nodes at T of the connections associated with v and b, respectively, then $w + \text{TRANSFER}(T) \prec c$ must hold, i.e., we respect the transfer time criterion at T.

Although the "blocked state" of each node is dynamic in the sense that it depends on the shortest path query, and therefore must be computed during the query, the set B_v of inducing blocked nodes can be precomputed for each node v by iterating through all departure nodes of the station and checking whether the above criteria apply to them.

Note that in contrast to the Route-Model, we do not have to deal with the transfer time criterion at S, since we only *block* nodes, and hence never allow a path to be taken which was forbidden by the transfer time criterion at S. In worst case, we block departure nodes which cannot be reached anyway due to the transfer time criterion of S. Moreover, all special cases are covered by our third condition.

Query. The modifications to standard DIJKSTRA algorithm are simple. We introduce an additional flag blocked(v) to all nodes of the graph, which is initialized to false. Then, whenever we try to insert a node v into the queue, we mark all nodes B_v as blocked. If v is marked as blocked, we prune the search.

Efficiently Storing Blocked Nodes. Storing the set B_v for each node is highly redundant, i.e., one particular departure node w at a station S is an element of many sets B_v of departure nodes at S. Thus, let for each departure node v the node $v_b \in B_v$ be the minimum node for which $v \prec v_b$ holds. In other words, v_b depicts the first blocked connection from the set B_v. Then for each node v we only store v_b as the *block target* of the node v, thus, reducing space consumption.

Regarding the query, when a node v is about to be inserted into the queue, we set blocked(v_b) to true, check for blocked(v), and do not insert v into the queue if and only if blocked(v) = *true*. By these means, we do not block the whole set B_v at once, but all nodes from B_v are eventually blocked bit by bit during the execution of the algorithm. The main loop of DIJKSTRA only needs the additions shown in red color at Algorithm 1.

Regarding correctness, we like to point out that we do not lose paths by this approach. Since it holds for each node v that by blocking v_b we always block a node that would have also been blocked by the original approach (because $v_b \in B_v$), our proof of correctness is applied to the stronger variant of blocking the whole set B_v.

Algorithm 1. Excerpt of DIJKSTRA's main loop

```
1  u ← a settled node
2  forall outgoing edges e = (u, v) do
3  |   if v is a new node then
4  |   |   if v_b ≠ nullnode then
5  |   |   └   blocked(v_b) ← 1
6  |   |   if blocked(v) = 1 then
7  |   |   └   continue
   |   |   // insert operation here
8  |   else
   |   └   // decreaseKey operation here
```

Combination with Route Model. Although our Route-Model and Node-Blocking follow the same ideas, the advantage of the Route-Model is the lower computational overhead during the query. However, as discussed in Section 3.2, it does not pay off to remodel major hubs. Hence, Node-Blocking harmonizes well with the Route-Model as we use Node-Blocking for pruning paths at such hubs.

Combination with Phase 1+ Models. Since from the Phase 1 model onwards departure nodes are removed, Node-Blocking has to be altered slightly to conform with these models. Instead of departure nodes blocking future departure nodes, we simply let the corresponding arrival nodes (belonging to the respective departure nodes) block each other. In this case, the arrival nodes assume the role of the previous departure nodes regarding blocking, which allows us to continue using the same query algorithm.

Correctness of Node-Blocking. We assume at this point that the reader is familiar with the notions introduced during the correctness proof of the Route-Model in Section 3.3, in particular with the terms sequence and cycle. However, we do not restrict ourselves to *simple cycles* here. The term *dispensable cycle* can be generalized to any cycle $\mathbf{S} = S_1 \to \cdots \to S_n \to S_1$ if it holds that

$$\sum_{i=1}^{n-1} \kappa_{S_i, S_{i+1}} + \kappa_{S_n, S_1} \geq \text{TRANSFER}(S_1).$$

Please note, that this condition does not contain transfers along the cycle, because we do not necessarily have a unique "turning point" that induces a transfer.

Theorem 2. *Applying Node-Blocking to* DIJKSTRA*'s algorithm yields correct solutions to the earliest arrival problem.*

Proof. We conduct this proof in two steps. First, we show that each shortest path with Node-Blocking enabled is also a path without Node-Blocking. Second, we show that a shortest path without Node-Blocking due to a minimal sequence of stations is also computable with Node-Blocking enabled.

Node-Blocking → Without Node-Blocking. This can easily be seen. Since Node-Blocking blocks nodes when they are about to be inserted into the priority queue, we can see this as dynamically deleting edges from the graph (namely the edges pointing to blocked nodes) during the query. Obviously, the graph emerging at the end of the query is a subgraph of the original graph, hence the computed path is also contained in the original graph without Node-Blocking enabled.

Without Node-Blocking → Node-Blocking. Let Π be a shortest path covering a minimal sequence of stations $\mathbf{S}$ computed by DIJKSTRA. Note again, that for *any* shortest path Π' covering a non-minimal sequence $\mathbf{S}'$ we can construct a minimal sequence $\mathbf{S}$ by removing each dispensable cycle. This directly induces the desired path Π. Then the following two statements hold.

1. *A path Π_B with Node-Blocking enabled covering the same sequence exists.* The default blocked-state of all departure nodes is *false*. Therefore, when we arrive at some station S_i along the sequence $\mathbf{S}$ on our path, the first departure node leading to S_{i+1} is not blocked when it is inserted into the priority queue (Note, a node never blocks itself, so this is even true if only one connection toward S_{i+1} existed). For that reason, there exists a path from S_i to S_{i+1}. Note, that due to the minimal nature of the sequence $\mathbf{S}$ the subsequence $S_i \rightarrow S_{i+1}$ is not contained again in $\mathbf{S}$ at a future point with one exception: The travel time of the cycle (beginning with S_{i+1}) is longer than TRANSFER(S_{i+1}), but in this case the departure node belonging to the respective connection arriving within the margin of TRANSFER(S_{i+1}) is not blocked.

2. *Π_B is a shortest path.* Assume $t_d(S_i)$ is the first time the DIJKSTRA algorithm discovers a departure node u along the route $S_i \rightarrow S_{i+1}$ in our sequence. Let furthermore $t_a(S_i)$ be the arrival time at S_{i+1} of the connection belonging to that departure node. Now assume further, that the optimal route continues at S_{i+1} at some point $t_d(S_{i+1}) \succ t_a(S_{i+1}) + \text{TRANSFER}(S_{i+1})$. Then taking the non-blocked connection through u and waiting at S_{i+1} yields an optimal subpath from S_i to S_{i+1}. If the optimal journey continues at S_{i+1} within the margin of transfer time, i.e. $t_d(S_{i+1}) \prec t_a(S_{i+1}) + \text{TRANSFER}(S_{i+1})$ then we ensure that the respective connections arriving within that margin are not blocked by u, hence the optimal subpath from S_i to S_{i+1} is prevailed as well. Since S_i and S_{i+1} were arbitrary sections along the (optimal) sequence $\mathbf{S}$, the computed path Π_B is a shortest path.

From this follows that Node-Blocking yields correct shortest path queries w.r.t. the earliest arrival problem.   □

4.2 Adapting Speed-Up Techniques

Although the adaption of many techniques may be promising, we choose basic goal-directed techniques for adaption. It turned out that adaption of more sophisticated techniques, e.g., Highway Hierarchies [21], Contraction Hierarchies [7], REAL [9], SHARC [1], is much more challenging than expected. The main reason are either the need of a bidirectional query algorithm or the bad performance of the contraction routine.

Arc-Flags. The classic Arc-Flag approach, introduced in [13, 14], first computes a partition $\mathcal{P}$ of the graph and then attaches a *label* to each edge e. A label contains, for each cell $P_i \in \mathcal{P}$, a flag $AF_{P_i}(e)$ which is *true* if a shortest path to at least one node in P_i starts with e. A modified DIJKSTRA—from now on called Arc-Flags DIJKSTRA—then only considers those edges for which the flag of the target node's cell is *true*. The big advantage of this approach is its easy query algorithm. However, preprocessing is very extensive. The original approach grows a full shortest path tree from each boundary node yielding preprocessing times of several weeks for instances like the Western European road network. Recently, a new *centralized* approach has been introduced [12]. However, it turns out that this centralized cannot be used in time-expanded transportation networks due to memory consumption. Hence, we use the original approach of growing full shortest path trees from each node.

Adaption. The query algorithm can be adapted to time expanded railway graphs very easily. We only have to consider that the exact target node is unknown (just the target station is known). For that reason we simply abort the DIJKSTRA algorithm as soon as a node belonging to the target station is settled. The preprocessing of Arc-Flags, however, needs some extra attention. Since we do not know the exact target node in advance, we have to ensure that all nodes belonging to the same station also get the same cell-id of the partition assigned. For that reason, we simply compute the partition on the condensed graph and map it to the expanded graph by assigning for each node $v \in V$ the cell-id due to cell(v) := cell(STATION(v)).

Computing the backwards-shortest path trees from each boundary node of each cell can then be done as described in [14]. However, this approach yields a problem specific on time expanded graphs. Since the length of any path in the graph always corresponds to the time needed to travel between the beginning and ending event (node) of that particular path, any two different paths between the same nodes *always* have the same length. Therefore, the number of shortest paths (in fact, there are *only* shortest paths in time expanded graphs) is tremendous. Unfortunately, if we set flags to true for every path, we do not observe any speed-up (cf. Section 5). In order to achieve a speed-up we have to prefer some paths over others. We examine the following four reasonable strategies for prefering paths:

- **Hop Minimization**
  For two paths of equal length, choose the one that has less hops (nodes) on it. This approach is often used in road networks [1].
- **Transfer Minimization**
  Choose the path that has less transfers between trains. While this is a good strategy for querying, it sets too many arc-flags to true, since for

different boundary nodes too many different paths lead a transfer-minimal connection.

- **Distance Minimization**
  Choose the path that is shorter (geographically).
- **Direct Geographical Distance**
  Choose the path whose direct geographical distance is closer to the source node of the shortest path tree, formally for some node v that is reached from u we choose the new predecessor according to

$$\text{pre}(v)_{\text{new}} := \underset{w \in \{u, \text{pre}(v)\}}{\text{argmin}} \{\sqrt{\left(\text{cd}_x(w) - \text{cd}_x(s)\right)^2 + \left(\text{cd}_y(w) - \text{cd}_y(s)\right)^2}\},$$

where s is the source node of the shortest path tree and cd depicts the x, respective y coordinate values for a given node. This optimization is very aggressive, as it leads to the same result for different boundary nodes of the same cell as often as possible.

Section 5 shows the huge difference in the query performance when the arc-flags are computed with different strategies. Note that we can optimize query times by setting as many flags as possible to false. However, we also loose the ability to choose the "best" path during the query (e.g. due to a minimal number of transfers, costs, etc.). This yields a trade-off between query time and the quality of the computed itineraries.

Arc-Flags and Node-Blocking. Unfortunately, Node-Blocking does not harmonize with Arc-Flags. This is due to the fact of Node-Blocking being a very aggressive technique, leaving only very few connection arcs per station and route accessible. The optimization criterion hereby, namely arriving as early as possible at the next station does not necessarily match with our path selection during Arc-Flags preprocessing. As a result, both techniques prune different shortest paths. A possible solution would be to adapt the path selection for Arc-Flags according to Node-Blocking. However, this turns out to be complicated as we have to grow shortest path trees on the reverse graph. Hence, this path selection strategy is not implemented yet.

ALT. Goal directed search, also called A^* [11], pushes the search towards a target by adding a *potential* to the priority of each node. The ALT algorithm, introduced in [8], uses a small number of nodes—so called *landmarks*—and the triangle inequality to compute such feasible potentials. Given a set $L \subseteq V$ of landmarks and distances $d(\ell, v), d(v, \ell)$ for all nodes $v \in V$ and landmarks $\ell \in L$, the following triangle inequations hold: $d(u, v) + d(v, \ell) \geq d(u, \ell)$ and $d(\ell, u) + d(u, v) \geq d(\ell, v)$. Therefore, $\pi(u, t) := \max_{\ell \in L} \max\{d(u, \ell) - d(t, \ell), d(\ell, t) - d(\ell, u)\}$ provides a lower bound for the distance $d(u, t)$ and, thus, can be used as a potential for u.

Adaption. The query algorithm is, again, straight forward to adapt to time-expanded railway graphs. Since the only difference to the standard DIJKSTRA algorithm is the key which is inserted into the priority queue, we can still simply abort the search as soon as a node of the target station gets settled. However, we cannot compute the landmarks on the expanded graph directly since then we would have to know the target node t in advance. Hence, we compute the landmarks on the much smaller condensed graph which still yields feasible potentials because the edge weights in the condensed graph are defined as the lower bounds regarding travel time. The potential function π during the query is then computed as follows:

$$\pi(v) = \max_{\ell \in L} \max\{ \operatorname{dist}(\operatorname{STATION}(v), \ell) - \operatorname{dist}(T, \ell),$$

$$\operatorname{dist}(\ell, T) - \operatorname{dist}(\ell, \operatorname{STATION}(v))\},$$

where T is the target station of the query. We can think of this as using a "lower bound of a lower bound" of the shortest path.

Former studies revealed that the selection of landmark nodes is crucial to the performance of ALT. The quality of the lower bounds highly depends on the quality of the selected landmarks. Thus, several selection strategies exist. To this point, no technique is known how to pick landmarks yielding the smallest search space for random queries. Thus, several heuristics exist. The best are *avoid* and *maxCover*. The first tries to identify regions that are not well covered by landmarks while the latter is basically the avoid routine followed by a local optimization. For details, we refer to [10].

Due to the small size of the condensed networks, another strategy for obtaining potentials seems promising. For each query, we use the target station T as landmark and compute the distances of all stations to T on-the-fly. The advantage of this *dynamic-landmark-selection* is a tighter lower bound. However, we have to run a complete DIJKSTRA in the condensed graph for each query which can take more time than using worse lower bounds from landmarks during the query. Note that this approach for obtaining lower bounds for A* was already proposed in [15].

Combining Arc-Flags and ALT. In [17], we observed that Arc-Flags (with the direct geographical distance strategy) and ALT optimize in two different ways. While Arc-Flags prunes paths that lead to the wrong direction geographically, ALT optimizes in time in the sense that fast trains are preferred over slow trains. Fast trains (having less stops in between) tend to get near the target station faster, yielding a lower key in the priority queue regarding the lower bound function. For that reason, it is suggestive to examine the combination of the two speed-up techniques. The implementation is straight-forward, since Arc-Flags does not interfere with ALT—Arc-Flags simply ignores edges that do not have their appropriate flag set, and ALT just alters the key in the priority queue.

5   Experiments

In this section, we present our experimental evaluation. Our implementation is written in C++ using solely the STL. As priority queue we use a binary heap.

Our tests were executed on one core of an AMD Opteron 2218 running SUSE Linux 10.3. The machine is clocked at 2.6 GHz, has 16 GB of RAM and 2 x 1 MB of L2 cache. The program was compiled with GCC 4.2, using optimization level 4.

Inputs. We use two inputs for our evaluation. The *railway* network of Central Europe and a local *bus* network of greater Berlin. Both networks have been provided by HAFAS for scientific use; the former network consists of 30,517 stations and 1,775,552 elementary connections. The corresponding figures for the latter are 2,874 and 744,005, respectively. While the network of Europe provides a good average structure for a railway network mixed of long-distance trains supported by short-distance trains, the bus network of Berlin consists of a very homogeneous structure, since there are almost no "long-distance" buses. Because of this and the very dense operations of buses with their short travel times between stations, it has already been shown [17] that this network seems to be a very hard instance for timetable information queries.

It should be noted that, while our timetable data is realistic, the transfer times at the stations were not available to us. Hence, we generated them at random and chose between 5 and 10 minutes for the railway and between 3 and 5 minutes for the bus network.

Default Settings. In the following, we report preprocessing times and the overhead of the preprocessed data in terms of *additional* bytes per node. We evaluate query performance by running 1 000 random s–t queries with source and target station picked uniformly at random. We *fix* the departure time to 7:00 am. We report the average number of settled nodes during the query as well as the average query time. The speed-up refers to the query time and is computed in reference to the classic time expanded model without any speed-up technique applied.

5.1   Models

Parameters. We start our experimental evaluation with parameter tests for our Route-Model. Recall that in the Route-Model we may affect the conversion process by the selection of γ which controls the maximum number of neighbors a station may have in order to become a Route-Model station. In the following we use values between 2 and 10 for γ. Table 1 reports for both our inputs: the resulting size (in terms of number of edges) and query performance. Note that we do not report number of nodes, as the remodeling routine does not add or remove any nodes. We also enabled Node-Blocking (see Section 4.1).

We observe that for both instances the Route-Model yields a speed-up. Increasing γ up to 5 increases performance, while values > 5 do not pay off. This is mostly due to the fact that for both graphs the majority of stations has less or equal than 5 neighbors (91% for the Europe and even 99% for the Berlin network).

Concerning Europe with $\gamma < 5$, we observe that the resulting graph has *less* edges than originally. Recall in the original graph the number of outgoing edges

Table 1. The effect of γ on the performance of the Route-Model with Node-Blocking enabled

	europe				bvb			
	SIZE	QUERY			SIZE	QUERY		
γ-value	#edges	#settled	[ms]	speed-up	#edges	#settled	[ms]	speed-up
reference	8,505,951	1,161,696	534.7	1.00	3,694,253	151,379	37.6	1.00
2	7,912,584	411,836	202.4	2.64	3,785,680	91,591	27.4	1.37
3	8,035,324	359,294	171.7	3.11	4,292,849	74,963	25.2	1.49
4	8,332,816	329,413	158.3	3.38	5,059,228	63,438	25.1	1.50
5	8,729,619	313,046	154.1	3.47	5,437,647	59,670	25.4	1.48
6	9,071,974	303,460	153.9	3.47	5,625,277	57,990	25.6	1.47
7	9,396,276	297,831	155.1	3.45	5,768,926	56,994	25.8	1.46
8	9,712,940	292,482	156.4	3.42	5,782,375	56,921	25.7	1.46
9	9,936,119	289,036	158.7	3.37	5,782,375	56,921	25.8	1.46
10	10,195,050	285,103	159.3	3.36	5,782,375	56,921	25.8	1.46

per arrival node is at most 2 (one toward the nearest transfer node and one toward the departure node of the same train). Hence, a decrease in number of the edges can only result from merely one edge being inserted for many arrival nodes at stations of degree 2. Interestingly, this observation of decreasing edges does not hold for our bus network which is due to the high density of the network: Because the stations are very close to each other, it often holds that the travel time to go forth and back between some stations S_1 and S_2 is less than TRANSFER(S_1), which results in back-edges being inserted for arrival nodes at S_2 (coming from S_1). Second, the operation frequency of the buses is very high, such that it may occur that edges toward more than the first bus of the route are inserted, when they arrive at the next station within the margin of its transfer time.

Summarizing, a value of $\gamma = 5$ yields the best results for railway inputs. The corresponding figure for the bus networks is 4.

Comparison to the Classic Time-Expanded Model. Next, we compare different contraction steps (Section 3) and our route model with the classic time expanded model. Table 2 shows the differences in graph size and query performance. While the overall graph size decreases when switching from the classic expanded to the phase 1 model, the number of edges significantely increases if applying our phase 2 model. Although the number of nodes decreases about 50%, this increase in number of edges leads to an *worse* query performance, since more edges are relaxed during the query. We hence conclude that the phase 2 model—and therefore the phase 3 model as well—is not the preferred choice for fast timetable queries.

Regarding the Route-Model, the increase in graph size is still reasonable while the query time decreases. However, we see that the query performance benefits from Node-Blocking as the speed-up more than doubles in the Europe network with Node-Blocking enabled. The reason for the weak performance without Node-Blocking is that paths through the graph that should be pruned by the Route-Model approach, are still relaxed when they are not blocked in non-converted big traffic hubs. In the bus network the general performance gain is

Table 2. Comparison of the different models. The Route-Model is computed with $\gamma = 5$ for *europe* and $\gamma = 4$ for *bvb*.

input	Model	SIZE		QUERY		
		#nodes	#edges	#settled	[ms]	spd-up
	Classic expanded	5,207,980	8,505,951	1,161,696	534.7	1.00
	Phase 1	3,472,022	6,769,991	768,181	426.5	1.25
	Phase 2	1,736,064	15,571,190	431,274	631.1	0.85
europe	Route	5,207,980	8,729,619	793,462	360.6	1.48
	Route w/ blocking	5,207,980	8,729,619	313,046	154.1	3.47
	Route + Phase 1	3,472,018	6,821,337	439,024	256.3	2.09
	Route + Phase 1 w/ blocking	3,472,018	6,821,337	200,213	122.8	4.35
	Classic expanded	2,232,016	3,694,253	151,379	37.6	1.00
	Phase 1	1,488,011	2,950,248	99,253	29.1	1.29
	Phase 2	744,006	13,229,482	60,218	56.8	0.66
bvb	Route	2,232,016	5,059,228	97,978	32.6	1.15
	Route w/ blocking	2,232,016	5,059,228	63,438	25.1	1.50
	Route + Phase 1	1,488,011	3,918,788	51,210	22.7	1.66
	Route + Phase 1 w/ blocking	1,488,011	3,918,788	34,032	18.6	2.02

not as big as with the railway network. Even Node-Blocking does not have such a great impact, which is mostly due to the very dense structure of this network.

Because the Route-Model can be combined well with the phase 1 model (departure nodes are simply removed after the conversion to the Route-Model), this gives us a gain in graph size while still keeping the advantages of the Route-Model. The query performance behaves as expected and increases by approximately one third compared to the Route-Model alone. If we then additionally apply Node-Blocking on the route + phase1 model, we get the best query performance of all the models which yields a speed-up of 4.35 in the railway network of Europe and 2.02 in the Berlin bus network.

5.2   Speedup Techniques

Up to now, we showed that by remodeling stations and using additional pruning techniques, we already achieve a speed-up of 4.35 over plain DIJKSTRA. Here, we now show that this approach harmonizes well with other speed-up techniques deriving from road networks.

Path-Selection during Arc-Flags Preprocessing. We already noted in Section 4.2 that in expanded timetable networks the number of shortest paths between two nodes is enormously high. It turns out that setting arc-flags for all paths yields a bad query performance. Hence, we have to favor some paths over the others. We proposed four different reasonable strategies: Minimize hops, minimize transfers, minimize accumulated geographic distance along the path and finally minimize the direct geographic distance from the preceding node to the source of the shortest path tree (see Section 4.2). Table 3 shows the impact of each strategy on the performance of Arc-Flags. Note that due to the long preprocessing times of Arc-Flags, we use a subnetwork of our European instance, namely the German railway network called *de_fern* (6822 stations and 554996 connections).

While minimizing hops is useful in road networks [1] (which can be interpreted there as preferring a route that has less road crossings) this results in a poor performance in railway network. Almost all flags are opened during preprocessing, thus the overhead of the Arc-Flags query algorithm outweighs the benefit from the few remaining pruned arcs. Interestingly, using minimal transfer or minimal distance strategies as path selection yields a poor query performance as well. This is mostly due to too many different paths of boundary nodes of the same cell being optimal, thus too many flags are set to *true*. Recall that the partition is computed on the condensed graph, hence for one station that is at the border of a cell, nodes belonging to all times of day are boundary nodes which may lead to very different transfer or distance minimal routes in the graph.

The minimal direct geographic distance strategy overcomes this issue by *always* choosing the same preceding node for *all* times of the day. For that reason, as many arc-flags as possible are kept *false*, which eventually yields a speed-up of 3.87 on the German railway network. Since all other strategies actually worsen the query performance, we choose the direct geographic distance strategy for further experiments involving Arc-Flags on time expanded railway networks.

Speed-Up Techniques on our Models. In the next experiment we compare the performance of the adapted speed-up techniques on the different models from Section 3. Because of the bad performance of the phase 2 model, we only compare the classic expanded model, the phase 1 model, the Route-Model and the combination of the route and phase 1 models.

Furthermore, we tested the effect of dynamic-landmark-selection against a precomputed set of landmarks. Table 4 shows our results. We show the query performance as well as preprocessing-costs by preprocessing time and additionally bytes per node required to store the preprocessed data. For each model we tested the following speed-up techniques:

- **BA:** Node-Blocking with ALT.
- **BdA:** Node-Blocking with ALT and dynamic-landmark-selection.
- **uFA:** Unidirectional Arc-Flags with ALT.
- **uFdA:** Unidirectional Arc-Flags with ALT and dynamic-landmark-selection.

Table 3. Arc-Flags. Evaluation of different path-selection strategies. For each strategy we apply a partition with 64 cells.

Strategy	PREPRO [h:m]	[B/n]	QUERY #settled	[ms]	speed-up
reference	—	0	152,998	58.1	1.00
hops	17:00	26.2	149,931	70.3	0.83
transfers	16:26	26.2	152,307	71.7	0.81
distance	20:53	26.2	134,462	61.8	0.94
geo. dist. to target	16:08	26.2	38,511	15.0	3.87

Table 4. Comparing different models in conjunction with the classic speed-up techniques. The parameter set used throughout: 128 cells, *geographic distance to target* path-selection-strategy for Arc-Flags and 8 landmarks using *maxCover* for the classic ALT algorithm.

	europe					*bvb*				
	PREPRO		QUERY			PREPRO		QUERY		
Model/Algo	[h:m]	[B/n]	#settled	[ms]	spd	[h:m]	[B/n]	#settled	[ms]	spd
Reference	—	0	1,161,696	534.7	1.00	—	0	151,379	37.6	1.00
Classic (BA)	≈ 4 s	4.0	261,151	162.7	3.29	≈ 2 s	4.1	96,533	33.6	1.12
Classic (BdA)	≈ 1 s	4.0	233,280	130.8	4.09	≈ 1 s	4.0	94,345	29.1	1.29
Classic (uFA)	106:11	106.5	71,937	32.7	16.35	45:30	108.0	49,921	17.0	2.21
Classic (uFdA)	106:11	106.5	65,143	33.9	15.77	45:30	107.9	49,014	15.2	2.47
Phase 1 (BA)	≈ 5 s	4.5	208,579	145.5	3.67	≈ 2 s	4.1	67,019	26.1	1.44
Phase 1 (BdA)	≈ 1 s	4.0	185,996	116.4	4.59	≈ 1 s	4.0	65,488	22.8	1.65
Phase 1 (uFA)	77:52	127.2	30,583	14.0	38.19	31:59	129.0	15,004	5.4	6.96
Phase 1 (uFdA)	77:52	126.7	27,310	18.5	29.06	31:59	128.9	14,713	5.1	7.37
Route (BA)	< 4 s	4.4	140,826	73.2	7.30	≈ 2 s	4.1	49,591	22.3	1.69
Route (BdA)	≈ 1 s	4.0	127,444	65.4	8.18	≈ 1 s	4.0	48,390	19.8	1.90
Route (uFA)	85:49	109.7	50,050	22.1	24.19	50:58	147.1	25,289	10.2	3.69
Route (uFdA)	85:49	109.3	45,180	25.3	21.13	50:58	147.0	24,785	9.3	4.04
Rt/Ph 1 (BA)	≈ 4 s	4.5	89,524	58.7	9.11	< 2 s	4.1	26,653	16.0	2.35
Rt/Ph 1 (BdA)	≈ 1 s	4.0	80,665	52.8	10.13	≈ 1 s	4.0	26,007	14.8	2.54
Rt/Ph 1 (uFA)	83:58	128.2	20,044	9.5	56.28	34:56	170.6	6,195	2.6	14.46
Rt/Ph 1 (uFdA)	83:58	127.7	17,805	15.2	35.18	34:56	170.5	6,053	2.8	13.43

Regarding classic ALT we always used a set of 8 precomputed landmarks by the *maxCover* [10] method. Arc-Flags were computed using a partition of 128 cells obtained from *SCOTCH* [18]. The strategy for path-selection was *geographic distance to target*. Note that for Arc-Flags, we turn off Node-Blocking (cf. Section 4.2).

We observe, that for all speed-up technique our modifications to the classic expanded model yield improvements regarding both query performance and preprocessing time. While the transition from the classic to the phase 1 model is more beneficial for Arc-Flags than ALT with Node-Blocking, the latter performs better on the Route-Model where Node-Blocking fits the model considerably better. The combination "Route + Phase 1" unifies the advantages of each model yielding the best speed-ups.

In general, Arc-Flags has a higher impact on the query time than ALT together with Node-Blocking (about 5.5 times faster on both networks) which is being paid for with very high preprocessing time and roughly 30 times more required space per node. Note that the dynamic ALT comes for free, as it does not require any preprocessing at all. With our modified models we can, however, still achieve a speed-up of 10.13 in Europe and 2.54 in Berlin with dynamic ALT and Node-Blocking, which is useful in a scenario where preprocessing is limited or not allowed.

Table 5. Performance of DIJKSTRAand uni-directional ALT using a *time-dependent* variant of our European input. For comparison, the corresponding figure for the time-expanded approach (route-model with phase 1) are given as well.

technique	time-dependent PREPRO time [h:m]	QUERIES #settled nodes	speed up	time [ms]	speed up	time-expanded PREPRO time [h:m]	QUERIES #settled nodes	speed up	time [ms]	speed up
Dijkstra	0:00	260 095	1.0	125.2	1.0	0:00	200 213	1.0	122.8	1.0
uni-ALT	0:02	127 103	2.0	75.3	1.7	0:01	89 524	2.2	58.7	2.1

Comparing the standard ALT against ALT with dynamic landmarks, we observe, that regarding query time dynamic ALT only pays off as long as the general speed-up (achieved through some other speed-up technique or model) does not exceed the cost we pay for computing the distance table on-the-fly. Since the condensed graph of Europe has about 11 times more stations than the Berlin graph, the cost for computing the dynamic distance table carries much more weight there—A one-to-all DIJKSTRA takes about 7 ms on the condensed graph of Europe. Hence, it never pays off using dynamic landmarks together with Arc-Flags here. The same effect can be observed in the Berlin network, however, only with the combination of the route and phase 1 models due to the much smaller condensed graph.

Summarizing, our modifications yield a speed-up of 3.5 if we apply ALT and Arc-Flags to both of our time-expanded graphs. The corresponding figure for our bus network is 5.5. This yields an overall speed-up of 56.28 for Europe and 14.46 for Berlin when compared to the classic model without any speed-up technique applied.

5.3 Comparison to the Time-Dependent Model

Table 5 compares the performance of DIJKSTRA's algorithm and ALT applied to our route+phase 1 time-expanded model and the time-dependent model. We observe that by the introduction of our Route-Model (and Node-Blocking) query performance of time-expanded queries are faster than for the time-dependent approach. Hence, we are able to close the performance-gap between both models. Analyzing the time-dependent approach, we notice that Node-Blocking is included implicitly: During a query we do not relax an edge more than once although it represents several connections running from one station to another. Hence, early connections *block* later ones. Our remodeling and Node-Blocking technique introduces these optimizations to the time-expanded approach. As a result the performance advantage of the time-dependent approach fades.

6 Conclusion

In this work, we introduced a local remodeling routine for the time-expanded approach based on the intuition that at many stations in a network, the number of reasonable choices is little. It turns out that this approach leads to a closely related speed-up technique harmonizing well with our remodeling. Moreover, we adapted speed-up techniques to the time-expanded model and show that they harmonize well with our new approach. Altogether, our approach yields query times up to 56.28 times faster than pure DIJKSTRA.

Regarding future work, we are optimistic that our approach would also work well for multi-criteria optimization. Although our pruning techniques may not work as strict as for single-criteria search, the number of reasonable choices is little in this scenario as well. Another very important problem is how to handle updates in case of delays. It seems as if updating a time-expanded graph is rather expensive, though possible [3, 16].

References

1. Bauer, R., Delling, D.: SHARC: Fast and Robust Unidirectional Routing. In: Munro, I., Wagner, D. (eds.) Proceedings of the 10th Workshop on Algorithm Engineering and Experiments (ALENEX 2008), pp. 13–26. SIAM, Philadelphia (2008)
2. Bauer, R., Delling, D., Wagner, D.: Experimental Study on Speed-Up Techniques for Timetable Information Systems. In: Liebchen, C., Ahuja, R.K., Mesa, J.A. (eds.) Proceedings of the 7th Workshop on Algorithmic Approaches for Transportation Modeling, Optimization, and Systems (ATMOS 2007). Internationales Begegnungs- und Forschungszentrum für Informatik (IBFI), pp. 209–225. Schloss Dagstuhl, Germany (2007)
3. Delling, D., Giannakopoulou, K., Wagner, D., Zaroliagis, C.: Timetable Information Updating in Case of Delays: Modeling Issues. Technical Report 133, Arrival Technical Report (2008)
4. Delling, D., Pajor, T., Wagner, D.: Engineering Time-Expanded Graphs for Faster Timetable Information. In: Proceedings of the 8th Workshop on Algorithmic Approaches for Transportation Modeling, Optimization, and Systems (ATMOS 2008), Dagstuhl Seminar Proceedings. Internationales Begegnungs- und Forschungszentrum für Informatik (IBFI), Schloss Dagstuhl, Germany (September 2008)
5. Delling, D., Sanders, P., Schultes, D., Wagner, D.: Engineering Route Planning Algorithms. In: Lerner, J., Wagner, D., Zweig, K.A. (eds.) Algorithmics of Large and Complex Networks. LNCS, vol. 5515, pp. 117–139. Springer, Heidelberg (2009)
6. Dijkstra, E.W.: A Note on Two Problems in Connexion with Graphs. Numerische Mathematik 1, 269–271 (1959)
7. Geisberger, R., Sanders, P., Schultes, D., Delling, D.: Contraction Hierarchies: Faster and Simpler Hierarchical Routing in Road Networks. In: McGeoch, C.C. (ed.) WEA 2008. LNCS, vol. 5038, pp. 319–333. Springer, Heidelberg (2008)
8. Goldberg, A.V., Harrelson, C.: Computing the Shortest Path: A* Search Meets Graph Theory. In: Proceedings of the 16th Annual ACM–SIAM Symposium on Discrete Algorithms (SODA 2005), pp. 156–165 (2005)
9. Goldberg, A.V., Kaplan, H., Werneck, R.F.: Better Landmarks Within Reach. In: Demetrescu, C. (ed.) WEA 2007. LNCS, vol. 4525, pp. 38–51. Springer, Heidelberg (2007)
10. Goldberg, A.V., Werneck, R.F.: Computing Point-to-Point Shortest Paths from External Memory. In: Proceedings of the 7th Workshop on Algorithm Engineering and Experiments (ALENEX 2005), pp. 26–40. SIAM, Philadelphia (2005)

11. Hart, P.E., Nilsson, N., Raphael, B.: A Formal Basis for the Heuristic Determination of Minimum Cost Paths. IEEE Transactions on Systems Science and Cybernetics 4, 100–107 (1968)
12. Hilger, M.: Accelerating Point-to-Point Shortest Path Computations in Large Scale Networks. Master's thesis, Technische Universität Berlin (2007)
13. Köhler, E., Möhring, R.H., Schilling, H.: Acceleration of Shortest Path and Constrained Shortest Path Computation. In: Nikoletseas, S.E. (ed.) WEA 2005. LNCS, vol. 3503, pp. 126–138. Springer, Heidelberg (2005)
14. Lauther, U.: An Extremely Fast, Exact Algorithm for Finding Shortest Paths in Static Networks with Geographical Background, vol. 22, pp. 219–230. IfGI prints (2004)
15. Müller-Hannemann, M., Schnee, M.: Finding All Attractive Train Connections by Multi-Criteria Pareto Search. In: Geraets, F., Kroon, L.G., Schoebel, A., Wagner, D., Zaroliagis, C.D. (eds.) Railway Optimization 2004. LNCS, vol. 4359, pp. 246–263. Springer, Heidelberg (2007)
16. Müller–Hannemann, M., Schnee, M., Frede, L.: Efficient On-Trip Timetable Information in the Presence of Delays. In: Proceedings of the 8th Workshop on Algorithmic Approaches for Transportation Modeling, Optimization, and Systems (ATMOS 2008), Dagstuhl Seminar Proceedings. Internationales Begegnungs- und Forschungszentrum für Informatik (IBFI), Schloss Dagstuhl, Germany (September 2008)
17. Pajor, T.: Goal Directed Speed-Up Techniques for Shortest Path Queries in Timetable Networks, Student Research Project (January 2008)
18. Pellegrini, F.: SCOTCH: Static Mapping, Graph, Mesh and Hypergraph Partitioning, and Parallel and Sequential Sparse Matrix Ordering Package (2007)
19. Pyrga, E., Schulz, F., Wagner, D., Zaroliagis, C.: Experimental Comparison of Shortest Path Approaches for Timetable Information. In: Proceedings of the 6th Workshop on Algorithm Engineering and Experiments (ALENEX 2004), pp. 88–99. SIAM, Philadelphia (2004)
20. Pyrga, E., Schulz, F., Wagner, D., Zaroliagis, C.: Efficient Models for Timetable Information in Public Transportation Systems. ACM Journal of Experimental Algorithmics 12, Article 2.4 (2007)
21. Sanders, P., Schultes, D.: Engineering Highway Hierarchies. In: Azar, Y., Erlebach, T. (eds.) ESA 2006. LNCS, vol. 4168, pp. 804–816. Springer, Heidelberg (2006)
22. Schulz, F., Wagner, D., Weihe, K.: Dijkstra's Algorithm On-Line: An Empirical Case Study from Public Railroad Transport. In: Vitter, J.S., Zaroliagis, C.D. (eds.) WAE 1999. LNCS, vol. 1668, pp. 110–123. Springer, Heidelberg (1999)

Time-Dependent Route Planning*

Daniel Delling and Dorothea Wagner

Universität Karlsruhe (TH), 76128 Karlsruhe, Germany
{delling,wagner}@ira.uka.de

Abstract. In this paper, we present an overview over existing speed-up techniques for time-dependent route planning. Apart from only explaining each technique one by one, we follow a more systematic approach. We identify basic ingredients of these recent techniques and show how they need to be augmented to guarantee correctness in time-dependent networks. With the ingredients adapted, three efficient speed-up techniques can be set up: Core-ALT, SHARC, and Contraction Hierarchies. Experiments on real-world data deriving from road networks and public transportation confirm that these techniques allow the fast computation of time-dependent shortest paths.

1 Introduction

Finding the quickest connection in transportation networks is a problem familiar to anybody who ever travelled. While in former times, route planning was done with maps at the kitchen's table, nowadays computer based route planning is established: Finding the best train connection is done via the Internet while route planning in road networks is often done using mobile devices.

An efficient approach to tackle this problem derives from graph theory. We model the transportation network as a graph and apply travel times as a metric on the edges. Computing the shortest path in such a graph then yields the provably quickest route in the corresponding transportation network. In principle, Dijkstra's classical algorithm [13] can solve this problem. However, for continental-sized transportation networks (consisting of up to 45 million road segments), Dijkstra's algorithm would take up to 10 seconds for finding a suitable connection, which is way too slow for practical applications. Roughly speaking, Dijkstra computes the distance to all possible locations in the network being closer than the target we are interested in. Clearly, it does not make sense to compute all these distances if we are only interested in the path between two points. Hence, many speed-up techniques have been developed within the last years. Such techniques split the work into two parts. During an *offline* phase, called preprocessing, we compute additional data that accelerates queries during the *online* phase. By exploiting several properties of a transportation network,

---

* Partially supported by the Future and Emerging Technologies Unit of EC (IST priority – 6th FP), under contract no. FP6-021235-2 (project ARRIVAL) and the DFG (project WA 654/16-1).

R.K. Ahuja et al. (Eds.): Robust and Online Large-Scale Optimization, LNCS 5868, pp. 207–230, 2009.

the fastest techniques can obtain the quickest path in road networks within microseconds for the price of few hours of preprocessing. See Fig. 1 for an example of the search space of a speed-up technique compared to Dijkstra's algorithm.

Up to the year 2008, research on route planning focused either on efficient speed-up techniques for *time-independent* route planning in road networks or on *modeling issues* (combined with basic algorithms for determining the best connection) in time-dependent networks deriving from public transportation. For an overview on time-independent route planning, see [10], while [28] presents the work for public transportation. Recently, the focus has shifted to the development of efficient route planning algorithms for time-dependent networks, both road networks and public transportation. It turned out that switching from a static to a time-dependent scenario is more challenging than one might expect: The input size increases drastically as travel times on time-dependent connections change frequently during the day. Moreover, shortest paths heavily depend on the time of departure, e.g., during rush hours it might pay off to avoid highways. On the technical side, the most efficient time-independent speed-up techniques rely on bidirectional search, i.e., a second search is started from the target. However, this concept is complicated in time-dependent scenarios as the arrival time would have to be known in advance for such a procedure.

Our Contributions. In this work, we recap the recent development on speed-up techniques for time-dependent route planning covering work from [1,6,7,8,9,26]. Apart from only explaining the techniques one by one we take a step back and reanalyze them. It turns out that the approach is the same for all time-dependent speed-up techniques: Augment the basic subroutines of preprocessing and the query algorithm such that correctness can still be guaranteed in time-dependent networks. Interestingly, all efficient techniques rely on four basic ingredients: Dijkstra's algorithm [13], landmarks [15,16], Arc-Flags [21,22], and contraction [29]. We here explain each ingredient in detail, how they are augmented, and how the recently developed speed-up techniques from combining some of these ingredients are obtained. Summarizing, in this paper we not only give a survey on time-dependent speed-up techniques but also reinterpret existing results so that the field on the whole becomes clearer to somebody who is new to time-dependent route planning.

Overview. This paper is organized as follows. First, we settle basic definitions in Section 2. In Section 3, we identify basic concepts for accelerating shortest path queries, show how they can be augmented so that correctness can be guaranteed in time-dependent networks, and analyze their drawbacks. Setting up efficient speed-up techniques from the (augmented) ingredients is done in Section 4. More precisely, we focus on three speed-up techniques: Core-ALT, SHARC, and Contraction Hierarchies. All three approaches are evaluated in Section 5 with real-world transportation networks and Europe. We conclude our work on time-dependent route planning by a summary and a discussion on future work in Section 6.

Fig. 1. Search space of different algorithms for the same sample query in a road network. The upper figure depicts the search space of Dijkstra, the lower one for a speed-up technique, i.e., SHARC [2]. Black edges are touched by algorithms, grey ones stay untouched. The shortest path is drawn thicker. We observe that the speed-up technique touches considerably fewer edges than Dijkstra.

2 Preliminaries

An (undirected) graph $G = (V, E)$ consists of a finite set V of *nodes* and a finite set E of *edges*. An edge is an unordered pair $\{u, v\}$ of nodes $u, v \in V$. If the edges are ordered pairs (u, v), we call the graph *directed*. In this case, the node u is called the *tail* of the edge, v the *head*. Throughout the whole work we restrict ourselves to directed graphs which are weighted by a length function *len*. The number of nodes $|V|$ is denoted by n, the number of edges $|E|$ by m. We say a graph is *sparse* if $m \in O(n)$. Given a set of edges H, $tails(H)$ / $heads(H)$ denotes the set of all tails / heads in H. With $\deg_{in}(v)$ / $\deg_{out}(v)$ we denote the number of edges whose head / tail is v. The reverse graph $\overleftarrow{G} = (V, \overleftarrow{E})$ is the graph obtained from G by substituting each $(u, v) \in E$ by (v, u). The 2-core of an undirected graph is the maximal node induced subgraph of minimum node degree 2. The 2-core of a directed graph is the 2-core of the corresponding simple, unweighted, undirected graph. A tree on a graph for which exactly the root lies in the 2-core is called an *attached tree*. All nodes not being part of the 2-core are called 1-shell nodes.

Time-Dependency. We model time-dependency by using functions for specifying edge weights. Throughout the whole work, we restrict ourselves to a function space $\mathbb{F}$ consisting of positive *periodic* functions $f : \Pi \to \mathbb{R}^+, \Pi = [0, p], p \in \mathbb{N}$ such that $f(0) = f(p)$ and $f(x) + x \le f(y) + y$ for any $x, y \in \Pi, x \le y$. Note that these functions respect the FIFO property (also called the *non-overtaking property*) which states that if A leaves the node u of an edge (u, v) before B, B cannot arrive at node v before A. Computation of shortest paths in FIFO networks is polynomially solvable [20]. In non-FIFO networks, complexity depends on the restriction whether waiting at nodes is allowed. If waiting is allowed, the problem stays polynomially solvable; otherwise, the problem is NP-hard [27].

In the following, we call Π the *period* of the input. We restrict ourselves to directed graphs $G = (V, E)$ with time-dependent length functions $len : E \to \mathbb{F}$. We use $len : E \times [0, p] \to \mathbb{R}^+$ to evaluate an edge for a specific departure time. Note that our networks fullfill the FIFO-property if we interpret the length of an edge as travel times due to our choice of $\mathbb{F}$. The composition of two functions $f, g \in \mathbb{F}$ is defined by $f \oplus g := g \circ f$. Moreover, we need to *merge* functions, which we define by $\min(f, g)$ with $\min(f, g)(x) := \min\{f(x), g(x)\}, x \in \Pi$. The upper bound of f is noted by $\overline{f} = \max_{x \in \Pi} f(x)$, the lower by $\underline{f} = \min_{x \in \Pi} f(x)$. An underapproximation $\downarrow f$ of a function f is a function such that $\downarrow f(x) \le f(x)$ holds for all $x \in \Pi$. An overapproximation $\uparrow f$ is defined analogously. Bounds and approximations of our time-dependent edge function *len* are given by analogous notations. Obviously, one can obtain a time-independent graph $\underline{G}$ from a time-dependent graph G by substituting the time-dependent length function by $\underline{len}$. We call $\underline{G}$ the *lower bound graph* of G.

We use piecewise linear functions for modeling time-dependency in transportation networks. Each edge gets a number of interpolation points assigned that depict the travel time on this edge at the specific time. Interestingly, evaluating

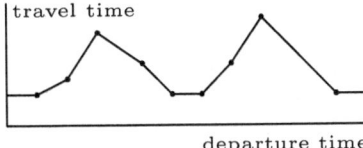
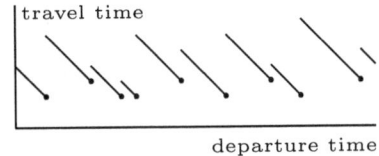

Fig. 2. Examples of piecewise linear travel time functions, the left figure shows a function used for road networks, while the right one is applied to railway networks. Interpolation points are depicted by dots. Note that the evaluation between two points is done in a different manner.

a function depends on the type of network we use. In road networks, evaluating a function at time τ is done by linear interpolation between the points left and right to τ. In railway networks, we identify the point p right to τ and return the travel time at p plus the waiting time. Figure 2 gives an example.

Paths. A path P in G is a sequence of nodes $(u_1, \ldots, u_k)$ such that $(u_i, u_{i+1}) \in E$ for all $1 \leq i < k$. In time-independent scenarios, the *length* of a path is given by $\sum_{i=1}^{k-1} len(u_i, u_{i+1})$. A path between two nodes s and t with minimum length is called a *shortest s–t path*. By $d(s,t)$ we denote the length of such a path. In time-dependent scenarios, the length $\gamma_\tau(P)$ of a path P departing from u_1 at time τ is recursively given by

$$\gamma_\tau\big((u_1, u_2)\big) = len\big((u_1, u_2), \tau\big)$$
$$\gamma_\tau\big((u_1, \ldots, u_j)\big) = \gamma_\tau\big((u_1, \ldots, u_{j-1})\big) + len\big((u_{j-1}, u_j), \gamma_\tau\big((u_1, \ldots, u_{j-1})\big)\big)$$

In other words, the length of the path depends on the departure time from s. In a time-dependent scenario, we are interested in two types of distances. On the one hand, we want to compute the shortest path between two nodes for a given departure time. On the other hand, we are also interested in retrieving the distance between two nodes for *all* possible departure times $\in \Pi$.

By $d(s, t, \tau)$ we denote the length of the shortest path $s, t \in V$ if departing from s at time τ. The distance-label, i.e., the distance between s and t for all possible departure times $\in \Pi$, is given by $d_*(s, t)$. Note that the distance-label is a function $\in \mathbb{F}$. In this work, we call a query for determining $d(s, t, \tau)$ an *s-t time-query*, while a query for computing $d_*(s, t)$ is denoted by *s-t profile-query*.

3   Ingredients and Their Augmentation

In this section, we identify basic ingredients all existing high-performance speed-up techniques for time-dependent route planning rely on. These are Dijkstra's algorithm, landmarks, Arc-Flags, and contraction. In the following, we explain each ingredient separately and show how they are augmented so that correctness can also be guaranteed in time-dependent networks.

3.1 Dijkstra's Algorithm

The classical algorithm for computing the shortest path from a given source to all other nodes in a directed graph with non-negative edge weights is due to Dijkstra [13]. The algorithm maintains, for each node u, a label $distance[u]$ with the tentative distance from s to u. A priority queue Q contains all nodes that depict the current search horizon around s. At each step, the algorithm removes (or *settles*) the node u from Q with minimum distance from s. Then, all outgoing edges (u, v) of u are relaxed, i.e., we check whether $d(s, u) + len(u, v) < distance[v]$ holds. If it holds, a shorter path to v via u has been found. Hence, v is either inserted to the priority queue or its priority is decreased.

Augmentation. Computing $d(s, t, \tau)$ can be solved by a modified Dijkstra [4]: when relaxing an edge (u, v) we have to evaluate the weight of it for time $\tau + d(s, u, \tau)$. In our scenario, the running time for evaluating functions is negligible, hence the additional effort for respecting the departure time is negligible as well.

However, computing $d_*(s, t)$ is more expensive but can be computed by a label-correcting algorithm [5], which can be implemented very similarly to Dijkstra. The source node s is initialized with a constant label $d_*(s, s) \equiv 0$, any other node u with a constant label $d_*(s, u) \equiv \infty$. Then, in each iteration step, a node u with minimum $\underline{d}_*(s, u)$ is removed from the priority queue. Then for all outgoing edges (u, v) a temporary label $l(v) = d_*(s, u) \oplus len(u, v)$ is created. If $l(v) \geq d_*(s, v)$ does *not* hold, $l(v)$ yields an improvement. Hence, $d_*(s, v)$ is updated to $\min\{l(v), d_*(s, v)\}$ and v is inserted into the queue. We may stop the routine if we remove a node u from the queue with $\underline{d}(s, u) \geq \overline{d}(s, t)$. If we want to compute $d_*(s, t)$ for many nodes $t \in V$, we apply a label-correcting algorithm and stop the routine as soon as our stopping criterion holds for all t. Note that we may reinsert nodes into the queue that have already been removed by this procedure. Also note that when applied to a graph with constant edge-functions, this algorithm equals a normal Dijkstra. An interesting result from [5] is the fact that the running time of label-correcting algorithms highly depends on the complexity of the edge-functions.

In the following, we construct *profile graphs (PG)*, i.e., compute $d_*(s, u)$ for a given source s and all nodes $u \in V$, with our label-correcting algorithm. We call an edge (u, v) a *PG-edge* if $d_*(s, u) \oplus len(u, v) > d_*(s, v)$ does *not* hold. In other words, (u, v) is a PG-edge iff it is part of a shortest path from s to v for at least one departure time.

Bidirectional Profile Search. As already mentioned, bidirectional search is prohibited for time-queries as the arrival time is unknown. However, we can directly apply bidirectional search for profile-queries since we investigate *all* arrival times. Compared to a time-independent bidirectional Dijkstra, we only need to adjust the stopping criterion. Stop the search if the lower bound of the minimum label in the forward queue added to the lower bound of the minimum label in the backward queue is larger than the upper bound of the tentative distance label.

3.2 A^* Search Using Landmarks (ALT)

Next, we explain the known technique of A^* search [17] in combination with landmarks, called ALT [15,16]. The search space of Dijkstra's algorithm can be visualized as a circle around the source. The idea of goal-directed or A^* search is to push the search towards the target. By adding a potential $\pi : V \to \mathbb{R}$ to the priority of each node, the order in which nodes are removed from the priority queue is altered. A 'good' potential lowers the priority of nodes that lie on a shortest path to the target. It is easy to see that A^* is equivalent to Dijkstra's algorithm on a graph with *reduced costs*, formally $len_\pi(u,v) = len(u,v) - \pi(u) + \pi(v)$. Since Dijkstra's algorithm works only on nonnegative edge costs, not all potentials are allowed. We call a potential π *feasible* if $len_\pi(u,v) \geq 0$ for all $(u,v) \in E$. The distance from each node v of G to the target t is the distance from v to t in the graph with reduced edge costs minus the potential of t plus the potential of v. So, if the potential $\pi(t)$ of the target t is zero, $\pi(v)$ provides a *lower bound* for the distance from v to the target t.

Preprocessing. There exist several techniques [31,32] to obtain feasible potentials using the layout of a graph. The ALT algorithm however, uses a small number of nodes—so called *landmarks*—and the triangle inequality to compute feasible potentials. Given a set $L \subseteq V$ of landmarks and distances $d(l,v), d(v,l)$ for all nodes $v \in V$. For a given landmark $l \in L$, the following triangle inequalities hold:

$$d(l,u) + d(u,v) \geq d(l,v) \quad \text{and} \quad d(u,v) + d(v,l) \geq d(u,l)$$

Therefore, $\underline{d}(u,v) := \max_{l \in L} \max\{d(u,l) - d(v,l), d(l,v) - d(l,u)\}$ provides a feasible lower bound for the distance $d(u,v)$. See Figure 3 for an illustration. The quality of the lower bounds highly depends on the quality of the selected landmarks.

Landmark Selection. A crucial point in the success of a high speed-up when using ALT is the quality of landmarks. Since finding good landmarks is difficult, several heuristics [15,16] exist. We focus on the best known techniques: *avoid* and *maxCover*.

Avoid [15]. This heuristic tries to identify regions of the graph that are not well covered by the current landmark set S. Therefore, a shortest-path tree T_r is grown from a random node r. The *weight* of each node v is the difference between $d(v,r)$ and the lower bound $\underline{d}(v,r)$ obtained by the given landmarks.

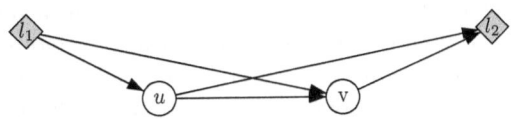

Fig. 3. Triangle inequalities for landmarks. The landmarks are l_1 and l_2.

The *size* of a node v is defined by the sum of its weight and the size of its children in T_r. If the subtree of T_r rooted at v contains a landmark, the size of v is set to zero. Starting from the node with maximum size, T_r is traversed following the child with highest size. The leaf obtained by this traversal is added to S. In this strategy, the first root is picked uniformly at random. The following roots are picked with a probability proportional to the square of the distance to its nearest landmark.

MaxCover [16]. The main disadvantage of avoid is the starting phase of the heuristic. The first root is picked at random and the following landmarks are highly dependent on the starting landmark. MaxCover improves on this by first choosing a candidate set of landmarks (using avoid) that is about four times larger than needed. The landmarks actually used are selected from the candidates using several attempts with a local search routine. Each attempt starts with a random initial selection.

Query. The unidirectional ALT-query is a modified Dijkstra operating on the input graph, the only difference to plain Dijkstra is that the key within the priority queue is not determined only by the distance to s but also by a lower bound of the distance to the target, given by the landmarks.

It turns out that unidirectional ALT only provides mild speed-ups over Dijkstra's algorithm [11]. The full potential of ALT is unleashed if applied bidirectionally. At a glance, combining ALT and bidirectional search seems easy. Simply use a feasible potential π_f for the forward and a feasible potential π_b for the backward search. However, such an approach does not work due to the fact that the searches might work on different reduced costs, so that the shortest path might not have been found when both searches meet. This can only be guaranteed if π_f and π_b are *consistent*, meaning $len_{\pi_f}(u, v)$ in G is equal to $len_{\pi_b}(v, u)$ in the reverse graph. We use the variant of an average potential function [19] defined as $p_f(v) = (\pi_f(v) - \pi_b(v))/2$ for the forward and $p_b(v) = (\pi_b(v) - \pi_f(v))/2 = -p_f(v)$ for the backward search. By adding $\pi_b(t)/2$ to the forward and $\pi_f(s)/2$ to the backward search, p_f and p_b provide lower bounds to the target and source, respectively. Note that these potentials are feasible and consistent but provide worse lower bounds than the original ones.

Augmentation. Based on observation that potentials stay feasible as long as edge weights only increase and do not drop below their initial values, we can adapt a unidirectional variant of the ALT algorithm to the time-dependent scenario: We perform both landmark selection and distance computation in the lower bound graph $\underline{G}$. It is obvious that we obtain a feasible potential. However, ALT implemented as bidirectional search is much faster than the unidirectional variant. As already mentioned, performing a bidirectional search in time-dependent networks is non-trivial. In [26], we showed how *bidirectional ALT* can be used in time-dependent networks anyway. The idea is as follows: A backward search is performed in $\underline{G}$ and is only used to restrict nodes that need to be visited by the forward search.

Bidirectional Query. The query algorithm is based on restricting the scope of a time-dependent A^* search from the source using a set of nodes defined by a time-*independent* A^* search from the destination, i.e., the backward search is a reverse search in $\underline{G}$, which corresponds to the graph G weighted by the lower bounding function $\underline{len}$. More precisely, it works in three phases:

1. A bidirectional ALT is applied to G, where the forward search is performed on the (time-dependent) graph, and the backward search is run on the lower bound graph $\underline{G}$. All nodes settled by the backward search are added to a set M. Phase 1 terminates as soon as the two search scopes meet.
2. Suppose that $v \in V$ is a node settled by both searches; then the time dependent cost $\mu = \gamma_\tau(p_v)$ of the path p_v going from s to t passing through v is an upper bound to $d(s, t, \tau)$. Let β be the key of the minimum element of the backward search queue; phase 2 terminates as soon as $\beta > \mu$. Again, all nodes settled by the backward search are added to M.
3. In the third phase, only the forward search continues, with the additional constraint that only nodes in M can be explored. The forward search terminates when t is settled.

Note that the time-dependent ALT algorithm also works in a dynamic time-dependent scenario: The algorithm still performs accurate queries as long as edge weights do not drop below their lower bound. Moreover, the bidirectional query algorithm can also be used to find a K approximation of the shortest path. Therefore, the second phase is already stopped as soon as $\beta > K\mu$ (cf. [26] for details).

3.3   Arc-Flags

The classic Arc-Flag approach, introduced in [21,22], first computes a partition $\mathcal{C}$ of the graph and then attaches a *label* to each edge e. A label contains, for each cell $C \in \mathcal{C}$, a flag $AF_C(e)$ which is true if a shortest path to at least one node in C starts with e. A modified Dijkstra—from now on called Arc-Flags Dijkstra—then only considers those edges for which the flag of the target node's cell is true. The big advantage of this approach is its easy query algorithm. Furthermore, we observed that for long-range queries in road networks, an Arc-Flags Dijkstra often is optimal in the sense that it *only* visits those edges that are on the shortest path. However, preprocessing is very extensive, either regarding preprocessing time or memory consumption.

Preprocessing. Preprocessing of Arc-Flags is divided into two parts. First, the graph is partitioned into k cells. The second step then computes k flags for each edge.

Partition. The first approach for obtaining a partition is based on a grid partition [22]. It turns out that the performance of an Arc-Flags query heavily depends on the partition used. In order to achieve good speed-ups, several requirements have to be fulfilled: cells should be connected, the size of the cells should be balanced, and the number of boundary nodes has to be low. A systematical experimental study of the impact of partitions on Arc-Flags has been published in [25].

Setting Arc-Flags. The second step of preprocessing is the computation of arc-flags. Throughout the years, several approaches have been introduced (see e.g., [18,21,22,23,24]). We here concentrate on two approaches which turned out to be the most efficient. For both approaches, we have to perform an initialization step, which sets the so-called *own-cell* flags of all edges not crossing borders to true. Note that the own-cell flag of an edge (u, v) in cell C, i.e., u and v both are in cell C, is $AF_C((u,v))$. If u and v are in different cells, no flag is set to true during the initialization phase.

Boundary Shortest Path Trees. A true arc-flag $AF_C(e)$ denotes whether e has to be considered for a shortest-path query targeting a node within C. The key observation of this approach is that all shortest paths ending in the cell C must pass any of the boundary nodes B_C of cell C. More precisely, a node $b \in C$ is called a boundary node of cell C if there exists an edge $(v, b) \in E$ with node v being part of a cell $C' \neq C$. With this observation, arc-flags can be computed as follows: Grow a shortest path tree in $\overleftarrow{G}$ from all boundary nodes $b \in B_C$ of all cells C. Then set $AF_C((u,v)) = $ true if (u, v) is a tree edge for at least one tree grown from all boundary nodes $b \in B_C$.

Centralized Approach. The drawback of the first approach is that we have to grow $|B|$ shortest path trees yielding long preprocessing times for large transportation networks. [18] introduces a new approach to computing flags. A label-correcting algorithm (also called centralized tree) is performed for each cell C. The algorithm propagates labels of size $|B_C|$ through the network depicting the distances to all boundary nodes of the cell. The algorithm terminates if no label can be improved any more. Then, $AF_C((u,v))$ is set to true if $len(u, v) + d(v, b) = d(u, b)$ holds for at least one $b \in B_C$.

Query. A unidirectional Arc-Flags query is a modified Dijkstra operating on the input graph. For any s–t query, it first determines the target cell T, and then relaxes only those edges e with $AF_T(e) = $ true. Note that compared to plain Dijkstra, an Arc-Flags query performs only one additional check.

Note that $AF_C(e)$ is true for almost all edges $e \in C$ due to the own-cell-flag. Due to these own-cell-flags an Arc-Flags Dijkstra yields no speed-up for queries within the same cell. Even worse, more and more edges become important when approaching the target cell (called the *coning effect*) and finally, all edges are considered as soon as the search enters the target cell.

Multi-Level Arc-Flags. While the coning effect can be weakened by a bidirectional approach, the problem of inner-cell queries persists also for bidirectional search. An approach to remedy this drawback is introduced in [25]: A second layer of arc-flags is computed for each cell. Therefore, each cell is again partitioned into several subcells and arc-flags are computed for each. A multi-level arc-flags query then first uses the flags on the topmost level and as soon as the query enters the target's cell on the topmost level, the low-level arc-flags are used for pruning.

Preprocessing in a time-independent scenario is done as follows. Arc-flags on the upper level are computed as described above. For the lower flags, grow a shortest path for all boundary nodes b on the lower level. Stop the growth as soon as all nodes in the supercell of C are settled. Then, we set a low-level arc-flag to true if the edge is a tree edge of at least one shortest path tree. Note that this approach can be extended to a multi-level approach in a straightforward manner. Also note that multi-level Arc-Flags can be applied bidirectionally as well.

Discusssion. The advantages of Arc-Flags is the easy concept combined with exceptional query performance: Preprocessing is based on Dijkstra-searches and the query algorithm performs only one additional check (per edge) compared to plain Dijkstra. Stunningly, bidirectional Arc-Flags long-range queries are often optimal—at least in road networks—in that sense that *only* shortest path edges are relaxed. However, the most crucial drawback of Arc-Flags is its time consuming preprocessing effort. Even the most advanced technique, i.e., the centralized approach, needs more than 17 hours to preprocess a continental-sized road network. Still, due to its superior undirectional query performance, Arc-Flags seemed to be a good starting point for time-dependent shortest path computations.

Augmentation. In time-independent scenarios, a set arc-flag $AF_C(e)$ denotes whether e has to be considered for a shortest-path query targeting a node within C. In other words, the flag is set if e is important for (at least one target node) in C. In a time-dependent scenario, we use the following intuition to set arc-flags: an arc-flag $AF_C(e)$ is set to true, if e is important for C at least once during Π. A straightforward adaption of computing arc-flags in a time-dependent graph is to construct a profile graph in $\overleftarrow{G}$ for all boundary nodes $b \in B_C$ of all cells C. Then we set $AF_C((u,v)) = $ true if (u,v) is a PG-edge for at least one PG built from all boundary nodes $b \in B_C$. In addition, we also set all own-cell flags to true as well. The time-dependent query is a normal time-dependent Dijkstra only relaxing edges with set flag for the target's cell.

Approximation. Computing arc-flags as described above requires to build a complete profile graph on the backward graph from each boundary node yielding too long preprocessing times for large networks. Recall that the running time of building PGs is dominated by the complexity of the function (cf. Section 3.1). Hence, we may construct two PGs for each boundary node, the first uses $\uparrow len$ as length functions, the second $\downarrow len$. Since we use approximations, we may use less interpolations points per label. By this, constructing two such PGs may be faster than building one exact one. We end up in two distance labels per node u, one being an overapproximation, the other being an underapproximation of the correct label. Then, for each $(u, v) \in E$, we set $\overline{AF}_C(u, v) = \mathsf{true}$ if $len(u, v) \oplus \uparrow\! d_*(v, b_C) > \downarrow\! d_*(u, b_C)$ does not hold.

If networks get so big that even setting approximate labels is prohibited due to running times, one can even use upper and lower bounds for the labels. This has the advantage that building two shortest-path trees per boundary node is sufficient for setting correct arc-flags. The first uses $\overline{len}$ as length function, the other $\underline{len}$. Note that by approximating arc-flags (denoted by $\overline{AF}$), their quality may decrease but correctness is untouched. Thus, queries remain correct but may become slower.

Heuristic Arc-Flags. In [8], we proposed a third approach for computing flags. The preprocessing is as follows: We grow $k + 2$ shortest-path trees from each boundary node. The first uses $\underline{len}$ as metric, the second one $\overline{len}$, and the remaining k trees are time-queries in $\overleftarrow{G}$ using a fixed arrival time at the boundary node. We set a flag of an edge for a cell C if the edge is part of at least one shortest path tree grown from the boundary nodes of C.

Unfortunately, this approach may yield incorrect queries as a shortest path for a specific departure time may have been missed. However, it is obvious that a path is found since at least for one departure time, flags are set to true for a shortest path to the target's cell. Experiments on the eventual error-rate can be found in Section 5.

3.4   Contraction

One reason for the success of hierarchical speed-up techniques is the iterative *contraction* of the input: Unimportant nodes are removed from the graph and additional *shortcuts* are inserted to preserve distances between non-removed nodes.

Node-Reduction. The number of nodes is reduced by iteratively *bypassing* nodes until no node is *bypassable* any more. To bypass a node v we first remove v, its incoming edges I and its outgoing edges O from the graph. Then, for each $u \in tails(I)$ and for each $w \in heads(O) \setminus \{u\}$ we introduce a new edge of the length $len(u, v) + len(v, w)$. If there already is an edge connecting u and w in the graph, we only keep the one with smaller length. All nodes not removed by the node-reduction are part of the so called *core* of the input.

Edge-Reduction. Note that this node-reduction routine potentially adds shortcuts not needed for keeping the distances in the core correct. See Figure 4 for an example. Hence, an edge-reduction is performed directly after node-reduction, similar to [30]. We grow a shortest-path tree from each node u of the core. We stop the growth as soon as all neighbors w of u have been settled. Then we check for all neighbors w whether u is the predecessor of w in the grown partial shortest path tree. If u is not the predecessor, we can remove (u, w) from the graph because the shortest path from u to w does not include (u, w). In order to remove as many edges as possible we favor paths with more hops over those with few hops.

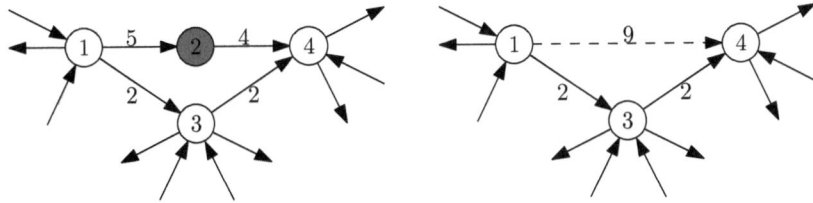

Fig. 4. Example of contraction. The figure on the upeer left depicts the input, edge labels indicate the weight of the edge. We contract, i.e., remove, node 2 and add an shortcut from node 1 to 4 with weight 9 (upper right). However, the shortest path from 1 to 4 is via node 3 with length 4. Hence, we can safely remove the shortcut (1,4) from the core in order to preserve distances between core nodes.

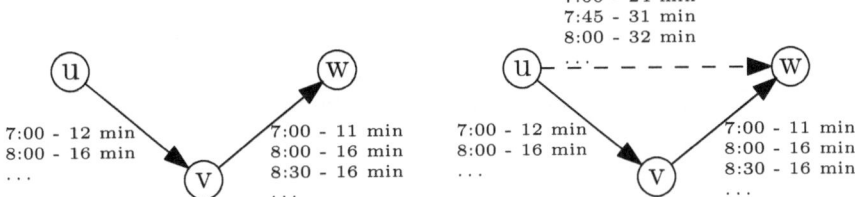

Fig. 5. Time-dependent contraction in road networks. Recall that we interpolate *linearly* between interpolation points, i.e., the travel time on edge (u,v) at 7:45 is 15 minutes. It is obvious that we have to add interpolation points at 7:00 and 8:00 to the function assigned to the shortcut (u,w). This would result in a travel time from u to w of 30 minutes when departing at 7:45. However, we arrive at v at 8:00 when departing from u at 7:45 and arrive at 8:16 at w. So, the travel time from u to w is 31 minutes instead of 30. Hence, we need to insert an additional interpolation point at 7:45. The reason for this is that the responsible interpolation points for evaluating $len(v,w)$ changes when departing from u at 7:45.

Augmentation. Time-dependent contraction is similar to a time-independent one. During node-reduction, new shortcuts (u,w), depicting the path from u via v to w, get the function $len(u,v) \oplus len(v,w)$ assigned. While this is straightforward in principle, one problem of node-reduction in time-dependent road networks is the following: Let $P(f)$ be the number of interpolation points of the function $f \in \mathbb{F}$. Then the composed function of $len(u,v) \oplus len(v,w)$, may have up to $P(len(u,v)) + P(len(v,w))$ number of interpolation points in the worst case. The main problem is that the interpolation points needed for evaluating $len(v,w)$ may change between two interpolation points of $len(u,v)$. Figure 5 gives an example, for details we refer the interested reader to [7]. This is one of the main problems when routing in time-dependent graphs: Almost all speed-up techniques developed for static scenarios rely on adding long shortcuts to the graph. While this is cheap for static scenarios, the insertion of time-dependent shortcuts yields a high amount of preprocessed data.

For edge-reduction, we build a PG (instead of a shortest path tree) from each node u of the core. We stop the growth as soon as all neighbors v of u have their final label assigned. Then we check for all neighbors whether $d_*(u, v) < len(u, v)$ holds. If it holds, we can remove (u, v) from the graph because for all possible departure times, the path from u to v does not include (u, v).

4   Speed-Up Techniques

In this section, we show how to assemble efficient speed-up techniques from the basic ingredients presented in Section 3. More precisely, we explain Core-ALT, SHARC, and Contraction Hierarchies. Due to their clear foundation on basic ingredients, the augmentation of these speed-up techniques is easier than for other approaches.

4.1   Core-ALT

Core-ALT was introduced in [3] and augmented to the time-dependent scenario in [9]. It is a combination of landmarks, bidirectional search, and contraction. As already discussed in Section 3, pure ALT suffers from two major drawbacks. Space consumption is rather high and—even more important—ALT cannot compete with hierarchical approaches—concerning query performance—in transportation networks. In [3], we showed how to remedy both drawbacks without violating the advantages of pure ALT, i.e., easy adaption to dynamic scenarios and robustness to the input. The key idea is to perform an initial contraction step prior to ALT preprocessing. Landmarks are then chosen from the core and landmark distances are also only stored for core nodes. This yields a 2-phase query. During the first phase, a plain bidirectional Dijkstra is performed until the core is reached. Within the core, bidirectional ALT is applied.

Preprocessing. At first, the input graph $G = (V, E)$ is contracted to a graph $G_C = (V_C, E_C)$, called the *core*. Then, we compute landmarks on the core and store the distances to and from the landmarks for all core nodes. After preprocessing the core, we store the preprocessed data and merge the core and the normal graph to a full graph $G_F = (V, E_F = E \cup E_C)$. Moreover, we mark the core-nodes with a flag.

Query. The s-t query is a modified bidirectional Dijkstra, consisting of two phases, both performed on G_F. During phase 1, we run a bidirectional Dijkstra rooted at s and t *not* relaxing edges belonging to the core. We add each core node, called *entrance point*, settled by the forward search to a set S (T for the backward search). The first phase terminates if one of the following two conditions hold: (1) either both priority queues are empty or (2) the sum of the distances to the closest entry points of s and t is larger than the length of the tentative shortest path. If case (2) holds, the whole query terminates. The second phase is an ALT-query, initialized by refilling the queues with the nodes belonging to S and T.

Augmentation. In [9], we augmented Core-ALT to time-dependent networks. The preprocessing is very similar to the time-independent variant. First, we extract a core $G_C = (V_C, E_C)$ with a time-dependent contraction routine. Then, we merge the core with the original graph to obtain $G_F = G_C \cup G = (V, E \cup E_C)$ since $V_C \subset V$. Finally, we select landmarks from G_C and compute landmark distances in G_C. The query algorithm again consists of two phases, performed on G_F. Due to the fact that the arrival time is unknown, the query algorithm is slightly more complicated than in the time-independent case.

1. Initialization phase: start a Dijkstra search from both the source and the destination node on G_F, using the time-dependent costs for the forward search and the time-independent costs *len* for the backward search, pruning the search (i.e., not relaxing outgoing edges) at nodes $\in V_C$. Add each node settled by the forward search to a set S, and each node settled by the backward search to a set T. Iterate between the two searches until: (*i*) $S \cap T \neq \emptyset$ or (*ii*) the priority queues are empty.

2. Main phase: (*i*) If $S \cap T \neq \emptyset$, then start a unidirectional Dijkstra search from the source on G_F until the target is settled. (*ii*) If the priority queues are empty and we still have $S \cap T = \emptyset$, then start a bidirectional time-dependent ALT (cf. Section 3) on the graph G_C, initializing the forward search queue with all leaves of S and the backward search queue with all leaves of T, using the distance labels computed during the initialization phase. The forward search is also allowed to explore any node $v \in T$, throughout the 3 phases of the algorithm. Stop when t is settled by the forward search.

In other words, the forward search "hops on" the core when it reaches a node $u \in S \cap V_C$, and "hops off" at all nodes $v \in T \cap V_C$. Moreover, we use time-dependent bidirectional ALT in case (*ii*) during the main phase. With the same arguments from Section 3.2, we can use Core-ALT to compute a K-approximation of the shortest path.

4.2   SHARC

SHARC Routing was introduced in [2] and augmented in [8]. It is based on contraction and Arc-Flags combined with a unidirectional query algorithm.

Preprocessing of static SHARC is divided into three sections. During the *initialization* phase, we extract the 2-core of the graph and perform a *multi-level* partition of G. Then, an *iterative* process starts. At each step i we first *contract* the graph by *bypassing* unimportant nodes and set the arc-flags *automatically* for each removed edge, depending on the tail u of the removed edge. If u is a core node, we only set the own-cell flag to true (and others to false) because this edge can only be relevant for a query targeting a node in this cell. Otherwise, all arc-flags are set to true as a query has to enter the core in order to reach a node outside this cell. See Fig. 6 for an example. For the remaining edges of the contracted graph we compute the arc-flags according to Section 3. In the *finalization* phase, we assemble the output-graph, refine arc-flags of edges

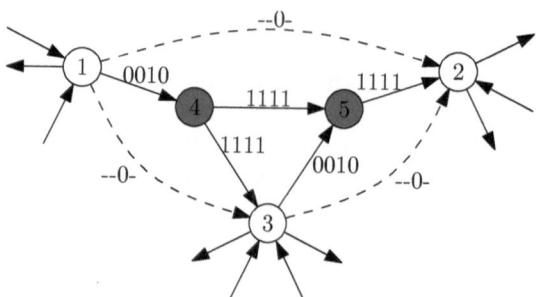

Fig. 6. Example for assigning arc-flags to removed edges during contraction for a partition having four cells. All nodes are in cell 3. The red nodes (4 and 5) are removed, the dashed shortcuts are added by the contraction. Arc-flags (edge labels) are indicated by a 1 for true and 0 for false. The edges heading a node removed by the contraction routine get *only* their own-cell flag set true. Any other removed edge gets all flags set to true. The added shortcuts get their own-cell flags fixed to false.

removed during contraction, and finally reattach the 1-shell nodes removed at the beginning.

Basically, the SHARC query is a modified Dijkstra that operates on the output graph. The modifications are the same as for a multi-level Arc-Flags query (cf. Section 3): When settling a node u, we compute the lowest level i on which u and the target node t are in the same supercell. When relaxing the edges outgoing from u, we consider only those edges having a set arc-flag on level i for the corresponding cell of t. Note that the SHARC query, compared to plain Dijkstra, only needs to perform two additional operations: computing the common level of the current node and the target and the arc-flags evaluation.

Augmentation. The adaption of SHARC [8] is done in a straightforward fashion. We use time-dependent contraction and time-dependent arc-flags computation during preprocessing instead of their time-independent counterparts.

Variants. In Section 3.3, we presented several ways of computing time-dependent arc-flags. The aggressive variant of SHARC uses exact flags during preprocessing, the economical version uses approximate flags, while heuristic SHARC uses heuristic flags. Hence, aggressive SHARC tends to have long preprocessing times combined with a better quality of flags, while economical SHARC has shorter preprocessing times for the price of worse flags. Heuristic SHARC however cannot guarantee correctness of the queries.

Landmarks. Approximate arc-flags yield worse results than exact ones. In order to partly remedy this loss in performance, we can add landmarks to SHARC. We can combine ALT with SHARC easily. We run a time-dependent ALT preprocessing consisting of selecting landmarks $L \subseteq V$ and computing $d(l, v), d(v, l)$ for all $v \in V, l \in L$. Then, we apply a normal SHARC-query but use $d(s, u, \tau) + \pi(u)$

(cf. Section 3) instead of $d(s, u, \tau)$ as priority key. We call this combination L-SHARC (**L**andmarks and **SHARC**).

4.3   Contraction Hierarchies

Contraction Hierarchies (abbreviated by CH) were introduced in [14] and augmented to the time-dependent scenario in [1]. This approach is solely based on contraction combined with a bidirectional query algorithm. Preprocessing is divided into two parts: node-ordering and contraction. Node-ordering assigns a priority to each node depicting its importance in an n-level hierarchy. Then, during contraction, the input graph G is transferred to two search graphs $G_{\uparrow}$ and $G_{\downarrow}$, which are called upward and downward graph, respectively. $G_{\uparrow}$ only stores edges directing from unimportant to important nodes, while $G_{\downarrow}$ contains only edges directing from important to unimportant nodes. These graphs can be constructed by running n node- and edge-reduction steps similarly to how it is explained in Section 3. However, each node-reduction step contracts exactly one node u, resulting in a very limited edge-reduction routine as unneeded shortcuts may only be added between neighbors of u. The query algorithm is conducted of two Dijkstra searches, a forward search (from s) operating on $G_{\uparrow}$ and a backward search (from t) on $G_{\downarrow}$.

Augmentation. Contraction Hierarchies is adapted by augmenting the contraction process with the process of node-ordering untouched. So, time-dependent preprocessing is straightforward; the main challenge is the adaption of the query algorithm.

The basic static query algorithm for CHs consists of a forward search in an upward graph $G_{\uparrow} = (V, E_{\uparrow})$ and a backward search in a downward graph $G_{\downarrow}$. Wherever these searches meet, we have a candidate for a shortest path. The shortest such candidate is a shortest path. Since the departure time is known, the forward search is easy to generalize. The easiest way to adapt the backward search is to explore *all* nodes that can *reach* t in $G_{\downarrow}$. During this exploration all edges connecting nodes that can reach t are marked. Let E_{marked} denote the set of marked edges. Then, an s–t-query can be performed by a forward search from s in $(V, E_{\uparrow} \cup E_{\text{marked}})$.

5   Experiments

In this section, we summarize experimental results on the performance of time-dependent ALT, Core-ALT, SHARC, and Contraction Hierarchies for road and railway networks. The experimental results are taken from [1,7].

All tests were executed on one core of two similar (with respect to performance) machines, both running SUSE Linux 10.3. The first machine is an AMD Opteron 2218 clocked at 2.6 GHz, has 16 GB of RAM and 2 x 1 MB of L2 cache. The second machine has a Xeon 5345 processors clocked at 2.33 GHz with 16 GByte of RAM and 2 x 4 MB of L2 cache. All programs were compiled with GCC 4.2.1 or 4.3.2, using optimization level 3.

Inputs. Two types of inputs are applied: Road and railway networks. For the former, we have access to a real-world time-dependent road network of Germany. It has approximately 4.7 million nodes and 10.8 million edges. In order to analyze the scalability of our approaches, we additionally use the available real-world *time-independent* network of Western Europe (18 million nodes and 42.6 million edges) and generate *synthetic* rush hours. All data has been provided by PTV AG for scientific use. The German data contains five different traffic scenarios, collected from historical data: Monday, midweek (Tuesday till Thursday), Friday, Saturday, and Sunday. As expected, congestion of roads is higher during the week than on the weekend: $\approx 8\%$ of edges are time-dependent for Monday, midweek, and Friday. The corresponding figures for Saturday and Sunday are $\approx 5\%$ and $\approx 3\%$, respectively. Our railways timetable data—provided by HaCon for scientific use—of Europe consists of 30 516 stations and 179 985 trains. The period is 24 hours. The resulting realistic, i.e., including transfer times, time-dependent network has about 0.5 million nodes and 1.4 million edges, and is fulfilling the FIFO-property.

Setup. In the following, we report preprocessing times and the overhead of the preprocessed data in terms of *additional* bytes per node. Moreover, we report two types of queries: *time-queries*, i.e., queries for a specific departure time, and *profile-queries*, i.e., queries for computing $d_*(s, t)$. For each type we provide the average number of settled nodes, i.e., the number of nodes taken from the priority queue, and the average query time. For *s-t* profile-queries, the nodes s and t are picked uniformly at random. Time-queries additionally need a departure time τ as well, which we pick uniformly at random as well. As all methods introduced in this chapter have approximate variants, we record four different statistics to characterize the solution quality: error rate, average relative error, maximum relative error, maximum absolute error. By *error rate* we denote the percentage of computed suboptimal paths over the total number of queries. By *relative error* on a particular query we denote the relative percentage increase of the approximated solution over the optimum, computed as $\omega/\omega^* - 1$, where ω is the cost of the approximated solution and ω^* is the cost of the optimum computed by Dijkstra's algorithm. We report *average* and *maximum* values of this quantity over the set of all queries. The maximum absolute error is given by $\omega - \omega^*$. All figures in this chapter are based on 100 000 random *s-t* queries and refer to the scenario that only the lengths of the shortest paths have to be determined, without outputting a complete description of the paths.

5.1   Road Networks

First, we compare all time-dependent algorithms discussed in this paper among each other. We hereby split our comparison in two parts. Exact queries and approximation. Table 1 reports query performance of time-dependent Dijkstra, uni-directional ALT, bidirectional ALT, Core-ALT (CALT), SHARC, and Contraction Hierarchies (CH) for our exact setup, while Tab. 2 depicts performance if suboptimal paths are allowed. As input we use our time-dependent road networks of Europe (high traffic) and Germany (midweek and Sunday). Note that no approximate variant of Contraction Hierarchies exists yet and that no results for Europe (high traffic) have been published. The reason for the latter is the high memory consumption making Contraction Hierarchies impractical for this input.

Exact Setup. Depending on the scenario, different algorithms perform best. While CALT is the technique with lowest preprocessing effort (both time and overhead), CH or SHARC win with respect to query performance. While CH tend to have fast query times, the space consumption is up to 1 000 bytes per node. For this reason, CH cannot be used for Europe (high traffic). Aggressive SHARC however, has the lowest query times but for the price of high preprocessing times. In fact, preprocessing times for aggressive SHARC are only practical for Germany on Sunday. As soon as the graph gets bigger or more edges are time-dependent, preprocessing takes more than 2 days. So, it seems as if economical L-SHARC and CALT are the techniques most robust to the input.

Summarizing, depending on the size of the graph and degree of perturbation, our presented speed-up techniques are 150 to 5 000 times faster than plain

Table 1. Performance of Dijkstra, uni- and bidirectional ALT, Core-ALT, SHARC, and Contraction Hierarchies (CH) in an exact setup. Note that no figures on the number of relaxed edges are given in [1].

| | | PREPRO | | QUERIES | | | | | |
		time [h:m]	space [B/n]	#delete mins	spd up	#relaxed edges	spd up	time [ms]	spd up
Ger mid	Dijkstra	0:00	0	2 305 440	1	5 311 600	1	1 502.88	1
	uni-ALT	0:23	128	200 236	12	239 112	22	148.36	10
	ALT	0:23	128	110 134	21	131 090	41	94.26	16
	CALT	0:09	50	3 190	723	12 255	433	5.36	280
	eco SHARC	1:16	155	19 425	119	104 947	51	25.06	60
	eco L-SHARC	1:18	219	2 776	831	19 005	279	6.31	238
	CH	0:25	1019	528	4 366	–	–	1.22	1231
Ger Sunday	Dijkstra	0:00	0	2 348 470	1	5 410 600	1	1 464.41	1
	uni-ALT	0:23	128	142 631	16	170 670	32	92.79	16
	ALT	0:23	128	58 956	40	70 333	77	42.96	34
	CALT	0:05	19	1 773	1 325	6 712	806	2.13	688
	eco SHARC	0:30	65	2 142	1 097	6 549	826	1.86	787
	eco L-SHARC	0:32	129	576	4 076	2 460	2 200	0.73	2 011
	agg SHARC	27:20	61	670	3 504	1 439	3 759	0.50	2 904
	agg L-SHARC	27:22	125	283	8 300	978	5 535	0.29	5 045
	CH	0:11	248	407	5 770	–	–	0.71	2 061
Europe high traffic	Dijkstra	0:00	0	8 877 158	1	21 006 800	1	5 757.45	1
	uni-ALT	1:15	128	2 143 160	4	2 613 994	8	1 520.83	4
	ALT	1:15	128	3 009 320	3	3 799 112	6	1 379.21	4
	CALT	1:00	61	60 961	146	356 527	59	121.47	47
	eco SHARC	6:44	134	66 908	133	480 768	44	82.12	70
	eco L-SHARC	6:49	198	18 289	485	165 382	127	38.29	150

Table 2. Performance of Dijkstra, uni- and bidirectional ALT, Core-ALT, and SHARC in an approximation setup

input	algorithm	PREPRO time [h:m]	space [B/n]	ERROR error-rate	max rel.	max abs[s]	TIME-QUERIES #del. mins	spd up	#rel. edges	time [ms]	spd up
Ger mid	ALT	0:23	128	12.4%	14.32%	1 892	50 764	45	60 398	36.92	41
	CALT	0:09	50	8.2%	13.84%	2 408	1 593	1 447	5 339	1.87	804
	SHARC	3:26	137	0.8%	0.61%	48	818	2 820	1 611	0.69	2 164
	L-SHARC	3:28	201	0.8%	0.61%	48	334	6 900	1 092	0.38	3 915
Ger Sun	ALT	0:23	128	10.4%	14.28%	1 753	50 349	47	59 994	36.04	41
	CALT	0:05	19	4.0%	12.72%	1 400	1 551	1 514	5 541	1.71	856
	SHARC	1:48	59	0.1%	0.36%	15	635	3 699	1 271	0.46	3 163
	L-SHARC	1:50	123	0.1%	0.36%	15	272	8 639	908	0.27	5 420
Eur high	ALT	1:15	128	35.4%	10.57%	5 789	311 209	29	382 061	214.24	27
	CALT	1:00	61	33.0%	8.69%	6 643	6 365	1 395	32 719	9.22	624
	SHARC	22:12	127	39.6%	1.60%	541	5 031	1 764	8 411	2.94	1 958
	L-SHARC	22:17	191	39.6%	1.60%	541	3 873	2 292	8 103	2.13	2 703

Dijkstra. For all evaluated networks, the query performance is sufficient for most real-world environments.

Approximation. In an approximate scenario, things are clearer. Performance of SHARC is boosted by more than an order of magnitude if we drop correctness combined with a reasonable preprocessing effort. This very good performance comes together with a very good quality of paths. Although ALT and Core-ALT also gain from allowing suboptimal paths, both query performance and quality of paths is (much) worse than for approximate SHARC. We conclude that SHARC is superior if we allow slightly suboptimal paths. Summarizing, approximate SHARC yields speed-ups between 2 700 to 5 420 over Dijkstra's algorithm combined with very low errors.

5.2   Timetable Information

Up to now, Contraction Hierarchies have not been evaluated on graphs deriving from public transportation. Hence, Table 3 shows the results of Dijkstra, uni- and bidirectional ALT, and SHARC for this input. We observe lower speed-ups for timetable information than for road networks in general. Unidirectional ALT is about 66% faster than plain Dijkstra. Even worse, switching from uni- to bidirectional ALT does not pay off. The bad performance of bidirectional ALT derives from the fact that the second phase of the algorithm is long. Hence, we have to explore a great part of the graph after the first path has been found. That is why speed-up over a unidirectional variant is—compared to road networks— rather low. We conclude that ALT works well for road networks but fails on graphs deriving from timetable information for railways.

For SHARC however, we observe a good performance in general. Queries for a specific departure times are up to 29.7 times faster than plain Dijkstra in terms

Table 3. Performance of time-dependent Dijkstra, uni- and bi-directional ALT and SHARC using our timetable data as input. Moreover, we report the increase in edge count over the input. #delete mins denotes the number of nodes removed from the priority queue, query times are given in milliseconds. Speed-up reports the speed-up over the corresponding value for plain Dijkstra.

| | PREPRO | | | TIME-QUERIES | | | | PROFILE-QUERIES | | | |
| | time | space | edge | #delete | speed | time | speed | #delete | speed | time | speed |
technique	[h:m]	[B/n]	inc.	mins	up	[ms]	up	mins	up	[ms]	up
Dijkstra	0:00	0	0%	260 095	1.0	125.2	1.0	1 919 662	1.0	5 327	1.0
uni-ALT	0:02	128	0%	127 103	2.0	75.3	1.7	1 434 112	1.3	4 384	1.2
ALT	0:02	128	0%	262 415	1.0	219.6	0.6	–	–	–	–
eco SHARC	1:30	113	74%	32 575	8.0	17.5	7.2	181 782	10.6	988	5.4
agg SHARC	12:15	120	74%	8 771	29.7	4.7	26.6	55 306	34.7	273	19.5

of search space. This lower search space yields a speed-up of a factor of 26.6. This gap originates from the fact that SHARC operates on a graph enriched by shortcuts. As shortcuts tend to have many interpolation points, evaluating them is more expensive than original edges. As expected, our economical variant is slower than the generous version but preprocessing is almost 8 times faster. Recall that the only difference between both version is the way arc-flags are computed during the last iteration step. Although the number of heap operations is nearly the same for running one label-correcting algorithm per boundary node as for growing two Dijkstra-trees, the former has to use functions as labels. As composing and merging functions is more expensive than adding and comparing integers, preprocessing times increase significantely.

Comparing time- and profile-queries, we observe that computing $d_*(s,t)$ instead of $d(s,t,\tau)$ yields an increase of about factor $4-7$ in terms of heap operations. Again, as composing and merging functions is more expensive than adding and comparing integers, the loss in terms of running times is much higher. Still, both our SHARC-variants are capable of computing d_* for two random stations in less than 1 second.

6 Conclusion

In this paper, we have given an overview over existing speed-up techniques for time-dependent route planning. We identified the basic ingredients these techniques are founded on. Since the speed-up techniques are based on basic ingredients, augmenting the ingredients yields time-dependent speed-up techniques. More precisely, three efficient speed-up techniques can be set up: Core-ALT, SHARC, and Contraction Hierarchies. Experiments on real-world data deriving from road networks and timetable information confirm that these techniques allow the fast computation of time-dependent shortest paths.

Regarding future work, one could think of faster ways of composing, merging, and approximating piece-wise linear functions as this would directly accelerate preprocessing. Aggressive SHARC is the superior technique with respect to query performance. Unfortunately, preprocessing times are impractical in high perturbation scenarios. Since preprocessing is based on building profile graphs being independent of each other, massive parallelization might be an option to preprocess aggressive SHARC in reasonable time for such networks. Another challenging task for the future is to reduce the space consumption of time-dependent Contraction Hierarchies.

Another open problem for route planning is that the quickest route is often not the best one. We might be willing to accept slightly longer travel times if the cost of the journey is less. Such better routes can be computed by running *multi-criteria* queries which take more than one metric into account. While SHARC works in such a scenario [12], it remains to be shown that other approaches can be augmented to such a scenario as well.

Acknowledgments. We would like to thank our coauthors on time-dependent route planning, G. Veit Batz, Leo Liberti, Giacomo Nannicini, Peter Sanders, Dominik Schultes, and Christian Vetter for their valuable contributions. We also had many interesting discussions with Andrew Goldberg, Riko Jacob, Matthias Müller-Hannemann, and Renato Werneck. Finally, we thank PTV AG and HaCon for providing us with real-world data for scientific use.

References

1. Batz, V., Delling, D., Sanders, P., Vetter, C.: Time-Dependent Contraction Hierarchies. In: Proceedings of the 11th Workshop on Algorithm Engineering and Experiments (ALENEX 2009), pp. 97–105. SIAM, Philadelphia (2009)
2. Bauer, R., Delling, D.: SHARC: Fast and Robust Unidirectional Routing. ACM Journal of Experimental Algorithmics, 2.4-2.29 14 (2009); special Section on Selected Papers from ALENEX 2008
3. Bauer, R., Delling, D., Sanders, P., Schieferdecker, D., Schultes, D., Wagner, D.: Combining Hierarchical and Goal-Directed Speed-Up Techniques for Dijkstra's Algorithm. In: McGeoch, C.C. (ed.) WEA 2008. LNCS, vol. 5038, pp. 303–318. Springer, Heidelberg (2008)
4. Cooke, K., Halsey, E.: The Shortest Route Through a Network with Time-Dependent Intermodal Transit Times. Journal of Mathematical Analysis and Applications 14, 493–498 (1966)
5. Dean, B.C.: Continuous-Time Dynamic Shortest Path Algorithms. Master's thesis, Massachusetts Institute of Technology (1999)
6. Delling, D.: Time-Dependent SHARC-Routing. In: Halperin, D., Mehlhorn, K. (eds.) Esa 2008. LNCS, vol. 5193, pp. 332–343. Springer, Heidelberg (2008)
7. Delling, D.: Engineering and Augmenting Route Planning Algorithms. PhD thesis, Universität Karlsruhe (TH), Fakultät für Informatik (2009)
8. Delling, D.: Time-Dependent SHARC-Routing. Algorithmica, July 2009. Special Issue: European Symposium on Algorithms (2008)
9. Delling, D., Nannicini, G.: Bidirectional Core-Based Routing in Dynamic Time-Dependent Road Networks. In: Hong, S.-H., Nagamochi, H., Fukunaga, T. (eds.) ISAAC 2008. LNCS, vol. 5369, pp. 812–824. Springer, Heidelberg (2008)

10. Delling, D., Sanders, P., Schultes, D., Wagner, D.: Engineering Route Planning Algorithms. In: Lerner, J., Wagner, D., Zweig, K.A. (eds.) Algorithmics of Large and Complex Networks. LNCS, vol. 5515, pp. 117–139. Springer, Heidelberg (2009)
11. Delling, D., Wagner, D.: Landmark-Based Routing in Dynamic Graphs. In: Demetrescu, C. (ed.) WEA 2007. LNCS, vol. 4525, pp. 52–65. Springer, Heidelberg (2007)
12. Delling, D., Wagner, D.: Pareto Paths with SHARC. In: Vahrenhold, J. (ed.) SEA 2009. LNCS, vol. 5526, pp. 125–136. Springer, Heidelberg (2009)
13. Dijkstra, E.W.: A Note on Two Problems in Connexion with Graphs. Numerische Mathematik 1, 269–271 (1959)
14. Geisberger, R., Sanders, P., Schultes, D., Delling, D.: Contraction Hierarchies: Faster and Simpler Hierarchical Routing in Road Networks. In: McGeoch, C.C. (ed.) WEA 2008. LNCS, vol. 5038, pp. 319–333. Springer, Heidelberg (2008)
15. Goldberg, A.V., Harrelson, C.: Computing the Shortest Path: A* Search Meets Graph Theory. In: Proceedings of the 16th Annual ACM–SIAM Symposium on Discrete Algorithms (SODA 2005), pp. 156–165 (2005)
16. Goldberg, A.V., Werneck, R.F.: Computing Point-to-Point Shortest Paths from External Memory. In: Proceedings of the 7th Workshop on Algorithm Engineering and Experiments (ALENEX 2005), pp. 26–40. SIAM, Philadelphia (2005)
17. Hart, P.E., Nilsson, N., Raphael, B.: A Formal Basis for the Heuristic Determination of Minimum Cost Paths. IEEE Transactions on Systems Science and Cybernetics 4, 100–107 (1968)
18. Hilger, M., Köhler, E., Möhring, R.H., Schilling, H.: Fast Point-to-Point Shortest Path Computations with Arc-Flags. In: Demetrescu, C., Goldberg, A.V., Johnson, D.S. (eds.) Shortest Path Computations: Ninth DIMACS Challenge. DIMACS Book, vol. 24. American Mathematical Society (to appear, 2009)
19. Ikeda, T., Hsu, M.-Y., Imai, H., Nishimura, S., Shimoura, H., Hashimoto, T., Tenmoku, K., Mitoh, K.: A fast algorithm for finding better routes by AI search techniques. In: Proceedings of the Vehicle Navigation and Information Systems Conference (VNSI 1994), pp. 291–296. ACM Press, New York (1994)
20. Kaufman, D.E., Smith, R.L.: Fastest Paths in Time-Dependent Networks for Intelligent Vehicle-Highway Systems Application. Journal of Intelligent Transportation Systems 1(1), 1–11 (1993)
21. Köhler, E., Möhring, R.H., Schilling, H.: Acceleration of Shortest Path and Constrained Shortest Path Computation. In: Nikoletseas, S.E. (ed.) WEA 2005. LNCS, vol. 3503, pp. 126–138. Springer, Heidelberg (2005)
22. Lauther, U.: An Extremely Fast, Exact Algorithm for Finding Shortest Paths in Static Networks with Geographical Background. In: Geoinformation und Mobilität - von der Forschung zur praktischen Anwendung, vol. 22, pp. 219–230. IfGI prints (2004)
23. Lauther, U.: An Experimental Evaluation of Point-To-Point Shortest Path Calculation on Roadnetworks with Precalculated Edge-Flags. In: Demetrescu, C., Goldberg, A.V., Johnson, D.S. (eds.) Shortest Path Computations: Ninth DIMACS Challenge. DIMACS Book, vol. 24, American Mathematical Society(to appear, 2009)
24. Möhring, R.H., Schilling, H., Schütz, B., Wagner, D., Willhalm, T.: artitioning Graphs to Speed Up Dijkstra's Algorithm. In: Nikoletseas, S.E. (ed.) WEA 2005. LNCS, vol. 3503, pp. 189–202. Springer, Heidelberg (2005)
25. Möhring, R.H., Schilling, H., Schütz, B., Wagner, D., Willhalm, T.: Partitioning Graphs to Speedup Dijkstra's Algorithm. ACM Journal of Experimental Algorithmics 11, 2.8 (2006)

26. Nannicini, G., Delling, D., Liberti, L., Schultes, D.: Bidirectional A* Search for Time-Dependent Fast Paths. In: McGeoch, C.C. (ed.) WEA 2008. LNCS, vol. 5038, pp. 334–346. Springer, Heidelberg (2008)
27. Orda, A., Rom, R.: Shortest-Path and Minimum Delay Algorithms in Networks with Time-Dependent Edge-Length. Journal of the ACM 37(3), 607–625 (1990)
28. Pyrga, E., Schulz, F., Wagner, D., Zaroliagis, C.: Efficient Models for Timetable Information in Public Transportation Systems. ACM Journal of Experimental Algorithmics 12, Article 2.4 (2007)
29. Sanders, P., Schultes, D.: Highway Hierarchies Hasten Exact Shortest Path Queries. In: Brodal, G.S., Leonardi, S. (eds.) ESA 2005. LNCS, vol. 3669, pp. 568–579. Springer, Heidelberg (2005)
30. Schulz, F., Wagner, D., Weihe, K.: Dijkstra's Algorithm On-Line: An Empirical Case Study from Public Railroad Transport. In: Vitter, J.S., Zaroliagis, C.D. (eds.) WAE 1999. LNCS, vol. 1668, pp. 110–123. Springer, Heidelberg (1999)
31. Sedgewick, R., Vitter, J.S.: Shortest Paths in Euclidean Graphs. Algorithmica 1(1), 31–48 (1986)
32. Wagner, D., Willhalm, T.: Drawing Graphs to Speed Up Shortest-Path Computations. In: Proceedings of the 7th Workshop on Algorithm Engineering and Experiments (ALENEX 2005), pp. 15–24. SIAM, Philadelphia (2005)

The Exact Subgraph Recoverable Robust Shortest Path Problem

Christina Büsing*

Institut für Mathematik, Technische Univeristät Berlin, Germany
cbuesing@math.tu-berlin.de

Abstract. Passengers of a public transportation system are often forced to change their planned route due to deviation in travel times. Rerouting is mostly done by simple means such as announcements. We introduce a model, in which the passenger computes his optimal route on his mobile device in a given subnetwork according to the actual travel times. Those travel times are sent to him as soon as a delay occurs.

The main focus of this paper is on the calculation of a small subnetwork. This subnetwork shall contain for every realization of travel times a shortest path of the original network and minimize the number of arcs. For this so called EXACT SUBGRAPH RECOVERABLE ROBUST SHORTEST PATH problem we introduce an approximation algorithm with an approximation factor of $\frac{m}{\ell}$, for any fixed constant $\ell \in \mathbb{N}$. This is the best possible approximation factor for the interval- and the Γ-scenario case, in which all realizations of travel times are given indirectly by lower and upper bounds on the arc cost. Unless $\mathbf{P} = \mathbf{NP}$, for those two scenario sets the problems is not approximable with a factor better than $m^{(1-\varepsilon)}$, where m is the number of arcs in the given graph and $\varepsilon > 0$.

1 Introduction

Motivation. The travel times in public transportation are subject to uncertainty. Major delays can occur by blockades of a track, the break-down of an engine or accidents. Whereas these are exceptional events, minor delays are frequent in all public transportation systems. These delays are caused for example by a jammed door of a vehicle, deviation in stopping times depending on the number of boarding passengers, or weather conditions. In a bus network where the vehicles of the public transportation system use the same infrastructure as the individual traffic, travel times, e.g. between two bus stops, must be considered as random data. Major and minor changes in the travel times can have an effect on the optimal route of an individual passenger through the system. For example, if the passenger was to change to a bus on a short but now crowded road it could be better to change at another station to a different bus line, reaching the same destination on a longer but currently faster road.

---

* Author's maiden name: Christina Puhl. Research supported by the Research Training Group "Methods for Discrete Structures" (DFG-GRK 1408) and the Berlin Mathematical School.

In a railway system classically the route of a passenger is communicated together with the purchase of a ticket. Rerouting is communicated by simple means. In case of delays new connections to main cities are announced by the conductor in the train or at the station. In addition at special information points passengers can get a printout of a new shortest route to their destination.

In the last ten years the electronic ticket has emerged in many public transportation systems. In the airline industry it has almost replaced paper tickets. An e-ticket is used to purchase a seat on an airline and can be ordered by telephone, mobile phone or over the Internet. This new technique is also widely used in urban public transportation systems, e.g. in several German regional transportation systems (Verkehrsverbund Rhein-Ruhr or Verkehrsverbund Berlin-Brandenburg), or in railway systems as managed by the Deutsche Bahn. Along with the ticket a fastest route is provided from the origin to the destination for the passenger. Yet, providing only the optimal itinerary for the planned travel times does not fully explore the potential of mobile devices. Providers now start to use the adherent possibilities for individual rerouting.

Some companies as the Verkehrsverbund Berlin-Brandenburg or the Verkehrsbetriebe Zürich already offer online-information on the actual arrival and departure times of their buses or trains and an updated traveling route, if needed. This information can be retrieved by the passenger via the Internet or with their mobile phone. Since the passenger does not know at what point in time data relevant to his journey has changed, he needs to check his way quite frequently. Multiplying this behavior with the increasing number and use of mobiles phones, the work load for the server can exceed capacity. For each request a shortest path in the complete transportation network has to be calculated and returned to the passenger. All shortest path calculations are done server sided.

Another service for mobile device users offered by airline companies and airports, e.g. Lufthansa or the airports München, Berlin and Frankfurt, is to send information on cancellations of a flight or its delay via SMS. The passenger registers for this service and if a delay is detected on his route, he is automatically informed. Updated routes are not provided.

Since the computational power of a mobile device is increasing, the calculation of a shortest path on small networks can be passed to the passengers. Thus, in a new approach along with an e-ticket a small enough map to be handled on a mobile device is provided. This map is chosen, such that for every likely realization of traveling times – i.e. for every scenario – a shortest path in the original network can already be found in the subnetwork. At the occurrence of a delay, the new distances are sent to the passengers mobile device, which can calculate the new shortest path on his own. Hence, the computational traffic on the main server is reduced since the shortest path calculation is transferred to the passenger.

In this paper, we address the question, which parts of the transportation network are important for the request of a passenger. We consider a set of likely scenarios, in which each scenario defines the travel times. The reduced map guarantees, that for each scenario a shortest path according to the original map is included. This map is recoverable robust against the considered set of likely scenarios. Furthermore, the size of the map is minimized. We call this problem the EXACT SUBGRAPH RECOVERABLE ROBUST SHORTEST PATH problem and provide an approximation algorithm with a best possible approximation factor.

Related Work. The EXACT SUBGRAPH RECOVERABLE ROBUST SHORTEST PATH problem belongs to the class of recoverable robust optimization problems. Recoverable robustness has been introduced by Liebchen et al. [8,9]. This concept combines and generalizes two classical approaches dealing with uncertainties: robustness and 2-stage stochastic programming.

In stochastic programming one assumes to have perfect knowledge about the probability distribution of the scenarios and seeks for a solution that optimizes some stochastic function. A special case, the 2-stage stochastic programming, defines a first stage decision, which is fix for all scenarios, and a second stage decision, taken after all data are known. Together they must form a feasible solution for the revealed scenario. The general aim is to minimize the costs for the first decision and the expected costs for the second part. Minkoff et al. [7] and Ravi et al. [11] applied this method to several combinatorial optimization problems. But in practice many problems tend to be solved once, therefore the expected value loses its relevance. Furthermore, a scenario might appear such that the total costs are much higher than the expected costs. This depends also on the fact that in many applications no stochastic information is given.

Robustness addresses those two problems by neglecting the distribution and using a min-max-criterion. A solution is chosen under the anticipation of a worst case scenario and may be overly conservative. The shortest path and other combinatorial optimization problems have been studied in the robust setting by Bertsimas and Sim [2], Yu and Yang [14] and Aissi et al. [1], among others. The drawback in those settings are the unacceptably high costs of an optimal solution. They also ignore the fact that in most problems a recovery involving at least a minor change to the previously determined solution is possible.

A different robust approach is to find a solution that minimizes the maximal deviation between the costs for the solution and the optimal costs in any scenario. This solution is less conservative. Zielinski [15] showed that this robust shortest path problem with interval scenarios is strongly **NP**-complete even for planar graphs. Karasan et. al [6] give a mixed integer program to solve the problem and introduce a methods for identifying arcs that are never part of a robust solution.

In a recoverable robust approach as defined in [8,9] a first stage decision has to be taken. This decision leads to first stage decision costs and limitations of the feasible solutions in the second stage. We call those the *recovery set* of a decision. In the second stage, when the scenario is known, from the recovery set any solution might be taken. For this solution the scenario costs have to be paid. An optimal recoverable robust solution is a first stage decision that minimizes the first stage costs and the maximal scenario costs by taking the best solution from its recovery set. In contrast to the robust approach, there exists no unique setting of recoverable robustness.

In [10] two different settings of a recoverable robust shortest path problem
have already been studied, the k-ARC-RRSP and the RENT-RRSP. In the
k-ARC-RRSP an (s,t)-path is fixed in the first stage, while in the second stage,
a path can be taken that differs by at most k arcs from the first stage path. In
the RENT-RRSP, as a first stage decision an (s,t)-path is chosen for which first
stage costs, depending on the rental factor α and the revealed scenario, have
to be payed. After the scenario is known any other path might be chosen as
recovery. For an arc a that was part of the first stage decision, we have to pay
in the second stage the difference between the scenario costs and the first stage
costs of this arc, i.e., $(1 - \alpha) \cdot c_a^S$. For any other arc we get extra inflation costs
given by the factor $(1 + \beta)$. Both problems are **NP**-hard for discrete scenario
sets and Γ-scenario sets. The RENT-RRSP is solvable in polynomial time for
interval scenarios and an approximation algorithm with an approximation factor
of $\gamma = \min\{\frac{1}{\alpha}, 2 + \beta\}$ is given for Γ-scenarios.

Definition and Results. In the two described recoverable robust shortest path
problems the scenario costs depend mostly on the scenario cost function and the
chosen recovery path. In the EXACT SUBGRAPH-RRSP setting this is different.
Here, we want to guarantee that for any given scenario a shortest path in the
original graph is also part of the chosen subgraph. This path is always chosen
as recovery and no second stage costs arise. The first stage costs just depend on
the size of the subgraph.

Definition 1 (Exact Subgraph-RRSP). *Let $G = (V, A)$ be a directed graph,
s, t two vertices in V and $\mathcal{S}$ a set of scenarios, each defining a cost function
$c^S : A \to \mathbb{R}^+$. We denote by $\mathcal{P}(G)$ the set of all (s,t)-paths in G. The scenario
costs $c^S(p)$ of a path $p \in \mathcal{P}(G)$ in the scenario $S \in \mathcal{S}$ are defined as*

$$c^S(p) := \sum_{a \in p} c^S(a).$$

A subgraph $G' \subseteq G$ is a feasible first decision, *if*

$$\min_{p' \in \mathcal{P}(G')} c^S(p') = \min_{p \in \mathcal{P}(G)} c^S(p) \quad \forall S \in \mathcal{S}.$$

An optimal solution to the EXACT SUBGRAPH RECOVERABLE ROBUST
SHORTEST PATH PROBLEM *is a feasible first decision with a minimal number of
arcs.*

The analysis of the problem depends on the given scenario set. We distinguish
three settings: the discrete scenario set $\mathcal{S}_D$, the interval scenario set $\mathcal{S}_I$ and
the Γ-scenario set $\mathcal{S}_\Gamma$. In the *discrete scenario set* every scenario is explicitly
given with its cost function [7,11,14]. Also covered in these articles is the *interval
scenario set*, an indirect description of all possible scenarios. For each arc a a
lower and an upper cost bound $\underline{c}_a$ and $\bar{c}_a$ with $0 \le \underline{c}_a \le \bar{c}_a$ is given. The cost
functions of all scenarios in $\mathcal{S}_I$ obey those bounds and for any cost function
$c : A \to \mathbb{R}^+$ with $c_a \in [\underline{c}_a, \bar{c}_a]$ for all $a \in A$, a scenario with this cost function

exists in $\mathcal{S}_I$. For the Γ-*scenario set* again lower and upper cost bounds are fixed. But a scenario $S \in \mathcal{S}_\Gamma$ is only allowed to have at most Γ arc costs deviating from the lower bound. This set has been introduced by Bertsimas and Sim [2].

The EXACT SUBGRAPH-RRSP with discrete scenario sets is **NP**-complete as we will show that 3SAT can be reduced to this setting. For interval scenarios we will show that the decision version of the EXACT SUBGRAPH-RRSP is in **CoNP**, when all arc costs suffer from deviation. This is done by introducing a criterion when an arc has to be part of the first decision. Furthermore, we prove that the optimization version is not approximable with a factor better than $m^{(1-\varepsilon)}$, unless $\mathbf{P} = \mathbf{NP}$, where m is the number of arcs in the graph and $\varepsilon > 0$. Here the reduction is from the DIRECTED 2-DISJOINT PATH PROBLEM.

Finally, we investigate the EXACT SUBGRAPH-RRSP with Γ-scenarios. This problem is **NP**-hard due to a reduction from EXACTLY-ONE-IN-TREE 3SAT. We also show that the problem is not approximable with a factor better than m, if $\mathbf{P} \neq \mathbf{NP}$. These results are stated in Section 2. In the last section we introduce an approximation algorithm with an approximation factor of $\frac{m}{\ell}$, for each $\ell \in \mathbb{N}$, and a running time of $\mathcal{O}(\ell \cdot 2^\ell \cdot m^\ell \cdot \mathrm{SP})$, where SP denotes the running time to solve a shortest path problem. The difficulty of an approximation algorithm lies in the verification of a feasible first decision.

2   The Complexity of the Exact Subgraph-RRSP

Discrete Scenario Sets. In a discrete scenario set $\mathcal{S}_D$, the cost function $c^S : A \to \mathbb{R}^+$ is explicitly given for every scenario S. Since we want to investigate the complexity of the EXACT SUBGRAPH-RRSP with $\mathcal{S}_D$, we consider in the following the decision problem whether a feasible first decision with less than K arcs exists.

Obviously the EXACT SUBGRAPH-RRSP is in **NP**: Let G' be a subgraph of a given graph G. By calculating for every $S \in \mathcal{S}_D$ a shortest path according to c^S in G and comparing its cost to the costs of a shortest path in G' we can decide in polynomial time if G' is a feasible first decision. Furthermore, the problem is solvable in polynomial time for constant K, since in that case the set of all subgraphs G' with K arcs has only $|A(G)|^K$ elements. If K is part of the input, however, the problem is strongly **NP**-complete.

Theorem 1. *The* EXACT SUBGRAPH-RRSP *with discrete scenario sets is strongly* **NP***-complete for* $K \leq \frac{\ell}{\ell+1} \cdot m$, *with* $\ell \in \mathbb{N}$ *arbitrary.*

Proof. We reduce from 3SAT. Let I be an instance of 3SAT with n variables $x_1, \ldots, x_n$ and m clauses $C_1, \ldots, C_m$, and let ℓ be some integer. We construct an instance I' of the EXACT SUBGRAPH-RRSP by defining a graph G' and a set of scenarios $\mathcal{S}_D$. For each variable x_i we introduce two parallel arcs between s and v_i and connect each node v_i via a simple path of length $\ell - 1$ with t (Fig. 1). We denote the (s,t)-path traversing the upper (s, v_i)-arc with p_{x_i} and the one traversing the lower arc with $p_{\overline{x}_i}$.

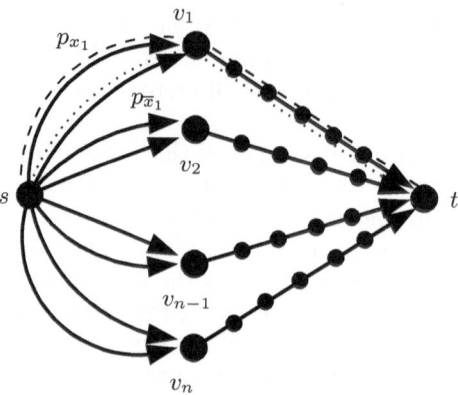

Fig. 1. Each variable x_i is represented by two paths p_{x_i} and $p_{\overline{x}_i}$

The set of scenarios $\mathcal{S}_D$ contains for every clause $C_j = y_{1j} \lor y_{2j} \lor y_{3j}$ a scenario S_j with the following cost function:

$$c^{S_j}(a) = \begin{cases} 0 & \text{if } a \in p_{y_{1j}} \cup p_{y_{2j}} \cup p_{y_{3j}} \\ 1 & \text{otherwise} \end{cases}.$$

Furthermore, we add for every variable x_i a scenario $\tilde{S}_i$ to $\mathcal{S}$ in which all arc costs of the paths p_{x_i} and $p_{\overline{x}_i}$ are set to 0 and all others to 1. Due to these extra scenarios each feasible first decision has to contain at least one of the two paths p_{x_i} and $p_{\overline{x}_i}$. Finally we set $K = \frac{\ell}{\ell+1} \cdot |A(G')|$. Since G' is a connected graph, $|A(G')| = (\ell+1) \cdot n$ and $|\mathcal{S}| = m+n$, the size of I' is polynomial in I.

In the next two paragraphs we will prove, that there exists a feasible first decision G^* with $|A(G^*)| = K = \frac{\ell}{\ell+1} \cdot |A(G')| = \ell \cdot n$ if and only if I is a yes-instance.

Let x^* be a feasible solution of I. We will show, that there exists a feasible first decision G^* with $|A(G^*)| = K$ of G'. This feasible first decision G^* is constructed by adding for each variable x_i the path p_{x_i} to G^* if $x_i^* = \text{true}$ and by adding the path $p_{\overline{x}_i}$ to G^* if $x_i^* = \text{false}$. Hence, $|A(G^*)| = \ell \cdot n = K \cdot |A(G')|$. For every scenario $\tilde{S}_i$ a path of length zero, i.e., a shortest path according to $c^{\tilde{S}_i}$, is contained in G^*. Since x^* is a feasible solution, G^* contains a path of length zero for every other scenario $S_j \in \mathcal{S}_D$. Therefore, G^* is a feasible first decision.

Let G^* be a feasible first decision with $|A(G^*)| = K = n \cdot \ell$. We will construct a feasible solution x^* for the instance I. Since for every variable x_i either p_{x_i} or $p_{\overline{x}_i}$ is part of G^*, those paths already add up to $n \cdot \ell$ arcs. Therefore, the assignment

$$x_i^* = \begin{cases} \text{true} & \text{if } p_{x_i} \in G^* \\ \text{false} & \text{otherwise} \end{cases}$$

is well-defined. Since for every scenario $S \in \mathcal{S}$ a path of Length zero is included in G^*, at least one literal of any clause is verified by x^*. Therefore, x^* is a feasible solution of I. □

Discrete scenario sets are very powerful, since there is in general no dependency between the different cost functions. To restrict the possible values of those cost functions, interval scenarios are often chosen.

Interval scenario sets. An interval scenario set $\mathcal{S}_I$ is defined indirectly by lower and upper bounds $\underline{c}_a$ and $\overline{c}_a$ on the costs for each arc a of the given graph, with $0 \leq \underline{c}_a \leq \overline{c}_a$. For each cost function $c : A \to \mathbb{R}^+$ with $c_a \in [\underline{c}_a, \overline{c}_a]$ there exists a scenario $S \in \mathcal{S}_I$ with $c^S = c$ and every scenario cost function obeys those bounds. The number of scenarios in $\mathcal{S}_I$ is infinite. We start by introducing a criterion for an (s,t)-path to be part of an optimal solution.

Definition 2. *Let p and p' be two (s,t)-paths. If*

$$c^S(p) \leq c^S(p') \quad \forall S \in \mathcal{S}_I,$$

we say p is dominating *p'. If an (s,t)-path is not dominated by any other (s,t)-path, we call this path* dominant.

This characterization of a path has already been defined by Demir et al. [4], but was called a weak path. Karasan et al. use this concept to show, that any robust deviation path with interval scenarios is a dominant path [6]. They also prove, that the dominance of a path can easily be tested.

Theorem 2. *A path p is dominant if and only if p is the unique shortest path according to the scenario S_p with*

$$c^{S_p}(a) = \begin{cases} \underline{c}(a) & \forall a \in p \\ \overline{c}(a) & \forall a \notin p \end{cases}.$$

Since a dominant path is the unique shortest path in the scenario S_p, it has to be part in any feasible first decision graph of an EXACT SUBGRAPH-RRSP instance. This criterion can be extended to a criterion for every arc a, if all arc costs suffer from deviation.

Lemma 1. *Let an EXACT SUBGRAPH-RRSP instance with interval scenario sets and $\underline{c} < \overline{c}$ be given and let G^* be an optimal first decision. Then $a \in G^*$ if and only if there exists a dominant path p with $a \in p$.*

Proof. Let G^* be a feasible first decision with minimal number of arcs. We assume, there is an arc $a' \in G^*$, but there exists no dominant path p with $a' \in p$. We will show that due to this assumption $G' = G^* \backslash \{a'\}$ is also a feasible first decision, a contradiction to the minimality of G^*. If G' is no feasible first decision, then there exists a scenario $S \in \mathcal{S}_I$, in which all shortest paths in G^* according to c^S cross a'. Let p^i with $i = 1$ be such a path. Due to our assumption p^i is not a dominant path. Hence, there exists a path $p^{i+1} \in \mathcal{P}(G)$ with $c^{S_{p^i}}(p^i) \geq c^{S_{p^i}}(p^{i+1})$. If p^{i+1} is not dominant, we set $i = i+1$ and can find another path that dominates p^i and and all other paths p^k with $k \leq i$. Repeating this scheme, we will eventually end up with a dominant path $\tilde{p}$, since the costs $c^{S_{p_i}}(p_i)$ are strictly monotonically decreasing in i due to $\underline{c}_a < \bar{c}_a$. This path $\tilde{p}$ dominates p^1 and is a shortest path according to c^S. Hence by Theorem 2, $\tilde{p}$ is part of G^*. Due to our assumption the arc $a' \notin \tilde{p}$ and therefore $\tilde{p} \in G'$. Thus G' is a feasible first decision. This is a contradiction to the optimality of G^*.   □

Corollary 1. *The* EXACT SUBGRAPH-RRSP *with interval scenario sets and* $\underline{c} < \bar{c}$ *has a unique optimal solution.*

With this local criterion we can show that the EXACT SUBGRAPH-RRSP decision problem is in **coNP**.

Theorem 3. *The* EXACT SUBGRAPH-RRSP *decision problem with interval scenarios and* $\underline{c} < \bar{c}$ *is in* **coNP**.

Proof. A problem is in **coNP**, if there is a polynomial certificate to verify the no-answer to the given decision problem. Let I be an instance of the EXACT SUBGRAPH-RRSP decision problem with a given K. If I is a no-instance, then the optimal feasible decision G^* of this instance has more than $K+1$ arcs. By Lemma 1 for every arc $a \in G^*$ there is a dominant path containing a. Hence, if I is a no-instance, there exists a set of at most $K+1$ dominant paths containing at least $K+1$ different arcs. Since for a set of $K+1$ paths it can be tested in polynomial time if every path is dominant, this results in a certificate to verify a no-answer.   □

Furthermore, the EXACT SUBGRAPH-RRSP optimization problem is strongly **NP**-hard, since it is already **NP**-hard to decide whether an arc is part of a dominant path. This was stated in [6]. Since the proof is not published, we prove the **NP**-hardness of the EXACT SUBGRAPH-RRSP in Theorem 4 by a reduction from DIRECTED 2-DISJOINT PATHS PROBLEM. The **NP**-completeness of this problem follows directly from a lemma published in 1980 by Fortune et al. [5].

Definition 3 (Directed 2-Disjoint Paths Problem). *Let $G = (V, A)$ be a directed graph and s_1, s_2, t_1, and t_2 four distinct vertices of G. The discrete 2-disjoint paths problem is to determine two disjoint paths, p_1 from s_1 to t_1 and p_2 from s_2 to t_2, if such paths exists.*

Notice, that for an EXACT SUBGRAPH-RRSP instance there is no polynomial certificate to test whether a given subgraph is a feasible first decision, unless **P = NP**.

Theorem 4. *The* EXACT SUBGRAPH-RRSP *optimization problem with interval scenarios is strongly* **NP**-*hard.*

Proof. We reduce from DIRECTED 2-DISJOINT PATHS PROBLEM. Let I be an instance of DIRECTED 2-DISJOINT PATHS PROBLEM, i.e., containing a directed

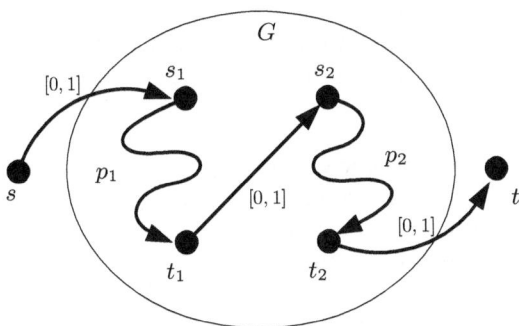

Fig. 2. Every dominant path is a simple path

graph $G = (V, A)$ and four nodes $s_1, t_1, s_2, t_2 \in V$. We transform this instance to an instance I' of the EXACT SUBGRAPH-RRSP by adding two node s and t and the arcs (s, s_1), (t_1, s_2) and (t_2, t) to G. The size of the new graph G' is polynomial in the original one. The lower and upper bounds on each arc are set to 0 and 1 (Fig. 2).

We will show, that there exist an (s_1, t_1)-path p_1 and a node-disjoint (s_2, t_2)-path p_2 in G, if and only if the optimal solution to I' contains the arc (t_1, s_2).

If the optimal solution to I' contains the arc (t_1, s_2), there exists a dominant path p with $(t_1, s_2) \in p$ (Lemma 1). Since p is dominant, p is a simple path. Hence, the parts of the path p connecting s_1 with t_1 and s_2 with t_2 are node disjoint and form an optimal solution to I.

On the other hand, let p_1 and p_2 be two node-disjoint paths in G. The path $p = (s, s_1) \cup p_1 \cup (t_1, s_2) \cup p_2 \cup (t_2, t)$ in G' is a simple path with $c^{S_p}(p) = 0$. The costs of any other (s, t)-path in G' in the scenario S_p are at least one. Thus, p is dominant by Theorem 2.   □

The reduction only works for directed graphs. In an undirected graph the 2-DISJOINT PATH PROBLEM can be solved in polynomial time [13]. We do not know, if the EXACT SUBGRAPH-RRSP remains **NP**-hard for undirected graphs. But we can extend the proof given for Theorem 4, to show the inapproximabiltiy for a factor better than m, with m being the number of arcs in the given graph.

Theorem 5. *There exists no efficient approximation algorithm with an approximation factor $m^{(1-\varepsilon)}$ for the* EXACT SUBGRAPH-RRSP *with interval scenario sets, unless* **P** = **NP**.

Proof. We assume there exists an approximation algorithm ALG with an approximation factor of $m^{(1-\frac{1}{\ell})}$ with $\ell \in \mathbb{N}$. Following the construction in the proof of Theorem 4, we just replace the arc (t_1, s_2) by $(|A(G)| + 2)^\ell - |A(G)| - 2$ parallel arcs. Hence, the new graph G' has $m'^\ell = (|A(G)| + 2)^\ell$ arcs. If the DIRECTED 2-DISJOINTS PATH instance I is a no-instance, $1 \le \text{OPT} \le m' - 1$ in the EXACT SUBGRAPH-RRSP instance I'. Hence, the graph G_{ALG} calculated by the algorithm can have at most $m'^{\frac{\ell-1}{\ell} \cdot \ell} \cdot (m' - 1) = m'^{\ell-1} \cdot (m' - 1)$ arcs.

Since the number of parallel (t_1, s_2) arcs exceeds this bound for $\ell \geq 3$, not all of these arcs can be contained in the solution G_{ALG}. On the other hand, if I is a yes-instance, all parallel arcs are part of G_{ALG}. This leads to a contradiction, unless $\mathbf{P} = \mathbf{NP}$. $\qquad\square$

Any algorithm just returning the given graph is already an approximation algorithm with a factor m. Before we introduce an approximation algorithm with a factor of $\frac{1}{\ell} \cdot m$, we finish the complexity study of the EXACT SUBGRAPH-RRSP by investigating the class of Γ-scenario sets.

$\boldsymbol{\Gamma}$-**scenario sets.** Recall, that the Γ-scenario set is defined as follows: Let $\underline{c}_a$ and $\bar{c}_a$ be lower and upper bounds on the arc costs. A scenario $S \in \mathcal{S}_\Gamma$ is only allowed to have at most Γ arc costs deviating from the lower bound. The concept of dominating paths cannot be transferred to the Γ-scenario sets. Let $G = (V, A)$ be a graph with $V = \{s, t\}$ and five parallel (s,t)-arcs, each of them having cost bounds of 0 and 1. If $\Gamma = 2$, there exists no dominant (s,t)-path, but any optimal solution contains three of the five arcs. Yet, it remains strongly **NP**-hard to compute an optimal solution to the EXACT SUBGRAPH-RRSP. The proof extends the reduction from EXACT-ONE-IN TREE 3SAT to the so called Max Scenario problem. An EXACT-ONE-IN TREE 3SAT instance is given by a set of variables $x_1, \ldots, x_n$ and a set of clauses $C_1, \ldots, C_m$ over the variables, such that each clause contains three literals. The question is, whether there exists a truth assignment of the variables, such that in each clause there is exactly one true literal. The problem is strongly **NP**-complete [12].

Definition 4 (Max Scenario Problem). *Let $G = (V, A)$ be a directed graph, $s, t \in V$ and $\underline{c}_a, \bar{c}_a$ lower and upper bounds for the arcs. The scenario costs for a scenario $S \in \mathcal{S}_\Gamma$ are defined as $c(S) = \min_{p \in \mathcal{P}} c^S(p)$. An optimal solution to the Max Scenario Problem is a scenario $S \in \mathcal{S}_\Gamma$ that maximizes the scenario costs.*

This problem is similar to the discrete time-cost tradeoff (DTCT) problem with negative processing times and the goal to maximize the makespan. The proof for the **NP**-hardness of the DTCT [3] can be transferred to the Max Scenario Problem.

Theorem 6. *The Max Scenario Problem is strongly **NP**-hard.*

A detailed proof is part of the appendix A. In the following we just fix some facts about the reduction from EXACT-ONE-IN-TREE 3SAT to the Max Scenario Problem: Let I be an instance of EXACT-ONE-IN-TREE 3SAT. We can construct a graph G_I with some cost uncertainties modeled by the intervals $[0, 2]$ and $[0, 4]$, such that

1. if there exists a scenario $S \in \mathcal{S}_\Gamma$ with $\min_{p \in \mathcal{P}} c^S(p) = 4$, the instance I is a yes-instance
2. if for every $S \in \mathcal{S}_\Gamma$ there exists a path with costs at most 2, the instance I is a no-instance.

In this reduction any scenario is allowed to have almost half of all uncertain values deviate from the lower bound. Using those facts and the reduction graph G_I, we show the **NP**-hardness of the EXACT SUBGRAPH-RRSP problem.

Theorem 7. *The* EXACT SUBGRAPH-RRSP *with Γ-scenario sets is strongly* **NP**-*hard.*

Proof. We reduce from EXACTLY-ONE-IN-TREE 3SAT. Let I be an instance of EXACTLY-ONE-IN-TREE 3SAT with n variables and m clauses, G_I the graph detailed described in the Appendix A in the proof of Theorem 12. We add an extra arc (s,t) to G_I with upper and lower arc costs of 3 and set $\Gamma = (2m+1) \cdot n + 2m$.

If I is a yes-instance, then there exists a scenario S^* in which all paths in G_I have length 4. Therefore, the arc (s,t) has to be part of the optimal solution. On the other hand, if I is a no-instance, then in every scenario a path of length 2 exists. Hence, the arc (s,t) is in no scenario a shortest path and can be deleted in any feasible first decision. □

Similar to the technique used for interval scenario sets, this proof can be extended to show that an approximation is hard to achieve.

Theorem 8. *There exists no approximation algorithm with a factor $m^{(1-\varepsilon)}$ for the* EXACT SUBGRAPH-RRSP *with Γ-scenario sets, unless* **P** = **NP**.

Proof. Let us assume there exists an algorithm ALG with an approximation factor $m^{(1-\frac{1}{\ell})}$, $\ell \in \mathbb{N}$. Let $G_I = (V, A_I)$ with $|A_I| = m'$ be the graph mentioned above constructed to a given EXACTLY-ONE-IN-TREE 3SAT instance I. Instead of a single (s,t)-arc as in Theorem 7, we add an (s,t)-path $\tilde{p}$ with length $(m'+1)^\ell - m'$ and fix its costs to 3. The new graph has $(m'+1)^\ell$ arcs. If I is a no-instance, the optimal solution is bounded by $1 \leq \text{OPT} \leq m'$. Hence, G_{ALG} has at most $(m'+1)^{\ell-1} \cdot m'$ arcs. Since the path $\tilde{p}$ for $\ell \geq 3$ is too long to be contained in G_{ALG}, this path cannot be part of the constructed graph. But if I is a no-instance, then $\tilde{p}$ has to be in G_{ALG}. Unless **P** = **NP**, this is a contradiction. □

In the next section we will give an algorithm that calculates an $\frac{m}{\ell}$-approximation for all three scenario types.

3   Approximation Algorithms

For any EXACT SUBGRAPH-RRSP instance independent of the type of scenarios sets, an algorithm returning the whole graph is an m-approximation. We improve this factor to $\frac{m}{\ell+1}$ for any $\ell \in \mathbb{N}$ and a running time of $\mathcal{O}(\ell \cdot 2^\ell \cdot m^\ell \cdot \text{SP})$. We start by introducing a criterion to test a given subgraph to its feasibility for interval scenario sets. The test is polynomial in the size of the given subgraph, if this size is bounded.

Theorem 9. *Let $G = (V, A)$ be a directed graph with lower and upper arc costs $\underline{c}$ and $\bar{c}$. Let $A' \subseteq A$ and*

$$\mathcal{S}_{A'}^I = \{S \in \mathcal{S} \mid c_a^S = \underline{c}_a \ \forall a \notin A' \text{ and } c_a^S \in \{\underline{c}_a, \bar{c}_a\} \ \forall a \in A'\}.$$

If for all scenario $\mathcal{S}_{A'}^I$ a shortest path in G is also contained in $G' = (V, A')$, then G' is a feasible first decision.

Proof. Assume G' is no feasible first decision. Then there exists a scenario $\bar{S}$, in which a shortest path $\bar{p}$ in G is shorter than any other path in G'. W.l.o.g. $c_a^{\bar{S}} \in \{\underline{c}_a, \bar{c}_a\}$ for all $a \in A$. Let p be any (s, t)-path in G'. Then

$$0 < c^{\bar{S}}(p) - c^{\bar{S}}(\bar{p})$$

$$= \sum_{a \in p \setminus \bar{p}} c_a^{\bar{S}} - \sum_{a \in \bar{p} \setminus p} c_a^{\bar{S}}$$

$$= \sum_{a \in p \setminus \bar{p}} c_a^{\bar{S}} - \sum_{\substack{a \in \bar{p} \setminus p \\ a \in A'}} c_a^{\bar{S}} - \sum_{\substack{a \in \bar{p} \setminus p \\ a \notin A'}} c_a^{\bar{S}}$$

$$\leq \sum_{a \in p \setminus \bar{p}} c_a^{\bar{S}} - \sum_{\substack{a \in \bar{p} \setminus p \\ a \in A'}} c_a^{\bar{S}} - \sum_{\substack{a \in \bar{p} \setminus p \\ a \notin A'}} \underline{c}_a.$$

Let us define a scenario S^* according to the cost function

$$c_a^{S^*} = \begin{cases} c_a^{\bar{S}} & \forall a \in A' \\ \underline{c}_a & \text{otherwise} \end{cases}.$$

Hence, $S^* \in \mathcal{S}_{A'}^I$ by definition. But due to the above calculations

$$c^{S^*}(p) - c^{S^*}(\bar{p}) > 0 \quad \forall p \in \mathcal{P}(G').$$

This is a contradiction.   □

A similar result can be achieved for Γ-scenario sets.

Theorem 10. *Let $G = (V, A)$ be a directed graph with lower and upper arc costs $\underline{c}$ and $\bar{c}$. Let $A' \subseteq A$ and*

$$\mathcal{S}_{A'}^{\Gamma} = \{S \in \mathcal{S}_{\Gamma} \mid c_a^S = \underline{c}_a \ \forall a \notin A' \text{ and } c_a^S \in \{\underline{c}_a, \bar{c}_a\} \ \forall a \in A'\}.$$

If for all scenario $\mathcal{S}_{A'}^{\Gamma}$ a shortest path in G is also contained in $G' = (V, A')$, then A' is feasible solution.

The proof works analogously to the one for Theorem 9. Both sets $\mathcal{S}_{A'}^I$ and $\mathcal{S}_{A'}^{\Gamma}$ contain exponentially many scenarios depending on the size of A'. If A'

is bounded, then the feasibility of $G' = (V, A')$ can be tested in polynomial time $(\mathcal{O}(2^{|A'|} \cdot \mathrm{SP}))$. This idea leads to the following algorithm:

Algorithm 1. An $\frac{m}{\ell+1}$-approximation for $\mathcal{S}_\Gamma$ and $\mathcal{S}_I$.

Data: Graph $G = (V, A)$, $s, t \in V$, $\ell \in \mathbb{N}$, $\Gamma \in \mathbb{N}$, $\underline{c}_a$, $\bar{c}_a$ $\forall a \in A$
Result: Feasible first decision $G' = (V, A')$

forall $i = \ell, \ldots, 1$ **do**
    forall $\overline{A} \subseteq A$ *with* $|\overline{A}| = i$ **do**
        forall $S \in \mathcal{S}_{\overline{A}}^\Gamma$ **do**
            Calculate $d^S = \min_{p \in \mathcal{P}(\overline{G})} c^S(p) - \min_{p' \in \mathcal{P}(G)} c^S(p')$
        if $d^S = 0$ *for all* $S \in \mathcal{S}_{\overline{A}}^\Gamma$ **then**
            Set $A' = \overline{A}$

if $A' = \emptyset$ **then**
    return $G' = (V, A)$
else
    return $G' = (V, A')$

Theorem 11. *Algorithm 1 calculates a feasible first decision G_{ALG} with*

$$\frac{|A(G_{\mathrm{ALG}})|}{|A(G_{\mathrm{OPT}})|} \leq \frac{m}{\ell+1}$$

for any $\ell \in \mathbb{N}$ and $\Gamma = |A(G)|$ in the interval case. The running time of the algorithm is $\mathcal{O}(\ell \cdot 2^\ell \cdot m^\ell \cdot \mathrm{SP})$.

Proof. For $A(G_{\mathrm{OPT}})$ the values d^S are zero for all $S \in \mathcal{S}_{A(G_{\mathrm{OPT}})}^\Gamma$. If $A(G_{\mathrm{OPT}}) \leq \ell$, the set $\overline{A} = A(G_{\mathrm{OPT}})$ is tested and A' is set to $A(G_{\mathrm{OPT}})$. Hence, $|A'| = |A(G_{\mathrm{ALG}})| = |A(G_{\mathrm{OPT}})|$. If the optimal solution contains more then $\ell + 1$ arcs,

$$\frac{|A(G_{\mathrm{ALG}})|}{|A(G_{\mathrm{OPT}})|} \leq \frac{m}{\ell+1}.$$

$\square$

Since no efficient approximation algorithm with a factor of $m^{(1-\varepsilon)}$ for the EXACT SUBGRAPH-RRSP with $\mathcal{S}_\Gamma$ and $\mathcal{S}_I$ exists (unless $\mathbf{P} = \mathbf{NP}$), no approximation algorithm with a better factor than the one for Algorithm 1 can be found.

4   Conclusion

We showed the **NP**-hardness for the EXACT SUBGRAPH-RRSP problem with scenario settings $\mathcal{S}_D$, $\mathcal{S}_I$ and $\mathcal{S}_\Gamma$. For $\mathcal{S}_I$ and $\mathcal{S}_\Gamma$ with just partial uncertainties on the arc costs the question whether the EXACT SUBGRAPH-RRSP decision problem is in **NP** or in **coNP** remains open. In the special case of $\mathcal{S}_I$, when all arc costs may change, we introduced a criterion to verify no-instances. This criterion is not valid for the general setting.

In the case of a discrete scenario set, Algorithm 1 also works by setting $\mathcal{S}_{A'}^{\Gamma} = \mathcal{S}_D$ and computes an $\frac{m}{\ell+1}$-approximation. So far we did not succeed in finding a lower bound for the approximation factor.

Acknowledgment

We thank Christian Liebchen, Rolf H. Möhring, Martin Skutella and Sebastian Stiller for fruitful discussions and motivation.

References

1. Aissi, H., Bazgan, C., Vanderpooten, D.: Approximation complexity of min-max (regret) versions of shortest path, spanning tree, and knapsack. In: Brodal, G.S., Leonardi, S. (eds.) ESA 2005. LNCS, vol. 3669, pp. 862–873. Springer, Heidelberg (2005)
2. Bertsimas, D., Sim, M.: Robust discrete optimization and network flows. Mathematical Programming B 98, 49–71 (2003)
3. De, P., Dunne, J., Wells, C.: Complexity of the discrete time-cost tradeoff problem for project networks. Operations Research 45(2), 302–306 (1997)
4. Demir, M.H., Kara, B.Y., Transel, B.C.: Tree network 1-median location problem with interval data: A parameter space based approach. IIE Transactions 37, 429–439 (2005)
5. Fortune, S., Hopcroft, J., Wyllie, J.: The directed subgraph homeomorphism problem. Theoretical Computer Science 10, 111–121 (1980)
6. Karasan, O.E., Pinar, M.C., Yaman, H.: The robust shortest path problem with interval data (2003)
7. Karger, D., Minkoff, M.: Building steiner trees with incomplete global knowledge. Proceedings of the 41st Annual IEEE Symposium on Foundations of Computer Science, 613–623 (2000)
8. Liebchen, C., Lübbecke, M., Möhring, R.H., Stiller, S.: Recoverable robustness. Tech. Report ARRIVAL-TR-0066, ARRIVAL-Project (2007)
9. Liebchen, C., Lüebbecke, M., Möhring, R.H., Stiller, S.: The concept of recoverable robustness, linear programming recovery, and railway applications. In: Ahuja, R.K., Möhring, R.H., Zaroliagis, C.D. (eds.) Robust and Online Large-Scale Optimization. LNCS, vol. 5868, pp. 1–27. Springer, Heidelberg (2009)
10. Puhl, C.: Recoverable shortest path problems. Preprint 034-2008, Institute of Mathematics, Technische Universität Berlin (2008)
11. Ravi, R., Sinha, A.: Hedging uncertainty: Approximation algorithms for stochastic optimization problems. Math. Program. 108, 97–114 (2006)
12. Schaefer, T.: The complexity of satisfiability problems. In: Proceedings of the tenth annual ACM symposium on Theory of computing, pp. 216–226 (1978)
13. Shiloach, Y.: A polynomial solution to the undirected two paths problem. Journal of Association for Computing Machinery 27(3), 445–456 (1980)
14. Yu, G., Yang, J.: On the robust shortest path problem. Computer and Operations Research 25(6), 457–468 (1998)
15. Zielinski, P.: The computational complexity of the relativ robust shortest paht problem. European Journal of Operational Research 158(3), 570–576 (2004)

A   The Max-Scenario-Problem

The Max-Scenario-problem is a sub-problem of the recoverable robust shortest path problems.

Definition 5 (Max-Scenario-problem). *Let $G = (V, A)$ be a directed graph, $s, t \in V$, and let $\mathcal{S}$ be a set of scenarios each defining a cost function $c^S : A \to \mathbb{R}_{\geq 0}$. The value* value$(S)$ *of a scenario S is determined through the shortest path according to c^S, i.e.*

$$\text{value}(S) = \min_{p \in \mathcal{P}} c^S(p).$$

An optimal solution to the Max-Scenario-problem *is a scenario $S \in \mathcal{S}$ with a maximal value.*

The Max-Scenario-problem is easy to solve for discrete scenarios and interval scenarios. For Γ-scenarios the problem is similar to the discrete time-cost tradeoff (DTCT-) problem with negative processing times and the goal to maximize the makespan. The proof for the **NP**-hardness of the DTCT [3] can be transferred to the Max-Scenario-problem with $\mathcal{S}_\Gamma$.

Theorem 12. *The   the   Max-Scenario-problem   with Γ-scenarios   is* **NP***-complete.*

Proof. For any scenario S its feasibility, i.e., $S \in \mathcal{S}_\Gamma$, and its value value(S) can be tested in polynomial time. Therefore, the decision version of the Max-Scenario-problem is in **NP**.

We reduce the **NP**-hard EXACT-ONE-IN-THREE 3SAT problem [12] to the Max-Scenario-problem with $\mathcal{S}_\Gamma$. Let I be an EXACT-ONE-IN-THREE 3SAT instance with $x_1, \ldots, x_n$ variables and $C_1, \ldots, C_m$ clauses. Each clause C_j consists of three literals $y_{j1}, \ldots, y_{j3} \in \{x_1, \overline{x}_1, \ldots, x_n, \overline{x}_n\}$, i.e.,

$$C_j = y_{j1} \vee y_{j2} \vee y_{j3}.$$

W.l.o.g. x_i or $\overline{x}_i$ is contained in a least one clause. A *feasible solution* to I is a vector $x \in \{\text{true}, \text{false}\}^n$, such that exactly one literal in every clause is fulfilled with true. We construct a Max-Scenario-instance I' with Γ-scenarios, i.e., we define a graph G, lower and upper cost-bounds and Γ. We start with the graph G. For each variable x_i the graph G contains a fork G_{x_i} with $s_i = s$, the origin node in G. A *fork* is a graph G_{x_i} defined by three arcs $a_i, a_{x_i}, a_{\overline{x}_i}$ and four nodes $s_i, y_i, v_{x_i}, v_{\overline{x}_i}$, with $a_i = (s_i, y_i)$, $a_{x_i} = (y_i, v_{x_i})$ and $a_{\overline{x}_i} = (y_i, v_{\overline{x}_i})$. The arcs a_i and $a_{\overline{x}_i}$ are block-arcs. A *block-arc* (v, w) is an arc representing M parallel (v, w) arcs each having the same properties, e.g., the lower and upper cost-bound. We call a_i the *handle* of a fork, a_{x_i} the *true arm* of a fork and $a_{\overline{x}_i}$ the *false arm* of a fork (Fig. 3).

Furthermore, G has three parallel arcs a_{j1}, a_{j2} and a_{j3} for each scenario C_j. Each arc represents a true assignment for C_j, where for a_{ji} the ith literal is true. We call those arcs the *clause-arcs*. Each clause-arc is connected with t,

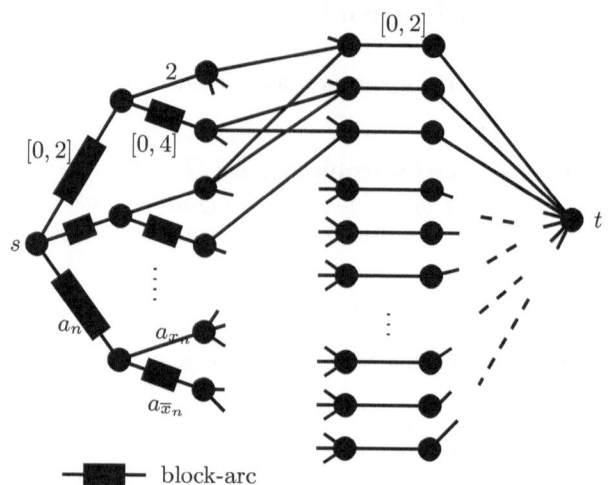

Fig. 3. The arcs a_n, a_{x_n} and $a_{\overline{x}_n}$ form the fork G_{x_n}. For every clause C_j, there exist three clause-arcs a_{j1}, a_{j2} and a_{j3}.

the destination node in G. We finish the construction of G by defining the arcs between the fork arms and clause-arcs. Let a_{ji} be a clause-arc to the clause $C_j = y_{j1} \vee y_{j2} \vee y_{j3}$. For $\ell \neq i$ and $y_{j\ell} = \overline{x}_k$, we connect the true arm of the fork G_{x_k} with a_{jk} and if $y_{j\ell} = x_k$ we connect the false arm of G_{x_k} with a_{ji}. For $\ell = i$, we add an arc between the true arm of G_{x_k} and a_{ji} if $y_{ij} = x_k$. If $y_{ij} = \overline{x}_k$, we connect the false arm of G_{x_k} with a_{ji} (Fig. 4).

We continue with the upper and the lower cost bounds in G. The handles, the true arms and the clause-arcs get upper cost bounds of 2 and the false arms get bounds of 4. Furthermore, the lower bounds of the true arms are set to 2, i.e., the costs of those arcs are not subject to uncertainties. Every other cost bound is set to 0 (Fig. 3). Note that the size of G is polynomial in the input for $M = 2m + 1$. We set $\Gamma = M \cdot n + 2m$.

We will prove, that there exists a Γ-scenario S^* with value$(S^*) = 4$ in I' if and only if there is a feasible solution for the instance I.

Let x^* be a feasible solution to I. We define the cost function of S^* for all arcs with uncertainty in the following way: If x_i^* is true, S^* assigns upper costs to the handle of G_{x_i} and lower costs to the false arm. If x_i^* is false, the false arm gets the upper costs and the handle the lower costs. Notice that any (s,t)-path already has a length of 2 due to this cost assignments. Since x^* is a feasible solution, in every clause C_j, there exists exactly one literal $i_j \in \{1, 2, 3\}$ which has a true assignment. Scenario S^* puts the costs of all clause-arcs a_{ji} with $i \neq i_j$ to their upper bounds and leaves the costs of a_{ji_j} at the lower bound (Fig. 4). In total S^* changes n block-arcs and $2m$ clause-arcs, i.e., $n \cdot M + 2m$ arc costs. Therefore, S^* is a Γ-scenario. It remains to show that any shortest path in G with c^{S^*} has a length of 4.

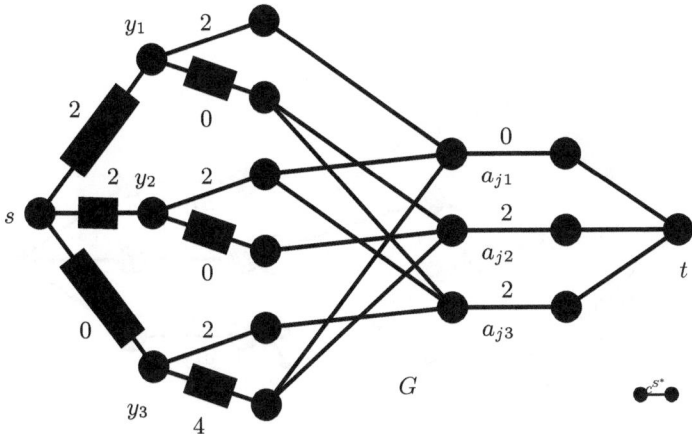

Fig. 4. This graph G is constructed for the instance I with $C_1 = x_1 \vee \bar{x}_2 \vee x_3$. The scenario S^* to a feasible solution $x^* = (\text{true, true, false})$ has value$(S^*) = 4$. In C_j the first variable x_1 verifies the clause. Therefore, the costs of a_{j1} are not raised.

Assume that there exists a path p with costs of 2. Then p has to cross a clause-arc a_{ji_j}, since all paths to a clause-arc have already a length of 2. If $y_{ji_j} = x_\ell$, then x_ℓ has a true assignment. Therefore, any path traversing the true arm of G_{x_ℓ} has, due to the definition of S^*, length of 4 or more. The same argument works for $y_{ji_j} = \bar{x}_\ell$. If $y_{ji} = x_\ell$ for $i \neq i_j$, then a_{ji_j} is connected to the false handle of G_{x_ℓ}. Since the literal y_{ji} is false, the variable x_ℓ^* is set to false. Therefore, any path crossing this arm, has length of at least 4. The same conclusions are valid for $y_{ji} = \bar{x}_\ell$. Hence, paths traversing a_{ji_j} have already a length of 4 before they pass the clause-arc. This is a contradiction.

Let S^* be a Γ-scenario in I' with value$(S^*) = 4$. Before we start with a construction of x^*, we need some observations.

1. Observation: The scenario S^* assigns in every fork exactly one block-arc to the upper cost bound.

Proof: Assume that there is a fork G_{x_ℓ} in which no block-arc is assigned to $\bar{c}$. Then in the handle block-arc and in the false arm block-arc exists an arc with costs of 0. An (s,t)-path traversing these two arcs has at most costs of 2. This is a contradiction to value$(S^*) = 4$. Since $2m < M$, at most n block-arcs can have upper bound costs. $\triangle$

2. Observation: Exactly two clause-arcs of each clause-are moved to their upper bounds.

Proof: Assume there exists a clause C_j, in which only one clause-arc is changed to the upper costs. Each one of the three clause-arcs a_{j1}, a_{j2} and a_{j3} is connected to the same forks G_{x_a}, G_{x_b} and G_{x_c}. Since in every fork one of the block-arcs has

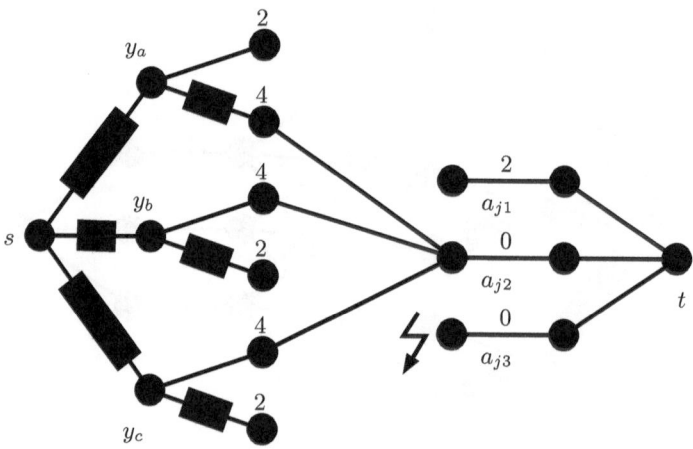

Fig. 5. If a scenario S moves just one of three clause-arcs, then there exists an (s,t)-path in G of length 2

been assigned to the upper costs, either a shortest path to the end of the true arm or a shortest path to the end of the false arm has length of 4. The other one has length of 2. Let a_{j1} w.l.o.g. be the one clause, in which the costs have been moved up. Since the shortest path from s to t has a length of 4 and the other two clause-arcs a_{j2} and a_{j3} have costs of 0, both must be connected to the three arms with the higher costs (Fig. 5). This is a contradiction to the construction of G.

Since S^* already changed $n \cdot M$ arc costs, there are just $2m$ possibilities left; two for every clause.   △

Now we define a solution x^* to the scenario S^*

$$x_i^* = \begin{cases} \text{true} & \text{if } c^{S^*}(a_i) = 2 \\ \text{false} & \text{otherwise} \end{cases}.$$

For every clause C_j, there is one clause-arc a_{ji_j} with costs 0. W.l.o.g. $i_j = 1$. If $y_{j1} = x_\ell$, then a_{j1} is connected to the true arm. Every path crossing this arm has to have a length of 4. Therefore, the handle arc a_ℓ has to have costs at the upper bound and hence $x_a^* = \text{true}$. The same argumentation works for $y_{j1} = \overline{x}_a$. Furthermore, for $y_{ji} = x_{b_i}$ or $y_{ji} = \overline{x}_{b_i}$ with $i \in \{2,3\}$ the two variables are set such that they neglect the clause. Hence, x^* is a feasible solution.

This completes the proof of the **NP**-completeness of the Max-Scenario-problem.   □

In the following we denote the graph G of the reduction from an EXACT-ONE-IN-TREE 3SAT instance I as G_I. Notice that G_I can be reduced by connecting all clause-arcs directly to t. Hence all simple (s,t)-paths have a length of 4.

Efficient Timetable Information in the Presence of Delays[*]

Matthias Müller-Hannemann[1] and Mathias Schnee[2]

[1] Martin-Luther-University Halle, Computer Science,
Von-Seckendorff-Platz 1, 06120 Halle, Germany
muellerh@informatik.uni-halle.de
[2] Darmstadt University of Technology, Computer Science,
Hochschulstraße 10, 64289 Darmstadt, Germany
schnee@algo.informatik.tu-darmstadt.de

Abstract. The search for train connections in state-of-the-art commercial timetable information systems is based on a static schedule. Unfortunately, public transportation systems suffer from delays for various reasons. Thus, dynamic changes of the planned schedule have to be taken into account. A system that has access to delay information about trains (and uses this information within search queries) can provide valid alternatives in case a connection does not work. Additionally, it can be used to actively guide passengers as these alternatives may be presented before the passenger is already stranded at a station due to an invalid transfer.

In this work, we present an approach which takes a stream of delay information and schedule changes on short notice (partial train cancellations, extra trains) into account. Primary delays of trains may cause a cascade of so-called secondary delays of other trains which have to wait according to certain policies for delays between connecting trains. We introduce the concept of a dependency graph to efficiently calculate and update all primary and secondary delays. This delay information is then incorporated into a time-expanded search graph which has to be updated dynamically. These update operations are quite complex, but turn out to be not time-critical in a fully realistic scenario.

We finally present a case study with data provided by Deutsche Bahn AG, showing that this approach has been successfully integrated into the multi-criteria timetable information system MOTIS and can handle massive delay data streams instantly.

Keywords: timetable information system, primary and secondary delays, dependency graph, dynamic graph update.

1 Introduction and Motivation

In recent years, the performance and quality of service of electronic timetable information systems has increased significantly. Unfortunately, not everything

---

[*] A preliminary version of this paper has appeared in Proceedings of ATMOS 2008 [1].

R.K. Ahuja et al. (Eds.): Robust and Online Large-Scale Optimization, LNCS 5868, pp. 249–272, 2009.
© Springer-Verlag Berlin Heidelberg 2009

runs smoothly in scheduled traffic and delays are the norm rather than the exception.

Delays can have various causes: Disruptions in the operations flow, accidents, malfunctioning or damaged equipment, construction work, repair work, and extreme weather conditions like snow and ice, floods, and landslides, to name just a few. On a typical day of operation in Germany, an online system has to handle about 6 million forecast messages about (mostly small) changes with respect to the planned schedule and the latest prediction of the current situation. Note that this high number of changes also includes cases where delayed trains catch up some of their delay.

A system that incorporates up-to-date train status information (most importantly, information about future delays based on the current situation) can provide a user with valid timetable information in the presence of disturbances.

Such an on-line system can additionally be utilized to verify the current status of a journey.

- Journeys can either be still valid (i.e., they can be executed as planned),
- they can be affected such that the arrival at the destination is delayed, or
- they may no longer be possible.

In the latter case, a connecting train will be missed, either because the connecting train cannot wait for a delayed train, or the connecting train may have been canceled. In a delay situation, such status information is very helpful. In the positive case – all planned train changes are still possible – passengers can be reassured that they do not have to worry about missing their connecting train(s). To learn that one will arrive x minutes late with the planned sequence of trains may allow a customer to make arrangements, e.g. inform someone to pick one up later. In the unfortunate case that a connecting train will be missed, this information can now be obtained well before the connection breaks and the passenger is stranded at some station. Therefore, valid alternatives may be presented while there are still more options to react. This situation is clearly preferable to missing a connecting train and then using any information system (ticket machine, service counter) to request an alternative.

Because up to now commercial systems do not take the current situation into account (even though estimated arrival times may be accessible for a given connection, these times are not used actively during the search), their recommendations may be impossible to use, as the proposed alternatives already suffer from delays and may even already be infeasible at the time they are delivered by the system.

From Static to Real-Time Timetable Information Systems. The standard approach to model static timetable information is as a shortest path problem in either a time-expanded or time-dependent graph. The recent survey [2] describes the models and suitable algorithms in detail. Previous research on timetable information systems has focused on the static case, where the timetable is considered as fixed.

Here we start a new thread of research on dynamically changing timetable data as a consequence of disruptions. Our contribution is

- the development of a first prototypal but yet completely realistic timetable information system that incorporates current train status information into a multi-criteria search for attractive train connections. Modeling issues have been discussed in the literature on a theoretical level [3] but no true-to-life system with real delay data has been studied and, to our knowledge, no such system that guarantees optimal results (with respect to even a single optimization criterion) exists. We provide results of implementing such a system for a real world scenario with no simplifying assumptions.
- We propose a system architecture intended for a multi-server environment, where the availability of search engines has to be guaranteed at all times. Our system consists of two main components, a *real-time information server* and one or several *search servers*. The real-time information server receives a massive stream of status messages about delayed trains. Its purpose is to integrate schedule changes into the "planned schedule". Moreover, it has to compute from the received messages (primary delays) all so-called secondary delays which result from trains waiting for each other according to certain waiting policies. The new overall status information is then sent to the search servers which incorporate all changes into their internal model. Search servers, in turn, are used to answer customer queries about train connections.
- Both servers require a specific graph model as the underlying basic data structure. We here introduce the concept of a *dependency graph* as a model to efficiently propagate primary delay information according to policies for delays in the real-time information server. Our dependency graph (introduced in Section 4) is similar to a simple time-expanded graph model with distinct nodes for each departure and arrival event in the entire schedule for the current and following days. This is a natural and efficient model, since every event has to store its own update information.

  For the search server we use a *search graph*. Here, we are free to use either the time-expanded or the time-dependent model. In this work, we have chosen to use the time-expanded model for the search graph, since our previous work, the timetable information server MOTIS [4], is based on this. Although update operations are quite complex in this model, it will turn out that they can be performed very efficiently, in $17\mu s$ per update message on average.
- To store a full timetable over a typical period of a year, static timetable systems are usually built on a compact data structure. For example, they identify the same events on different days of operation and use bitfields to specify valid days. This space saving technique does not work in a dynamic environment since the members of such an equivalence class of events have to be treated individually, as they will have, in general, different delays. We will show how a static time-expanded graph model can be extended to a dynamic graph model without undue increase in space consumption.

Related Work. Delling et al. [3] independently of us came up with ideas on how to regard delays in timetabling systems. In contrast to their work we do not primarily work on edge weights, but consider nodes with timestamps. The edge weight for time follows, whereas edge weights for transfers and cost do not change during the update procedures. This is important for the ability to do multi-criteria search. Due to a number of low-level optimizations we achieve a considerable speed-up over the preliminary work in Frede et al. [1].

A related field of current research is disposition and delay management. Gatto et al. [5,6] have studied the complexity of delay management for different scenarios and have developed efficient algorithms for certain special cases using dynamic programming and minimum cut computations. Various policies for delays have been discussed, for example by Ginkel and Schöbel [7]. Schöbel [8] also proposed integer programming models for delay management. Stochastic models for the propagation of delays are studied, for example, by Meester and Muns [9]. Policies for delays in a stochastic context are treated in [10].

Overview. The remainder of this paper is organized as follows. In Section 2, we will discuss primary and secondary delays. We introduce our system architecture in Section 3, and describe its two main components afterwards. First, we explain our dependency graph model and the propagation algorithm for delays (in Section 4). Then, we briefly review the time-expanded graph model and our search server MOTIS (Section 5). Afterwards, we present the update of the search graph (in Section 6). A major issue for a real system, the correct treatment of days of operation, will be discussed in Section 7. Afterwards, we provide our experimental results in Section 8. Finally, we conclude and give an outlook.

2   Up-to-Date Status Information

2.1   Primary Delay Information

First of all, the input stream of status messages consists of reports that a certain train departed or arrived at some station at time τ either on time or delayed by x minutes. In case of a delay, such a message is followed by further messages about predicted arrival and departure times for all upcoming stations on the train route.

Besides, there can be information about additional trains (specified by a list of departure and arrival times at stations plus category, attribute and name information). Furthermore, we have (partial) train cancellations, which include a list of departure and arrival times of the canceled stops (either all stops of the train or from some intermediate station to the last station).

Moreover, we have manual decisions by the transport management of the form: "Change from train t to t' will be possible" or "will not be possible". In the first case it is guaranteed that train t' will wait as long as necessary to receive passengers from train t. In the latter case the connection is definitely going to break although the current prediction might still indicate otherwise. This information may depend on local knowledge, e.g. that not enough tracks are available to wait or that additional delays are likely to occur, or may be based on global considerations about the overall traffic flow. We call messages of this type *connection status decisions*.

2.2 Secondary Delays

Secondary delays occur when trains have to wait for other delayed trains. Two simple, but extreme examples for policies for delays are:

- *never wait.* In this policy, no secondary delays occur at all in our model. This causes many broken connections and in the late evening it may imply that customers do not arrive at their destination on the same travel day. However, nobody will be delayed who is not in a delayed train.
- *always wait as long as necessary.* In this strategy, there are no broken connections at all, but massive delays are caused for many people, especially for those whose trains wait and have no delay on their own.

Both of these policies seem to be unacceptable in practice. Therefore, train companies usually apply a more sophisticated rule system specifying which trains have to wait for others and for how long. For example, the German railways Deutsche Bahn AG employ a complex set of rules, dependent on train type and local specifics.

In essence, this works as follows: There is a set of rules describing the maximum amount of time a train t may be delayed to wait for passengers from a feeding train f. Basically, these rules depend on train categories and stations. But there are also more involved rules, like if t is the last train of the day in its direction, the waiting time is increased, or during peak hours, when trains operate more frequently, the waiting time may be decreased.

The *waiting time* $wt_s(t, f)$ is the maximum delay acceptable for train t at station s waiting for a feeding train f. Let $dep_s^{sched}(t)$ and $dep_s(t)$ be the departure time according to the schedule resp. the new departure time of train t at station s, $arr_s(t)$ the arrival time of a train and $minct_s(f, t)$ the *minimum change time* needed from train f to train t at station s. Note that in a delayed scenario the change time can be reduced, as guides may be available that show changing passengers the way to their connecting train. Train t waits for train f at station s if

$$arr_s(f) + minct_s(f, t) - dep_s^{sched}(t) < wt_s(t, f).$$

In this case, train t will incur a secondary delay. Its new departure time is determined by the following equation

$$dep_s(t) = \begin{cases} arr_s(f) + minct_s(f, t) & \text{if } t \text{ waits} \\ dep_s^{sched}(t) & \text{otherwise.} \end{cases}$$

In case of several delayed feeding trains, the new departure time will be determined as the maximum over these values.

During day-to-day operations these rules are always applied automatically. If the required waiting time of a train lies within the bounds defined by the rule set, trains will wait. Otherwise they will not. All exceptions from these rules have to be given as connection status decisions.

3   System Architecture

Our system consists of two main components, see Figure 1 for a sketch. One part is responsible for the propagation of delays from the status information and for the calculation of secondary delays, while the other component handles connection queries. The core of the first part, our *real-time information server*, is a *dependency graph* which models all the dependencies between different trains and between the stops of the same train and is used to compute secondary delays (in Section 4 we introduce in detail the dependency graph and propagation algorithm). The dependency graph stores the obtained information needed to update the search servers and transmits this information in a suitable format to them. The search servers in turn update their internal graph representation whenever they receive these changes. This decoupling of dependency and search graph allows us to use any graph model for the search graph.

In a distributed scenario this architecture can be realized with one server running as the real-time information server that continuously receives new status information and broadcasts it. We will present some details in the following subsection. Load balancing can schedule the update phases for each server. If this is done in a round-robin fashion, the availability of service is guaranteed.

Multi-server Approach

The *search server* mainly consists of a search graph, an update component for the search graph, and a query algorithm.

In a multi-server environment, updates of a search server are either triggered by a load balancer or an internal clock after a maximum amount of time without

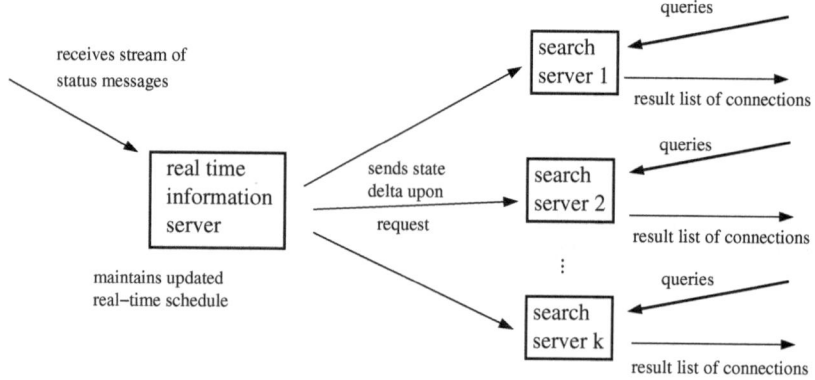

Fig. 1. Sketch of the system architecture

update. The data it receives (called *state delta* for the remainder of this work) are lists of changed departure and arrival times as well as meta-information about additional and canceled trains and connection status decisions. Subsequently, it adjusts the search graph accordingly and thereafter the graph looks exactly as if it were constructed from a schedule with all these updated departure and arrival times. Thus, the search algorithm does not need to know whether it is working on a graph with updated times or not.

The *real-time information server* receives all the up-to-date status information, uses its internal *dependency graph* to compute updated departure and arrival times (cf. Section 4) and stores these and the meta-information in a data structure UDS (*update data structure*). UDS maintains for every event with a changed timestamp a 3-tuple consisting of (1) a reference to the event itself, (2) the latest updated timestamp of this event, and (3) the release time when the last update of this event took place. Whenever a search server requests an update, it receives all events with a release time later than the last update of that server. If the timestamp of an event (or node in the graph model) changes, we call the necessary update a *(node) shift*.

For a true multi-server architecture with multiple search servers we basically have two update scenarios:

- An additional search server joins in and has to be initialized to the current time: We iterate over all existing event entries in UDS and transmit all those with times differing from the scheduled time.
- A search server has answered a number of queries and now enters update mode: We could simply transmit all events with release time greater than the last update time of the search server (referenced as *iterator version*). As this requires iterating over all stored events even to calculate a small delta, we can do more efficiently utilizing a map (referenced as *map version*).

In the map version a map of all changed events and their previous event time is maintained for each search server individually. Whenever a new event time is released, we look for that event in the map. Only if it is not already present, we store the event itself and its event time before the last change. This is the current timestamp of the event in the search server. To answer an update request we simply return all events in this map, whose new event time differs from the event time in the map (and thus the time in the current server), and clear the map afterwards. Using this technique we not only save iterating over all entries to determine the set of changed events (our state delta) but also avoid transmitting events that have been changed more than once and do not require a shift, since their new event time is the same as in the last update.

Our UDS data structure enables us to transmit only consistent state deltas on demand. Thereby, we can decrease both the time spent in communication and updating the graphs (e.g. if between two update phases more than one information for a single event is processed in the dependency graph, it is not required to transmit the intermediate state and adjust the graph accordingly).

4   Dependency Graph

4.1   Graph Model

Our *dependency graph* (see Figure 2) models the dependencies between different trains and between the stops of the same train. Its node set consists of four types of nodes:

– departure nodes,
– arrival nodes,
– forecast nodes, and
– schedule nodes.

Each node has a timestamp which can dynamically change. Departure and arrival nodes are in one-to-one correspondence with departure and arrival events. Their timestamps reflect the current situation, i.e. the expected departure or arrival time subject to all delay information known up to this point.

Schedule nodes are marked with the planned time of an arrival or departure event, whereas the timestamp of a forecast node is the current external prediction for its departure or arrival time.

The nodes are connected by five different types of edges. The purpose of an edge is to model a constraint on the timestamp of its head node. Each edge $e = (v, w)$ has two attributes. One attribute is a Boolean value, signifying whether this edge is currently active or not. The other attribute $\tau(e)$ denotes a point in time which basically can be interpreted as a lower bound on the timestamp of its head node w, provided that the edge is currently active.

– *Schedule edges* connect schedule nodes to departure or arrival nodes. They carry the planned time for the corresponding event of the head node (according to the published schedule). Edges leading to departure nodes are always active, since a train will never depart prior to the published schedule.
– *Forecast edges* connect forecast nodes to departure or arrival nodes. They represent the time stored in the associated forecast node. If no forecast for the node exists, the edge is inactive.
– *Standing edges* connect arrival events at a certain station to the following departure event of the same train.
    They model the condition that the arrival time of train t at station s plus its minimum standing time $stand_s(t)$ must be respected before the train can depart (to allow for boarding and disembarking of passengers). Thus, for a standing edge e, we set $\tau(e) = arr_s(t) + stand_s(t)$. Standing edges are always active.
– *Traveling edges* connect a departure node of some train t at a certain station s to the very next arrival node of this train at station s'. Let $dep_s(t)$ denote the departure time of train t at station s and $tt(s, s', t)$ the travel time for train t between these two stations. Then, for edge $e = (s, s')$, we set $\tau(e) = dep_s(t) + tt(s, s', t)$. These edges are only active if the train currently has a secondary delay (otherwise the schedule or forecast edges provide the necessary conditions for its head node).

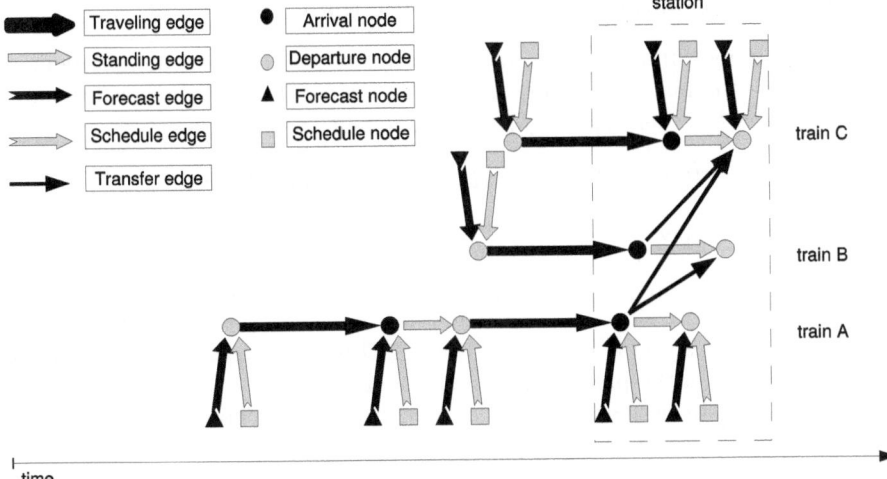

Fig. 2. Illustration of the dependency graph model

Due to various, mostly unknown factors determining the travel time of trains in a delayed scenario, e.g. speed of train, condition of the track, track usage (by other trains and freight trains that are not in the available schedule), used engines with acceleration/deceleration profiles, signals along the track etc. we assume for simplicity that $tt(s, s', t)$ is the time given in the planned schedule. However, if a more sophisticated, but efficiently computable oracle for $tt(s, s', t)$ taking the mentioned factors into account were available, it could be used without changing our model.

- *Transfer edges* connect arrival nodes to departure nodes of other trains at the same station, if there is a planned transfer between these trains. Thus, if f is a potential feeder train for train t at station s, we set $\tau(e) - wait_s(t, f)$, where

$$wait_s(t, f) = \begin{cases} arr_s(f) + minct_s(f, t) & \text{if } t \text{ waits for } f \\ 0 & \text{otherwise} \end{cases}$$

(cf. Section 2.2) if we respect the waiting rules. Recall that t waits for f only if the following inequality holds

$$arr_s(f) + minct_s(f, t) - dep_s^{sched}(t) < wt_s(t, f)$$

or if we have an explicit connection status decision that t will wait.

By default these edges are active. In case of an explicit connection status decision "will not wait" we mark the edge in the dependency graph as not active and ignore it in the computation.

For an "always wait" or "never wait" scenario we may simply always set $\tau(e)$ to the resulting delayed departure time or to zero, respectively.

4.2   Computation on the Dependency Graph

The current timestamp for each departure or arrival node can now be defined recursively as the maximum over all deciding factors: For a departure of train t at station s with feeders $f_1, \ldots, f_n$ we have $dep_s(t) =$

$$\max\{dep_s^{sched}(t), dep_s^{for}(t), arr_s(t) + stand_s(t), max_{i=1}^{n}\{wait_s(t, f_i)\}\}.$$

For an arrival we have

$$arr_s(t) = \max \left\{ arr_s^{sched}(t), arr_s^{for}(t), dep_{s'}(t) + tt(s', s, t) \right\}$$

with the previous stop of train t at station s'. Inactive edges do not contribute to the maximum in the preceding two equations.

If we have a status message that a train has finally departed or arrived at some given time dep^{fin} resp. arr^{fin}, we do not longer compute the maximum as described above. Instead we use this value for future computations involving this node.

We maintain a priority queue (ordered by increasing timestamps) of all nodes whose timestamps have changed since the last computation was finished. Whenever we have new forecast messages, we update the timestamps of the forecast nodes and, if they have changed, insert them into the queue. For a connection status decision we modify the corresponding transfer edge and update its head node. If its timestamp changes, it is inserted into the queue. As long as the queue is not empty, we extract a node from the queue and update the timestamps of the dependent nodes (which have an incoming edge from this node). If the timestamp of a node has changed in this process, we add it to the queue as well.

For each node we keep track of the edge e_{max} which currently determines the maximum so that we do not need to recompute our maxima over all incoming edges every time a timestamp changes. Only if $\tau(e_{max})$ was decreased or $\tau(e)$ for some $e \neq e_{max}$ increases above $\tau(e_{max})$ the maximum has to be recomputed.

- If $\tau(e)$ decreases and $e \neq e_{max}$ nothing needs to be done.
- If $\tau(e)$ increases and $e \neq e_{max}$ but $\tau(e) < \tau(e_{max})$ nothing needs to be done.
- If $\tau(e)$ increases and $e = e_{max}$ the new maximum is again determined by e_{max} and the new value is given by the new $\tau(e_{max})$.

When the queue is empty, all new timestamps have been computed and the nodes with changed timestamps can be sent to the search graph update routine or, in the multi server architecture, to the UDS data structure.

A Note on the Implementation. For ease of exposition we have introduced all kinds of nodes and edges in the dependency graph as being real nodes and edges. Of course, in our implementation we do not use a node and an edge to encode nothing more than a single timestamp for schedule and forecast times. Only arrival and departure nodes are real nodes with entering and leaving edges plus two integer variables representing the scheduled and forecast time. The latter is set to some predefined value to specify "no real-time information available (yet)". An arrival node has a container of leaving transfer edges, one entering traveling edge and one leaving standing edge. Analogously, a departure node has a container of entering transfer edges, one entering standing edge and one leaving traveling edge. Iterators over incoming dependencies and markers for the current input determining the timestamp of the node (the incoming edge or schedule or forecast time with maximum timestamp) have to be able to traverse resp. point to the different representations. We deemed the much more elegant version of the update routines - pretending the existence of nodes and edges for schedule and forecast times as well - better suited for presentation.

5   Time-Expanded Search Graph Model

5.1   The Static Model

Let us briefly describe the realistic time-expanded search graph model used in this work. Its basic idea is — as in the dependency graph before — to model each departure and arrival event of some train as a node with a timestamp. Each timestamp here represents the time after midnight in minutes.

Again, for each departure event of some train there is a *traveling edge* to its very next arrival event. With each traveling edge we associate a number of additional attributes: a bitfield representing traffic days, with one bit for each day of the schedule period, and several train attributes (train category, train number and name, availability of extra services, and the like).

The difference between the dependency graph and the search graph comes from the need to model the transfer between trains more explicitly in the latter case so that a Dijkstra-like shortest path algorithm can be used. Non-constant transfer times between pairs of connecting trains are modeled with the help of additional *change nodes*. For every departure time at a station there is a change node which is connected via *entering edges* to all departure nodes at that time. Change nodes at the same station are ordered increasingly by their timestamp, and subsequent change nodes (in this order) are interconnected with *waiting edges*. Moreover, at each station the last change node before midnight is linked to the first change node after midnight. For each arrival node there is a *leaving edge* connecting it to the corresponding first change node which is reachable in the time needed for a transfer from this train to any other. All possible shorter transfer times (e.g. for trains at the same platform) are realized using *special transfer edges*. The subgraph formed by all edges incident to change nodes of a certain station will be referred to as the *change level* of this station.

Finally, we have *stay-in-train edges* each of which connects the arrival node of some train to the corresponding departure node at the same station, provided the latter exists. For each optimization criterion, a certain length is associated with each edge.

Traffic days, possible attribute requirements and train class restrictions with respect to a given query can be handled quite easily. We simply mark traveling edges as *invisible* for the search if they do not meet all requirements of the given query. With respect to this visibility of edges, there is a one-to-one correspondence between feasible train connections and paths in the graph. More details of the graph model can be found in [4].

5.2   Search Server MOTIS

Over the last years, the authors developed the timetable information system MOTIS (multi-objective train information system) which performs a multi-criteria search for train connections in a realistic environment based on the above described time-expanded graph. To be more precise, MOTIS uses an extension of this model to incorporate additional features from practice which we omit here for clarity of the presentation.

Our underlying model ensures that each proposed connection is indeed feasible, i.e. can be used in reality. MOTIS is parameterized to search with respect to a selection of optimization criteria, most importantly travel time, number of interchanges, ticket cost, and reliability of all interchanges of a connection. The search algorithm used within MOTIS is a generalized multi-criteria Dijkstra-like algorithm enhanced with some additional speed-up techniques like goal-directed search [4]. The system is designed not only to present the true Pareto-optima, but more generally "all attractive" connections to customers, see also [4]. MOTIS has successfully been extended to search for low-cost connections [11] and night trains [12].

5.3   The Dynamic Model

The static time-expanded graph model has been slightly adapted for the dynamic scenario. Compared to the standard search graph we have to store additional information, namely status decisions, a second timestamp for each node to report actual and scheduled time in query results, additional strings containing reasons for the delays, and the like. Moreover, we need a slightly different representation of trains with identical schedules on multiple days. We defer details of this modification to Section 7.

6   Updating the Search Graph

The update in the search graph does not simply consist of setting new timestamps for nodes (primary and secondary delays), insertions (additional trains) and deletions (cancellations) of nodes and resorting lists of nodes afterwards. All the edges modeling the changing of trains at the affected stations have to be recomputed respecting the changed timestamps, additional and deleted nodes, and connection status information. The following adjustments are required on the change level (see Figure 3):

- Updating the leaving edges pointing to the first node reachable after a train change.
- Updating the nodes reachable from a change node via entering edges.

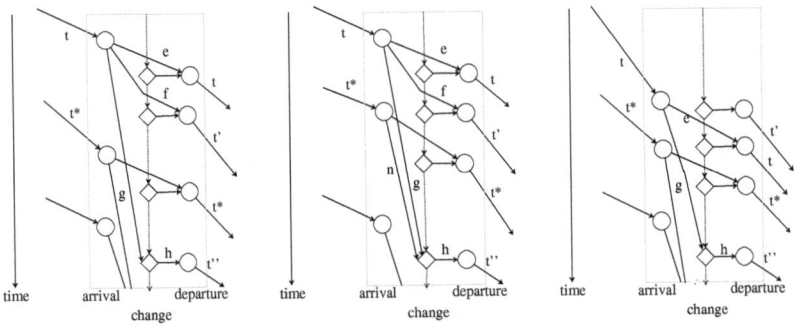

Fig. 3. The change level at a station (left) and necessary changes if train t^* arrives earlier (middle) or train t arrives later (right)

- Inserting additional change nodes or unhooking them from the chain of waiting edges at times where a new event is the only one or the only event is moved away or canceled.
- Recalculating special interchange edges from resp. to arrival resp. departure nodes with a changed timestamp (either remove, adjust or insert special interchange edges).

The result of the update phase is a graph that looks and behaves exactly as if it was constructed from a schedule describing the current situation. Additionally, it contains information about the original schedule and reasons for the delays.

Next, we give two examples for updating the search graph. In Figure 3 (left) it is possible to change from train t to all trains departing not earlier than t'' using leaving edge g, any number of consecutive waiting edges and an entering edge (e.g. h to enter t''). A change to train t' on the same platform is also feasible using special interchange edge f and, of course, to stay in train t via stay-in-train edge e. However, it is impossible to change to train t^* although it departs later than t', because it requires more time to reach it. Suppose train t^* manages to get rid of some previous delay and now arrives and departs earlier than previously predicted (see Figure 3, middle part). In the new situation it is now possible, to change from t^* to train t'' using the new leaving edge n and the existing entering edge h.

In our second example let train t arrive delayed as depicted in Figure 3 (right). As it now departs after t', it is not only impossible to change to t' (special interchange edge f is deleted), but also the departure nodes for the departures of t' and t are in reverse order. Therefore, the waiting edges have to be re-linked.

7   Traffic Days

A common simplification in theoretical work on timetable information systems is the assumption that trains operate periodically. Often even a periodicity of one hour is used. In real schedules, however, there is a considerable difference between peak hours, late evenings and "quiet" nights. For our timetable server MOTIS we take time modulo a single day in order to have a better manageable graph size as opposed to full time expansion. Recall that traveling edges carry traffic day flags (stored in bitfields) to model the days of operation, e.g. trains operating only on weekdays, or weekends, different schedules for school days and non school-days, trains operating on public holidays according to the weekend schedule etc.

In our scenario with delay information we have to take care of multiple traffic days as well. To be able to present the customer with updated information we need to model not only "today" (the current day) but also tomorrow as some connections might pass the midnight border (have a so-called "*night jump*"), especially if we query with a departure in the afternoon or evening. Resulting alternative journeys, requested after a delay on a journey, may even end on the next day due to delays, although no night jump was present in the original connection.

Therefore, we chose a schedule length of two days. In our time-expanded graph we represent all the trains operating today and tomorrow. However, trains that have the same schedule on both days can no longer be represented just once with two traffic day flags set. To be able to shift today's train without affecting the version of tomorrow, thus not incorrectly cloning delays, or vice versa, we need two distinct versions of such trains.

7.1   Memory Consumption Issues

The simplest version to attain separate nodes for today's and tomorrow's events is to use full time expansion on all our schedule days and not take time modulo 1440 and use traffic day flags on the train edges. Unfortunately, this would not only increase the number of event nodes and edges as well as the change edges, it would also significantly increase the number of change nodes and waiting edges. Whereas there is no way to avoid the increase for the former type of nodes and edges, we found a way to keep the size of the rest the same: We only use full time expansion for departure and arrival events and link all events to a change level with only one node per necessary timestamp, regardless of the day of that event, i.e. the number of change nodes and waiting edges remains the same, only the number of adjacent edges to the change nodes increases. Three different models for the search graph arise.

- Model (A) is the static model where the same events on two subsequent days are represented only once but two traffic day flags are set.
- Model (B) treats each arrival and departure event individually and uses the sparse change level implementation as described above.
- Model (C) also treats each arrival and departure event individually but uses full change level expansion.

Note that in the dependency graph we opted for full time expansion. There is no change level with waiting edges and all the change representation is between the

Table 1. Sizes of the search graph for two days, Wednesday and Thursday resp. Sunday and Monday and the increase when changing between the models (A), (B), and (C) as described in the text

model	unit	event train/std nodes	edges	change nodes	change edges	waiting edges	total nodes	total edges
We & Th (A)	(in k)	988	950	459	988	459	1447	2397
We & Th (B)	(in k)	1956	1878	459	1954	459	2415	4291
We & Th (C)	(in k)	1956	1878	912	1954	912	2868	4744
increase (A → B)	(in %)	98.0	97.7	0	97.8	0	66.9	79.0
increase (A → C)	(in %)	98.0	97.7	98.7	97.8	98.7	98.2	97.9
Su & Mo (A)	(in k)	1181	1134	498	1180	498	1679	2812
Su & Mo (B)	(in k)	1702	1634	498	1701	498	2200	3833
Su & Mo (C)	(in k)	1702	1634	798	1701	798	2500	4133
increase (A → B)	(in %)	44.1	44.1	0	44.2	0	31.0	36.3
increase (A → C)	(in %)	44.1	44.1	60.2	44.2	60.2	48.9	47.0

trains itself and only necessary to decide whether trains wait for others or not and compute the resulting changes. In this model a source delay propagation may or may not delay events on the following day. There is no need for case distinctions due to day changes.

Test Data. To study the effect of these models on the space consumption, we use the train schedule of Germany in 2008. The schedule contains 68,300 trains for the whole year with over 5,000 distinct bitfields for the days of operation. We look at the graphs prepared for two subsequent days, either two weekdays, Wednesday and Thursday (We & Th) with 38,600 trains with distinct schedules or for Sunday and Monday (Su & Mo) with 46,600 trains with distinct schedules.

Comparison of Models. In Table 1, we compare our three different models for the search graph. For the more homogeneous case of two weekdays version (C) requires double the amount of space while for our variant (B) we manage to increase the number of nodes by two thirds and the number of edges by four fifths. The tremendous increase of (C) is due to the large number of trains operating identically each weekday. If we look at the graph for Sunday and Monday the increase is much smaller as many of the trains operate either on Monday or on Sunday, therefore the increase in nodes and edges for the trains is below 50%. Still our model improves the additional required memory space from nearly one half to about one third.

During the actual search for train connections, variant (B) has a slight running time overhead in comparison with full time expansion (C). This overhead turns out to be negligible if a look-ahead in the search process categorizes entering edges as not allowed if they lead to a departure node for a train not operating on the required day.

7.2   Moving from One Day to the Next

At midnight we have to change the current day for our real-time information server as well as the search servers. Now, information about yesterday is no longer relevant as tomorrow becomes today and we need to load the "new tomorrow".

The real-time information server loads the dependency graph for tomorrow and "forgets" yesterday. With the fully time-expanded model there is no hassle in doing so. Note that we still have to keep yesterday's events that are delayed to today and have not happened yet, but nothing more about yesterday is needed any longer. Thus, we can delete all information about yesterday's events in the data structure. With our prototype, this whole procedure is finished in less than 35 seconds for the complete German timetable.

The search servers need a longer update phase than usual as they have to be restarted with the now current day and the following day. Afterwards, they request an update for a new server (exactly as described for a new server in Section 3). In this update they receive all information for today currently available. These updates typically take less than ten seconds. Together with the restart time of about 20 seconds a single search server is down for about half a minute. Even a server that has not yet changed days can still be updated after midnight and produce correct search results, as only the information about the next day is missing, not the current day. So there is no problem with the last server updating at say 01:00 a.m. Since midnight is not a peak hour for timetable information systems a number of servers might change days concurrently without compromising the availability of service. In summary, within a multi-server solution down-times of individual servers can easily be hidden from the customer.

8   Evaluation of the Prototype

We implemented the dependency graph and the update algorithm described in Section 4 and extended our time table information system MOTIS to support updating the search graph (cf. Section 6). Although these update operations are quite costly, we give a proof of concept and show that they can be performed sufficiently fast for a system with real-time capabilities.

Our computational study uses the German train schedule of 2008. During each operating day all trains that pass various trigger points (stations and important points on tracks) generate status messages. There are roughly 5000 stations and 1500 additional trigger points. Whenever a train generates a status message on its way, new predictions for the departure and arrival times of all its future stops are computed and fed into a data base. German railways Deutsche Bahn AG provided delay and forecast data from this data base for a number of operation days. The simulation results for these days look rather similar without too much fluctuation neither in the properties of the messages nor in the resulting computational effort.

In the following subsection, we present results for a standard operating day with an average delay profile testing various waiting profiles broadcasting the update information as soon as it becomes available. In the succeeding subsection we will present first results for the multi-server architecture (as described in Section 3) and test different update intervals. All experiments were run on an Intel Xeon 2.6 GHz with 8GB of RAM.

As no system with the capabilities of our prototype exists, we cannot compare our results to others. To ensure the correctness of our approach we used automated regression tests continuously checking the status of a large number of connections and determining alternatives, collecting meta-information about the encountered delays in the process. Furthermore, we intensively investigated isolated test cases (e.g. explicit search for trains known to us that they were delayed, search for trains departing next to a delay, searches for which the off-line optimum was affected by a delay).

8.1   Overall Performance and Waiting Profiles

To test our system, we used five sets of waiting profiles. Basically, the train categories were divided into five classes: high speed trains, night trains, regional trains, urban trains, and class "all others." Waiting times are then defined between the different classes as follows:

- *standard* High speed trains wait for each other 3 minutes, other trains wait for high speed trains, night trains, and trains of class "all others" 5 minutes, night trains wait for high speed and other night trains 10 minutes, and 5 minutes for class "all others."
- *half* All times of scenario standard are halved.
- *double* All times of scenario standard are doubled.
- *all5* All times of scenario standard are set to five minutes, and in addition regional trains wait 5 minutes for all but urban trains.
- *all10* All times of the previous scenario are doubled.

It is important to keep in mind that the last two policies are far from reality and are intended to strain the system beyond the limits it was designed to handle.

Our dependency graph model assumes that we know at each station which pairs of trains have potentially to wait for each other, i.e., which transfer edges are present. In our implementation we use the pragmatic rule, that if the difference between the departure event of train t_1 and the arrival event of another train t_2 at the same station does not exceed a parameter δ, then there is a transfer edge between these two events.

For each of these different waiting profiles we tested different maximum distances (in minutes) of feeding and connecting trains $\delta \in \{5, 15, 30, 45, 60\}$, and compare them to a variant without waiting for other trains (policy *no wait*). In this reference scenario it is still necessary to propagate delays in the dependency graph to correctly update the train runs. Thus, the same computations as with policies for delays is carried out, only the terms for feeding trains are always zero.

We constructed search and dependency graphs from the real schedule consisting of 36,700 trains operating on the selected day. There are 8,817 stations in

Table 2. Properties of the search graph (left) and dependency graph (right) for one day

search graph			dependency graph	
event nodes	0.99 mil		events	0.97 mil
change nodes	0.46 mil		standing edges	0.45 mil
edges	2.40 mil		traveling edges	0.49 mil

Table 3. The number of transfer edges depending on the waiting policy and the maximum allowed time difference δ between feeding and connecting train

transfer edges	5min	15min	30min	45min	60min
std / half / double	7.1k	54.7k	123.8k	207.8k	267.8k
all5 / all 10	14.6k	168.3k	399.6k	665.4k	874.3k

the data. The number of nodes and edges in both graphs are given in Table 2. The number of standing and traveling edges are in one-to-one correspondence to the stay-in-train and traveling edges of the search graph. The number of transfer edges depends on the waiting policy and parameter δ and can be found in Table 3. Note that, whether a transfer edge exists or not, depends on the classes that wait for each other and not on the actual number of minutes they wait. Therefore, the number of edges are identical for the policies *half*, *standard*, and *double* as well as for the policies *all5* and *all10*. There is a monotonous growth in the number of transfer edges depending on the parameter δ. Additionally, the number of these edges increases as more trains wait for other trains because of additional rules.

In Table 4, we give the results for our test runs for the different policies and values of δ. Running times are averages over 25 test runs. For the chosen simulation day we have a large stream of real forecast messages. Whenever a complete sequence of messages for a train has arrived, we send them to the dependency graph for processing. 336,840 sequences are handled. In total we had 6,340,480 forecast messages, 562,209 messages of the type "this train is now here" and 4,926 connection status decisions. Of all forecast messages 2,701,277 forecasts are identical to the last message already processed for the corresponding nodes. The remaining messages either trigger computations in the dependency graph or match the current timestamp of the node. The latter require neither shifting of nodes nor a propagation in the dependency graph. The resulting number of node shifts is given in the seventh column of Table 4. Depending on the policy we have a different number of nodes that were shifted and stations that have at least one delayed event (last two columns of the table).

The key figures for the computational efficiency (required CPU times in seconds, operation counts for the number of touched stations and node shifts in multiples of thousand) increase when changing to policies for which trains wait longer or more trains have to wait. Increasing δ yields a higher effect the more trains wait. The overall small impact of changing δ is due to the majority of delays being rather small. We notice a significant growth in all key criteria when

Table 4. Computation time for the whole day (propagation in the dependency graph (DG) and update of the search graph (SG), IO and total) and key figures (in multiples of thousand) for the executed node shifts in the search graph and the number of nodes and stations with changed status information with respect to different policies for delays

Instance policy	δ in min	Computation time for SG in s	DG in s	IO in s	total in s	Node shifts in k	With delay nodes in k	With delay stations
no wait	-	59.8	6.4	39.4	105.6	3,410	396.2	5,385
half	5	59.1	6.2	40.0	105.3	3,432	396.6	5,397
	15	60.7	6.4	39.7	106.8	3,525	400.1	5,483
	30	60.8	6.4	40.4	107.7	3,535	400.4	5,494
	45	61.2	6.5	40.0	107.8	3,539	400.6	5,494
	60	62.3	6.8	39.7	108.8	3,540	400.7	5,496
standard	5	59.1	6.2	39.3	104.6	3,443	396.8	5,408
	15	62.6	6.5	39.5	108.5	3,614	402.5	5,532
	30	63.4	6.7	40.1	110.2	3,636	403.2	5,541
	45	63.6	6.8	39.9	110.2	3,646	403.6	5,541
	60	63.6	6.7	40.3	110.7	3,651	403.7	5,545
double	5	58.9	6.3	39.7	104.9	3,447	396.8	5,419
	15	66.4	6.6	40.4	113.4	3,835	406.2	5,590
	30	67.9	6.9	40.5	115.3	3,908	407.5	5,639
	45	69.4	7.2	40.1	116.7	3,945	408.0	5,642
	60	69.0	7.3	39.9	116.2	3,959	408.1	5,642
all5	5	60.7	6.4	40.3	107.4	3,623	403.5	5,588
	15	123.1	11.5	40.0	174.6	7,603	440.5	6,051
	30	124.9	13.0	40.4	178.3	7,670	442.8	6,064
	45	124.9	14.7	40.6	180.2	7,687	443.4	6,064
	60	126.0	16.5	40.4	182.9	7,689	443.7	6,070
all10	5	60.7	6.4	40.4	107.5	3,651	404.0	5,608
	15	193.8	19.0	39.8	252.6	13,052	457.9	6,118
	30	195.2	21.6	40.9	257.7	13,231	463.0	6,145
	45	198.0	24.6	40.6	263.2	13,346	464.4	6,148
	60	200.7	27.3	40.7	268.7	13,466	465.3	6,162

increasing δ from 5 to 15. All policies behave rather similarly for $\delta = 5$, whereas the differences between the realistic policies and the extreme versions and even from *all5* to *all10* for higher values of δ is apparent.

Amongst the plausible policies there is only a 16% difference in the number of moved nodes. It little more than doubles going to policy *all5* and even increases by a factor of 3.8 towards policy *all10*. Roughly 40 seconds of our simulation time are spent extracting and preprocessing the messages from the forecast stream. This IO time is obviously independent of the test scenario. The increase in running time spent in the search graph update is no more than 3 seconds for $\delta > 5$ for all policies except *all10* with 7 seconds and differs by at most 10 seconds or 17% among the realistic scenarios. The running time scales with the number of shifts. An increase of factor 1.9 resp. 3.4 of node shifts results in a factor of 1.8 resp. 3.3 in running time (compare policies *double* to *all5* and *all10* with $\delta = 60$). The time spent in the dependency graph differs by at most 1 second (about 16%) for realistic scenarios and stays below 30 seconds even for the most extreme policy.

Even for the most extreme scenario a whole day can be simulated in less than 5 minutes. The overall simulation time for realistic policies lies around 2 minutes. For the policy *standard* with $\delta = 45$, we require on average $17\mu s$ reconstruction work in the search graph per executed node shift. By incident, also the overall runtime per computed message is $17\mu s$.

Worst-Case Considerations (Based on Policy Standard with $\delta = 45$). The highest number of messages received per minute is 15,627 resulting in 29,632 node shifts and a computation time of 0.66 seconds for this minute. However, the largest amount of reconstruction work occurred in a minute with 5,808 messages. It required 172,432 node shifts and took 2.38 seconds; this is the worst case minute which we observed in the simulation. Thus, at our current performance we could easily handle 25 times the load without a need for event buffering. This clearly qualifies for live performance.

8.2   Multi-server Performance

As we have seen in the previous subsection most of the time is spent in reconstructing the search graph. Applying sophisticated software engineering the update process has been sped up considerably. Additionally, a big potential lies in doing less reconstruction work. In a real-time environment it is not necessary to update multiple times per minute as soon as new information is available (as we did in the previous subsection). It clearly suffices to update each minute. Depending on the load balancing scheme every 2 or 3 minutes might still produce results of high quality.

To be able to compare the numbers to the previous section we tested the two servers as introduced in Section 3 "in line", i.e. one waited for the other to finish computation before continuing his own work. We use our waiting profile "standard" with $\delta = 45$ for all versions. The *baseline* version does not use the UDS and immediately updates the search graph. The version *split* additionally inserts and retrieves events into/from the UDS. Our code spends about 47 seconds on extracting and preprocessing the messages from the forecast stream and propagation in the dependency graph. Pushing all the events through the UDS data structure in the split architecture only requires an additional 7.2 seconds.

As we do not see a need for update intervals shorter than one minute, we now read all incoming messages for a particular minute and calculate the resulting changed event times in the dependency graph. These are transmitted to the data structure UDS in our real-time information server part. Meanwhile the search graph requests an update every 1, 2, 3, 4, or 5 minutes, using either the *iterator* or *map* version. The results can be found in Table 5. The numbers are averages over 25 runs.

By sending the state delta of the last x minutes as a batch job to the search graph we save a lot of reconstruction work due to mutually interacting messages

Table 5. The number of transmitted events, node shifts and execution time for simulating the whole day. We compare version with and without two server architecture using an iterator or a map to determine the relevant events (cf. Section 3) for different update intervals.

Instance		Transmissions		Computation time			
		needed	unnec-	SG	UDS		total
Version	interval		essary		ins	ext	
	in min	in k	in k	in s	in s	in s	in s
baseline	-	3646	0	63.5	0.0	0.0	110.3
split	-	3646	0	63.7	3.7	3.5	117.7
iterator	1	3143	0	53.9	3.1	55.6	159.0
	2	2809	149	45.5	2.9	29.3	124.4
	3	2447	284	38.3	2.9	20.4	108.2
	4	2177	360	33.3	2.8	15.8	98.6
	5	1954	404	29.3	2.8	13.1	91.4
map	1	3143	0	54.3	4.9	2.1	107.4
	2	2809	0	45.4	4.9	1.9	98.5
	3	2447	0	38.3	4.8	1.8	91.2
	4	2177	0	33.4	4.7	1.7	86.3
	5	1954	0	29.2	4.7	1.5	81.7

arriving between two subsequent updates, e.g. oscillating forecasts for trains, or reconstruction is done for a train but later it is shifted again due to a changed arrival time of one of its feeding trains.

With increasing interval size the number of messages to transmit significantly decreases. The resulting time required for updating the search graph is sped up by nearly 10 seconds when changing from immediate update to an interval of one minute. The increase of the interval size by one additional minute within the range of [1-5] reduces the execution time by a few seconds.

The *iterator* version of detecting events to transmit (cf. Section 3) only uses the release time information and cannot detect that an event does not require shifting, therefore it transmits 149k to 404k (depending on the update interval) of these irrelevant messages demanding a node "shift" to the node's current position. On the other hand, the *map* version only transmits events with changed timestamp, even if the release time is newer, therefore we do not have unnecessary transmissions. As shifts to the same position are never executed we only have the unnecessary transmission and no extra work, as we can see with the identical run times for the search graph update (column SG).

Inserting the information (column UDS ins) about changed event times into the UDS takes between 2.8 and 3.7 seconds, depending on the number of insertions (and thus the interval size). For the *map* version the bookkeeping requires an additional 1.8 to 2.0 seconds for the whole day. While the extraction (column UDS ext) using the *map* version requires 1.5 to 2.1 seconds, iterating for each update over all stored events to find the relevant new information in the *iterator* version is very costly and takes 13.1 to 55.6 seconds. Obviously, these times do not depend on the number of transmissions but on the number of iterations, as we observe that the extraction time is inversely proportional to the interval size.

The improvement in run time of 3 seconds (from 110.3 to 107.4 seconds), when changing from the baseline version to the split version with an interval of one minute, does not seem like much. However, it enabled us to do load balancing and handle updates on demand with our multi-server approach. The update time for the search servers consists of the time for receiving events from the UDS plus the time for the search graph update. Therefore, instead of taking 110.3 seconds to read messages, propagate delays and update the search in our baseline version, we only need 56.4 seconds in the split architecture for keeping the search graph up-to-date. Thus, we gain more than 50 seconds of available computation time per search server (about half the time required by the baseline version that does all the work on its own). Together with the initial startup phase and the first update with all relevant information for today depending on yesterday's data of about half a minute (cf. Subsection 7.2) a search server is only 60 to 90 seconds per day busy with startup and updating. This means that each search server can use 99.9% of its time for answering search queries.

The real-time information server spends about 47 seconds for reading messages and propagation in the dependency graph and additional 3 seconds storing the data. For each registered server (in our tests just one) it takes 2 seconds maintaining the map of relevant events and 2 seconds to extract and transmit those events. Thus, we have by far enough time to update a multitude of search servers.

9   Conclusions and Future Work

We have built a first prototype which can be used for efficient off-line simulation with massive streams of delay and forecast messages for typical days of operation within Germany. Using the presented multi-server solution, the correct handling of all necessary updates is so fast that each search server can use almost all of its time for answering search queries. Stress tests with extreme policies for delays showed that the update time scales linearly with the amount of work. So even for cases of major disruptions we expect a sufficient performance of such a multi-server solution. Compared to typical stream profiles, we are able to handle about 25 times as much reconstruction work.

It remains an interesting task to implement a live feed of delay messages for our timetable information system and actually test real-time performance of the resulting system. Since update operations in the time-dependent graph model are somewhat easier than in the time-expanded graph model, we also plan to integrate the update information from our dependency graph into a multi-criteria time-dependent search approach developed in our group (Disser et al. [13]).

A true real-time timetable information system as demonstrated by our prototype opens the door for a new service to passengers who want their travel plans supervised. The provider of such a service would constantly check the validity of planned connections, and in case of necessary changes due to delays inform the affected passenger and propose new alternative connections by sending messages to a mobile phone or an email address.

Acknowledgments

This work was partially supported by the DFG Focus Program Algorithm Engineering, grant Mu 1482/4-1. Thanks go to our students Lennart Frede and Mohammad Keyhani who contributed to the software design and implementation. We wish to thank Deutsche Bahn AG for providing us with timetable data and up-to-date status information for scientific use, and Christoph Blendinger and Wolfgang Sprick for many fruitful discussions.

References

1. Frede, L., Müller-Hannemann, M., Schnee, M.: Efficient on-trip timetable information in the presence of delays. In: Fischetti, M., Widmayer, P. (eds.) Proceedings of ATMOS 2008 - 8th Workshop on Algorithmic Approaches for Transportation Modeling, Optimization, and Systems, Schloss Dagstuhl - Leibniz-Zentrum für Informatik, Germany (2008)
2. Müller-Hannemann, M., Schulz, F., Wagner, D., Zaroliagis, C.: Timetable information: Models and algorithms. In: Geraets, F., Kroon, L.G., Schoebel, A., Wagner, D., Zaroliagis, C.D. (eds.) Railway Optimization 2004. LNCS, vol. 4359, pp. 67–90. Springer, Heidelberg (2007)
3. Delling, D., Giannakopoulou, K., Wagner, D., Zaroliagis, C.: Timetable Information Updating in Case of Delays: Modeling Issues. Technical report ARRIVAL-TR-0133, ARRIVAL Project (2008)
4. Müller-Hannemann, M., Schnee, M.: Finding all attractive train connections by multi-criteria Pareto search. In: Geraets, F., Kroon, L.G., Schoebel, A., Wagner, D., Zaroliagis, C.D. (eds.) Railway Optimization 2004. LNCS, vol. 4359, pp. 246–263. Springer, Heidelberg (2007)
5. Gatto, M., Glaus, B., Jacob, R., Peeters, L., Widmayer, P.: Railway delay management: Exploring its algorithmic complexity. In: Hagerup, T., Katajainen, J. (eds.) SWAT 2004. LNCS, vol. 3111, pp. 199–211. Springer, Heidelberg (2004)
6. Gatto, M., Jacob, R., Peeters, L., Schöbel, A.: The computational complexity of delay management. In: Kratsch, D. (ed.) WG 2005. LNCS, vol. 3787, pp. 227–238. Springer, Heidelberg (2005)
7. Ginkel, A., Schöbel, A.: The bicriteria delay management problem. Transportation Science 41, 527–538 (2007)
8. Schöbel, A.: Integer programming approaches for solving the delay management problem. In: Geraets, F., Kroon, L.G., Schoebel, A., Wagner, D., Zaroliagis, C.D. (eds.) Railway Optimization 2004. LNCS, vol. 4359, pp. 145–170. Springer, Heidelberg (2007)
9. Meester, L.E., Muns, S.: Stochastic delay propagation in railway networks and phase-type distributions. Transportation Research Part B 41, 218–230 (2007)
10. Anderegg, L., Penna, P., Widmayer, P.: Online train disposition: to wait or not to wait? ATMOS 2002. In: ICALP 2002 Satellite Workshop on Algorithmic Methods and Models for Optimization of Railways, Electronic Notes in Theoretical Computer Science, vol. 66 (2002)

11. Müller-Hannemann, M., Schnee, M.: Paying less for train connections with MO-TIS. In: Kroon, L.G., Möhring, R.H. (eds.) 5th Workshop on Algorithmic Methods and Models for Optimization of Railways, Internationales Begegnungs- und Forschungszentrum für Informatik (IBFI), Schloss Dagstuhl, Germany (2006)
12. Gunkel, T., Müller-Hannemann, M., Schnee, M.: Improved search for night train connections. In: Liebchen, C., Ahuja, R.K., Mesa, J.A. (eds.) Proceedings of AT-MOS 2007 - 7th Workshop on Algorithmic Approaches for Transportation Modeling, Optimization, and Systems, Internationales Begegnungs- und Forschungszentrum für Informatik (IBFI), Schloss Dagstuhl, Germany (2007); Extended journal version appears in Networks
13. Disser, Y., Müller-Hannemann, M., Schnee, M.: Multi-criteria shortest paths in time-dependent train networks. In: McGeoch, C.C. (ed.) WEA 2008. LNCS, vol. 5038, pp. 347–361. Springer, Heidelberg (2008)

Integrating Robust Railway Network Design and Line Planning under Failures

Ángel Marín[1], Juan A. Mesa[2], and Federico Perea[2]

[1] Department of Applied Mathematics and Statistics,
Madrid Polytechnic University, Spain
angel.marin@upm.es
[2] Department of Applied Mathematics II,
University of Seville, Spain
{jmesa,perea}@us.es

Abstract. Traditionally, when designing robust transportation systems, one wants to increase the functionality of the system in presence of failures, even though they might not work optimally when no failures occur, which is the usual case. In this paper we make an attempt to integrate robust network design and line planning without decreasing the efficiency of the system when no failures occur. Therefore, extra costs must be met (price of robustness). Two different concepts of robustness are considered: one from the user's point of view, which aims at minimizing total travel time, and one from the operator's point of view, which aims at minimizing extra costs, both assuming possible disruptions.

1 Introduction

Designing a Railway Network (RN), or even extending one that is already functioning, is a vital subject due to the fact that they reduce traffic congestion, travel time and pollution. The interest on this is reflected both in the number of papers on this subject that one can find in the literature and in the special effort that governments are making in order to improve their transportation (usually railway) networks. Examples of the research community's interest on this topic are the papers by Gendreau et al. [10], Dufourd et al. [9] and Bruno et al. [4]. On the other hand, governmental interest is reflected in the number of research projects in this area that are being supported by public institutions, for instance *ARRIVAL*.

When facing a Railway Network Design (RND) problem there is usually another transportation system already operating in the area where the RN is to be built or extended. Therefore competition between the RN to be constructed and the alternative mode must be taken into account. In Laporte et al. [13], an Integer Linear Programming (ILP) problem is used to design a RN in the presence of a competing mode. Indeed, in that paper, the robustness of the network is taken into account by adding certain extra constraints which provide

R.K. Ahuja et al. (Eds.): Robust and Online Large-Scale Optimization, LNCS 5868, pp. 273–292, 2009.
© Springer-Verlag Berlin Heidelberg 2009

potential users with different routes so that in case of failure they still find the RN attractive with respect to the alternative mode. The robustness in optimization models has been addressed from different angles. For instance, Malcom and Zenios [15] choose different parameters that are uncertain whereas Bertsimas and Sim, [2] and [3], propose models that control the degree of conservatism, therefore avoiding the classic "worst case scenario".

The following step after designing a RN is planning its lines (origin and destination stations, stops and frequencies), from now on called Railway Line Planning (RLP) problem. In passenger transportation the train fleet is operated in a cyclic timetable, and a service or line is defined by trains with the same route and stop stations. The frequency of a service is the number of trains that run in each direction per cycle on their common route. The line design considers the demand satisfaction and some capacity constraints. There are two main conflicting objectives to be pursued when planning a line system, namely (i) optimizing passenger service, and (ii) minimizing operational costs of the railway system. The improvement of the passenger service may be defined from different points of view; minimizing transfers, minimizing total travel time or maximizing comfort. Maximizing the number of direct connections usually results in long lines, however, long lines may transfer delays more easily and provide an inefficient allocation of rolling stock, because it is usually allocated according to the peak demand along the line, see Abbink et al. [1]. Therefore, in a robust system the lines are relatively short, which may force passengers to transfer from one train to another too often.

The references about RLP are focused on interurban trains. They may be classified into two groups: operator's point of view and user's point of view. In the first group, Claessens et al. [6] consider the minimization of service costs. They define a non-linear mixed integer model involving binary decision variables for the selection of services and additional variables for the frequencies and train lengths, considering the type vehicle. They solve the problem by linearization and make use of Branch and Bound techniques. Cordeau et al. [8] made a good state-of-the-art survey on line planning but oriented to freight transportation. Cordeau et al. [7] consider different types of trains and use Benders Decomposition. From the user's point of view, Bussieck et al. [5] maximize the number of passengers without transfers. In order to reduce the number of variables they use aggregated variables considering the relaxation of capacity constraints. Scholl [18] minimizes the number of transfers, which implies the use of passenger routes in the model, for which the concept of "Switch-and-Ride" is defined. Lagrangian Relaxation is used to obtain lower bounds and heuristics are needed to generate feasible solutions. Schöbel and Scholl [17] minimize the travel time with the inclusion of route and transfer times and considering budget constraints.

In this paper we wish to design (or extend) a robust RN and provide a line planning assuming that there is a competing transportation mode already operating in the area and budget is limited. Our robustness definitions are regarded to link failures, assuming that the utility of a network is adversely affected by such failures. We will consider two definitions of robustness:

- from the user's point of view. A RN will be considered robust if failure on edges affect the total travel time of the network the least possible.
- from the operator's point of view. A RN will be considered robust if the extra operator costs caused by failures are as low as possible, that is, the number of vehicles affected are as few as possible.

Both types of robustness will be defined in detail in Section 3.

Attempts to combine the different steps in transportation planning have already been addressed in the literature, see Guihaire and Hao [11] and the references therein. Robust RND and Robust RLP have been addressed separately in the literature (see [13] and Kontogiannis and Zaroliagis [12]). This paper includes the novelty of combining both problems into one, adding robustness constraints as well. In Laporte et al. [13] and [14] it was shown that the Robust RND is extremely complex, and only small instances could be optimally solved. The latter paper also shows the theoretical NP-hardness of our RND problem. As for the complexity of RLP, in [17] it is proved that the RLP is NP-hard. Therefore a combined ILP problem would be intractable from the computational point of view. Thus, in this work we propose an iterative process that adds robustness to the network (from both definitions) sequentially and is computationally more affordable.

The rest of the paper is structured as follows. In Section 2 both the RND problem and the RLP problem are summarized. In Section 3, the indexes of robustness we are to use and heuristics to solve the robust counterparts of the problems defined in the previous section are presented. Section 4 is devoted to showing an illustrative example of the algorithm proposed. The paper finishes with some conclusions and two appendixes where RND and RLP models are presented in more detail in order to make this paper self-contained.

2   Network Design and Line Planning

In this section we briefly introduce the two problems under consideration in this paper: the RND problem and the RLP problem. Only the necessary data for the iterative process that will be defined in the next section are introduced, both models being explained in more detail in the appendixes.

RND is possibly the first step in the complete process for creating a railway system. In this paper we assume that a RN is to be built in a given area, connecting certain points, knowing the mobility patterns between them, and without exceeding a maximum budget. It is also assumed that an alternative transportation system is already operating in the area. Therefore, the input data are:

- A graph (N, A) where $N = \{n_1, \ldots, n_I\}$ are the potential sites for locating stations and A is a set of feasible (bidirectional) arcs linking the elements in N, from which we can choose the links of the railway network. Costs of building stations and links (c_i is the cost for building a station at node n_i and c_{ij} is the cost for building the link directly connecting stations n_i and n_j) are known.
- The demand patterns are given by a matrix G, where g_w denotes the number of trips from n_i to n_j.
- The necessary time to go from n_i to n_j using the alternative transportation mode is known, for every pair of points of N, and denoted by u_{ij}^{COM}.

In [13], an ILP model for the RND problem is presented. The output of this model can be summarized as follows:

- $x_{ij} = 1$ if arc $(n_i, n_j) \in A$ belongs to the railway network; 0 otherwise.
- $y_i = 1$ if node $n_i \in N$ is a station of the railway network; 0 otherwise.
- $p_w = 1$ if the origin-destination (OD) pair w chooses the railway system; 0 otherwise.
- $f_{ij}^w = 1$ if the OD pair w uses the arc (n_i, n_j) of the railway system in a shortest path; 0 otherwise.
- u_w is the necessary time for the pair w to complete its journey by using its fastest alternative.

Note that such variables define the topological network of the railway system. Appendix A gives a detailed summary of the RND model considered.

Once the topology of the railway network is defined (stations and links) one should face the RLP problem in which the lines of the network including the stops of each line and their frequencies are designed. The RLP model takes as input data the demand patterns and the edges and nodes chosen by the RND model. The utility of the network is obtained after solving the RND problem, that is, the demand patterns are redefined as $g_w = g_w p_w$. The lines are elaborated from the topological network (defined after solving the RND problem). In our model we will consider all the service alternatives that may be defined from the nodes and edges of topological network, but in practice it is usual to find that only a line pool is considered.

In RLP, considering the routing of the demand through the lines (and their capacities) and constraints on the capacities of the service system (number of trains available, capacity of the section of the services, etc.), one wants to find the configuration of services (with their frequencies) that minimizes the total cost of the network (viewed as travel times or operator extra costs) and is able to satisfy all the demand that is to use it. The output of a RLP model can be summarized as:

1. $x_{ij}^l = 1$ if arc $(n_i, n_j) \in A$ belongs to service l; 0 otherwise.
2. $y_i^l = 1$ if node $n_i \in N$ is a stop of service l; 0 otherwise.
3. v_l, the frequency of trains on the service $l \in L$.

4. h_r is the flow of the route r, $\forall\, r \in R_w$, $\forall\, w \in N \times N$, where R_w is the set of possible routes in the network for the demand pair w.

In Appendix B a more detailed summary of the RLP problem is given.

3   Robust Railway Network Design and Line Planning

In this section we present an iterative process that aims to mix robust RND and robust RLP by iterating from one problem to another.

A model that includes both robust RND and robust RLP could have been proposed. The main drawback that one would find in such a model is its computational complexity. Both models separately are very complex, and it is logical to think that when joining them into one unique model, plus adding robustness constraints, the computational costs might become intractable. Therefore we propose an iterative process that, although it solves each problem separately, it connects them to each other with the aim of making it affordable from the computational point of view.

Two robust approaches may be considered: the first one uses the passenger's time (user robustness), whereas the second one takes into account the distribution of the line services (operator robustness). In the rest of the paper, the concept of network that will be used is:

Definition 1. *Let (N, A) be the graph on which we are to build our transportation network. A network r on (N, A) is a set of lines and their frequencies, each stop of each line being a node in N, and each link of each line being an edge in A. The set of all possible networks on G shall be denoted by $\mathcal{R}(N, A)$, or just $\mathcal{R}$ in case there is no confusion about the graph under consideration.*

3.1   User Robust Railway Network Design and Line Planning

First, the concept of User Robustness RND and RLP (UR-RND-RLP) used in this section must be defined. Roughly speaking, we consider that a network is robust from the user point of view when, if the route of a passenger is affected by a failure, there are other possible routes in the network, and those alternative routes allow that the total travel time does not increase dramatically.

Definition 2. *Let (N, A) be a graph, let r be a network over (N, A), and W the set of OD pairs that are going to use the network. The following robustness index is defined:*

$$R^U(r) = \frac{\min_{\bar{r} \in \mathcal{R}} \max_{(i,j) \in A} \{ \sum_{w \in W} T_w^{\bar{r}}(i,j) g_w \}}{\max_{(i,j) \in A} \{ \sum_{w \in W} T_w^{r}(i,j) g_w \}}, \tag{1}$$

where $T_w^r(i,j)$ denotes the length of the fastest route in network r for pair w when arc (i,j) is not operative. If there is no possible route in r for pair w when arc (i,j) fails, we set $T_w^r(i,j) = M$, for M sufficiently large.

Sometimes, it is not logical to test the robustness of certain areas of the network (for instance, if the graph (N, A) has branches). Therefore, sometimes in the definition of R^U we do not calculate the maximum on all arcs of A but only on a subset of it.

Index R^U satisfies three basic properties: it lays within 0 and 1, it is invariant against scale changes, and it is monotone, as stated in the following proposition.

Proposition 1. *Let $G = (N, A)$ be a graph and let W be the set of OD pairs that are going to use the network, respectively. Let r and r' be two networks over G, satisfying that $r \subset r'$, that is, all stations and links of r are in r' as well. Then, it holds that:*

1. *(Within [0,1] property)*
   $0 \leq R^U(r) \leq 1$. *Besides, $R^U(r) = 0$ if and only if there are some OD pairs of W such that the failure in an edge of r leaves them unconnected. Besides, $R^U(r^*) = 1 \; \forall \; r^* \in \arg\min_{\bar{r} \in \mathcal{R}} \max_{(i,j) \in A} \{\sum_{w \in W} T_w^{\bar{r}}(i,j) g_w\}$.*
2. *(Scale invariance)*
   If the scales in which the time and/or the number of passengers change, the index $R^U(r)$ remains invariant.
3. *(Monotonicity)*
   $R^U(r) \leq R^U(r')$.

Proof. 1. (Within [0,1] property)
   Trivially, $0 \leq R^U(r) \leq 1$. Note that $R^U(r)$ is always strictly positive and it can reach zero value if and only if its denominator tends to infinite, that is, if and only if there is a pair w and an arc (i,j) such that there is no path joining w when arc (i,j) fails.

2. (Scale invariance)
   Assume that the distance is now measured on a different scale, given by constant k_1, and the number of passengers is also measured on a different scale, given by constant k_2. Let $\tilde{g}_w = k_2 g_w$ the new demand patterns. Let $\tilde{T}_w^r(i,j)$ denote the length of the fastest route in the network r for the pair w when arc (i,j) is not operative, measured on the new scale. One has that $\tilde{T}_w^r(i,j) = k_1 T_w^r(i,j)$. Therefore,

$$\tilde{R}^U(r) = \frac{\min_{\bar{r} \in \mathcal{R}} \max_{(i,j) \in A} \{\sum_{w \in W} \tilde{T}_w^{\bar{r}}(i,j) \tilde{g}_w\}}{\max_{(i,j) \in A} \{\sum_{w \in W} \tilde{T}_w^r(i,j) \tilde{g}_w\}}$$

$$= \frac{\min_{\bar{r} \in \mathcal{R}} \max_{(i,j) \in A} \{\sum_{w \in W} k_1 k_2 T_w^{\bar{r}}(i,j) g_w\}}{\max_{(i,j) \in A} \{\sum_{w \in W} k_1 k_2 T_w^r(i,j) g_w\}} = R^U(r) \; .$$

3. (Monotonicity)
   Since $T_w^{r'}(i,j) \leq T_w^r(i,j)$ for all $(i,j) \in A$, and for all $w \in W$, $R^U(r') \geq R^U(r)$. $\qquad \square$

Those three properties justify the use of index R^U to compare the robustness of different networks.

Definition 3. *Let r^1 and r^2 be two networks covering the same set of OD pairs. Network r^1 is said to be more robust than r^2 from the user's point of view if $R^U(r^1) > R^U(r^2)$.*

Once we have defined our concept of robustness and the index we are to use to measure it, let us describe the sequential process we propose to join user Robust RND-RLP.

First, the RND problem is solved, as in Section 2 of [13]. This model optimizes a certain utility function of the railway system to be built in competition with an alternative mode of transportation. In this section we consider the total travel time as utility, which results in the following ILP problem (see Appendix A for more details):

$$
\begin{aligned}
RND : \text{min } & \text{Total Travel Time} \\
\text{s.t.: } & \text{Budget Constraints,} \\
& \text{Alignment Location Constraints,} \\
& \text{Routing Demand Conservation Constraints,} \\
& \text{Location-Allocation Constraints,} \\
& \text{Splitting Demand Constraints.}
\end{aligned}
\tag{2}
$$

The solution to this problem will yield a topological network minimizing the total travel time, that is, a set of edges and nodes that will constitute the actual transportation network. From this topological network, we have to calculate the optimal configuration of lines (with their frequencies), that is, we have to solve the RLP problem minimizing user costs. The vehicle fleet is assigned to the topological network in a competitive way, solving the following ILP problem:

$$
\begin{aligned}
RLP : \text{min } & \text{User Costs} \\
\text{s.t.: } & \text{Routing Demand Conservation Constraints,} \\
& \text{Arc User Capacity Constraints,} \\
& \text{Arc Vehicle Capacity Constraints,} \\
& \text{Fleet Capacity Constraints.}
\end{aligned}
\tag{3}
$$

The solution to the RLP problem will yield an optimal network, that is, a set of active lines and their frequencies. This would be the end of the initial iteration, which results in:

- the configuration of lines minimizing total travel time, r^0.
- the set of covered OD pairs by r^0, W^0.
- $T_w^{r^0}(i,j)$, for all $w \in W^0$, $(i,j) \in A$.
- $\arg\max_{(i,j)\in A}\{\sum_{w\in W} T_w^{r^0}(i,j)g_w\}$.

In following iterations we intend to extend r^0 so that it covers (at least) the same OD pairs and is more robust, at a minimum extra cost. That is, we want to build a network r^1 so that $R^U(r^0) \leq R^U(r^1)$. Since we do not want to adversely affect the utility of network r^0, the budget must be increased. Therefore, the second iteration of our process will solve the RND problem fixing the arcs of r^0. This way, both the trip coverage and the total travel time of the network will not be negatively affected. Besides, we will introduce the robustness constraints defined in Section 3.1 of [13], for (i^*, j^*) and for all $w \in W^0$, thus ensuring alternative routes for those pairs. Therefore, we will build a new network with a total travel time not higher than r^0 that still attracts (at least) all OD pairs in W^0, and is more robust. This is done by solving the following ILP problem:

$$ROB - RND : \min z_{costs} = \sum_{(i,j)} c_{ij} x_{ij} + \sum_i c_i y_i$$
$$\text{s.t. Alignment Location Constraints,}$$
$$\text{Routing Demand Conservation Constraints,}$$
$$\text{Location-Allocation Constraints,}$$
$$\text{Splitting Demand Constraints,}$$
$$x_{ij} = 1, \quad \forall \, (i,j) \in r^0,$$
$$f^w_{(i^*, j^*)} \leq \tfrac{1}{2}, \quad \forall \, w \in W^0,$$
$$(i^*, j^*) \in \arg\max_{(i,j) \in A}\{\sum_{w \in W} T^{r^0}_w(i,j) g_w\},$$

where z_{costs} is the cost of building the RN and f^w_{ij} are now allowed to lie in $[0, 1]$. Note that in this problem both the OD pairs covered W^0 and the total travel time of the network that had been previously computed can only be improved. Note as well that we minimize the construction costs so the robustness constraints are met at a minimum extra cost. The final process of this iteration is to solve the RLP for the topological network just obtained, which yields the network r^1. The following theorem states that the second network is more robust than the first one over the set W^0.

Theorem 1. $R^U(r^0) \leq R^U(r^1)$.

Proof. Since r^1 has the same arcs as r^0 plus (possibly) some more, and from the monotonicity of our index of robustness, the result follows.   □

A pseudocode of this process is given in Algorithm 1.

3.2   Operator Robust Railway Network Design and Line Planning

The concept of Operator Robust RND and RLP (OR-RND-RLP) used in this subsection is roughly defined as follows. A service is robust when the vehicles of a fleet are more distributed in the arcs of a given network. Thus, if a section of a line is affected by a failure less vehicles of the fleet are affected by it (and so less users). A more formal definition is:

Definition 4. *Let r be a network and let $v_r(i,j)$ be the number of vehicles of r using the arc (i,j), $v_r(i,j) = \sum_{l \in r} v_l \delta^l_{ij}$, $\delta^l_{ij} = 1$ if line $l \in r$ is defined using the edge (i,j), and zero otherwise. The following service robustness index is defined:*

Algorithm 1

Input data

User Robust Railway Railway Network Design and Line Planning

1. Initialization
   (a) **RND**: Solve RND minimizing total travel time. Let W^0 be the OD pairs covered by this optimal network and let T^* be its total travel time.
   (b) **RLP**: From W^0 and the topological network obtained in 1a, solve RLP minimizing user costs. Let r^0 be the resulting network.
2. Iterations. Set $k = 1$.
   (a) **Robust RND**: Solve ROB-RND, minimizing construction costs and with the following extra constraints: $x_{ij} = 1 \ \forall \ (i,j) \in r^{k-1}$ (keep the edges of r^{k-1}), $f_{i^*j^*}^w \leq \frac{1}{2} \ \forall \ w \in W^0$ (it opens new routes for users crossing (i^*, j^*)), where $(i^*, j^*) \in \arg\max_{(i,j) \in r^{k-1}} \{\sum_{w \in W^0} T_w^{r^{k-1}}(i,j)g_w\}$.
   (b) **RLP**: From the topological network obtained in 2a, solve RLP, minimizing user costs. Let r^k be the resulting network.
   (c) $k \rightarrow k + 1$. **Go to step 2a (until we run out of budget).**

$$R^O(r) = \frac{\min_{\bar{r} \in \mathcal{R}} \max_{(i,j) \in \bar{r}} v_r(i,j)}{\max_{(i,j) \in r} v_r(i,j)} . \tag{4}$$

Network r_1 is more robust than network r_2 if $R^O(r_1) > R^O(r_2)$.

Following analogous proofs as we did for the user robustness index R^U, the following properties are proven.

Proposition 2. *Let $G = (N, A)$ be a graph and let W be the set of OD pairs that are going to use the network, respectively. Let r and r' be two networks over G, satisfying that $r \sqsubset r'$, that is, all stations and links of r are in r' as well. Then, it holds that:*

1. *(Within [0,1] property)*
   $0 \leq R^O(r) \leq 1$.
2. *(Scale invariance)*
   If the scales in which the time and/or the number of passengers change, the index $R^O(r)$ remains invariant.
3. *(Monotonicity)*
   $R^O(r) \leq R^O(r')$

The iterative process to obtain the Operator Robustness Network Design and Line Planning is similar to the previous UR-RND-RLP, but changing the objective functions in problems (2) and (3) and the robustness index. Now the utility used in the RND problem is to maximize the trip coverage, the utility used in the RLP problem is to minimize the operational costs and the robustness index is R^O. This process is detailed in Algorithm 2.

Algorithm 2

Input data

Operator Robust Railway Network Design and Line Planning

1. Initialization
   (a) **RND**: Solve RND maximizing trip coverage. Let W^0 be the OD pairs covered.
   (b) **RLP**: From W^0 and the topological network obtained in 1a, solve RLP minimizing operational costs. Let r^0 be the resulting network.
2. Iterations. Set $k = 1$.
   (a) **Robust RND**: Solve ROB-RND, minimizing construction costs and with the following extra constraints: $x_{ij} = 1 \ \forall \ (i,j) \in r^{k-1}$ (keep the edges of r^{k-1}), $f_{i^*j^*}^w \le \frac{1}{2} \ \forall \ w \in W^0$ (it opens new routes for users crossing (i^*,j^*)), where $(i^*,j^*) \in \arg\max_{(i,j) \in r^{k-1}} \{v_{r^{k-1}}(i,j)\}$.
   (b) **RLP**: Solve RLP, minimizing operational costs. Let r^k be the resulting network.
   (c) $k \to k + 1$. **Go to step 2a (until we run out of budget).**

---

4   An Illustrative Example

For this example we will consider Andalucía, a region in the South of Spain, see Figure 1. A high-speed train network wants to be built in the area. Its potential

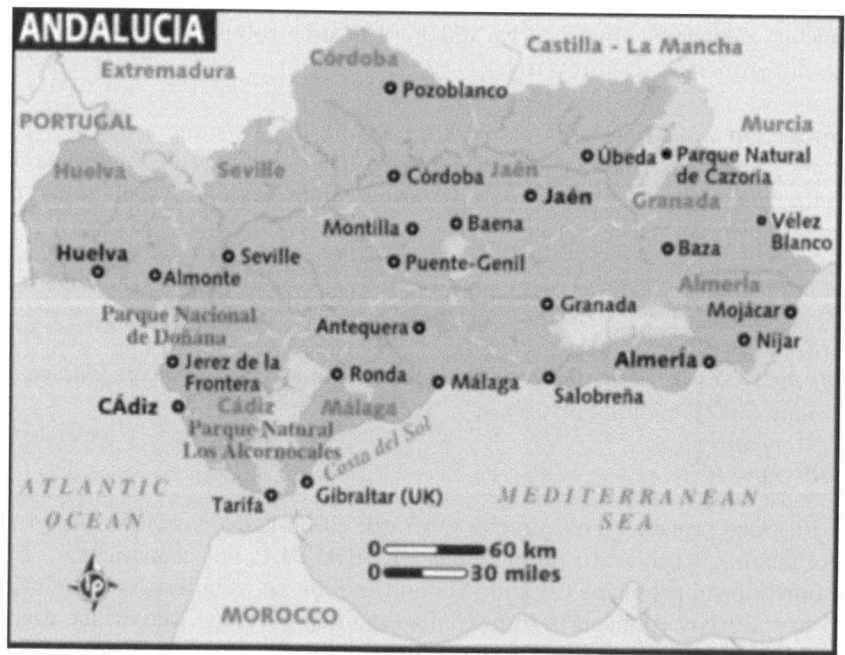

Fig. 1. Map of Andalucía

stations are the capitals of the eight provinces of the region, namely: Huelva (1), Seville (2), Cádiz (3), Córdoba (4), Málaga (5), Granada (6), Jaén (7) and Almería (8). Besides, two other nodes have been added to the problem, Antequera (9) and Guadix (10), for their importance as transfer stations. The set of potential stations and possible links (based on the orography and forbidden areas of the region) is depicted in Figure 2. Each node has an associated construction cost c_i and each possible edge a number d_{ij}, which is the time needed to traverse it by the railway network to be built. Such times are an approximation, which has been made as follows. There already exists a high-speed connection between Sevilla and Córdoba, which takes 41 minutes. The trip on the road takes approximately 107 minutes. Therefore, the links between different cities have been approximated from the road trip times using the same proportion: $\frac{41}{107}$ (results have been rounded to the closest integer). The construction costs of each edge (c_{ij}) are assumed to be proportional to their length, and for the sake of simplicity we have set $c_{ij} = d_{ij} \ \forall \ (i,j)$. A maximum budget of 300 units has been considered. The OD demands g_w and the cost u_w^{COM} for each demand pair $w \in W$ are given by the following matrices:

$$G = \begin{pmatrix}
0 & 2974718 & 576797 & 70924 & 131169 & 44198 & 17131 & 35264 & 0 & 0 \\
2974718 & 0 & 2977353 & 1146110 & 773926 & 315251 & 295295 & 21089 & 0 & 0 \\
576797 & 2977353 & 0 & 307126 & 1736095 & 844903 & 24032 & 56680 & 0 & 0 \\
70924 & 1146110 & 307126 & 0 & 448310 & 258062 & 133859 & 26604 & 0 & 0 \\
131169 & 773926 & 1736095 & 448310 & 0 & 1383515 & 215241 & 62623 & 0 & 0 \\
44198 & 315251 & 844903 & 258062 & 1383515 & 0 & 572196 & 650714 & 0 & 0 \\
17131 & 295295 & 24032 & 133859 & 215241 & 572196 & 0 & 72667 & 0 & 0 \\
35264 & 21089 & 56680 & 26604 & 62623 & 650714 & 72667 & 0 & 0 & 0 \\
0 & 0 & 0 & 0 & 0 & 0 & 0 & 0 & 0 & 0 \\
0 & 0 & 0 & 0 & 0 & 0 & 0 & 0 & 0 & 0
\end{pmatrix} ;$$

$$u_w^{COM} = \begin{pmatrix}
0 & 61 & 135 & 162 & 206 & 232 & 244 & 324 & 171 & 257 \\
61 & 0 & 83 & 107 & 149 & 175 & 184 & 267 & 114 & 200 \\
135 & 83 & 0 & 174 & 170 & 231 & 256 & 323 & 171 & 256 \\
162 & 107 & 174 & 0 & 114 & 141 & 99 & 233 & 85 & 166 \\
206 & 149 & 170 & 114 & 0 & 98 & 159 & 172 & 52 & 123 \\
232 & 175 & 231 & 141 & 98 & 0 & 69 & 113 & 79 & 49 \\
244 & 184 & 256 & 99 & 159 & 69 & 0 & 157 & 122 & 90 \\
324 & 267 & 323 & 233 & 172 & 113 & 157 & 0 & 173 & 78 \\
171 & 114 & 171 & 85 & 52 & 79 & 122 & 173 & 0 & 105 \\
257 & 200 & 256 & 166 & 123 & 49 & 90 & 78 & 105 & 0
\end{pmatrix} .$$

The demand pairs shown refer to number of trips per year, and therefore for our problem we have divided them by 365 in order to give a daily line planning. The distances of the complementary mode have been calculated using the fastest route on the road network, given by *www.google.com*.

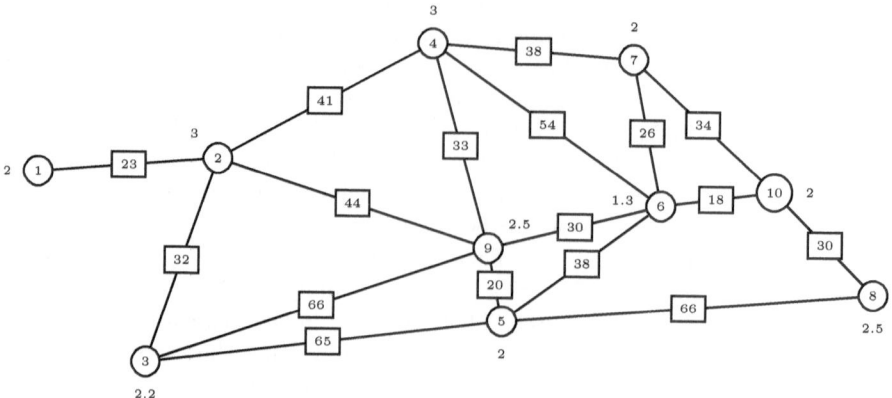

Fig. 2. Andalucía: potential stations and potential links

4.1   User Robustness

Let us now calculate a Robust RN from the user's point of view. Note that, since a failure in edge $(1, 2)$ would leave the graph disconnected, we will not take this edge into account when calculating index R^U, that is, rename $A = A \setminus \{(1, 2)\}$ when executing Algorithm 1. The initialization phase yields the following line configuration:

$r^0 = \{l_1, l_2, l_3, l_4\}$,
$l_1 = (1 - 2 - 3)$, $l_2 = (4 - 9 - 5)$, $l_3 = (2 - 9 - 6 - 7)$, $l_4 = (1 - 2 - 9 - 6 - 10 - 8)$
$u_1 = 30$, $u_2 = 22$, $u_3 = u_4 = 27$.

with a construction cost of 285.8 units.

From network r^0, one has that: $(i^*, j^*) = (2, 3)$, with a value of $17871M$. Note that there are no alternative routes in case any of the arcs fail, therefore $T_w^{r^0}(i, j) = M$, $\forall w \ (i, j) \in r^0$. Thus, the chosen arc for the next iteration is that holding the maximum flow.

For the first iteration, $k = 1$, we have to extend network r^0 so that there are alternative routes for all covered pairs in case arc $(2, 3)$ fails, and minimizing the extra costs. Such an optimization problem yields the topological network consti-tuted by the edges in r^0 plus edge $(3, 5)$. With this topological network, the RLP problem generates the optimal configuration of lines of the first iteration r^1:

$r^1 = \{l_1, \ldots, l_7\}$,
$l_1 = (1 - 2 - 3)$, $l_2 = (2 - 9 - 6)$, $l_3 = (3 - 5 - 9)$, $l_4 = (5 - 9 - 6)$,
$l_5 = (4 - 9 - 6)$, $l_6 = (6 - 10 - 8)$, $l_7 = (7 - 6 - 10)$
$u_1 = 22$, $u_2 = 19$, $u_3 = 8$, $u_4 = 6$, $u_5 = 11$, $u_6 = 5$, $u_7 = 7$.

Our process guarantees that this network is more robust than network r^0. The construction costs are 350.8 units. In Figure 3 both networks r^0 and r^1 are

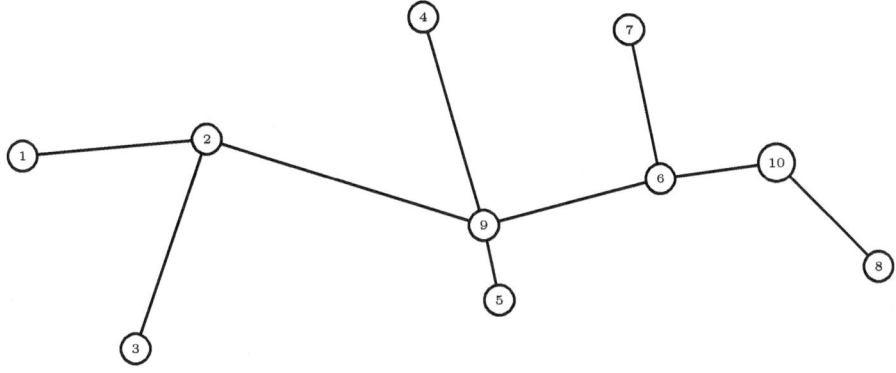

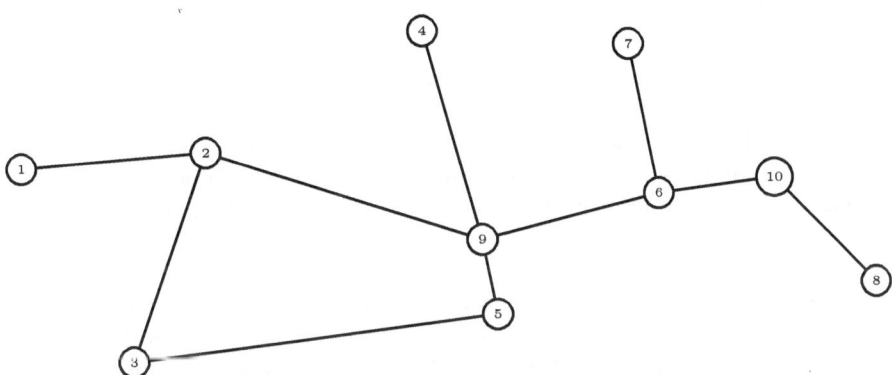

Fig. 3. Topological networks obtained in iteration 0 and iteration 1 of the user robustness algorithm

depicted (in order to not make the picture too messy we have just drawn the topological network). Notice the loop that is built in r^1 as opposed to r^0 where there are no loops. This loop improves the robustness of the network. For the following iteration, $(i^*, j^*) = (2, 9)$, and the algorithm continues (if there is available budget) from this edge.

If the index applied is the operator robustness, the algorithm is analogous. We just have to take into account that the objective function in the RND is trip coverage and in RLP we have to include only operator costs in the objective function. In this case the first network results:

$r^0 = \{l_1, \ldots, l_9\}$,
$l_1 = (1 - 2 - 3)$, $l_2 = (1 - 2 - 9)$, $l_3 = (2 - 9 - 4)$,
$l_4 = (3 - 2 - 9)$, $l_5 = (4 - 9 - 5)$, $l_6 = (3 - 2 - 9 - 5)$,
$l_7 = (5 - 9 - 6)$, $l_8 = (6 - 10 - 8)$, $l_9 = (7 - 6 - 9)$,
$u_1 = 11$, $u_2 = 7$, $u_3 = 1$, $u_4 = 18$, $u_5 = 10$, $u_6 = 1$, $u_7 = 11$, $u_8 = 5$, $u_9 = 7$.

The following iteration continues from the arc with the maximum flow, which is the same as in the User Robustness case. Imposing the corresponding robustness constraints, this results in network r^1:

$r^1 = \{l_1, \ldots, l_9\}$,
$l_1 = (1 - 2 - 3)$, $l_2 = (1 - 2 - 9)$, $l_3 = (2 - 9 - 4)$,
$l_4 = (3 - 2 - 9 - 5)$, $l_5 = (3 - 2 - 9 - 6)$, $l_6 = (4 - 9 - 6)$,
$l_7 = (6 - 10 - 8)$, $l_8 = (2 - 3 - 5 - 9)$, $l_9 = (7 - 6 - 9)$,
$u_1 = 11$, $u_2 = 7$, $u_3 = 1$, $u_4 = 18$, $u_5 = 10$, $u_6 = 1$, $u_7 = 11$, $u_8 = 5$, $u_9 = 7$.

Note that the resulting networks r^0 and r^1 have the same topological configuration when applying both robustness concepts but different line configurations.

5 Conclusions

The main contribution of this paper is the introduction of a heuristic procedure for integrating two different stages of the railway planning at the same time: Network Design and Line Planning. This sequential process has the following steps: 1) design a network without robustness constraints; 2) design a line plan on this network with no respect to robustness; 3) fix the OD pairs served in the current plans and the arcs used; 4) construct a more robust network (keeping what is fixed in 3); 5) construct a line plan on the new network; and 6) goto 3 unless budget for total cost is exceeded. We considered two different definitions of robustness, one from the user's point of view and the other from the operator point's of view. For each robustness concept we give an index that measures how robust a RN with a line configuration is for each definition of robustness (and the corresponding versions of the heuristic proposed are separately introduced). Both indexes satisfy three properties: scale invariance, monotonicity and membership in $[0,1]$. The paper concludes with an illustrative example that shows how the proposed procedures work.

Acknowledgments

The authors would like to thank two anonymous referees for their comments and suggestions. This work was partially supported by the Future and Emerging Technologies Unit of European Council (IST priority - 6th FP), under contract no. FP6-021235-2 (project ARRIVAL), by Ministerio de Fomento (Spain) under project PT2007-003-08CCPP, by Ministerio de Educación y Ciencia (Spain) under project MTM06-15054, by Ministerio de Ciencia e Innovación (Spain) under projects TRA2005-09068-C03-01 and TRA2008-06782-C02-01, and by Junta de

Andalucía (Spain) under project P06-FQM-01366. We also want to thank Professor García-Benítez for providing us with some necessary data for the Example section.

Appendix A: Railway Network Design

In the model for RND problem we assume that the mobility patterns in a metropolitan area are known. This implies that the number of potential passengers from each origin to each destination is given. We also assume that the location of the potential stations is known. There already exists a different mode of transportation and the RN will compete with it. For deciding which mode each demand is allocated to, the comparison between the generalized costs of the travellers is used. The aim of the model is to design a network (i.e. to decide in which nodes stations are to be located and how to connect them) consisting of lines so that it covers as many trips as possible. Since the resources are limited we also impose some budget constraints, which depend on the construction costs both of edges and nodes.

- Data and notation
  - A set $N = \{n_i;\ i = 1, 2, \ldots, I\}$ of potential sites for locating stations is given.
  - A set A' of feasible (bidirectional) arcs linking the elements in N is known. Therefore, we have a graph $\mathcal{G}' = (N, A')$, from which arcs are to be selected to form lines. Furthermore, there exists a graph $\mathcal{G}'' = (N, A'')$, representing the network used by the complementary mode (e.g. the street network). Let $G = (N, A)$, where $A = A' \cup A''$, be the whole network. Let us denote by $N(i) = \{n_j : \exists\, a \in A',\ a = (n_i, n_j)\}$ the set of adjacent nodes to n_i.
  - Each feasible arc $a = (n_i, n_j) \in A'$ has an associated length d_{ij}. The lengths of the arcs in A' usually correspond to approximate Euclidean distances if the system to be designed is underground and street network distances if it is at grade. However, forbidden regions or streets will increase the distances, and d_{ij} can also be interpreted as the generalized cost for traversing arc $a = (n_i, n_j) \in A'$.
  - For each node n_i and each arc $a \in A'$ there is an associated cost for constructing the corresponding infrastructure: c_i is the cost for building a station at node n_i, c_a being the cost for building the link a. A bound C_{max} on the available budget is also given.
  - The mobility pattern is given by a matrix $G = (g_w) :\ w \in W$, where W is the ordered index pair set: $W = \{w = (i, j);\ n_i, n_j \in N\}$, also referred to as the set of demands.
  - The generalized cost for satisfying each demand by the RN and the complementary modes are u_w^{RTN} and u_w^{COM}, respectively. While costs using the complementary mode depend on its (street) network and therefore are input data, the RN ones will depend on the topology of the network

to be constructed. The computation of RN costs u_w^{RTN} can be done by adding the lengths of the arcs of the path that demand w will use in the RN. Let u_w be the generalized cost for the demand w either by $\mathcal{G}'$ or by $\mathcal{G}''$

The aim of the model is to design a RN of at most L lines, $|L|$ being a low number (in reality, networks designed from scratch usually contain 3, 4 or 5 lines). Since constraints on the total cost will be imposed, we will allow some lines in L to not be included in the RN.

- The following variables will be used in the model.

  - $y_i^l = 1$, if node n_i is a station of line l; 0 otherwise.
  - $x_{ij}^l = 1$, if the arc $a = (n_i, n_j) \in A'$ belongs to line $l \in L$; 0 otherwise.
  - $x_{ij} = 1$, if the arc $a = (n_i, n_j) \in A'$ belongs to the RN; 0 otherwise.
  - f_{ij}^w denotes the normalized flow of the demand $w \in W$ through arc $(n_i, n_j) \in A'$, from n_i to n_j, $f_{ij}^w \in \{0,1\}$ if no failure occurs. Note that such variables will define the fastest route for the demand w in the network to be built.
  - φ_{ij}^w denotes the normalized flow of the demand $w \in W$ through arc $(n_i, n_j) \in A''$, from n_i to n_j, $\varphi_{ij}^w \in \{0,1\}$ if no failure occurs.
  - $h_l = 1$, if line l is included; 0 otherwise.
  - $p_w = 1$, if demand w is allocated to the RN, that is, if its fastest route in the network takes less time than the alternative mode; 0, otherwise.

- Objective functions and constraints
  The objective of our model is to maximize the RN trip coverage in case that everything works fine

$$z_1 = \sum_{w=(p,q)\in W} g_w p_w \ ,$$

or to minimize its total travel time

$$z_2 = \sum_{w=(p,q)\in W} g_w u_w \ .$$

The constraints have been grouped according to their aims as follows:

- Budget constraints

$$\sum_{n_i,n_j\in A',i<j} c_{ij} x_{ij} + \sum_{l\in L} \sum_{n_i\in N} c_i y_i^l \leq C_{\max} \tag{1}$$

- Alignment location constraints

$$x_{ij}^l \leq y_i^l, \ (n_i, n_j) \in A', \ i < j, \ l \in L \tag{2}$$

$$x_{ij}^l \leq y_j^l, \ (n_i, n_j) \in A', \ i < j, \ l \in L \tag{3}$$

$$x_{ij}^l = x_{ji}^l, \ (n_i, n_j) \in A', \ i < j, \ l \in L \tag{4}$$

$$x_{ij}^l \leq x_{ij}, \ (n_i, n_j) \in A', \ i < j, \ l \in L \tag{5}$$

$$x_{ij} \leq \sum_{l \in L} x_{ij}^l, \ (n_i, n_j) \in A', \ i < j \tag{6}$$

$$\sum_{j \in N(i)} x_{ij}^l \leq 2, n_i \in N, \ l \in L \tag{7}$$

$$h_l + \sum_{(n_i, n_j) \in A' \ i < j} x_{ij}^l = \sum_{n_i \in N} y_i^l, \ l \in L \tag{8}$$

$$\tfrac{1}{2} - \sum_{(n_i, n_j) \in A' \ i < j} x_{ij}^l + M(h_l - 1) \leq 0, \ l \in L \tag{9}$$

$$\tfrac{1}{2} - \sum_{(n_i, n_j) \in A' \ i < j} x_{ij}^l + M h_l \geq 0, \ l \in L \tag{10}$$

$$\sum_{n_i \in B} \sum_{n_j \in B} x_{ij}^l \leq |B| - 1, B \subseteq N, |B| \geq 2, \ l \in L \tag{11}$$

• Routing demand conservation constraints

$$\sum_{(n_i, n_p) \in A'} f_{ip}^w + \sum_{(n_i, n_p) \in A''} \varphi_{ip}^w = 0, \ w = (p, q) \in W \tag{12}$$

$$\sum_{(n_p, n_j) \in A'} f_{pj}^w + \sum_{(n_p, n_j) \in A''} \varphi_{pj}^w = 1, \ w = (p, q) \in W \tag{13}$$

$$\sum_{(n_i, n_q) \in A'} f_{iq}^w + \sum_{(n_i, n_q) \in A''} \varphi_{iq}^w = 1, \ w = (p, q) \in W \tag{14}$$

$$\sum_{(n_q, n_j) \in A'} f_{qj}^w + \sum_{(n_q, n_j) \in A''} \varphi_{qj}^w = 0, \ w = (p, q) \in W \tag{15}$$

$$\sum_{(n_i, n_k) \in A'} f_{ik}^w - \sum_{(n_k, n_j) \in A'} f_{kj}^w = 0, \ \text{if } k \notin \{p, q\}, \ w = (p, q) \in W \tag{16}$$

$$\sum_{(n_i, n_k) \in A''} \varphi_{ik}^w - \sum_{(n_k, n_j) \in A''} \varphi_{kj}^w = 0, \ \text{if } k \notin \{p, q\}, \ w = (p, q) \in W \tag{17}$$

$$f_{ij}^w + \varphi_{ij}^w \leq 1, \ (n_i, n_j) \in A, \ w \in W \tag{17}$$

• Location-Allocation constraints

$$f_{ij}^w + p_w - 1 \leq \sum_{l \in L} x_{ij}^l, \ (n_i, n_j) \in A', \ w \in W \tag{18}$$

• Splitting demand constraints

$$\varepsilon + u_w - \mu u_w^{COM} - M(1 - p_w) \leq 0, \ w = (p, q) \in W \tag{19}$$

where $u_w = \sum_{(n_i, n_j) \in A'} d_{ij} f_{ij}^w + \sum_{(n_i, n_j) \in A''} u_{ij}^{COM} \varphi_{ij}^w$, M is a big enough real number and $\varepsilon > 0$ small enough, and u_{ij}^{COM} is the generalized cost for traversing arc(i, j) by the complementary mode.

$$x_{ij}^l, \ x_{ij}, \ y_i^l, \ h_l, \ f_{ij}^w, \ \varphi_{ij}^w, p_w \in \{0, 1\}. \tag{20}$$

Constraint (1) takes into account budget availability. Constraints (2) and (3) ensure that a link is included in the RN only if the nodes that the link is incident to are also selected. Constraints (4) allow edges to be used in both directions. Constraints (5) and (6) impose that the arc (n_i, n_j) is to be built if and only if there is a line that uses it.

Constraints (7) require that each node has at most two associated edges of each line. Constraints (11) ensure that a line does not contain a cyclic subgraph. Let us note that these constraints along with (8) guarantee that lines to be constructed are path subgraphs. However, a line must have at least one edge; this is ensured by constraints (9). If a line l is not considered in the design then it does not have any edge (constraints (10)). (12) to (17) are flow conservation constraints in each node. The incoming flow equals the outgoing flow in both pairs of variables. Separately, we require that both outgoing flows and both incoming flows are equal to 1 at the beginning of the paths and at the end of the paths, respectively. Constraints (18) guarantee that a demand is routed on an edge only if this edge belongs to the rapid transit network. Finally, constraints (19) force demands to be assigned to the rapid transit mode if the associated cost for using this network (taking the fastest route) is less than or equal to the corresponding cost of the complementary mode and the opposite. Note the importance of the variable ε in these constraints to break possible ties.

For more details see [13].

Appendix B: Rapid Transit Line Planning

The RLP formulation is based on the modeling of the physical network, the services and the demand. These elements are briefly described in the rest of this appendix. The reader should note that some of the input data or variables are the same as in the RND model defined in Appendix A.

- Physical network: The physical network is formed by the stations and the arcs linking them. It is obtained from the RND output.
- Service network: A service is characterized by an origin station, a set of stop stations and a destination station. The arcs linking these stations is part of the definition of the service l. L is the service set. A section s of a service is the arc between each pair of stations. Each service l is characterized by a section set S_l including the sections used by the service l. Each service has other technical characteristics such as the vehicle speed and the number of passengers that can be moved in a train of this service q_l.
- Demand: Each O-D pair w is characterized by three parameters: the origin and destination stations and the demand of passengers g_w between them. The pair set is W. The demand is realized by route flow h_r, that is the use variable. Each demand w has a route set R_w. R_s is the route set using the section s.

- Costs: We will consider costs associated to users on the routes and costs associated to operators on the services. The routing cost rc is assumed to be a linear term that depends on route r. The operative cost oc is also assumed to be a linear term that depends on section s.
- Note that each train can run a given service several times during the planning time if the service time is sufficiently inferior. For this reason we consider the inclusion of another parameter in the model: t_l is the time to run each service l. Thus, if u_l is the frequency in the service, the number of needed trains for a given service l is $t_l u_l$. The fleet capacity is defined by fc. The operative variable is the frequency u_l of the service l.

The RLP model is defined by:

$$\min \Theta \sum_{w \in W,\ r \in R_w} rc_r h_r + (1 - \Theta) \sum_{s \in S_l,\ l \in L} oc_s u_l$$
$$\text{subject to } \sum_{r \in R_s} h_r - \sum_{l \in S_l} q_l u_l \leq 0, \quad \forall\, s \in S$$
$$\sum_{l \in S_l} u_l \leq q_s, \quad \forall\, s \in S$$
$$\sum_{r \in R_w} h_r = g_w, \quad \forall\, w \in W$$
$$\sum_{l \in L} t_l u_l \leq fc$$
$$y_l \in \mathbb{Z}^+,\ h_r \in \mathbb{R}^+.$$

RLP minimizes the user costs weighted by Θ, that represents the monetary cost of the user time, and the operative costs. The first constraints (Routing Demand Conservation Constraints) ensure that there are enough active services for the passenger flow in each route. The second group (Arc User Capacity Constraints) are capacity constraints that avoid sections to be saturated. The third group (Arc Vehicle Capacity Constraints) ensures the meeting of the demand. The fourth set of constraints (Fleet Capacity Constraints) ensure that the fleet capacity is sufficient for the frequency required. For more details the reader is referred to Marín and Salmerón [16].

The *RLP* is short defined by:

$$RLP : \min \text{ User costs, Operative costs}$$
$$\text{s.t.: Routing Demand Conservation Constraints,}$$
$$\text{Arc User vehicle Capacity Constraints,}$$
$$\text{Arc Vehicle Capacity Constraints,}$$
$$\text{Fleet Capacity Constraints.}$$

References

1. Abbink, E., van den Berg, B., Kroon, L., Salomon, M.: Allocation of railway rolling stock for passenger trains. Transportation Science 38, 33–41 (2005)
2. Bertsimas, D., Sim, M.: Robust Discrete Optimization and Network Flows. Mathematical Programming Series B 98, 49–71 (2003)
3. Bertsimas, D., Sim, M.: The Price of Robustness. Operations Research 52, 35–53 (2004)
4. Bruno, G., Gendreau, M., Laporte, G.: A heuristic for the location of a rapid transit line. Computers & Operations Research 29, 1–12 (2002)

5. Bussieck, M.R., Kreuzer, P., Zimmermann, U.T.: Optimal lines for railway systems. European Journal of Operational Research 96, 54–63 (1996)
6. Claessens, M.T., van Dijk, N.M., Zwaneveld, P.J.: Cost Optimal Allocation of Passenger Lines. European Journal of Operational Research 110, 474–489 (1998)
7. Cordeau, J.F., Soumis, F., Desrosiers, J.: A Benders Decomposition Approach for the Locomotive and Car Assignment Problem. Transportation Science 34, 133–149 (2000)
8. Cordeau, J.F., Toth, P., Vigo, D.: A Survey of Optimization Models for Train Routing and Scheduling. Transportation Science 32, 380–404 (1998)
9. Dufourd, H., Gendreau, M., Laporte, G.: Locating a Transit Line Using Tabu Search. Location Science 4, 1–19 (1996)
10. Gendreau, M., Laporte, G., Mesa, J.A.: Locating Rapid Transit Lines. Journal of Advanced Transportation 29, 145–162 (1995)
11. Guihaire, V., Hao, J.K.: Transit network design and scheduling: A global review. Transportation Research Part A 42, 1251–1273 (2008)
12. Kontogiannis, S., Zaroliagis, C.: Robust Line Planning under Unknown Incentives and Elasticity of Frequencies. In: Fischetti, M., Widmayer, P. (eds.) ATMOS 2008. Schloss Dagstuhl - Leibniz-Zentrum fuer Informatik, Germany (2008)
13. Laporte, G., Marín, A., Mesa, J.A., Perea, F.: Designing Robust Rapid Transit Networks with Alternative Routes. Journal of Advanced Transportation (to appear, 2010)
14. Laporte, G., Mesa, J.A., Perea, F.: A Game Theoretic Framework for the Robust Railway Transit Network Design Problem. Transportation Research B (to appear, 2010)
15. Malcom, S., Zenios, S.A.: Robust Optimization for Power Systems Expansion under Uncertainty. Journal of the Operational Research Society 45, 1040–1049 (1994)
16. Marín, A., Salmerón, J.: Tactical Design of Rail Freight Networks. Part I: Exact and Heuristic Methods. European Journal of Operational Research 90, 26–44 (1996)
17. Schöbel, A., Scholl, S.: Line Planning with Minimal Transfers. In: 5th Workshop on Algorithmic Methods and Models for Optimization of Railways, vol. 06901, Dagstuhl Seminar Proceedings (2006)
18. Scholl, S.: Customer-Oriented Line Planning. Ph.D. Thesis, University of Kaiserslautern (2005)

Effective Allocation of Fleet Frequencies by Reducing Intermediate Stops and Short Turning in Transit Systems

Juan A. Mesa[1], Francisco A. Ortega[2], and Miguel A. Pozo[3]

[1] Department of Applied Mathematics II, University of Seville
jmesa@us.es
[2] Department of Applied Mathematics I, University of Seville
riejos@us.es
[3] Department of Applied Mathematics I, University of Seville
miguelpozo@us.es

Abstract. A rapid transit system is called robust if it maintains its functionality under perturbations. To be robust, strategies for producing less vulnerable plans and procedures of recovery actions in case of disruptions (including timetable adjustment, rolling stock re-scheduling and crew re-scheduling) are used. This paper deals with the first task and develops an effective plan for allocating fleet frequencies at stops along a line based on three objectives: minimizing passenger overload, maximizing passenger mobility and minimizing passenger loss. Schedules for decongesting and recovering the line are determined by means of optimization models. Heuristic approaches are discussed and computational results for a real case study are provided.

Keywords: rapid transit system, planning, robustness, frequencies.

1 Introduction

The global transit planning process consists of determining a set of lines and their associated timetables to which vehicles and drivers are assigned, taking a complex scenario of constraints into consideration. According to Ceder and Wilson [2], the global transit planning process can be decomposed into a sequence of five components; namely, design of routes, setting of frequencies, timetabling, vehicle scheduling and crew scheduling. Ideally, all those steps should be treated simultaneously in order to ensure a desirable level of interaction and feedback, thus leading to better results. However, due to the exceptional complexity of the process, this global approach appears intractable in practice. As a result, various sub-problems have been identified in order to solve the planning problem in a sequential manner, although the global optimality cannot be guaranteed at the end of the process (Guihaire and Hao [8]).

Perturbations can appear due to an increase of demand (even in predictable situations, like peak hours or crowded social events) or as a consequence of fleet

R.K. Ahuja et al. (Eds.): Robust and Online Large-Scale Optimization, LNCS 5868, pp. 293–309, 2009.

size reduction (as in the case of a driver's strike). Both scenarios give rise to un-supplied demand zones that generate passenger overloads in the available vehicles. A robust Rapid Transit System (RTS) should be able to absorb such demand peaks through suitable strategies so that users' waiting time does not exceed an admissible length of time and the system managers avoid, at the same time, high operational costs to satisfy this demand.

The fleet itinerary along a line can be determined following different strategies previously planned (*off-line*). Operating normally, each train runs according to a timetable and stops in each station; when the vehicle arrives at the last station, it repeats the same itinerary in the opposite direction. Until the unit reaches the last station, the cycle is not restarted. Short-turning is a tactical decision which is useful when high demand zones need to be attended (this situation is typical in lines which connect to distant residential areas with the city centre or economic centres). For this strategy some vehicles can perform short cycles in order to increase frequency in specific zones of the line (Furth [6]). Another strategy for alleviating those overloaded stops faster consists of skipping stops (deadheading) at those stations with less demand. A deadheaded vehicle runs empty through a number of stations, after all passengers have already alighted, in order to start the new cycle earlier. In railway systems, deadheading at intermediate stops can be seen as expressing a vehicle through a segment of the line, once the stations to be skipped have been previously announced. Every time an intermediate station is skipped, time spent in decelerating, alighting and boarding, and accelerating the vehicle again is saved for users. As a disadvantage, bad use of express service restricts transport supply and can provoke passenger overloads in those vehicles which perform the full cycle. Other control strategies include holding a vehicle at a station, adding vehicles held in reserve or splitting a train (Wilson et al [14], Soeldner [11]).

Among the actions addressed to recovering system functionality, this paper deals with the problem of reallocating vehicle frequencies in order to distribute transport supply in a more efficient manner. A model (of tactical planning) is formulated for integrating strategies of Short-Turning And Expressing (STAE) in order to obtain appropriated frequencies which preserve the supply-demand equilibrium as well as maintain a minimum service level. For this purpose, fleet frequency will be increased in high demand zones in exchange for reducing service at stops with less demand, but without fully canceling it. Schedules for routes and stops, associated to the frequency allocation, will be obtained by means of optimization models. As an application, the methodology proposed has been applied with real data of commuter trains of Madrid (Spain), provided by RENFE operator. The paper is organized as follows: the next section reviews some approaches related to our problem. In Section 3, the formulation is built step by step and the paper contribution is specified. A greedy solution algorithm is considered for solving this problem in Section 4. In Section 5, the methodology is applied to a real case, making use of real data from the Line C10 commuter trains of Madrid.

2   Background

Strategies for managing perturbations, based on controlling the frequency of vehicles at stops have attracted increasing interest the past few years. In an approach where the concept of deficit function is introduced, Ceder and Stern [3] suggest the use of deadheading trips between various critical terminals to reduce fleet requirements. Eberlein et al. [4] deal with the real-time deadheading problem (RTDP) in bus transit operations control. Although the considered transit system consists of a one-way loop network (representative of, for example, a single rail transit line, or a simple bus or trolley bus line), the RTDP is shown to be a difficult nonlinear integer programming problem, computationally intractable. Therefore, an algorithm is proposed for solving a simplified version of the problem.

Guentari and Codina [9] exhibit a bilevel formulation for modeling congested public transport networks in an urban context, considering the feasibility of the asymmetry in the frequencies. In order to solve the problem, two heuristics are proposed, given the difficulty of applying exact algorithms for real scenarios.

Tirachini el al. [12] develop a model that integrates short turns and deadheading (at terminal stops) for an isolated transit line where the variables to be optimized are the frequencies within and outside of the high demand zone (of a continuous nature) and the stations where the strategies begins and ends (discrete variables). The objective is to decrease the total costs for users and operators in the public transport corridor by means of skipping stops (empty nodes in Figure 1 and Figure 2). With this objective, the fleet is divided into a service that cyclically runs stopping in all stations, and into another service that performs a short asymmetric cycle (Figure 1) between the nodes corresponding to the zone where a higher demand of trips exists.

In the following section, the model of Tirachini et al. [12] will be generalized for the case in which every vehicle of the fleet can perform a different short asymmetric cycle including expressing at intermediate stops (STAE) (Figure 2). Moreover, the notation used in the model will be able to optimism three relevant objectives: minimizing passenger overload, maximizing passenger mobility and minimizing passenger loss.

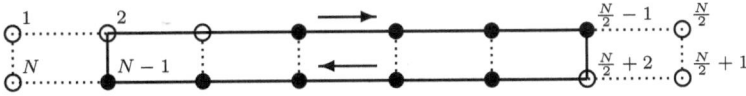

Fig. 1. Asymmetric short cycle in a linear corridor of transit

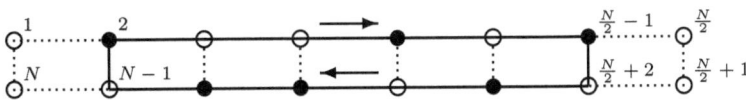

Fig. 2. STAE in a linear corridor of transit

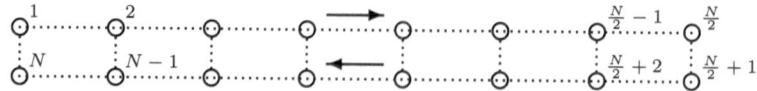

Fig. 3. Linear corridor of transit

3   Problem Formulation

As was previously noted, the transit system in consideration consists of a one-way loop network, as is shown in Figure 3. Such a network is representative of a single rail transit line. Since a one-to-one correspondence can be established between the same stations for both directions, N is always even and $N/2$ is always integer. The temporary scenario for determining strategies will be established by an operational time T.

Throughout this paper, we assume the following assumptions if not otherwise noted:

A1. A station can accommodate only one vehicle at a time and no overtaking between vehicles can occur at any point in the network.

A2. A vehicle is allowed to change its direction at any station of the line, without spending additional time during handling.

A3. In presence of a fleet size reduction, the new train schedule is efficiently announced to riders well in advance, in order for them to fit their interests to the new service.

A4. Passengers between each pair of stations will be homogeneously distributed in the set of all trips supplied by the system.

A5. When the frequency of vehicles, φ_{ij} between an origin i and a destination j decreases with regard to the initial frequency of vehicles, φ_{ij}^0, the number of users od_{ij}, who travel from origin i to destination j, also decreases according to the result given by $od_{ij}g_{ij}$, where

$$g_{ij} = \frac{1 - e^{-\lambda \frac{\varphi_{ij}}{\varphi_{ij}^0}}}{1 - e^{-\lambda}} \in (0,1) \tag{1}$$

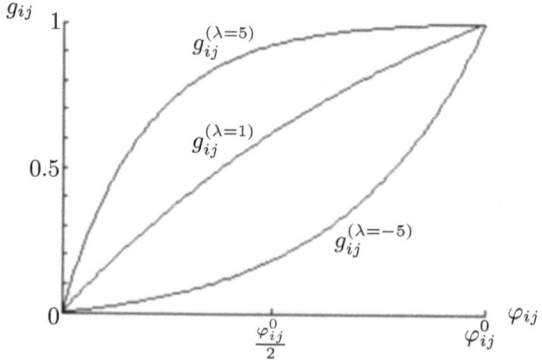

Fig. 4. Different slopes of g_{ij} for values of $\lambda = \{5, 1, -5\}$

Note that values g_{ij} are between 0 and 1 and the increasing behavior of its drawing can be calibrated by means of the parameter λ, as is shown in Figure 4.

A6. When an episode of overloading takes place in vehicle k at station i, a number of users $(\Delta\phi_{ki}^{++})$ cannot board the train. The number of users that was initially going to alight at current station i (ϕ_{ki}^{-}) will have to be decreased, taking the previous values of $\Delta\phi_{kj}^{++}$ $(j < i)$ into consideration (as shown in Equation 10).

3.1 Notation

In order to formulate the approach as a programming model, the following parameters and variables are defined:

N	number of stations of the line
$i, j \in \{1, ..., N\}$	indices associated to stations of the line
od_{ij}	number of passengers going from origin i to destination j
d_{ij}	distance (through the line) from i to j
K	fleet size
$k \in \{1, ..., K\}$	index associated to vehicles of the line
T	operation time
t_e	standard time at stops required in operations of acceleration and deceleration, boarding and alighting
v	average speed of vehicles without considering stops
μ	maximum capacity of vehicles

The decision of stopping at station i for vehicle k generates a set of $K \times N$ binary variables, defined as follows:

$$x_{ki} = \begin{cases} 1, \text{ if vehicle } k \text{ stops at station } i \\ 0, \text{ otherwise} \end{cases}$$

In order to make more comprehensive the formulation, a set of functions are defined (w.c.c. $\equiv$ without capacity constraints):

g_{ij}	demand reduction factor
φ_{ij}	frequency of vehicles going from i to j
ϕ_{ki}^{+}	users expected to board vehicle k at station i (w.c.c.)
ϕ_{ki}^{-}	users expected to alight from vehicle k at station i (w.c.c.)
C_{ki}^{*}	users at vehicle k departing from i (w.c.c.)
ϕ_{ki}^{++}	users actually boarding vehicle k at station i
$\Delta\phi_{ki}^{++}$	users unable to board vehicle k at station i
ϕ_{ki}^{--}	users actually alighting from vehicle k at station i
C_{ki}	users of vehicle k departing from i

3.2 Defining Frequencies, Flows and Load

The time used by vehicle k to complete a full cycle of its itinerary without stopping is given by:

$$\tau_k^d = \frac{2 max_{i,j}\{x_{ki}x_{kj}d_{ij}\}}{v} \tag{2}$$

i.e., twice the distance between the farthest stations, where vehicle k makes stops, divided by the average speed of vehicles without considering stops. On the other hand, the time spent in stops due to deceleration, alights, boards and acceleration,

$$\tau_k^p = \sum_{i \in N} t_e x_{ki} \tag{3}$$

could be considered taking, instead of the constant t_e, a function t_i dependent on the time required in each station for boarding and alights.

The frequency of trips between stations i and j is given by:

$$\varphi_{ij} = \sum_{k \in K} \frac{T}{\tau_k^d + \tau_k^p} x_{ki}x_{kj} \tag{4}$$

Combining the previous equation and assumption [A4.], the number of users, homogeneously distributed, which boards vehicle k at station i can be obtain:

$$\phi_{ki}^+ = \sum_{j=i+1}^{N} \frac{od_{ij}g_{ij}x_{ki}x_{kj}}{\varphi_{ij}} \tag{5}$$

Following an analogous reasoning the number of passengers which alights from vehicle k at station i can be deduced:

$$\phi_{ki}^- = \sum_{j=1}^{i-1} \frac{od_{ji}g_{ij}x_{ki}x_{kj}}{\varphi_{ij}} \tag{6}$$

By using the previous expressions the passenger load of vehicle k after departing from station i can be determined:

$$C_{ki}^* = \sum_{j=1}^{i} (\phi_{kj}^+ - \phi_{kj}^-) \tag{7}$$

3.3 Capacity Constraints

The previous three expressions have been defined in absence of capacity constraints. If the vehicle capacity is limited to μ, a recursive system of definitions is required to determine C_{ki}, the actual load of passengers in vehicle k after departing from station i.

The number of passengers that actually can board vehicle k at station i is given by

$$\phi_{ki}^{++} = \phi_{ki}^{+}\chi_{(\phi_{ki}^{+}+C_{k(i-1)}-\phi_{ki}^{--}<\mu)} + (\mu - C_{k(i-1)} + \phi_{ki}^{--})\chi_{(\phi_{ki}^{+}+C_{k(i-1)}-\phi_{ki}^{--}>\mu)}$$
(8)

where $\chi(\cdot)$ is the usual characteristic function. By means of this expression, only the number of passengers that actually fit in the vehicle is computed as boarded riders. Hence we can obtain the number of passengers that cannot board at station i:

$$\Delta\phi_{ki}^{++} = (\phi_{ki}^{+} - \phi_{ki}^{++})\chi_{(\phi_{ki}^{+}+C_{k(i-1)}-\phi_{ki}^{--}>\mu)}$$
(9)

The real number of passengers that alight at a station depends on the number of passengers that could have boarded on previous stations, so by using assumption [A6.] it is possible its evaluation:

$$\phi_{ki}^{--} = \phi_{ki}^{-} - \sum_{j=1}^{i-1}\Delta\phi_{kj}^{++}\frac{\phi_{ki}^{-}}{\sum_{i'=j+1}^{N}\phi_{ki'}^{-}}$$
(10)

Consequently, the real passenger load of a vehicle can be determined:

$$C_{ki} = \sum_{j=1}^{i-1}(\phi_{kj}^{++} - \phi_{kj}^{--})$$
(11)

3.4 Objectives

The objectives of users (typically expressed in terms of travel time and waiting time) and operators (in terms of fleet size) are commonly minimized simultaneously (Israeli and Ceder [10]). This involves a multi-objective formulation whose resolution leads to finding a set of solutions which represent different degrees of commitment between the conflicting objectives. From the operator's point of view, it is usual to minimize costs corresponding to vehicle/hour and vehicle/kilometer. However, if the fleet size is assumed to be fixed and all the vehicles remain available during all the operation time then the operator's objectives can be omitted in the optimization. From the user's point of view, three objectives are studied and described next.

Objective 1: Minimize Passenger Overload. A sudden cancelation of the service at an stop or a noticeable increase of passengers at a station can give rise to overloads, impeding that an important number of users can aboard. Looking after the convenience of ridership, it is of special interest to minimize the number of customers unable to board in vehicles (total overload), quantified through the function:

$$Z_1 = \sum_{k\in K}\sum_{i\in N}\Delta\phi_{ki}^{++}\varphi_k$$
(12)

where φ_k is the number of times that each vehicle performs its itinerary:

$$\varphi_k = \frac{T}{\tau_k^d + \tau_k^p} \tag{13}$$

Objective 2: Maximize Mobility. From the users' point of view, another objective consists of having a travel frequency distribution that provides higher frequencies between those o-d pairs with higher number of users. This criterion, that facilitates mobility, is well modeled by minimizing the sum of ratios passengers/trips between origin-destination pairs:

$$Z_2 = \sum_{i,j,i\neq j} \frac{od_{ij}}{\varphi_{ij}} \tag{14}$$

Objective 3: Minimize Passenger Loss. Given an initial transport supply, expressed by means of $\sum_{ij} \varphi_{ij}^0$, which satisfies an o-d demand, represented by matrix od_{ij}, a recommendable objective is the minimization of the number of users loss after a fleet size reduction, which can be described as was pointed out in assumption [A5.]:

$$Z_3 = \sum_{i\in N}\sum_{j\in N} od_{ij} - \sum_{i\in N}\sum_{j\in N} od_{ij}g_{ij} = \sum_{i\in N}\sum_{j\in N} od_{ij}(1 - g_{ij}) \tag{15}$$

3.5   Formulating Models

Scenario 1: Network Decongestion. If the demand is assumed inelastic, the objectives of interest will be to minimize the passenger overload (Z_1) and maximize the mobility (i.e. minimize Z_2), in order to prevent that the existing fleet size is unable to supply the demand.

For this network decongestion scenario, the problem of setting frequencies including Short Turning and Expressing at intermediate stops (STAE) can be formulated as follows:

$$\text{minimize } Z_1 = \sum_{k\in K}\sum_{i\in N} \Delta\phi_{ki}^{++} \frac{T}{\tau_k^d + \tau_k^p}$$

$$\text{minimize } Z_2 = \sum_{i,j} \frac{od_{ij}}{\varphi_{ij}}$$

subject to

$$\tau_k^d = \frac{2max_{i,j}\{x_{ki}x_{kj}d_{ij}\}}{v}, \qquad \forall k \tag{C1}$$

$$\tau_k^p = \sum_{i\in N} t_e x_{ki}, \qquad \forall k \tag{C2}$$

$$\varphi_{ij} = \sum_{k\in K} \frac{T}{\tau_k^d + \tau_k^p} x_{ki}x_{kj}, \qquad \forall i,j \tag{C3}$$

$$\phi_{ki}^{+} = \sum_{j=i+1}^{N} \frac{od_{ij}x_{ki}x_{kj}}{\varphi_{ij}} \frac{1 - e^{-\lambda \frac{\varphi_{ij}}{\varphi_{ij}^0}}}{1 - e^{-\lambda}}, \qquad \forall k, i \tag{C4}$$

$$\phi_{ki}^{-} = \sum_{j=1}^{i-1} \frac{od_{ij}x_{ki}x_{kj}}{\varphi_{ij}} \frac{1 - e^{-\lambda \frac{\varphi_{ij}}{\varphi_{ij}^0}}}{1 - e^{-\lambda}}, \qquad \forall k, i \tag{C5}$$

$$y_{ki} = \chi_{(\phi_{ki}^{+} + C_{k(i-1)} - \phi_{ki}^{--} < \mu)}, \qquad \forall k, i \tag{C6}$$

$$z_{ki} = \chi_{(\phi_{ki}^{+} + C_{k(i-1)} - \phi_{ki}^{--} > \mu)}, \qquad \forall k, i \tag{C7}$$

$$\phi_{ki}^{++} = \phi_{ki}^{+} y_{ki} + (\mu - C_{k(i-1)} + \phi_{ki}^{--}) z_{ki}, \quad \forall k, i \tag{C8}$$

$$\Delta\phi_{ki}^{++} = (\phi_{ki}^{+} - \phi_{ki}^{++}) z_{ki}, \qquad \forall k, i \tag{C9}$$

$$\phi_{ki}^{--} = \phi_{ki}^{-} - \sum_{j=1}^{i-1} \Delta\phi_{kj}^{++} \frac{\phi_{ki}^{-}}{\sum_{i'=j+1}^{N} \phi_{ki'}^{-}}, \qquad \forall k, i \tag{C10}$$

$$C_{ki} = \sum_{j=1}^{i} (\phi_{kj}^{++} - \phi_{kj}^{--}), \qquad \forall k, i \tag{C11}$$

$$x_{ki} \in \{0, 1\} \qquad \forall k, i \tag{C12}$$

Constraints C1-C5 correspond to the definitions made for frequencies, flows and load. Capacity constraints (C8 - C11) have been formulated by previously introducing two binary functions (y_{ki}, z_{ki}) associated to the characteristic functions which are used to correctly evaluate the real passenger load of vehicles.

Scenario 2: Network Restoration. In a context of an increasing elastic demand in front of a rigid supply of fleet, it is logical to suppose that a number of users will be deviated to another means of transport, as was previously established in assumption [A5.]. Therefore, the strategic objectives will now be to minimize the passenger overload (Z_1) and minimize the passenger loss (Z_3), in order to effective and efficiently (respectively) recover the functionality of the network after episodes of overloading.

For this scenario of network restoration, the STAE problem can be formulated as follows:

$$\text{minimize } Z_1 = \sum_{k \in K} \sum_{i \in N} \Delta\phi_{ki}^{++} \frac{T}{\tau_k^d + \tau_k^p}$$

$$\text{minimize } Z_3 = \sum_{i \in N} \sum_{j \in N} od_{ij} (1 - \frac{1 - e^{-\lambda \frac{\varphi_{ij}}{\varphi_{ij}^0}}}{1 - e^{-\lambda}})$$

subject to the same set of constraints [(C1)-(C10)] associated to the previous model.

4   Resolution Algorithm

Owing to the features of both approaches of the STAE problem, they can be identified as combined cases of the known problems of Transit Network Design and Frequencies Setting (TNDFSP), where the objectives and the constraints are non-linear.

In general, the TNDFSP is a special type of integer optimization problem, where its space of solutions is a set of combinations of subsets of integer numbers (Israeli and Ceder [10]). In our approach, each solution can be assumed as a frequency set between every o-d pair and a route set that vehicles have to follow, once variables $\mathcal{X} = [x_{ki}], k \in K, i \in N$ have been determined in the model.

The enumeration of feasible solutions for variables x_{ki} may be extremely large and complex, even for small size instances of the problem, needing the utilization of commercial software to solve it (see Wan and Lo [13], and Barra et al. [1]). In particular, both formulations of the STAE problem present involve KN decision variables and $N^2 + 8KN + 2K$ constraints; for the example shown in Section 5, with 26 stations and 10 vehicles, the approach include 260 decision variables and 2776 constraints.

Since exact methods seem to be unable for solving the routes and frequencies optimization problem, approximate methods appear to be an admissible alternative. Metaheuristics (Glover y Kochemberg [7]) provide approximating methods that implement efficient and effective approaches for the exploration of the solution space. Their multi-objective variant, represents specific mechanisms to carry out the resolution of optimization models with several objectives in conflict (Ehrgott y Gandibleux [5]).

The easiest approximate method for solving the model on real life instances consists of designing ad hoc a greedy algorithm. In general, a greedy algorithms obtains a local optimum at each stage which is iteratively improved until reaching or approximating to a global optimum.

The structure of the algorithm proposed is the following:

INPUT: Obtaining an initial solution

Set $x_{ki} = 1, \forall k \in K, \forall i \in N$ (i.e., an schedule where all the K vehicles stops in the N stations). Let X=X^0 be this initial solution. Evaluate the objective function for this solution [Z(X)].

STEP1: Selecting a subset Θ of new candidate solutions

From route X new possible solutions are considered, for different strategies, and collected in set Θ. Namely,(1) removing one or two stops, making shorter the cycle by the left, (2) idem by the right, (3) removing the station with the lowest demand (i.e., lowest value of $\phi_{ki}^+ + \phi_{ki}^-$)

STEP2: Selecting the best new candidate solution

Select the best solution X* of Θ according to the objective Z under consideration. Determine frequencies, flows and loads of X*.

TEST: Comparing solutions

IF Z(X*) provides a better value than Z(X), then reassign X=X* and **GO TO** STEP1; otherwise **END**

In spite of its simplicity, the described algorithm is flexible enough to start from any initial solution provided by the system operator (INPUT), to include other strategies of improving solutions (STEP 2) and to compare the quality of the solutions obtained for both scenarios. Although this algorithm has been designed to solve the problem in a context of off-line information, it is worthwhile to see its adaptability to the on-line problems where managers decide to apply immediate actions by canceling (or not) specific stops. In broad sense, this approach may be considered as "robus", since the formulation assumes that disruptions can change the known input data and, nevertheless, the model validity remains.

5   Computational Results

5.1   Validating the Algorithm

The methodology proposed in this work and the effectiveness of the algorithm was applied to a single line of 7 nodes. The diagram of boards and alights along the line is described in Figure 5.

The number of users boarding and alighting at every station differs notice-ably in both directions of the line giving rise to a heterogeneous distribution of origin-destination flows:

$$od_{ij} = \begin{pmatrix} 0 & 34 & 31 & 21 & 14 & 12 & 21 \\ 24 & 0 & 46 & 40 & 132 & 65 & 12 \\ 44 & 24 & 0 & 22 & 262 & 279 & 14 \\ 32 & 289 & 30 & 0 & 34 & 28 & 17 \\ 66 & 21 & 36 & 19 & 0 & 98 & 13 \\ 48 & 197 & 97 & 245 & 42 & 0 & 32 \\ 34 & 21 & 25 & 34 & 19 & 28 & 0 \end{pmatrix}$$

The values of o-d pairs $(2,5)$, $(3,5)$, $(3,6)$, $(4,2)$, $(6,2)$, $(6,4)$, suggest the need of providing these pairs with a higher trip frequency (objective 2). On the other hand, this criterion should be complemented by avoiding passenger overloads at each station of the corridor. The assumption that demand could be elastic or inelastic, requires studying the objectives of Section 3.4 in scenarios 3.5.1 and 3.5.2.

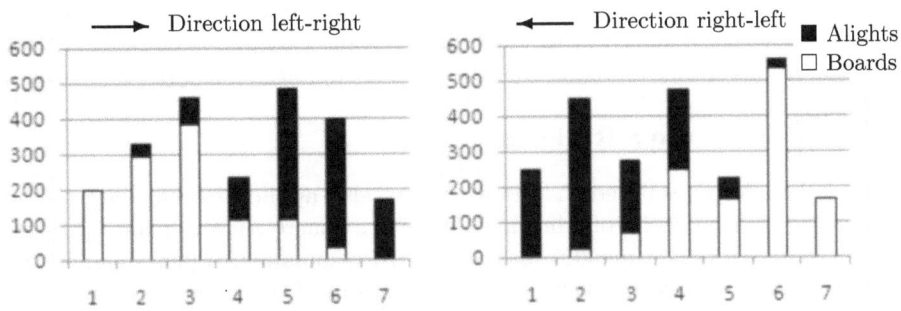

Fig. 5. Boards and alights at stations

Table 1. Results obtained for Scenarios 1 y 2

Operation	$Z_{1,2}$	$Z_{1,3}$
without strategy $(Z_{1,2}^0, Z_{1,3}^0)$	1007	634
STAE with $\alpha = 0.5$	836	411

Table 2. Schedules of routes and stops

Direction left-right ($\rightarrow$)						
k=1	-	2	3	-	5	6	-
k=2	-	2	3	4	5	6	-
k=3	1	2	3	4	5	6	7

Direction right-left ($\leftarrow$)						
k=1	-	2	-	4	-	6	-
k=2	-	2	3	4	5	6	-
k=3	1	2	3	4	5	6	7

The simultaneous minimization of objectives Z_1 and Z_2 in Scenario 1, is carried out through the minimization of the weighted sum $Z_{1,2} = \alpha Z_1 + (1 - \alpha)Z_2$ for values of $\alpha \in (0, 1)$. This will give rise to a non-dominated pareto optimum solutions set for each value of α. Analogously, the simultaneous minimization, $Z_{1,3}$, of objectives Z_1 and Z_3 in Scenario 2 is formulated.

All possible configurations $\mathcal{X} = [x_{ki}], k \in K, i \in N$ were analyzed in order to obtain an optimum solution. Such process needed a computational time of 417 seconds for each value of α, considering constraints of minimum trip frequencies ($\varphi_{ij} \geq 1$), minimum itinerary performance ($\varphi_k \geq 1$), and fixed stops for every vehicle in the stations of the main o-d pairs (3,6),(4,2). However, the same optimum solution set were obtained, using the algorithm proposed, in a computational time of 0.01 seconds, without any consideration of fixed stops.

As an example, Table 1 shows the values of objective functions $Z_{1,2}$ and $Z_{1,3}$ when the fleet operates without strategy $(Z_{1,2}^0, Z_{1,3}^0)$, and with STAE strategy for $\alpha = 0.5$.

Each one of the solutions obtained from the set $\mathcal{X} = [x_{ki}], k \in K, i \in N$ has a schedule of routes and stops assigned to each vehicle. As an example, the one obtained for case $\alpha = 0.5$ in both directions is provided in Table 2. For instance, vehicle 2 runs from station 2 to 6 skipping station 4 (superior table) and then changes the direction running from station 6 to 2 skipping 5 and 3 (inferior table).

5.2   Application to a Real Case

The methodology proposed in this work was also applied to real data of the commuter train systems of Madrid. Through a macro-survey made to users, an o-d matrix for the time slot [6:00, 8:59] (corresponding to peak-hour in the morning) was obtained. Focusing on the 26 stations of the linear corridor of line C10, a disequilibrium (in the same terms of section 5.1) was observed between directions left-right and right-left (Figure 6).

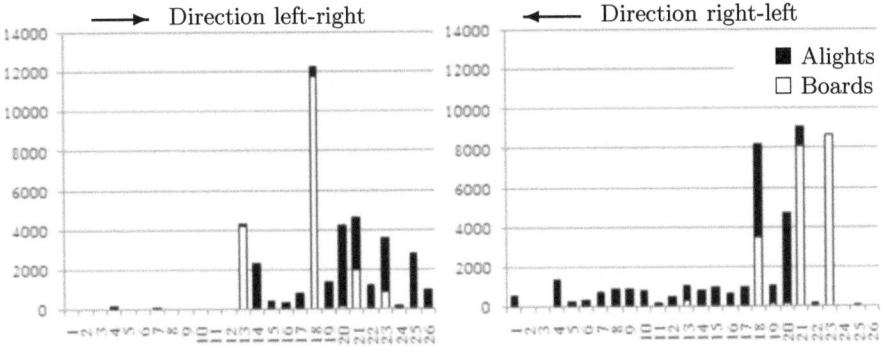

Fig. 6. Boards and alights at the stations of line C10

First, we assume that an initial fleet size of $K = 10$ vehicles (with capacity $\mu = 1150$ users/vehicle) is provided to serve this corridor, and no strategies are adopted in their itinerary (vehicles serve all corridor from one side to another, stopping at every station). A fleet size reduction to $K = 5$ vehicles give rise to the distribution $\frac{od_{ij}}{\varphi_{ij}}$ (users/trip) for the 26×26 o-d pairs (Figure 7).

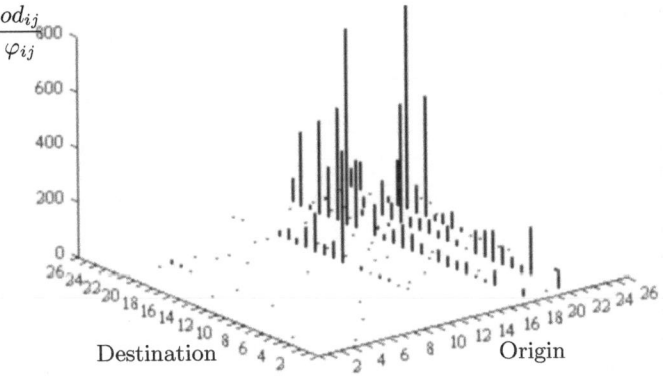

Fig. 7. Distribution *users/trip* for the 26×26 o-d pairs

The set of non dominated solutions for the investigated values of parameter α, can be identified along the Pareto frontier in Figure 8 and Figure 9, where four of those solutions have been pointed up.

Tables 3 and 4 shows the values of objective functions when the fleet operates without strategy (Z_1^0, Z_2^0, Z_3^0), and also the values obtained from four representative efficient solutions. The improvement respect to the case without strategy is exhibited as well.

Finally, the schedule of routes and stops assigned to each vehicle obtained for case $\alpha = 0.458$ in Scenario 1 is provided in Table 5 and analogously for case $\alpha = 0.483$ in Scenario 2 in Table 6.

Table 3. Results obtained in Scenario 1

Operation	Z_1	Z_2	ΔZ_1^0	ΔZ_2^0
without strategy (Z_1^0, Z_2^0)	16654	7901	-	-
STAE with $\alpha = 0.134$	6969	6440	-58.1%	-18.5%
STAE with $\alpha = 0.458$	5512	6578	-66.9%	-16.7%
STAE with $\alpha = 0.605$	4519	7238	-72.9%	-8.4%
STAE with $\alpha = 0.875$	4025	7489	-75.8%	-5.2%

Table 4. Results obtained in Scenario 2

Operation	Z_1	Z_3	ΔZ_1^0	ΔZ_3^0
without strategy (Z_1^0, Z_3^0)	13208	4897	-	-
STAE with $\alpha = 0.336$	3905	2209	-70.4%	-54.9%
STAE with $\alpha = 0.483$	3069	2763	-76.8%	-43.5%
STAE with $\alpha = 0.539$	2309	3291	-82.5%	-32.8%
STAE with $\alpha = 0.603$	1625	3392	-87.7%	-30.7%

Table 5. Schedules of routes and stops for Scenario 1

Direction left-right ($\rightarrow$)

	1	2	3	4	5	6	7	8	9	10	11	12	13	14	15	16	17	18	19	20	21	22	23	24	25	26
k=1	-	-	-	-	-	-	-	-	-	-	-	-	-	-	-	-	-	18	19	20	21	22	23	-	-	-
k=2	-	-	-	-	-	-	-	-	-	-	-	-	13	14	-	-	17	18	19	20	21	22	23	-	25	-
k=3	-	-	-	4	-	-	-	-	-	-	-	-	13	14	15	16	17	18	19	20	21	22	23	-	25	-
k=4	-	-	-	4	-	-	7	-	-	-	-	-	13	14	15	16	17	18	19	20	21	22	23	24	25	26
k=5	1	2	3	4	5	6	7	8	9	10	11	12	13	14	15	16	17	18	19	20	21	22	23	24	25	26

Direction right-left ($\leftarrow$)

	1	2	3	4	5	6	7	8	9	10	11	12	13	14	15	16	17	18	19	20	21	22	23	24	25	26
k=1	-	-	-	-	-	-	-	-	-	-	-	-	-	-	-	-	-	18	19	20	21	-	23	-	-	-
k=2	-	-	-	-	-	-	-	-	-	-	-	-	13	14	15	16	17	18	-	20	21	-	23	-	25	-
k=3	-	-	-	4	5	6	7	8	9	10	-	12	13	14	15	16	17	18	19	20	21	22	23	-	25	-
k=4	-	-	-	4	5	6	7	8	9	10	11	12	13	14	15	16	17	18	19	20	21	22	23	-	-	26
k=5	1	2	3	4	5	6	7	8	9	10	11	12	13	14	15	16	17	18	19	20	21	22	23	24	25	26

Table 6. Schedules of routes and stops for Scenario 2

Direction left-right ($\rightarrow$)

	1	2	3	4	5	6	7	8	9	10	11	12	13	14	15	16	17	18	19	20	21	22	23	24	25	26
k=1	-	-	-	-	-	-	-	-	-	-	-	-	-	-	-	-	-	18	19	20	21	22	23	-	-	-
k=2	-	-	-	-	-	-	-	-	-	-	-	-	13	14	15	16	17	18	19	20	21	22	23	-	25	-
k=3	-	-	-	-	-	-	-	-	-	-	-	-	13	14	15	16	17	18	19	20	21	22	23	24	25	-
k=4	-	-	-	-	-	-	7	-	-	-	-	-	13	14	15	16	17	18	19	20	21	-	-	-	-	-
k=5	1	2	3	4	5	6	7	8	9	10	11	12	13	14	15	16	17	18	19	20	21	22	23	24	25	26

Direction right-left ($\leftarrow$)

	1	2	3	4	5	6	7	8	9	10	11	12	13	14	15	16	17	18	19	20	21	22	23	24	25	26
k=1	-	-	-	-	-	-	-	-	-	-	-	-	-	-	-	-	-	18	19	20	21	-	23	-	-	-
k=2	-	-	-	-	-	-	-	-	-	-	-	-	13	14	15	16	17	18	19	20	21	22	23	-	25	-
k=3	-	-	-	-	-	-	-	-	-	-	-	12	13	14	15	16	17	18	19	20	21	22	23	-	25	-
k=4	-	-	-	-	-	-	7	8	9	10	11	12	13	14	15	16	17	18	19	20	21	-	-	-	-	-
k=5	1	2	3	4	5	6	7	8	9	10	11	12	13	14	15	16	17	18	19	20	21	22	23	24	25	26

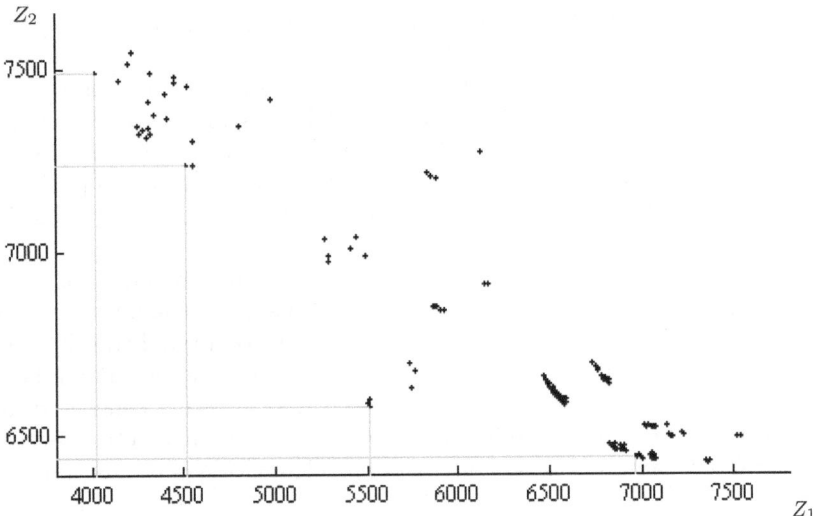

Fig. 8. Selection of four representative efficient solutions in Scenario 1

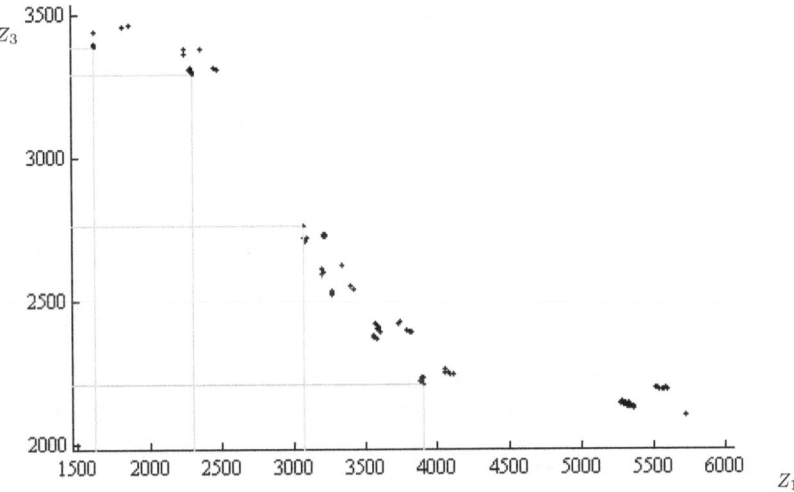

Fig. 9. Selection of four representative efficient solutions in Scenario 2

6   Conclusions

This paper formulates a model for the application of the integrated strategy Short Turning and Expressing (STAE) in a railway linear corridor. The model develops a frequency distribution and a route configuration for a fleet size unable to supply the existing demand in the corridor. Multicriteria optimization is applied taking as objectives to raise the service quality (maximizing the mobility and minimizing the passenger loss) at the same time that a system smooth-running is provided (minimizing the passenger overload).

The simultaneous utilization of several traffic strategies for each vehicle involves a mathematical formulation which achieves combinatorial complexity. Therefore, the determination of solutions requires the use of heuristics algorithms. This work provides a greedy algorithm that removes stops from the schedule throughout an iterative process, increasing the quality of the candidate solutions at each stage.

The methodology proposed in this work was applied to real data of the commuter train systems of Madrid, where eventual massive demand of this means of transport could make inefficient the operativeness of a rapid transit network, giving rise to overloads, passenger loss, and a bad system service. From the analysis of the results obtained, it can be stressed that applying control strategies of frequencies. The computational experience indicates that better results are obtained when criteria of good performance and service quality are combined.

Acknowledgments. This work was partially supported by the Future and Emerging Technologies Unit of EC (IST priority-6th FP), under contract no. FP6-021235-2 (project ARRIVAL), and by the Spanish research Project PT-2007-003-08CCPP (Centro de Experimentación de Obras Públicas, Ministerio de Fomento - Gobierno de España).

References

1. Barra, A., Carvalho, L., Teypaz, N., Cung, V.D., Balassiano, R.: Solving the Transit Network Design Problem with Constraint Programming. In: Proceedings of the 11th World Conference in Transport Research, University of California, Berkeley, USA, June 24-28 (2007)
2. Ceder, A., Wilson, N.H.M.: Bus Network Design. Transportation Research B 20, 331–334 (1986)
3. Ceder, A., Stern, H.I.: Deficit Function Bus Scheduling with Deadheading Trip Insertion for Fleet Size Reduction. Transportation Science 15(4), 338–363 (1981)
4. Eberlein, X.J., Wilson, N.H.M., Barnhart, C.: The Real Time Deadheading Problem. Transportation Research B 32, 77–100 (1997)
5. Ehrgott, M., Gandibleu, X.: Approximative Solution Methods for Multiobjective Combinatorial Optimization. TOP 12(1), 1–89 (2004)
6. Furth, P.G.: Short Turning on Transit Routes. Transportation Research Record 1108, 42–52 (1987)
7. Glover, F.W., Kochemberger, G.A.: Handbook of Metaheuristics. Springer, Heidelberg (2003)
8. Guihaire, V., Hao, J.-K.: Transit Network Design and Scheduling: A Global Review. Transport. Res. Part A. 42, 1251–1273 (2008)
9. Guentari, M.S., Codina, E.: Optimización de Frecuencias Combinadas y Asignación de Vehículos en Redes de Transporte Público Congestionadas. VIII Congreso de Ingeniería de Transporte, A Coruña, Spain (2008)

10. Israeli, Y., Ceder, A.: Transit Route Design Using Scheduling and Multiobjective Programming Techniques. In: Proceedings of the Sixth International Workshop on Computer Aided Scheduling of Public Transport, pp. 56–75. Springer, Heidelberg (1993)
11. Soeldner, D.W.: A Comparasion of Control Options on the MBTA green line. Master thesis, Civil Engineering, MIT (1993)
12. Tirachini, A., Cortés, C.E., Jara-Díaz, S.: Estrategia Integrada de Asignación de Flota en un Corredor de Transporte Público. XIII Congreso Chileno de Ingeniería de Transporte. Santiago, Chile (2007)
13. Wan, Q.K., Lo, H.K.: A Mixed Integer Formulation for Multiple-route Transit Network Design. Journal of Mathematical Modelling and Algorithms 2(4), 299–308 (2003)
14. Wilson, N.H.M., Macchi, R.A., Fellows, R.E., Deckoff, A.A.: Improving Service on the MBTA Green Line through Better Operations Control. Transportation Research Record 1361 (1992)

Shunting for Dummies:
An Introductory Algorithmic Survey*

Michael Gatto, Jens Maue, Matúš Mihalák, and Peter Widmayer

Institute of Theoretical Computer Science, ETH Zurich, Switzerland
firstname.lastname@inf.ethz.ch

Abstract. In this survey we present a selection of commonly used and new train classification methods from an algorithmic perspective.

1 Introduction

Railway optimization has started to evolve from manual to computerized solutions over the past decades. Problems of one particular class within the area of railway optimization ask how to arrange train cars into specific sequences so as to form desired trains. These problems, in railway terms called *shunting, marshalling,* or *classification* problems, have been considered for a variety of settings, with different assumptions about possible elementary operations and operational constraints, and for several objectives. The goal of this survey is to serve as an entry point into the field, with an emphasis on fully explaining key ideas in a typical setting so that they are easily accessible to a computer scientist. Our detailed explanations do not aim for encyclopedic completeness (problem descriptions of this nature can be found in [1,2]), but we do point out the state of the art in algorithmic shunting research.

Shunting problems arise in railway networks in many circumstances. We explain shunting at the example of freight trains. The purpose of freight trains is to transport rail cars from origin stations to destination stations. For the purpose of this car delivery, the routes to travel of freight trains have to be decided, along with what cars the train pick up along their routs and bring to a marshalling yard. In the yard, trains meet, exchange and rearrange cars, and travel along another route to distribute their cars. Picking up or delivering a car is easy if the car can simply be attached at or detached from the end of the train. This imposes conditions on the sequence of cars in both trains that enter and trains that leave a yard. In real life, the whole rearrangement process in yards occurs over time, has to obey a number of constraints due to physics and to operational requirements of the yard, and incurs costs with several components. In this introductory survey, we base our explanations on a specific setting as for the constraints and costs, and we ignore time entirely.

More specifically, we study a marshalling yard that consists of a set of parallel tracks, called *classification tracks*, connected on one end in the form of a binary

---

* This work was partially supported by the Future and Emerging Technologies Unit of EC (IST priority - 6th FP) under contract no. FP6-021235-2 (project ARRIVAL).

R.K. Ahuja et al. (Eds.): Robust and Online Large-Scale Optimization, LNCS 5868, pp. 310–337, 2009.

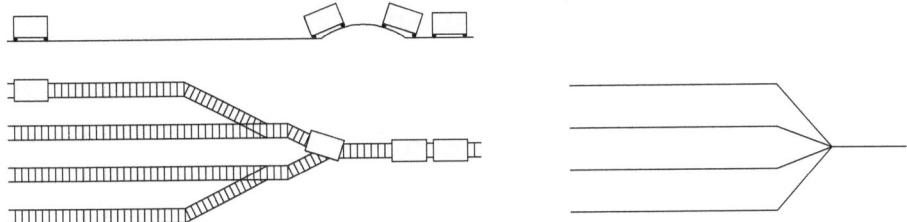

Fig. 1. Schematic representation of a hump yard. Left, seen from the side and the top, featuring the tree-like connectivity to the hump track (bottom), and a schematic representation of the hump (top), shown during a roll in of the same cars. Right, the abstract representation we use throughout the paper, shown for the same yard.

tree, with no connection at the other end (that is, a dead end), see Figure 1. Each connection between two tracks merges them into one track that takes part in further merging (unless it is the only remaining track). Cars leaving the classification track have no choice where to go, but for entering cars (that is, cars traveling in the other direction) such a connection serves as a switch: A car going over the switch in the direction of the dead ends can go either one of both ways, depending on how the switch has been set. In effect, a car that enters the tree of switches from the outside can be guided to any track, by setting all relevant switches accordingly. Sending a car to a track, hence, amounts to choosing a path in the tree from the root to a leaf. Switches are programmable, in the sense that they can read the identity of a passing car (a little in front of the switch) and then set the switch according to plan.

In such a marshalling yard, the most important elementary operation for an incoming train is to distribute all its cars, in the order in which they arrive within the train, one by one onto the classification tracks. Since cars have no engines, marshalling yards often use gravity to make individual cars travel onto the classification tracks. To that aim, the track leading the incoming trains to the classification tracks is built on a small artificial hill, such that cars can roll from the top of the hill to the classification tracks. This hill is called the *hump* of such a *hump yard*, and the track for incoming trains the *hump track*. In more detail, the cars of an incoming train are decoupled in front of the hump, and are then pushed over the hump by a shunting engine at the end of the train. This operation is called a *roll-in*. All cars on a classification track can be coupled together, so that a shunting engine can attach to them and pull them over the hump, a *pull-out* operation that puts the corresponding sequence of cars in the same situation as an incoming train. In particular, roll-in and pull-out operations have to alternate. This mode of operation is well known in computing: A classification track serves as a special kind of stack (a linear data structure which stores elements on the *last-in first-out* basis), where individual elements can be pushed on top of the stack. A pull-out operation is a sequence of pop operations, each removing a top element from the stack, until the stack is empty; each removed element is pushed on the stack that reflects the hump track.

Fig. 2. Schematic representation of a double ended hump yard. The hump is present on just one of the two tracks connecting to the classification tracks (e.g., on the right hand side).

For the purpose of this survey, we mostly focus on a single train entering the yard, with several trains (or even just a single one in particular) leaving it. The most important parameters that determine how to create the outgoing trains from the incoming train refer to the number of classification tracks, the lengths of these tracks, the number of roll-in and pull-out operations, and to a lesser degree the number of times that a car goes over the hump (so as to limit wear and tear on cars). Some of the presented methods will disregard the arrangement of cars in the incoming train and work just as well for any such order. This has the advantage of a certain robustness against differences in train composition of incoming trains, but suffers from the disadvantage that a certain sorting that may be inherent in the incoming train cannot be exploited to make shunting even more efficient and better adapted. Practical shunting rests on non-adaptive methods, one reason being that adaptive methods can be quite complicated and require computational support (as against a fixed shunting scheme). It is interesting to study tradeoffs between simplicity and robustness on one side, and optimality on the other side; we will explain some tradeoffs in this survey, but some others have still not been understood precisely.

We now give an example to illustrate the ideas and questions we are interested in. Consider a train built by n cars arriving at a double ended hump yard, that is, a classification yard where the set of parallel classification tracks are connected at and are accessible from both sides, and a hump is present on just one side of the yard, see Figure 2. Here, we want to reverse the order of the cars of the incoming train. For this example, we number the cars from 1 to n in their outgoing sequence. Thus, the inbound train T_{in} is represented by the ordered sequence of cars $T_{in} = (n, n-1, n-2, \ldots, 2, 1)$, and the goal is to build an outbound train $T_{out} = (1, 2, \ldots, n-2, n-1, n)$. One option to achieve such an ordering is to apply single-stage sorting (we elaborate on this method in Section 3). Here, the cars are pushed over the hump, and each car is rolled onto a different track. In a subsequent step, the cars are collected from the opposite side of the yard in their outgoing order, thus building the train T_{out}. To achieve the ordering with this method, each car is rolled over the hump exactly once, and in total n cars are rolled over the hump, which is best-possible. However, this approach requires n tracks to perform the sorting, which is as bad as it gets. As an alternative, the output train can also be built using just two tracks and repeatedly pushing cars over the hump (a so-called multistage method, see Section 4 and Section 5). The train is pushed over the hump n times. At the first roll-in operation, all cars roll onto the second track, except for car number one, which is rolled onto the first track. Then, all cars of the second track are pulled out over the hump again, and the sorting continues iteratively as follows. At the i-th roll-in operation, the car with number i is rolled onto the first track, while the remaining cars roll onto the second track. Then, the cars on the second track are pulled over the hump. After n roll-in operations, the outgoing train T_{out} has been formed on the first track. Here, we sort the train using very few tracks (just two), but using n roll-in operations; moreover, the total number of times the cars roll over the hump (most cars will roll repeatedly) is roughly $\frac{n^2}{2}$. Under the assumption that the first track is only used to build the outbound train, and thus its cars cannot be re-humped (i.e., pulled out and subsequently rolled in), this method is optimal for the given input sequence.

Observe that the two aforementioned methods can be applied to sort cars appearing in any order in the inbound train T_{in}, and thus these methods can be viewed as robust, as the order of the cars in the incoming trains does not influence the sorting (shunting) steps. However, for several important criteria that we did not specify yet (but will do later) these methods may perform poorly. Consider, e.g., any two cars with consecutive numbers in the outbound train which are already in the correct relative order in T_{in}. For the first method, these cars can be assigned to the same classification track, thus reducing the number of tracks needed to sort T_{in}; for the second method, these cars can be rolled onto the first track in the same roll-in phase, thus reducing the number of pull-out and roll-in operations. For example, the inbound train $T_{\text{in}} = (3, 1, 2, 4, 5)$ can be sorted using two tracks with both methods. For the first, cars 1 and 2 can be rolled onto the first track, and cars $3, 4, 5$ onto the second track, thus already restoring the order. For the second, the sorting can be achieved requiring two roll-ins instead of five as follows. In the first roll-in, cars 1 and 2 roll onto the first track, all other cars onto the second track. The cars on the second track are now also consecutive in number, and can be put onto the first track in one roll-in. In this example less tracks or less roll-in steps can be used because some of the cars of the incoming train were in the correct relative order, a concept that can be formalized and is often called *presortedness* of the input, and which we talk about later in this paper.

Moreover, the two extremes shown above naturally raise the question of the trade-off between the number of tracks available for shunting, and the number of roll-in operations required to sort the cars: what is the minimum number of tracks necessary to sort cars in a single stage for any given input sequence, and what is the minimum required number of sorting steps (i.e., roll-ins) if there are W tracks available for sorting?

In computer science, the shunting problem can be seen as a sorting problem for a sequence of numbers (identifying rail cars) using quite special sorting operations (pulling all the cars out of one classification track to a hump track, rolling every car from the hump track onto an arbitrary classification track). In algorithmic terms, a track (both a classification and the hump track) operate as a *stack* (i.e., as a linear data structure which stores and retrieves elements in the *last-in first-out* order). The operations *push* and *pop* allow to store an element and to remove the top element of the stack, respectively. The problem of sorting an input sequence of numbers with a network of stacks and queues (recall that a queue stores and retrieves elements in the *first-in first-out* order) was first considered in the seventies. Knuth introduced the problem (motivated by shunting problems of railway cars) in the first volume of his book *The Art of Computer Programming* (see [3] for its 3rd edition) that was first published in 1968. We will briefly discuss the main findings along these lines in Section 6.

Earlier Surveys

For the sake of perspective, we briefly discuss earlier surveys [1,2,4] with a different emphasis.

The work of Siddiquee from 1972 [4] promotes the use of computer-aided systematic methods for car sorting and train formation in the United States. In view of developing tailor-made schemes for the U.S. environment, Siddiquee describes four currently used systematic methods that are in use worldwide – the *sorting by groups*, the *sorting by train*, the *triangular sorting*, and the *geometric sorting* schemes. Additionally to the description of the methods, Siddiquee also points out basic characteristics of the methods, and outlines when a method can be superior to other methods. A major part of our survey has the same general goal. In order to keep our survey self contained, in Section 4 we also describe the four sorting methods and characterize them in a similar way. Additionally, we put these methods in the context of the newer developments – methods that take the *presortedness* of the incoming trains into consideration, and therefore sort different incoming trains in potentially different ways.

Hansmann and Zimmermann dedicate the first part of their paper [1] to a thorough characterization and general description of variants of the shunting problem. They name the class of shunting problems under consideration the SORTINGOFROLLINGSTOCK problems – the class SRSP. Loosely speaking, SRSP consists of problems which ask to find an optimum schedule for rearranging units of rolling stock (railcars, trams, trains,...) under certain constraints and objectives. A shunting problem is characterized according to the *track topology* at the shunting yard, according to the *sorting mode* (e.g., the allowed shunting operations, or the separation of time when all cars arrive from the time when they depart), and according to the desired structure of the *output sequence* (e.g., cars are asked to be grouped in a consecutive subsequence, or groups are asked to depart in a certain order). Combining any of these three characteristics results in one particular shunting problem. Hansmann and Zimmermann also give references to known results.

Similar in spirit to Hansmann and Zimmermann is the report of Di Stefano et al. [2], which presents various models for car rearrangement, reflecting the real-world limitations of how cars can move, shove and turn at train stations or shunting yards. They define a train as a sequence of symbols, and a configuration of a train as its permutation. The output constraints on the shunting are modeled as a set of permutations of the input train. For every described class of shunting problems, they define allowed shunting operations, and the set of feasible outputs, together with an objective function which they want to optimize while transforming the input sequence into one of the feasible sequences (using the allowed shunting operations).

2   Problem Definition

In this section, we give a mathematical description of the shunting problems at hand.

Definition 1 (Tracks, Shunting Yard). *A track works like a stack. A stack supports two types of operations: the addition of one element on the top of the stack (the classical push operation), and the removal of the top element from the stack (the classical pop operation). A shunting yard is a set of W classification tracks $\theta_i, i \in \{1, \ldots, W\}, W \in \mathbb{N}$, and one hump track.*

Definition 2 (Roll-in, Pull-out, Sorting Step). *A pull-out operation of a classification track is the removal of all elements from the classification track, and addition of the elements in reversed order to the hump track (using the stack jargon, a pull-out is an operation that repeats the following two stack operations until the track is empty: pop the top element from the classification track, push the element to the hump track).*

A roll-in operation is the removal of all elements of the hump track, and addition of these elements to classification tracks (using the stack jargon, a roll-in repeats the following two stack operations until the hump track is empty: pop an element from the hump track, push the element to a classification track).

A sorting step is a sequence of one pull-out operation followed by one roll-in operation.

Observe that in a roll-in operation any element can be assigned to any track. In general, we will concentrate on the setting where a pull-out is followed by a roll-in.

Definition 3 (Cars, Train). *A car is identified with a unique positive integer. A train T is defined as a sequence of cars, i.e., as a sequence of integers $T = (\tau_1, \ldots, \tau_n), \tau_i \in \mathbb{N}, i \in \{1, \ldots, n\}$.*

Definition 4 (Train Realization). *A classification track θ realizes a train T if the sequence of cars in θ, in the order the cars were pushed to θ, equals T. A hump track realizes a train T if the sequence of cars in the hump track, in the reverse order the cars were pushed to the hump track, equals T.*

The problems we consider ask to rearrange one train called the *inbound* train T_{in}, using only the roll-in and pull-out operations, to a set of *outbound* trains $T_{\text{out}} = \{T_{\text{out}}^1, T_{\text{out}}^2, \ldots\}$, where the outbound trains satisfy a given *sorting constraint*

$\mathcal{C}_{\text{out}}$ on the structure of the outbound trains. We will assume that the cars of the inbound train T_{in} form the set $\{1, 2, \ldots, n\}$. The inbound train $T_{\text{in}} = (\tau_1, \tau_2, \ldots, \tau_n)$ is initially on the hump track such that τ_1 is on top, and τ_n is on the bottom of the stack (that is, τ_1 is closest to the hump). The sorting constraint $\mathcal{C}_{\text{out}}$ specifies how many outbound trains are to be built (we denote it by m), what cars belong to which train, and what are the requirements on the order of cars in each outbound train. For example, a simple sorting constraint may ask for $m = 1$ outbound train T_{out} in which the order of the cars is $(1, 2, \ldots, n)$, i.e., it asks for a sorted sequence of the cars of T_{in}.

Definition 5 (Sorting Schedule). *A sorting schedule for an inbound train T_{in} and sorting constraint $\mathcal{C}_{\text{out}}$ is a sequence of a roll-in operation and h sorting steps after which the (non-empty) sorting tracks realize trains that satisfy the sorting constraint $\mathcal{C}_{\text{out}}$. We call h the* length *of the sorting schedule.*

A sorting schedule thus defines the set of operations which allow to build the outbound trains from the inbound train.

   We are now ready to define the class of shunting problems we consider within this survey.

Problem: TRAINSORTING
Input: A shunting yard with W tracks, an inbound train T_{in} realized on the hump track, and a sorting constraint $\mathcal{C}_{\text{out}}$.
Output: A sorting schedule S leading to a realization of the outbound trains satisfying the constraint $\mathcal{C}_{\text{out}}$.

Problem: TRAINSORTINGMINSTEP
Input and *Output* as in TRAINSORTING
Objective: Minimize the number of sorting steps.

Problem: TRAINSORTINGMINTRACK
Input and *Output* as in TRAINSORTING, but with $W = \infty$
Objective: Minimize the number of used tracks.

In some of the methods we consider hereafter, the focus is to achieve a feasible sorting schedule for any input, and no rigorous optimization goal is considered.
   A commonly considered sorting constraint partitions cars into groups according to the cars' travel destination, and requires cars from the same group to appear consecutively (but in arbitrary order) in an outbound train.

Definition 6 (Groups, Group Sorting Constraint). *Let $\mathcal{G} = \{G_1, \ldots, G_g\}$ be a partition of the cars $\{1, 2, \ldots, n\}$ of an inbound train T_{in}. We call every G_i, $i \in \{1, 2, \ldots, g\}$, a* group. *The group sorting constraint $\mathcal{C}_{\text{out}}^{\mathcal{G}}$ for the partition $\mathcal{G}$ requires m outbound trains, specifies for every group the outbound train it belongs to, and requires that all cars of a group appear consecutively in the corresponding outbound train.*

Let us finally note that in the literature a classification track is sometimes called a *sorting* track, a sorting scheme is also called a *classification schedule*, a sorting problem is sometimes called a *shunting* problem, and a group is also called a *block*. Blocks and groups are, however, not always used as synonyms (see e.g. [5] for more on this topic).

3 Single-Stage Sorting

To get the flavor of the shunting problems, we first consider a special case (and simpler case in a way) of the more general sorting problems – the problem where, besides the initial roll-in, only one additional roll-in operation is allowed. The problem we consider is a special case of the TRAINSORTINGMINTRACK problem with a group sorting constraint $C_{out}^{\mathcal{G}}$ which asks for $m = 1$ outbound train. The specialty here is that after the initial roll-in operation, all tracks are pulled out (one track after another, without a roll-in operation), and then the cars from the hump track are rolled onto an arbitrary track, where they are asked to realize a train that satisfies the requirements of $T_{out}^{\mathcal{G}}$. Thus, in this problem, there are only two roll-ins allowed.

Dahlhaus et al. [6] study this problem under a different name, the so called *train marshalling problem*. They assume that the cars of the inbound train are sorted, i.e., $T = (1, 2, \cdots, n)$. We now outline their considerations and results for the problem.

Observe first that for a train with g groups, we do not need more than g tracks. We can just send every car from G_j, $j = 1, \cdots, g$, to track j and then form an outbound train by an arbitrary order of pull-outs of the tracks and one roll-in. However, it is not always necessary to use g tracks. Consider for example the situation with $n = 11$ cars, and $g = 5$ groups, where $\mathcal{G} = \{G_1, G_2, G_3, G_4, G_5\}$, and $G_1 = \{1, 6, 11\}$, $G_2 = \{2, 7\}$, $G_3 = \{3, 8\}$, $G_4 = \{4, 9\}$, and $G_5 = \{5, 10\}$, and recall that the cars appear in sorted order in the inbound train. Then, one can roll cars $(1, 6, 11)$ onto the first track, cars $(2, 7, 8)$ onto the second track, cars $(3, 4, 9, 10)$ onto the third track, and car 5 onto the fourth track. The outbound train that is built by pulling out track 4, track 3, track 2 and track 1 (in this order), and then rolling the cars from the hump track onto any classification track, satisfies the condition that all cars from every G_j, $j = 1, 2, \cdots, g$, appear consecutively (i.e., the outbound train is a concatenation of trains that are realized in tracks 1, 2, 3 and 4).

Dahlhaus et al. further show that the decision variant of this special version of the TRAINSORTINGMINTRACK problem is NP-complete. Furthermore the authors show that for any instance at most $\lceil n/4 + 2 \rceil$ tracks are needed. Notably, the number of necessary tracks is independent from the number of destinations. They also conjecture that the number of needed tracks is at most $\lceil (m-1)d^2/n + 1/m \rceil$, where $m = \min_i\{|G_i|\}$, and $d = \lfloor n/m \rfloor$.

4 Multistage Sorting

While single-stage sorting can be performed also in yards with no hump, with not much more additional effort, the multistage sorting, which uses many roll-in operations, conceptually relies on the existence of a hump, where no engine is needed to bring single cars to chosen classification tracks. In contrast to single-stage sorting, in multistage sorting cars are repeatedly pulled from the classification tracks back to the hump track to be rolled in again.

In this section we first present the objectives of the multistage methods, and then describe and discuss various multistage sorting methods: sorting by train in Section 4.1, and simultaneous sorting in Section 4.2, for which we describe four particular variants.

We will also take into account the existence of groups, and constraints posed upon them. Recall that m denotes the number of outbound trains, and $\mathcal{G} = \{G_1, G_2, \cdots, G_g\}$ is the partition of the cars of the inbound train into groups. The *strict group sorting constraint* $\mathcal{C}_{\mathrm{out}}^{\mathcal{G}}$ requires cars of the same group to appear consecutively in the outbound trains, and also requires the groups to appear in order of their increasing index. We will denote the number of groups in the ith outbound train by g_i, $i = 1, 2, \cdots, m$. The biggest and smallest number among g_i, $i = 1, 2, \cdots, m$, will be referred to by $g_{\max}$ and $g_{\min}$, respectively.

Restrictions and Objectives in Real Life. On the one hand, train sorting schedules should do the sorting quickly. The time required to carry out a classification schedule is mainly determined by the number h of sorting steps and the total number r of cars that rolled in, and it can be approximated by the expression $h(c_{\mathrm{pull}} + c_{\mathrm{roll}}) + r c_{\mathrm{push}}$, where c_{pull}, c_{roll}, and c_{push} are constants denoting the time required to pull out a track, the time the last car needs to roll from the hump onto a track, and the time for decoupling and pushing one car over the hump, respectively. Increasing h allows reducing r and vice versa; this correlation is described in more detail in [7], which also includes an example with experimentally derived data for h and r. In a common classification yard, the value of $h(c_{\mathrm{pull}} + c_{\mathrm{roll}})$ usually dominates $r c_{\mathrm{push}}$, so h can be regarded as the main objective. This approach is taken, e.g., in [8].

On the other hand, we want to use as little track space as possible for a classification process. For an existing classification yard, we want to use as few classification tracks for the multistage method as possible since the remaining tracks can be used for other sorting activities. Usually some number W of tracks is reserved only for multistage sorting in real-world classification yards, and their (say uniform) length C clearly determines how many cars fit on each track. Hence, the values of W and C usually constrain our optimization problem, while they would become objectives in the design process of a new classification yard.

4.1 Sorting by Train

The sorting scheme *sorting by train* [8,9,10], which is also called *initial grouping according to outbound trains* [4], works in two stages. First, cars are rolled in according to their respective outbound trains, where one track – or several if one track is too short – is reserved for each outbound train. In the second stage, there is one step for every track containing cars of an outbound train T_{out}^i: the track(s) with cars of this train are pulled out (sequentially), and the cars are rolled back in, guiding cars of the same group to the same track (each group

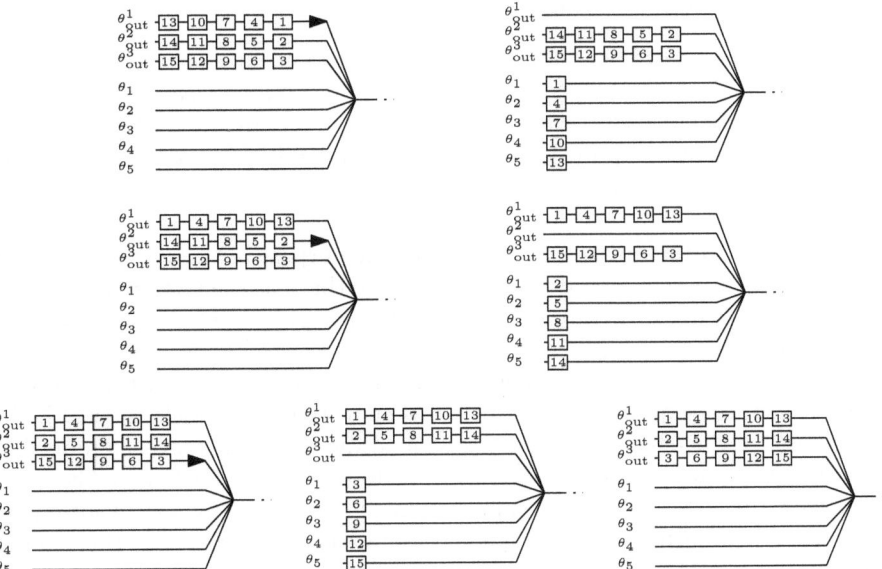

Fig. 3. Example for sorting by train. The input train is built by 15 cars, numbered from 1 to 15, and arrives at the yard in the inverted numbered sequence $T_{in} = (15, 14, 13, 12, 11, 10, 9, 8, 7, 6, 5, 4, 3, 2, 1)$. The cars are to be rearranged to form the following three outbound trains, with each car forming one group, and car numbers appearing in each train in increasing order: $T_{out}^1 = \{1, 4, 7, 10, 13\}, T_{out}^2 = \{2, 5, 8, 11, 14\}, T_{out}^3 = \{3, 6, 9, 12, 15\}$. The sorting procedure is shown in the figures (the order of the figures is, row-wise, left to right). First, the cars are rolled in and sorted onto three tracks, according to their outbound train (top left figure); then, the cars of the first train are pulled out from track θ_{out}^1, and rolled in onto 5 tracks (top right figure); the train is then realized on track θ_{out}^1 (middle left figure). The cars of the remaining two trains are sorted in the same way (middle right and bottom left figures for T_{out}^2, bottom center and bottom right figures for T_{out}^3). At the end (bottom right figure) all three trains have been sorted.

thus has a dedicated track). The groups are then hauled in the correct order to some outgoing track, where they are finally coupled together. This procedure is repeated for every outbound train. An example for this method is illustrated in Figure 3.

The length of a schedule following this sorting scheme is given by $h = m + \sum_{i=1}^{m} g_i$, and every car is rolled in exactly two times – or three times if the separated groups are connected via the hump track. After the first stage, exactly m tracks are occupied, so the number of required tracks ranges between $m + g_{min} - 1$ and $m + g_{max} - 1$, while the latter value is attained if the first train processed in the second stage has a number of g_{max} groups. Regarding the capacity constraint, no track needs to hold more cars than those of the longest outbound train.

4.2   Simultaneous Sorting

Different to sorting by train, in *simultaneous sorting* [9,11], also called the *simultaneous method, simultaneous marshalling* [12], *sorting by block* [10], or *initial grouping according to subscript* [4], the incoming cars are first sorted according to the group membership in their respective outbound trains, and then by train membership in further steps.

Elementary Simultaneous Sorting. For the purpose of this method we refer to the groups also by a second identifier: a group that has the k-th smallest index among all groups assigned to an outbound train T_{out}^j, $j = 1, 2, \cdots, m$, is called the *k-th group of train T_{out}^j*.

In the first stage of the elementary version of simultaneous sorting, all cars of the same group of different trains are sent to the same track. Thus, all the first groups of the outbound trains are sent to the same track, every group that appears second in its outbound train is sent to a common track, etc. Then, the track with the cars of the first groups is pulled out, and the cars are sorted according to their outbound trains. This procedure is repeated for all other groups taken in increasing number. Here, one track is reserved for each outbound train, and in every step every such train grows by one group. An example illustrating this method is shown in Figure 4.

The number of sorting steps is $h = g_{\max}$, which is a great improvement over sorting by train. Compared to other variants of simultaneous sorting, this method

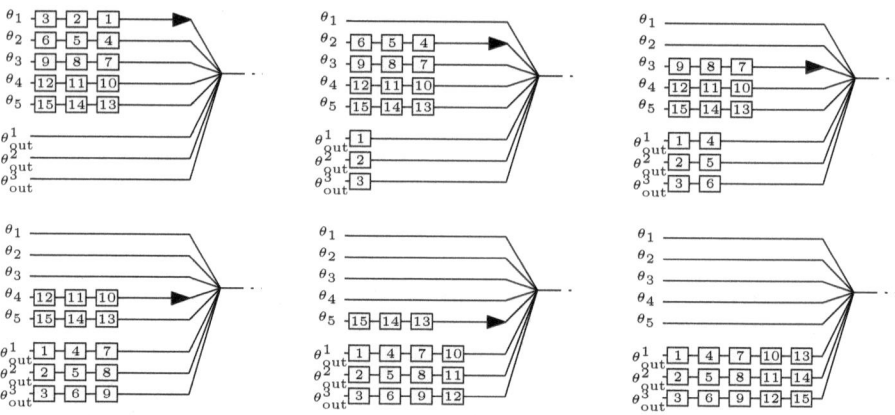

Fig. 4. Example for elementary simultaneous sorting. The structure of the inbound and outbound trains is as in Figure 3. The sorting procedure is shown row-wise from left to right. First, the cars of the inbound train are rolled in and collected group-wise onto 5 different classification tracks (top left figure). Next, the cars of the first group are pulled out from track θ_1; in the subsequent roll-in, each car rolls onto the track where the respective outbound train is built (top middle figure). The remaining groups are pulled out and rolled in sequentially following the same procedure, continuing with tracks θ_2 through θ_5. As a result, the outbound trains are sorted, one on each track (bottom right figure).

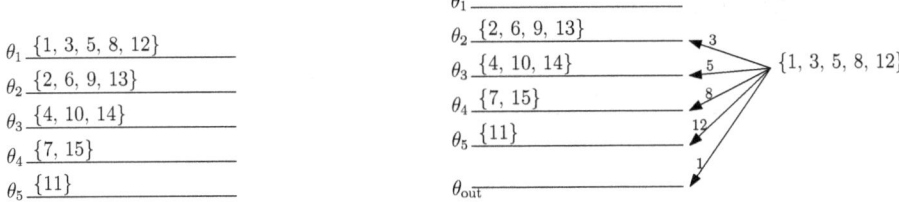

Fig. 5. Left, the assignment scheme of groups to tracks in the initial roll in stage of triangular sorting, for 15 groups. Right, the assignment of the groups to tracks during the first roll in of the second stage. Note that the actual order of the cars assigned to a track may be arbitrary, depending on the order of the cars in the initial roll-in.

minimizes the number of cars rolled in, with every car rolled in exactly twice. However, it has the highest value of h compared to the methods reported in the subsequent sections. The track requirement between $m+g_{\min}-1$ and $m+g_{\max}-1$ is the same as for sorting by train: exactly $g_{\max}$ tracks are occupied after the first stage, and up to $m-1$ further tracks are required. Furthermore, even if the groups can be renamed in such a way that all outbound trains have a group indexed with $g_{\max}$ and use the highest possible group indices for their remaining groups, the assembly of every outbound train is forced as soon as the first of the last $g_{\min}$ tracks is rolled out. This explains the lower bound on the track use.

Triangular Sorting. *Triangular sorting* is an extension of the basic simultaneous sorting scheme. In triangular sorting every car is rolled in at most three times. This method has been considered in [4,8,9,10,11,12,13].

Triangular sorting can be subdivided into two main stages. In the first stage, cars are initially rolled in and assigned to classification tracks according to a certain scheme. In the second stage, all cars on the sorting tracks are pulled out track by track, and the cars are re-assigned either to one of the other used classification tracks, or to the track θ_{out} where the outbound train is being built.

In the first stage, the groups of cars are assigned to classification tracks by iteratively filling an imaginary diagonal: Group 1 is assigned to track θ_1, and forms the first diagonal. Group 2 is assigned to track θ_2, and group 3 to track θ_1, filling the second diagonal. The third diagonal is built by group 4, assigned to track θ_3, group 5, assigned to track θ_1, group 6 assigned to track θ_2. As a general scheme, the i-th diagonal is built by assigning to track θ_j the group number $1+\sum_{i=1}^{j-1} i = 1 + \frac{j^2-j}{2}$, and each group numbered from $2+\frac{j^2-j}{2}$ to $\frac{j^2+j}{2}$ is assigned to track θ_1 to θ_{j-1}, respectively. The structure of this assignment is given in Figure 5. Clearly, each group is assigned to exactly one track.

In the second stage, the tracks are pulled out sequentially, each time processing the track containing the cars with smallest group number, which results in pulling out tracks with increasing index θ_i. In each roll in phase, the car with (group) number i is rolled onto the classification track holding car $i-1$, or the track θ_{out} if no classification track holds group $i-1$. The structure of this assignment is shown in Figure 5 for the roll in corresponding to the pull out of track θ_1. Note

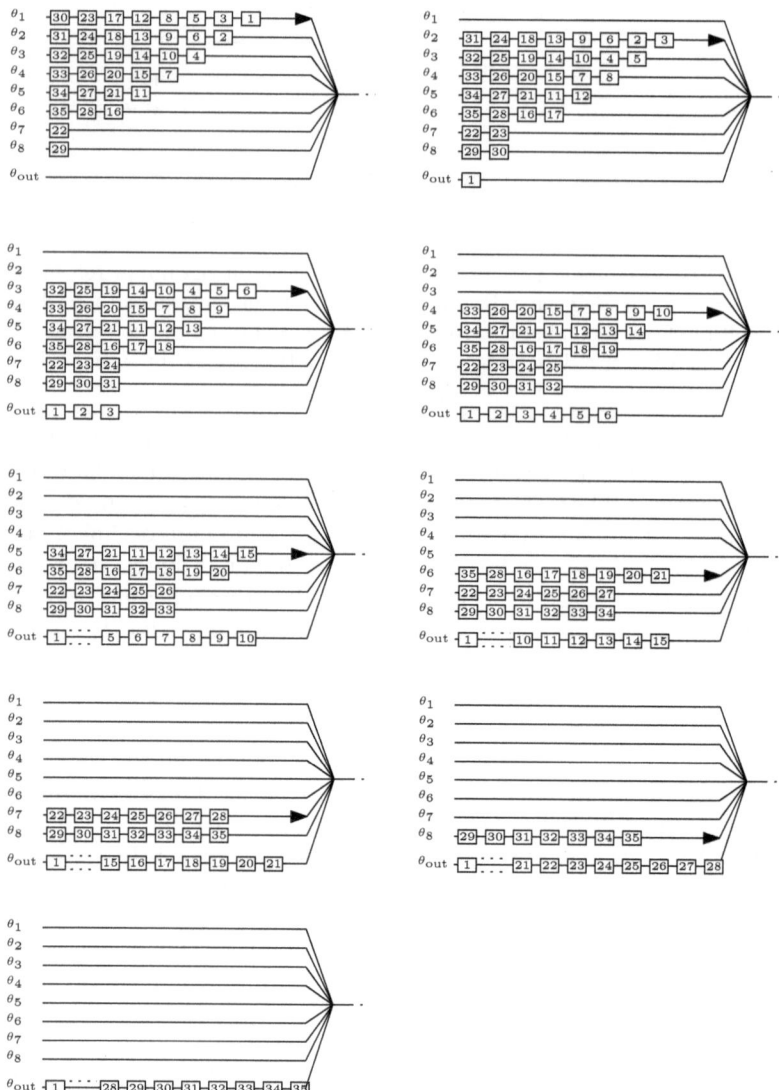

Fig. 6. Example for triangular sorting. The input train is built by 35 cars, numbered from 1 to 35, and arrive at the yard in the inverted numbered sequence $T_{in} = (35, 34, \ldots, 2, 1)$. The cars are to be rearranged to form one outbound train T_{out}. Each car forms one group, and groups must appear in increasing order of the cars: $T_{out} = (1, 2, \ldots, 34, 35)$. The sorting procedure is shown row-wise from left to right. The top left figure shows the yard configuration after the initial roll in, with the assignment according to triangular sorting. Next, track θ_1 is pulled out; the car with smallest number (i.e., car 1) is assigned to the track where the train is being realized. The other cars are assigned as follows: car number i is assigned to the track containing car $i-1$. The remaining tracks (which now contain some cars assigned during previous pull outs) are pulled out sequentially, continuing from θ_2 and ending with θ_6, and classified according to the same scheme. After 6 pull-out/roll-in steps, the ordered train has been formed on θ_{out} (bottom figure).

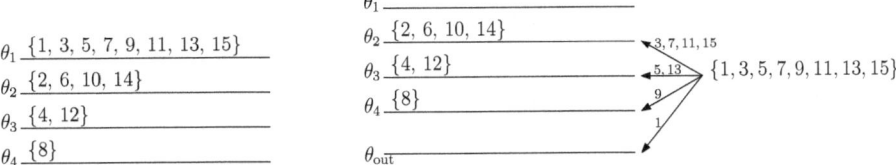

Fig. 7. Left, the assignment scheme of groups to tracks in the initial roll in stage of geometric sorting, for 15 groups. Right, the assignment of the groups to tracks during the first roll-in of the second stage.

that because of this second stage, each car added to a classification track during the second stage is in correct relative order with the cars with preceding group number. Moreover, the cars added to a classification track during the second stage appear on it in consecutive, increasing order.

An example of triangular sorting of an inbound train to be rearranged in one outbound train is given in Figure 6 for the special case where each group consists of one car.

The number of groups g_{max} of an outbound train that can be sorted with this method within h steps is bounded by $g_{max} \leq \frac{1}{2}h(h+1)$ [9,8]. Conversely, the number of sorting steps is $h = \lceil \sqrt{2g_{max}} - \frac{1}{2} \rceil$, and the number of required classification tracks ranges between $\lceil \sqrt{2g_{max}} - \frac{1}{2} \rceil$ and $m + \lceil \sqrt{2g_{max}} - \frac{1}{2} \rceil - 1$. We recall that every car is rolled in at most three times, so the total number of car roll-ins is at most $3n$.

Geometric Sorting. A further generalization of triangular sorting is achieved by the sorting scheme *geometric sorting* [4,8,9,11,12]. This scheme uses less classification tracks than the previous ones, but a car can roll in up to $\log_2 n$ times. Similar to triangular sorting, geometric sorting can be subdivided into two main stages. During the first stage, the cars of the inbound train are initially rolled in. The assignment of groups to tracks is defined as follows. All groups with an odd number (of the form $(2k-1), k \in \mathbb{N}$) are assigned to the first track θ_1. The track $\theta_j, j \in \mathbb{N}$, contains all groups with number of the form $(2k-1) \cdot 2^{j-1}$. It is easy to see that each group is assigned to exactly one track. A schematic representation of the assignment for group (car) number up to 15 is given in Figure 7. Observe that the number of groups in the tracks form a geometric series.

In the second stage, the classification tracks are pulled out sequentially. The track containing the group with smallest number is pulled out first. After each pull out, the pulled out cars are rolled in again. Cars of group number i are assigned to the track θ_j which holds cars of group number $i - 1$, or, if the cars with group number $i - 1$ are on the track θ_{out} where the outbound train is being realized, to track θ_{out}. The structure of this assignment is shown in Figure 7 for the pull-out of track θ_1. In the second phase, the sorting method ensures that each group of cars newly assigned to a classification track is in the correct relative order with the group with previous number. The ordering of the initial roll-in ensures that at the end the groups are in the correct order.

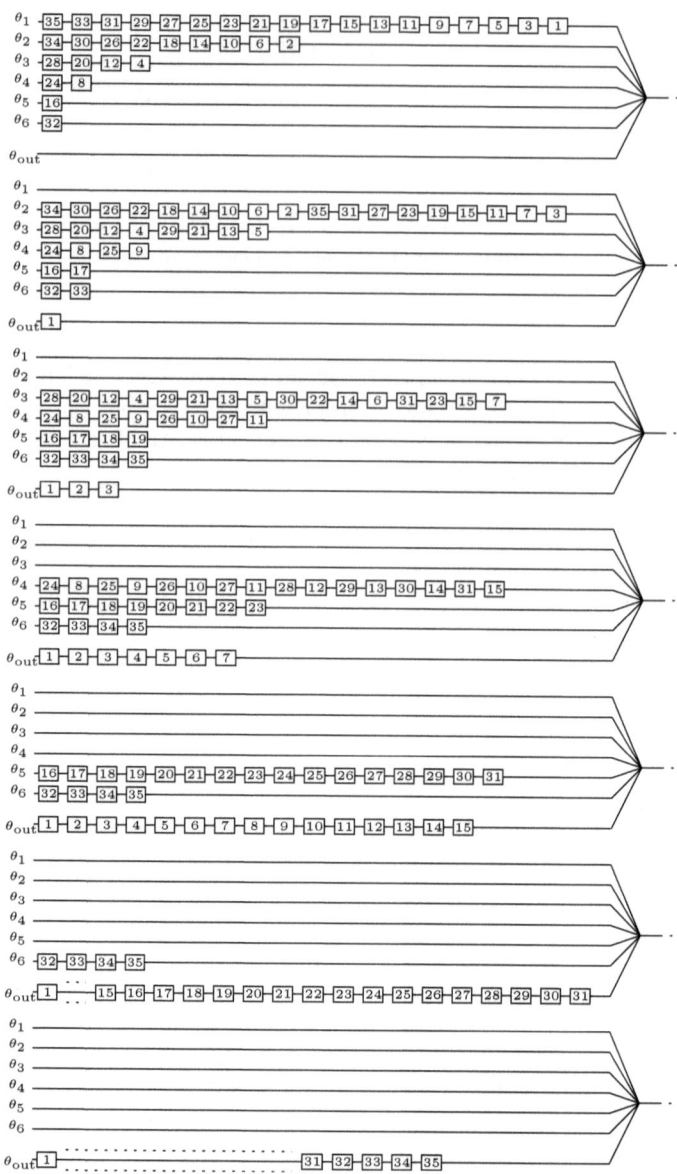

Fig. 8. Example for geometric sorting. The structure of the inbound and outbound train is as in Figure 6. During the initial roll-in, the cars are assigned to a track according to the geometric scheme (top figure). Note the geometric distribution of the total number of cars assigned to the tracks. Next, track θ_1 is pulled out. During the roll in, in a similar way to the triangular sorting scheme, each car is assigned to the track containing the car labeled with the preceding number (second figure from top). Thus, for example, car number 9 is assigned to track θ_4, which contains car number 8. Note that because of this, these two cars are in the correct relative order. Car number 1 is assigned to θ_{out}. The remaining tracks are pulled out and rolled in sequentially, continuing with θ_2 and ending with θ_6. After the roll-in of track θ_6, the complete train has been assembled onto θ_{out}.

An example of this method applied to one inbound and one outbound train is given in Figure 8.

For this method, as stated in [9], $g_{max} \leq 2^h - 1$ for h sorting steps. Put differently, the number of sorting steps of this method is given by $h = \lceil \log_2(g_{max}+1) \rceil$, which is proven in [8]. Similar to triangular sorting, the track requirement of this method ranges between $\lceil \log_2(g_{max} + 1) \rceil$ and $m + \lceil \log_2(g_{max} + 1) \rceil - 1$. In total, there are $\frac{n \log_2 n}{2}$ car roll-ins, so each car rolls in $\frac{\log_2 n}{2}$ times on average.

5   Simultaneous Sorting with Presorted Input

All methods we have considered till now simply take the inbound train and transform it using one uniform way into the desired outbound trains. In particular, for a fixed number of cars the methods use the same number of steps, and roll the same number of cars regardless of the structure of the input. In this section we describe approaches that consider the order of the cars in the inbound train, and make use of the fact that some cars come in the correct relative order to produce sorting schedules that need less sorting steps, or use less classification tracks. The basic idea is to handle two cars that should appear consecutively in an outbound train, and that are in the correct relative order in the inbound train, in the "same way". In the literature this is often described as the *presortedness* of the input. We would like to discover how much of the input is already sorted in the correct order, and to subsequently use it in our favor.

5.1   Parallel Pull-Outs

Dahlhaus et al. consider a special variant of the sorting problem in which after a roll-in operation *all* tracks are pulled out [14]. For our purposes we call the act of pulling out all tracks a *parallel* pull-out. The authors study the case where only W tracks are available, and ask how many parallel pull-outs are needed to sort the inbound train of n cars to single outbound train $T_{out} = (1, 2, 3, \ldots, n)$. In the following we describe some of their results.

Inspired by radix sort, one can use the following technique to sort the inbound train. The sorting scheme performs a sequence of sorting steps $i = 1, 2, 3, \ldots$, where one sorting step is a roll-in operation followed by a parallel pull-out. We will call such a sorting step a *parallel sorting step*. After every parallel sorting step i, the algorithm maintains the following invariant (observe that at this moment the cars are on the hump track): Consider the cars $1, 2, 3, \ldots, n$ in batches, where each batch contains W^i consecutive numbers; Let us denote by B_b^i the batch of cars $\{\tau \mid (b - 1)W^i + 1 \leq \tau \leq bW^i\}$, $b = 1, 2, \ldots$; (Observe that each batch, but possibly the last, contains W^i cars;) Then all cars within one batch are in the correct relative order. It is now a simple observation that after $\log_W n$ parallel sorting steps the cars on the hump track realize the outbound train $T_{out} = (1, 2, \ldots, n)$ and it can be rolled onto an arbitrary classification track. If we consider the initial situation when no sorting step has been performed (i.e., $i = 0$), then every batch is a singleton and the invariant trivially holds. Now, in the first step, the cars are assigned to W classification tracks such that after one parallel pull-out (where the tracks are pulled out in the order $\theta_W, \theta_{W-1}, \ldots, \theta_1$) every batch of W consecutively numbered cars are in the correct relative order. It is easy to see that if the car 1 is rolled onto θ_1, car 2 onto θ_2, ..., car W onto θ_W, then after a parallel pull-out these cars are in a correct relative order. The other cars are assigned to tracks in similar manner, in first step car τ is rolled onto track θ_ℓ where $\ell = 1 + ((\tau - 1) \mod W)$. In general, in the $(i+1)$-th step, all cars in batch B_b^i are sent to track θ_j, where $j = 1 + ((b-1) \mod W)$. Observe that after the subsequent parallel pull-out, the cars in every batch B_b^{i+1}, $b = 1, 2, \ldots$, are in correct relative order. Thus, the invariant is maintained and the sorting procedure needs $\log_W n$ steps.

However, we do not have to start the first roll-in with every single car representing a batch. Instead, we can identify batches of cars that are in a correct relative order already, and handle them as one unit. Thus, we can partition the cars into batches $B_1^0, B_2^0, \ldots, B_k^0$, where batch B_1^0 contains the maximal sequence of cars $1, 2, 3, \ldots, c_1$ such that all cars of B_1^0 are in correct relative order in T_{in}, B_2^0 contains the maximal sequence of cars $c_1 + 1, c_1 + 2, \ldots, c_2$ which are in correct relative order in T_{in}, etc. This results in a certain number of batches, say k. Similarly to the described sorting procedure, cars from batch B_b^i are sent to track number $1 + ((b-1) \mod W)$, and then new batches are formed as in the previous approach. A new batch B_b^{i+1} is formed by the union of W old batches B_j^i, $j = (b-1)W + 1, \ldots, bW$. With every step, the number of batches is decreased by a factor of W, and thus in $\log_W k$ parallel sorting steps the hump yard realizes the train $(1, 2, 3, \ldots, n)$.

5.2   Sequential Pull-Outs

We now concentrate on the original sorting problem in which the sorting scheme operates with sorting steps, i.e., every pull-out of a track is followed by a roll-in operation. In this section we describe approaches that aim to design, for a given inbound train T_{in}, an optimum sorting scheme for various optimization goals – minimizing the number of sorting steps, or minimizing the number of used tracks. As before, sorting is required to build an outbound train with cars representing numbers in increasing, sorted order.

This problem is studied by Jacob et al. in [5,8]. We describe their ideas for the case of minimizing the number of sorting steps. In particular, in this case, we can assume there are arbitrarily many classification tracks at our disposal (the effect of this assumption, as compared with dropping it, are shown in [5]). Assuming this, we can, for simplicity, assign a new track θ_i to each sorting step i, $i = 1, 2, 3, \ldots$, i.e., we can assume that in sorting step i a track θ_i is pulled out. Observe that now a sorting scheme is fully specified by prescribing which tracks each car visits during the shunting.

Intuitively, to minimize the number of sorting steps, the cars shall share the classification tracks as much as possible. For example, if all cars come sorted in the inbound train T_{in}, all cars can roll in to one classification track with no subsequent sorting steps. We will call the classification track on which the outbound train is built a *destination* track. As no car is pulled out from this track, it does not add to the number of sorting steps. Observe now that if two cars τ_i and τ_j from the inbound train T_{in}, $i < j$, are in the correct order, i.e., $\tau_i < \tau_j$, the cars can be assigned to the same tracks in a classification schedule. Doing so, car τ_i will be always in front of τ_j, regardless of where other cars roll in. If the cars τ_i and τ_j appear in the wrong order in T_{in}, i.e., $\tau_i > \tau_j$, the cars cannot use the same tracks, i.e., the cars have to "split" at some point (before they meet again, the latest on the destination track). In particular, car τ_j has to "overtake" the car τ_i at some point. This can be done, in a simple way, by rolling in the cars τ_i, τ_j to tracks k and l, where $l < k$, and then in the next sorting steps, to roll the cars onto the destination track. In general, however, more than one such overtaking may be necessary, and the course of a car may access more tracks. In this case we need to make sure that τ_i does not overtake τ_j after the proper order has been restored.

Jacob et al. devised a neat description of sorting schemes which includes all aspects of the previous discussion. The description encodes in a binary form the course of each car through the classification tracks (recall that the classification tracks correspond one-to-one with the sorting steps). As we shall see, this encoding allows a simple analysis of sorting schemes (in terms of needed sorting steps), and characterization of optimal sorting schemes. For a sorting scheme with $h > 0$ sorting steps, the bitstring $b^j = b^j_{h-1} b^j_{h-2} \cdots b^j_0$ of h bits assigned to the car j, $j = 1, 2, \cdots, n$, represents which classification tracks out of $\theta_0, \theta_1, \cdots, \theta_{h-1}$, besides the destination track, the car visits. The bitstring b^j is thus read from left to right with the interpretation that $b^j_k = 1$ if and only if the car j visits the track θ_k that is pulled out in sorting step k. The subsequent roll-in of the car j in sorting step i is to the track ℓ which is the index of the next bit set to one, i.e., $\ell = \min_{k < i \leq h-1} \{i \mid b^j_i = 1\}$. If such a bit does not exist, i.e., $b^j_i = 0$ for all $i > k$, the car j is rolled onto the destination track. Figure 9 is the example given in [8] that illustrates the encoding of sorting schemes with bitstrings.

Deriving Optimal Schedules using the Bitstring Encoding. Given sorting scheme, we can describe it with the bitstring encoding, which immediately tells the number of sorting steps of the sorting scheme. Conversely, using the encoding and the conditions posed on the "overtaking" of cars within a sorting scheme, we can derive an optimum sorting scheme for a given inbound train T_{in} [8]: if two consecutively numbered cars τ and $\tau + 1$ are in the correct order in the inbound train, the same bitstring can be assigned to both cars as they can take the same course and never need to change their relative order during the classification. However, if two consecutive cars τ and $\tau + 1$ are in the reverse order in the inbound train, $\tau + 1$ must be assigned to a bitstring $b' = b'_{h'-1} \ldots b'_0$ that, if regarded as the binary representation of the integer number $\sum_{i=0}^{h'-1} (2^i b'_i)$, is strictly greater than the bitstring $b = b_{h-1} \ldots b_0$ assigned to τ. Then, with $b < b'$, let k be the most significant index (i.e., the leftmost index) with $b_k = 0$ and $b'_k = 1$. Car $\tau + 1$ is sent to some track θ_{next} in sorting step k. As $b_i = b'_i$ for all $i > k$ (i.e., the cars are handled the same way in the remaining sorting

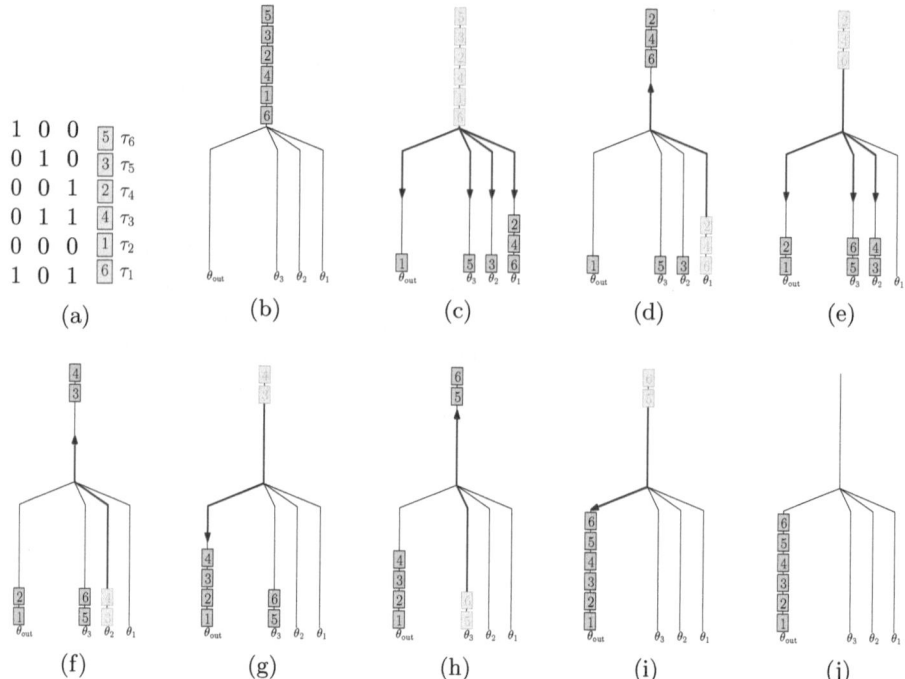

Fig. 9. A classification procedure for $h = 4$ and $n = 6$, using track θ_{out} for the output train. The encoding is shown in (a), the input train in (b). (c)–(j) show the consecutive situations during the procedure, always pulling out the cars of the rightmost occupied track.

steps) car τ was already sent to θ_{next} in a previous step. Hence, τ appears at a position in front of $\tau + 1$ on track θ_{next}. The two cars will not swap their relative order at any later step, so τ ends up on the output track at a position in front of $\tau + 1$ in the outbound train.

This insight directly yields a process to derive feasible schedules: all cars of an ordered subsequence of consecutive cars in the inbound train can get the same bitstring. Two cars of different maximal such subsequences, however, are in reversed order and must thus be assigned different bitstrings as explained above. Therefore, if c denotes the number of maximal ordered subsequences of consecutive cars in the inbound train, c different bitstrings are required to produce the outbound order. The number of steps h equals the length of the longest used bitstring, so the number of steps required to sort the train is given by $h = \lceil \log_2 c \rceil$. This sorting scheme thus needs less pull-out operations than the sorting scheme of Dahlhaus et al. [14], which needs $\log_W c$ parallel pull-outs, which corresponds to $\frac{W \log_2 c}{\log_2 W}$ single pull-outs. The number of roll-ins per car of this sorting scheme is clearly at most h.

Approximate Schedules for Restricted Capacity. As mentioned before, the classification tracks have a restricted capacity in practice, i.e. there is an upper bound on the number of cars that fit on each track. Even for an unrestricted number of classification tracks and the same upper bound C for each track, finding a length-optimal schedule is an NP-hard problem as shown in [8,5]. Nevertheless, there is a polynomial-time 2-approximation [5], which is outlined in the following.

The *capacity constraint* requires that, for every track at every step of the classification procedure, the number of cars on this track does not exceed C. In terms of bitstrings, for each i this translates to

$$\sum_{j=1}^{n} b_i^j \leq C \tag{1}$$

The 2-approximation also uses a constraint called the *weight constraint*, which is a relaxation that considers only the sum of all capacity constraints instead of the individual capacity constraint per track:

$$\sum_{i=0}^{h-1} \sum_{j=1}^{n} b_i^j \leq hC \tag{2}$$

The 2-approximation works in two phases: first, a length-optimal schedule is derived that has minimum weight among all schedules that satisfy the weight constraint. This is done with a dynamic programming approach. The resulting schedule may violate the capacity constraint. Then, this schedule is successively modified until it satisfies the capacity constraint. This modification is done by introducing additional steps for every column of the schedule that violates the capacity constraint, but the details go too far to be described here.

Finding Optimal Schedules Using Integer Programming. The aforementioned encoding is an elegant way to describe sorting schemes, and its binary character allows for a straightforward modeling of various train sorting optimization problems as an integer program (IP) with binary variables. This approach was followed in [15] to derive a basic integer program. This work also showed how various real-world restrictions can be implemented by linear constraints using the binary encoding. This approach is taken for the purpose of deriving actual sorting schedules for real-world instances.

A basic IP model to compute a sorting schedule is shown in Figure 10. Given a fixed number of sorting steps h, the objective is to minimize the total number r of cars rolled in. The constants $\text{rev}(x, y)$ indicate whether two cars x and y are in the correct order: $\text{rev}(j-1, j) = 1$ if the jth and $(j-1)$th car are in reversed order in the inbound train; otherwise, $\text{rev}(j-1, j) = 0$. Furthermore, F denotes the subset of cars that are the first in their respective outbound train. Constraint (3) ensures that the bitstrings assigned to consecutive cars of the same outbound train that appear in reversed order in the inbound train are different, and that the car with smaller number gets the smaller bitstring. In other words, (3) enforces

$$\min \quad \sum_{\substack{1 \le j \le n \\ 0 \le i < h}} b_i^j$$

$$\text{s.t.} \quad \sum_{0 \le i < h} 2^i b_i^j \ge \text{rev}(j-1, j) + \sum_{0 \le i < h} 2^i b_i^{j-1} \; \forall j \in \{2, \ldots, n\} \setminus F \qquad (3)$$

$$\sum_{1 \le j \le n} b_i^j \le C \qquad\qquad\qquad \forall i \in \{0, \ldots, h-1\} \qquad (4)$$

$$b_i^j \in \{0, 1\} \qquad\qquad\qquad \forall i \in \{0, \ldots, h-1\}, j \in \{1, \ldots, n\} \quad (5)$$

Fig. 10. Basic IP model [15] for deriving a feasible schedule of length h on h sorting tracks of restricted capacity C and a total of n cars

a *feasible* schedule. Constraint (4) is a capacity constraint, and enforces that at most C cars are assigned to a track in the sorting schedule (recall that as defined earlier, the constant C denotes the number of cars that fit on each track). As previously stated, the objective is to minimize the number of cars rolled in. In order to minimize the length h of the sorting schedule, a sequence of IPs with increasing values of h can be solved.

The IP model shown in Figure 10 actually describes the method for sequential pull-outs of Section 5 that considers presorted input generalized to multiple outbound trains. The basic version of this IP model can be extended to capture various real-world restrictions as described in [15]. Therefore, this integer programming model can be regarded as a generalized representation and solution technique for train classification.

Robustness of Sorting Schedules. While a schedule that is optimal for a given order of cars (in an arriving train) makes sure that resources are used in the best possible way, it can cause problems in reality, when the actual order of the cars within the arriving train happens to differ from what was expected. In this case, the schedule may not produce the required, sorted order any more, and as a consequence one may need to adjust the schedule. We note that the inherently robust (but not optimal) sorting schemes such as the triangular sorting or the geometric sorting do not depend on a particular input order, so they are resilient against any changes in the input order. In this section we review the robustness issues that were studied with respect to optimal schedules as described in the previous paragraphs.

Cicerone et al. [16] were the first to consider robustness questions in train sorting problems when optimality or near-optimality is the primary goal. Cicerone et al. deal with several kinds of disruptions for the special case of a single incoming and a single outgoing train in a hump yard. Some of these results can also be found in [17].

In [16], Cicerone et al. apply the concept of *recoverable robustness*, which was originally introduced in [18,19], to the problem of train classification. The original schedule may become infeasible as a consequence of a disruption such as a delayed train arrival. After such a disruption, the concept of recoverable robustness allows taking some action that alters the original schedule in order to obtain a feasible schedule for the new situation. This action is called recovery, and the way and the extent to which the schedule may be altered is called *recovery strategy*. An initial solution that can be recovered, naturally within the limits allowed by the recovery strategy, for every possible disruption scenario is called *recovery robust*. The worst case ratio between the cost of a recovery robust solution and the cost of an optimal (non-robust) solution is called the *price of robustness*.

The disturbances with regard to train classification considered in [16] are small deviations of key resources: a single car occurs at a different position in the incoming train than planned, there is one additional or one missing car, and one track is unavailable. Cicerone et al. estimate the price of robustness with respect to the following three basic recovery strategies: zero-recovery, which means that no recovery action may be taken at all; assigning a limited number of cars to a bitstring different from the one in the original schedule; and complete reassignment of all cars to bitstrings without increasing the length of the schedule beyond the original length. For all these cases, the paper gives characteristics of the binary code-representation of a classification schedule. The authors also show that some algorithms that are optimal (for the robustness problem) for some disruptions are infeasible for others. This approach can be seen as an interesting first step to apply robustness to train classification when presortedness is considered.

6   Other Models and Related Problems

In this survey we have presented a selection of algorithmic problems related to rearranging railway units. In this section we would like to point out some variants or generalizations of the TRAINSORTING problem that have been previously studied in the literature.

In the introduction we briefly outlined that the TRAINSORTING problem can be viewed as a specific instance of the problem of sorting an input sequence of numbers with a network of stacks and queues, with a restricted operational modus. A network of stacks and queues is a directed graph where every node is (associated with) a stack or a queue. The network has two special nodes s and t, the *source* node and the *target* node, where s has no incoming edge, and t has no outgoing edge. The source node s is a stack containing initially the input sequence of the first n natural numbers. The elements of s (i.e., the numbers) shall traverse through the network and arrive at t in the sorted order $1, 2, 3, \cdots$. The traversal of the elements obeys the network topology and the stack/queue storage rules: If an element is popped from a stack/queue x then the element is pushed into a stack/queue y that is a neighbor of x in the network. This problem was first introduced by Knuth in 1968 and generalized by Even and Itai, Tarjan, and Pratt, immediately afterwards [3,20,21,22]. These early studies mainly considered various network topologies with the aim of classifying the permutations that can be sorted by the given topology, counting the number of such permutations, and finding minimal unsortable permutations. As an example, a network with one stack only (besides the two nodes s and t) can sort a permutation Π if and only if it avoids the pattern $(2, 3, 1)$, i.e., Π is a permutation which does not contain a subsequence (a, b, c) for which $c < a < b$ holds [3]. It can be shown that in this network the number of sortable permutations of length n is $\binom{2n}{n}/(n+1)$. Also, permutation $(2, 3, 1)$ is the smallest unsortable permutation, as clearly every permutation of length two can be sorted with one stack. This result, combinatorial in its nature, has been then generalized and refined in various ways. Bóna summarizes the subsequent results (again, mainly combinatorial) in his survey [23]. Noticeably, the new results appeared only after a gap of twenty years, with the work of Atkinson et al., Bóna et al., or West (see the survey of Bóna [23] for references). The only early algorithmic considerations are by Even and Itai [20] who reduce the problem of deciding whether a given sequence is sortable by a network of parallel stacks to a graph-coloring problem (a network of parallel stacks is a network where s is connected to k stacks, $k \geq 1$, and these stacks are connected to t). A first solely algorithmic focus on the problem appeared only recently in the work by König and Lübbecke [24], who studied the problem of minimizing the number of "moves" in the network that sort a given sequence in a complete network of stacks. They show that this problem, and even the problem to find a sublinear approximation, is NP-hard for networks with at least 4 stacks. It is rather surprising that there has been no earlier study of this nature, as both sorting and data structures as stacks are a natural concern of a theoretical computer scientist.

An interesting and important variation of single-stage sorting is the problem of parking *autonomous* rolling stock (such as buses, trams, or trains with non-separable engines, i.e., units that are capable of autonomous motion – unlike the cars of trains we have considered till now) in a depot (which can be modelled as a set of classification tracks) in the evening such that, if possible, in the morning the rolling stock can leave from the depot in a prescribed order without "blocking" each other on the tracks. The goal is to use the minimum number of tracks, or, if there is a limited number of tracks, to minimize the "disturbance caused by the blockings".

Blasum et al. [25] and Winter and Zimmerman [26] considered a special version of this problem in the context of trams used in local public transportation. The depot they consider is one sided, i.e., the entrance and exit is from the same side. Thus, each track can be seen as a stack with *last-in first-out* rule (LIFO). In their setting, there are n trams, each tram is of a certain type $t_i \in \{t_1, t_2, \cdots, t_c\}$, and the number of tracks in the depot is W, where $l_j, j = 1, \cdots, W$, denotes the capacity of track j, i.e., the maximum number of trams that fit on that track. The morning departure schedule requires a specific type of tram to depart (rather than a specific tram), i.e., it prescribes the order in which trams of certain type should depart from the depot.

The work of Blasum et al. [25] assumes that the trams are already positioned on the tracks in the depot. The problem is to decide whether the trams can leave in an order (obeying the LIFO rule of each track) that preserves the required types. Blasum et al. show that this problem is NP-complete even if the number of types c is 2, and they present a dynamic programming algorithm that solves the problem in time $\mathcal{O}\left(n^W\right)$ (recall that W is the number of tracks). The authors also report experimental results.

Winter and Zimmermann [26] consider the more general problem where no assignment of incoming trams to tracks has been made yet. That is, for a given order of arriving trams, they ask for a good assignment of the trams to the tracks such that in the morning, for a given departure schedule of tram types, the trams from the tracks can: (1) depart from the depot satisfying the departure schedule, using the minimum number of shunting operations in the depot, or (2) depart without a single shunting operation, minimizing the number of type-mismatches. Not surprisingly, the decision variant of the problems remains NP-complete. The authors model the problem as a quadratic integer program with binary variables, and also derive its linearized version. The experiments using a commercial MIP solver show that only relatively small instances can be solved using this approach. The authors then present various heuristics for the problem, together with an experimental evaluation of the presented solution approaches.

Di Stefano and Koči [27] study the problem of parking trains in a depot under various models of the depots. For a given sequence $T = (\tau_1, \tau_2, \cdots, \tau_n)$ of arriving trains (train τ_1 arrives first, train τ_n arrives last) to the depot, the problem is to assign each train to a track such that the trains can depart in the order $1, 2, 3, \cdots, n$. The tracks are assumed to have sufficient capacity to accommodate every train. The objective is to minimize the number of used tracks. Di Stefano and Koči study various models of tracks. When each track is modeled as a *First-in First-out* (FIFO) queue, the problem is solvable in polynomial time by a straightforward application of Dilworth's theorem (see [28] for details) and the underlying polynomial algorithm. Alternatively, the problem is shown to be equivalent to the coloring problem on a graph where each node is a train, and the edges represent conflicts, i.e., the fact that two trains cannot be stored on the same track. Secondly, for the track being modeled as a queue where trains can be added from both ends, and depart from one end only, the problem is shown to be equivalent to a coloring problem of a hypergraph, where an edge corresponds to three trains that cannot be stored on one track – a hyper-edge $e = \{\tau_i, \tau_j, \tau_k\}$, where τ_j is the smallest number among trains in e, and the train τ_j appears in the arriving sequence T in between the trains τ_i and τ_k. The complexity of this coloring problem is left open (but shown later in [29] to be NP-hard). The authors show, however, that the problem is always solvable with at most $\lfloor \frac{\sqrt{8n+1}-1}{2} \rfloor$ tracks, show that this bound is tight (in the worst case), and also present a polynomial algorithm that never uses more tracks than this bound. The authors also show that this problem is equivalent to the problem with a queue where the trains can enter from one side only, but can leave from both sides. Finally, if the tracks are modelled as a queue where trains can enter and depart from both ends, the problem is shown to be equivalent to a coloring problem of a hyper-graph where a hyper-edge encodes which trains cannot be stored on the same track. Every hyper-edge contains four vertices. The complexity of the coloring problem is left open (and shown later in [29] to be NP-hard). At the end, the paper considers the case where the trains do not necessarily have to depart after all trains arrived, i.e., the arrival times and the departure times of the trains can be arbitrary, and especially they can interleave. For the tracks modelled as a FIFO queue (with one end only), the problem of computing an optimum assignment of trains to tracks is shown to be equivalent to a coloring problem of *circle graphs* (see e.g. [30] for more details on this problem in circle graphs). It is an easy observation that for tracks modelled as FIFO it does not matter whether the departures and arrivals are interleaved or not.

Demange and others [31,32,33] study several variants of the online version of the problem of assigning tracks to autonomous vehicles at a depot, the so-called track assignment problem. In their setting, trains appear online at the depot, and disclose a time interval representing their arrival time and their departure time from the depot. The goal is to assign the trains to tracks online, such that each train can leave the depot at its departure time without requiring any shunting activity. This setting is analyzed for different models of tracks and constraints on times, and also for capacity constraints on the tracks, given by a maximum number of train units that a track can hold. The authors tackle the online problem by constructing conflict graphs as described above: each train is represented by a node, and an edge represents the infeasibility of assigning both corresponding trains to the same track. Now, the problem of finding a valid assignment is equivalent to coloring the resulting conflict graph in an online fashion: the nodes of the graph are presented online, together with the edges adjacent to the already disclosed nodes. The different settings of the track assignment problem lead to specific graph classes, which have special characteristic with respect to their coloring. For instance, if tracks are modeled as stacks (which implies that trains must arrive and leave from the same direction), and all time intervals of the trains intersect a specific time point (the so-called *midnight constraint*), the resulting graph is a permutation graph. For the resulting coloring problems on the graph classes Demange et al. show lower bounds on the competitive ratio. The First-Fit algorithm, which in every iteration assigns to the newly presented node the color with lowest possible number among the feasible ones, is analyzed for different settings. For example, for the case mentioned above, and with tracks holding at most b train units, the authors show a lower bound of $2 - \frac{1}{\min\{b,k\}}$ on the competitive ratio for any online algorithm finding a b-bounded coloring, given that the graph admits a coloring with k colors and that at most b nodes are allowed to share the same color. For the same case, they also show that the bounded-First-Fit algorithm matches this bound if the nodes are presented in increasing order of arrival time of the corresponding trains. Results of this flavor are derived for several online coloring problems, with different orders of presentation of the nodes. For some variants, it is also shown that the First-Fit algorithm produces an optimal coloring.

7 Conclusion

In this work we have surveyed the main algorithmic approaches to a specific class of shunting problems that arise in railway transportation – the TRAINSORTING problems. While the early methods focused on simplicity and robustness (such as the triangular method), the latest research focuses on algorithms that consider the order of cars in the inbound train and design sorting schemes accordingly. These methods are obviously not robust, as a small change in the input may result in outbound trains that do not satisfy a given sorting constraint if the sorting schedule is left unchanged. It is however a challenging problem to consider also the robustness issues, which may require the development of new adaptive methods. In this survey we have also outlined the first steps in this directions.

The theoretical problems that have attracted most of the attention in the research community reflect the needs of practitioners only partially. This is especially the case with adaptive multistage sorting schemes, where we assume that the input sequence of the incoming train(s) is known in advance. This is, however, not the case in practice. The operation center of a shunting yard indeed has a plan of the expected arrivals of trains with cars and their order, but the trains may in reality arrive later (due to delays) or not at all (an engine break; a strike on a part of the network, etc.). Thus, it may happen that in the middle of the shunting process the sorting schedule expects to start with a roll-in of a certain inbound train T_{in}^i which did not yet arrive. Any practical sorting schedule therefore has to adapt to the new situation. Delays of the trains are not the only cause of troubles in real-world operation of shunting yards. Other difficulties appear when the order of cars in an inbound train is different from the expected one. This can be detected before the first roll-in operation, but it may also happen that the identity of a car is discovered only when an outbound train is ready and inspected for the departure. An interesting question here is how to adapt the sorting scheme such that it re-sorts this outbound train, together with the sorting work it was planning to do. Last but not least, one is interested in online scenarios where certain tracks remain temporarily unavailable (due to a breakdown of a switch), or the number of available engines is limited (in which case placing more cars on tracks such that fewer tracks are used is preferred).

Acknowledgments

We thank Marc Nunkesser and Riko Jacob for many interesting discussions, and the anonymous reviewers for their helpful suggestions.

References

1. Hansmann, R.S., Zimmermann, U.T.: Optimal sorting of rolling stock at hump yards. In: Mathematics - Key Technology for the Future: Joint Projects Between Universities and Industry. Springer, Heidelberg (2007)

2. Di Stefano, G., Maue, J., Modelski, M., Navarra, A., Nunkesser, M., van den Broek, J.: Models for rearranging train cars. Technical Report TR-0089, ARRIVAL (2007)
3. Knuth, D.E.: The Art of Computer Programming, 3rd edn. vol. 1. Addison-Wesley, Reading (1997)
4. Siddiqee, M.W.: Investigation of sorting and train formation schemes for a railroad hump yard. In: Proceedings of the 5th International Symposium on the Theory of Traffic Flow and Transportation, pp. 377–387 (1972)
5. Jacob, R., Márton, P., Maue, J., Nunkesser, M.: Multistage methods for freight train classification. Networks (2009)
6. Dahlhaus, E., Horak, P., Miller, M., Ryan, J.F.: The train marshalling problem. Discrete Applied Mathematics 103(1-3), 41–54 (2000)
7. Maue, J., Nunkesser, M.: Evaluation of computational methods for freight train classification schedules. Technical Report TR-0184, ARRIVAL (2009)
8. Jacob, R., Marton, P., Maue, J., Nunkesser, M.: Multistage methods for freight train classification. In: Proceedings of the 7th Workshop on Algorithmic Methods and Models for Optimization of Railways (ATMOS), IBFI Schloss Dagstuhl, pp. 158–174 (2007)
9. Krell, K.: Grundgedanken des Simultanverfahrens. ETR RT 22, 15–23 (1962)
10. Daganzo, C.F.: Static blocking at railyards: Sorting implications and track requirements. Transportation Science 20(3), 189–199 (1986)
11. Flandorffer, H.: Vereinfachte Güterzugbildung. ETR RT 13, 114–118 (1953)
12. Pentinga, K.J.: Teaching simultaneous marshalling. The Railway Gazette (1959)
13. Daganzo, C.F., Dowling, R.G., Hall, R.W.: Railroad classification yard throughput: The case of multistage triangular sorting. Transportation Research, Part A 17(2), 95–106 (1983)
14. Dahlhaus, E., Manne, F., Miller, M., Ryan, J.: Algorithms for combinatorial problems related to train marshalling. In: Proceedings of the Eleventh Australasian Workshop on Combinatorial Algorithms (AWOCA), pp. 7–16 (2000)
15. Márton, P., Maue, J., Nunkesser, M.: An improved classification procedure for the hump yard Lausanne Triage. In: Proceedings of the 9th Workshop on Algorithmic Methods and Models for Optimization of Railways (ATMOS), Wadern, Germany, IBFI Schloss Dagstuhl (2009)
16. Cicerone, S., D'Angelo, G., Stefano, G.D., Frigioni, D., Navarra, A.: Robust algorithms and price of robustness in shunting problems. In: Proceedings of the 7th Workshop on Algorithmic Approaches for Transportation Modeling, Optimization, and Systems (ATMOS), Wadern, Germany, IBFI Schloss Dagstuhl, pp. 175–190 (2007)
17. Cicerone, S., D'Angelo, G., Di Stefano, G., Frigioni, D., Navarra, A., Schachtebeck, M., Schöbel, A.: Recoverable robustness in shunting and timetabling. In: Ahuja, R.K., Möhring, R.H., Zaroliagis, C.D. (eds.) Robust and Online Large-Scale Optimization. LNCS, vol. 5868, pp. 28–60. Springer, Heidelberg (2009)
18. Liebchen, C., Lübbecke, M., Möhring, R.H., Stiller, S.: Recoverable robustness. Technical Report TR-0066, ARRIVAL (2007)
19. Liebchen, C., Lüebbecke, M., Möhring, R.H., Stiller, S.: The concept of recoverable robustness, linear programming recovery, and railway applications. In: Ahuja, R.K., Möhring, R.H., Zaroliagis, C.D. (eds.) Robust and Online Large-Scale Optimization. LNCS, vol. 5868, pp. 1–27. Springer, Heidelberg (2009)
20. Even, S., Itai, A.: Queues, stacks and graphs. In: Proceedings of an International Symposium on the Theory of Machines and Computations, pp. 71–86 (1971)
21. Tarjan, R.: Sorting using networks of queues and stacks. Journal of the ACM 19(2), 341–346 (1972)

22. Pratt, V.R.: Computing permutations with double-ended queues, parallel stacks and parallel queues. In: Proceedings of the fifth annual ACM symposium on Theory of computing (STOC), pp. 268–277 (1973)
23. Bóna, M.: A survey of stack-sorting disciplines. The Electronic Journal of Combinatorics 9(2) (2003)
24. König, F.G., Lübbecke, M.E.: Sorting with complete networks of stacks. In: Hong, S.-H., Nagamochi, H., Fukunaga, T. (eds.) ISAAC 2008. LNCS, vol. 5369, pp. 895–906. Springer, Heidelberg (2008)
25. Blasum, U., Bussieck, M.R., Hochstättler, W., Moll, C., Scheel, H.H., Winter, T.: Scheduling trams in the morning. Mathematical Methods of Operations Research 49(1), 137–148 (1999)
26. Winter, T., Zimmermann, U.T.: Real-time dispatch in storage yards. Annals of Operations Research 96(1-4), 287–315 (2000)
27. Di Stefano, G., Koči, M.L.: A graph theoretical approach to the shunting problem. Electronic Notes in Theoretical Computer Science 92, 16–33 (2004)
28. Dilworth, R.P.: A decomposition theorem for partially ordered sets. The Annals of Mathematics 51(1), 161–166 (1950)
29. Di Stefano, G., Krause, S., Lübbecke, M.E., Zimmermann, U.T.: On minimum k-modal partitions of permutations. Journal of Discrete Algorithms 6, 381–392 (2008)
30. Gavril, F.: Algorithms for a maximum clique and a maximum independent set of a circle graph. Networks 3(3), 261–273 (1973)
31. Demange, M., Di Stefano, G., Leroy-Beaulieu, B.: Online bounded coloring of permutation and overlap graphs. Electronic Notes in Discrete Mathematics 30, 213–218 (2008); (Proceedings of the IV Latin-American Algorithms, Graphs, and Optimization Symposium (LAGOS))
32. Leroy-Beaulieu, B.: Some coloring and walking problems in graphs. PhD thesis, Ecole Polytechnique Federale de Lausanne (EPFL), Switzerland (2008)
33. Demange, M., Di Stefano, G., Leroy-Beaulieu, B.: On the online track assignment problem. Technical report, ARRIVAL (2006)

Integrated Gate and Bus Assignment at Amsterdam Airport Schiphol

Guido Diepen[1,*,**], J.M. van den Akker[2,*], and J.A. Hoogeveen[2,*]

[1] Paragon Decision Technology
Schipholweg 1, 2034 LS Haarlem, The Netherlands
Guido.Diepen@aimms.com
[2] Department for Information and Computing Sciences
Utrecht University
P.O. Box 80089, 3508 TB Utrecht, The Netherlands
{marjan,slam}@cs.uu.nl

Abstract. At an airport a series of assignment problems need to be solved before aircraft can arrive and depart and passengers can embark and disembark. A lot of different parties are involved with this, each of which having to plan their own schedule. Two of the assignment problems that the 'Regie' at Amsterdam Airport Schiphol (AAS) is responsible for, are the gate assignment problem (i.e. where to place which aircraft) and the bus assignment problem (i.e. which bus will transport which passengers to or from the aircraft). Currently these two problems are solved in a sequential fashion, the output of the gate assignment problem is used as input for the bus assignment problem. We look at integrating these two sequential problems into one larger problem that considers both problems at the same time. This creates the possibility of using information regarding the bus assignment problem while solving the gate assignment problem. We developed a column generation algorithm for this problem and have implemented a prototype. To make the algorithm efficient we used a special technique called stabilized column generation and also column deletion. Computational experiments with data based on real-life data from AAS indicate that our algorithm is able to compute a planning for one day at Schiphol in a reasonable time.

Keywords: gate assigment, integrated planning, airports, column generation, stabilized column generation, integer linear programming.

1 Introduction

Between the time an aircraft lands at an airport and the time it departs again many things must happen. One of the most obvious things is that passengers needs to disembark the aircraft. Moreover, the aircraft need to be refueled, new

---

* Supported by BSIK grant 03018 (BRICKS: Basic Research in Informatics for Creating the Knowledge Society).
** This research was performed while the author was at Utrecht University.

R.K. Ahuja et al. (Eds.): Robust and Online Large-Scale Optimization, LNCS 5868, pp. 338–353, 2009.
© Springer-Verlag Berlin Heidelberg 2009

passengers need to board, new supplies have to be put on board, the aircraft has to get cleaned. All of the actions take place while the aircraft is standing at a *gate*. We will consider the arrival of an aircraft until the following departure of the same aircraft as one *stay*. The *gate assignment problem* deals with assigning a given set of stays to a set of gates such that certain criteria are met.

In this paper, we consider the gate assignment at Amsterdam Airport Schiphol (AAS). We investigate the daily planning, i.e. the creation of a planning for the upcoming day on the basis of the available information about the stays of that day. In Diepen et al. [4], we have presented a column generation algorithm to create an assignment for aircraft to gates that is robust from a practical point of view, meaning that any small deviation from the scheduled arrival and departure times can be incorporated without lots of rescheduling.

Some stays are not assigned to a gate with an air bridge but to a so-called remote stand. This implies that passengers have to be transported to and from the aircraft by buses. We have shown how we can create a robust schedule for these platform buses by a similar type of column generation algorithm (see Diepen [3]) in case the gate assignment is given.

This approach resembles the way AAS is actually solving these two problems currently. First the gate assignment problem is solved, the solution of which is then used as input for the bus planning problem. Although the bus planner has the possibility to influence the gate planning by providing preferences, in general the two problems are solved in a sequential way.

Observe that this could imply that a schedule for the gate assignment results in an instance for the bus planning problem for which only poor solutions are possible. In many cases minor changes to the original solution for the gate assignment problem would allow better assignments for the buses. So although this would mean a sub-optimal solution for the gate assignment problem to be used, the solution for both the gate and bus planning as a whole would improve.

In this paper, we focus on *the integration of gate assignment and bus planning*. Our goal is to achieve better overall robustness and a more efficient bus planning without too much negative effects on the gate assignment. The airport authorities at AAS have indicated that robustness is very important for them, in order to limit the amount of gate changes during the day of operations.

During the last years, a significant amount of research has been performed on the integration of real-life scheduling problems. For example Freling, Huisman, and Wagelmans [6] look into the integration of solving the combination of the vehicle and crew scheduling problems that arise in the public transport scheduling. They present two different models and algorithms for solving the integrated version of the two problems, and compare the results to the results obtained by using the standard sequential approach.

One of the areas where the integration of real-life scheduling problems is investigated a lot, is in the *airline* industry. Cordeau et al. [2] investigate the integration of the aircraft routing problem with the crew scheduling problem. They propose a solution approach based on Benders decomposition and show that solving these two problems as one integrated problem yields significant cost savings. Other integrations that have been considered are schedule assignment and the fleet assignment problems (see Rexing et al. [9] and Lohatepanont and Barnhart [8]) and the integration of the fleet assignment and the crew scheduling problems (see Gao [7], Clarke et al. [1], and Sandhu and Klabjan [10]).

At Amsterdam Airport Schiphol, the software package currently in use for solving the gate assignment problem, uses a rule based approach for optimizing the assignment. It includes many aspects, however, it does not support the main thing we aim for: robustness. The software package is also capable of scheduling additional processes besides the assignment of aircraft to gates. For instance, in Vancouver the same program is used and there the scheduling of the push-back trucks is also handled by the program.

The purpose of the research described in this paper is to enable the use of information regarding the bus planning problem while solving the gate assignment problem. Instead of an iterative method in which the separate problems are solved in turns and are allowed to send constraints or preferences to each other, our approach is to combine the two assignment problems into *one big problem* and to solve this one big problem as a whole, where the objective is to maximize overall robustness.

The outline for the remainder of the paper is as follows: In Section 2 we will give the problem formulation and our model and in Section 3 we present solution method. Furthermore, in Section 4 we will report on the results of the experiments that we performed and finally, in Section 5 we give our conclusions.

2   Problem Formulation

In this section, we describe the problem and present an integer linear programming formulation. For the upcoming day we want to create a gate assignment for a given set of stays and a planning for the platform buses transporting passengers to and from stays at a remote stand.

For the gate assignment several properties of the stays are important:

- Arrival and departure time
- Region of origin and destination (Schengen/EU/Non-EU)
- Size category
- Ground handler

At AAS the ground handlers are divided into two groups: KLM Ground Services and other companies. Clearly, two stays cannot be assigned to the same gate at the same time. At AAS the minimum amount of idle time between two consecutive stays at a gate is 20 minutes. For each gate it is known which regions (because of safety regulations), size categories, and ground handlers it can serve. This results in constraints to ensure that at a gate there are only stays matching the properties of the gate with respect to region, size of the aircraft and ground handler.

Moreover, certain preferences might be taken into account. For example, some airlines such as KLM have their 'own' gates or want their stays to be grouped as much as possible on certain gates, for example we could require that at least 5 out of 7 Swiss stays are on a specific gate.

Flights that stay on the ground for a longer period, e.g. 3 hours, may have to be split. This means that after some time the stay is removed from the gate and later is moved back to some (possibly other) gate.

According to operational rules used at AAS this proceeds as follows:

- *Arrival part.* After the aircraft lands, it will stay at the gate for 65 minutes, after which it is towed to some buffer stand.
- *Intermediate part.* During this part the aircraft resides on a buffer stand, where it does not use precious gate capacity.
- *Departure part.* The aircraft is taken from the buffer to the appropriate gate, 95 minutes before the aircraft will depart.

We have included this option in our algorithm. We have omitted it from the upcoming LP and ILP formulations for reasons of simplicity. For the full model, we refer to [4].

Our objective is to create an assignment schedule that is robust from a practical point of view, meaning that the resulting schedule is able to cope with minor disturbances during the actual day as well as possible. The following picture shows an example of a schedule that is typically non-robust and can be improved by interchanging stays 3 and 4.

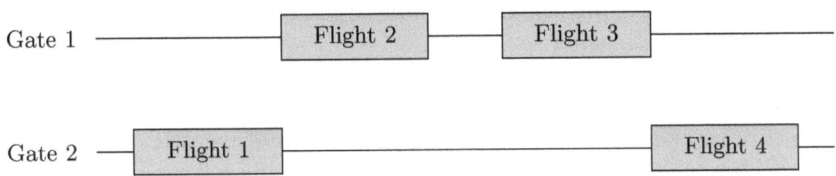

Fig. 1. Example of a non-robust schedule

Observe that a schedule is best able to cope with disturbances if all idle times between each pair of consecutive stays on a gate are as large as possible. We model this with a cost function that greatly penalizes short idle times, while giving very low cost to large, and thus favorable idle times.

For the cost of the idle time t between two consecutive stays v and w on a gate we use the same cost function presented in [4]

$$c^G(t) = conv(v,w)1000(\arctan(0.21(-t)) + \frac{\pi}{2}),$$

where $conv(v,w)$ denotes the convenience multiplier expressing the preference of stay w directly succeeding stay v on a gate. For example, this multiplier is small if v and w belong to the same airline since in this case the airline has a clear incentive to make v depart on time. This cost function is devised such that it mimics the preferences of the planners: a solution with low cost is always one of the solutions preferred by the planners.

If a stay is handled at a remote stand, the passengers are moved to and from the terminal by bus. The number of buses needed depends on the number of passengers. In this way, each stay assigned to a remote stand results in a number of bus trips. At arrival all trips take place at the same time, and for the departure there by rule have to be at least two trips, where the first trip starts already some time before the departure of the stay. When ordering buses and drivers, AAS can specify the amount buses required per 15 minutes. As a result the bus drivers (about 60) on a day work in about 20 types of shifts, where shifts longer than 4.5 hours contain a mandatory break.

To maximize robustness, we make use of a similar cost function of the idle times t between consecutive trips of the same bus. The exception is that at each given time we have significantly lowered the total cost, this to resemble the fact that the gate assignment is still the more important problems of the two:

$$c^B(t) = 50(\arctan(0.21(-t)) + \frac{\pi}{2}),$$

By taking the sum of the total cost of both sub problems, we now have a representation for the quality of the robustness of a solution as a whole. This is a quite standard representation of robustness, which works well in practice.

The ILP Formulation. The model is obtained by combination and extension of the separate models presented in [4] and [3] to solve the gate assignment and the bus planning problems respectively.

Our model is based on so-called *gate plans*, which consist of a set of stays that will be assigned to one gate. The cost of such a gate plan is equal to the total cost of the idle times between each pair of successive stays. We aggregate gates with the same properties into groups of gates and each such group we refer to as a *gate type*. These properties contain at least the origin/destination, size and ground handler. However, a trivial aggregation in which each separate gate (except for the platform stands) is considered a single type is also possible.

We define the decision variable

$$x_i = \begin{cases} 1 \text{ if gate plan } i \text{ is selected} \\ 0 \text{ otherwise,} \end{cases}$$

Since it might be non-trivial to assign all stays to a gate, we allow a stay to be unassigned at high cost. This is modelled by the binary variable UAF_v. Let V denote the number of stays, A the number of gate types, S_a the number of gates of type a, and K the number of preferences. Now the robustness cost of the gate assignment are given by:

$$\text{Min } \sum_{i=1}^{N} c_i^G x_i + \sum_{v=1}^{V} Q_v UAF_v$$

and the gate plans have to satisfy the following constraints:

$$\text{UAF}_v + \sum_{i=1}^{N} g_{vi}x_i = 1 \qquad v = 1 \ldots V \tag{1}$$

$$\sum_{i=1}^{N} e_{ia}x_i \leq S_a \qquad a = 1 \ldots A \tag{2}$$

$$\sum_{i=1}^{N}\sum_{v=1}^{V}\sum_{a=1}^{A} p_{vak}e_{ia}g_{vi}x_i \geq P_k \qquad k = 1, \ldots, K \tag{3}$$

where

$$g_{vi} = \begin{cases} 1 \text{ if stay } v \text{ is in gate plan } i \\ 0 \text{ otherwise,} \end{cases}$$

$$e_{ia} = \begin{cases} 1 \text{ if gate plan } i \text{ is for gate type } a \\ 0 \text{ otherwise,} \end{cases}$$

$$p_{vak} = \begin{cases} 1 \text{ if stay } v \text{ has preference for gate of type } a \text{ in preference } k \\ 0 \text{ otherwise,} \end{cases}$$

Constraint (1) ensures that all stays are either present in one of the selected gate plans, or the unassignment variable for the stay will have the value 1, resulting in a penalty in the objective function.

Constraint (2) ensures that we select as many gate plans of a certain type as there are gates of the type and Constraint (3) ensures that we fulfill all of the preferences that are given with respect to the gate assignment. Here P_k is the minimum required number of stays with a preference for gate type a that we have to assign to a gate of type a to meet the preference constraints, e.g. the constraint can be that at least 7 out of the 10 stays of a certain airline are at a given gate.

In case we need to solve the bus planning problem for a given solution of the gate assignment problem, we know exactly which stays are assigned to which platform. With this information we can create the trips needed to transport all the passengers; in the model we must ensure that each of these trips is either driven by a bus, or it is left unassigned with a penalty cost.

Similar to the gate assignment, we define bus plans as the set of trips performed by one bus. We define

$$y_j = \begin{cases} 1 \text{ if bus plan } j \text{ is selected} \\ 0 \text{ otherwise,} \end{cases}$$

and the binary variable UAT_t to signal if trip t is unassigned. Let T be the number of trips, B be the number of shift types and T_b be the number of buses with drivers available is shift b. We now obtain the following model:

$$\text{Minimize} \sum_{j=1}^{M} c_j^B y_j + \sum_{t=1}^{T} R_t \text{UAT}_t$$

subject to

$$\text{UAT}_t + \sum_{j=1}^{M} h_{tj} y_j = 1 \ \text{ for } t = 1, \ldots, T \tag{4}$$

$$\sum_{j=1}^{M} f_{jb} y_j = T_b \ \text{ for } b = 1, \ldots, B \tag{5}$$

$$y_j \in \{0,1\} \ \text{ for } j = 1, \ldots, n. \tag{6}$$

We can solve this problem in the same way we solved the gate assignment problem. The only additional complication is that in the pricing problem we have to ensure that bus plans which last more than 4.5 hours require a mandatory break.

When we look at the combination of the two problems, we do not yet know which stays will be placed on the platform (and also, on which platform) and therefore we have to find a way to determine which trips we actually need to assign to buses.

To handle this problem, we generate all possible trips for stays that could be assigned to the remote stands. This means that for each of these stays we create the trips for each of the platforms that it can be assigned to. For example, if an arriving stay requires two trips because of the number of passengers and it can be assigned to the D/E platform, as well as the B platform it means that we will create two trips from the D/E platform and two trips from the B platform to the terminal building. Similarly, not only different platforms, but also different destinations in the terminal building must be considered. For each possible combination we would have to create the trips also. To allow for this coupling we will work with all possible trips and determine which of these are needed in a solution and which are not. For this purpose we will use the variables NNT_t for each trip t to denote whether the trip t needs to be assigned to a bus or that the trip is irrelevant for the assignment problem.

$$\text{Min} \ \sum_{i=1}^{N} c_i^G x_i + \sum_{v=1}^{V} Q_v \text{UAF}_v + \sum_{j=1}^{M} c_j^B y_j + \sum_{t=1}^{T} R_t \text{UAT}_t$$

subject to

$$(1) - (3)$$

$$\sum_{j=1}^{M} f_{jb} y_j \le T_b \qquad b = 1 \ldots B \tag{7}$$

$$\text{NNT}_t + \text{UAT}_t + \sum_{j=1}^{M} h_{tj} y_j = 1 \qquad t = 1 \ldots T \tag{8}$$

$$\text{NNT}_t + \sum_{i=1}^{N} \sum_{v=1}^{V} t_{tvi} g_{vi} r_i x_i = 1 \qquad t = 1 \ldots T \tag{9}$$

$$x_i \in \{0,1\} \qquad i = 1 \ldots N \tag{10}$$

$$y_j \in \{0,1\} \qquad j = 1 \ldots M \tag{11}$$

where

$$f_{jb} = \begin{cases} 1 \text{ if bus plan } j \text{ is for a shift of type } b \\ 0 \text{ otherwise,} \end{cases}$$

$$h_{tj} = \begin{cases} 1 \text{ if trip } t \text{ is in bus plan } j \\ 0 \text{ otherwise,} \end{cases}$$

$$t_{tvi} = \begin{cases} 1 \text{ if assigning stay } v \text{ to gate plan } i \text{ implies trip } t \text{ must be driven} \\ 0 \text{ otherwise,} \end{cases}$$

$$r_i = \begin{cases} 1 \text{ if gate plan } i \text{ is for a remote stand} \\ 0 \text{ otherwise,} \end{cases}$$

Constraint (7) ensures that for each bus shift we select at most the number of buses present in that shift.

Constraint (8) states that trip t is either not needed, or, in case it is needed, must either be assigned to a bus plan or it must be explicitly become unassigned at high cost.

Without any further constraints on the NNT_t variables, the easiest solution would be to set the value of all of these variables to 1 and all of the trip constraints would be satisfied right away. Constraint (9) ensures that this cannot happen for trips that are defined for stays assigned to the remote stands. It is also this constraint that actually links the gate and bus model into one large model.

3   Solving the Problem

3.1   Assigning Stays to Gate Plans and Trips to Bus Plans

Observe that the above model determines for each group of gates and each group of shifts an equal sized set of gate plans and bus plans respectively. To approximate the optimal solution of the above ILP-formulation, we will first relax the integrality constraints (10) and (11). After that we will solve the resulting LP relaxation to optimality by making use of column generation.

The Pricing Problem. After each iteration of the column generation process, we need to determine whether other columns exist that might improve the value of the objective function, the so-called *pricing problem*. In our case we have to solve two types of pricing problems, one for finding gate plans and one for finding bus plans.

The pricing problem for the gate assignment part boils down to a set of shortest path problems. For each gate type a we define a graph G^a, the nodes of which are the stays that are allowed to be assigned to gate type a, and there is an arc between each pair of stays (v, w) such that w can directly succeed v on that gate, i.e., the difference between the arrival time T_w^{arr} of stay w and the departure time T_v^{dep} is at least 20 minutes. Furthermore we add a source vertex s with an arc to every node v and a sink t and an arc from every node to t. Now every path in G_a corresponds to a feasible gate plan and vice versa. To be able to solve the pricing problem as a shortest path problem, we assign to each arc leaving v a cost component equal to

$$-\pi_v - \sum_{k=1}^{K} p_{vak}\psi_k - \sum_{t=1}^{T} t_{tv}\rho_t.$$

Moreover, for each arc (v, w) we add a cost component equal to the cost of the idle time between v and w, which amounts to

$$c^G(T_w^{\mathrm{arr}} - T_v^{\mathrm{dep}}).$$

Here the dual multipliers π_v for stay v and ψ_k for preference k follow from Constraint (1) and Constraint (3) respectively. Moreover, ρ_t is the dual multiplier of Constraint (9), which only applies to gate plans that are for remote stands (because only then $r_i = 1$). The last term which is due to the 'coupling' constraint is the only difference with the pricing problem for the gate assignment problem. Finally, we include the dual multiplier corresponding to Constraint (2), which is constant, as the gate type is given.

We may assume that the stays are sorted by their arrival times, which implies a topological order on the vertices of the graph. Because we now have a DAG with a topological order it is possible to find the shortest path in $O(|V| + |E|)$ time.

The pricing problem with respect to the bus problem boils down to a similar type of shortest path problem and is the same as the pricing problem for solving only the bus planning problem separately (see Diepen [3]). The only difference is that the size of the individual graphs is larger due to the increased number of trips.

Because solving all of the pricing problems in each iteration may be rather time consuming, we have tried out different strategies with respect to which of the pricing problems we solve during each iteration. One possible approach is to interleave the solving of the pricing problems; one iteration we solve the pricing problems for the buses and the other iteration we solve the pricing problems for the gates.

Although, after some initial tests we found that searching for both gate and bus plans with negative reduced cost from the beginning on turned out to work better than the other possibilities.

In [4] and [3], we generated a pool of additional columns that can be added to the ILP and enable us to solve the ILP in a reasonable amount of time. For the gate assignment problem we obtain these by removing from the current DAG a stay that appears in the optimal gate plan, after which we resolve the shortest path problem. We perform this step for every stay in the optimal solution of the pricing problem. For bus planning we generate additional columns in the same way. When solving the problems separately, the columns are added when we start solving the ILP. However, when solving the integrated problem all additional columns with negative reduced cost are already added during the column generation process.

Improvements in Solving the LP. During our first experiments, it turned out that the LP problem is very degenerate and tends to require a long solution time. This degeneracy appears in two ways during the column generation process. First, resolving the restricted master problem after new columns have been added takes quite many iterations and second, new columns that are generated with negative reduced cost do not improve the objective function after they have been added to the restricted master problem.

We have applied two different approaches to improve the solving of the LP. The first approach we used is *column deletion* and consists of the removal of columns with too large positive reduced cost after every given number of iterations. The effect of this removal is not only that the model is simplified and some degeneracy is removed, but also that the resulting model is smaller and therefore it can be solved more quickly. For solving the problems separately, this approach showed promising results for decreasing the computation time.

The second approach we implemented is so-called **stabilized column generation**. This technique was introduced in du Merle et al. [5] and consists of a combination of two techniques. One technique is the addition of extra perturbation variables with a component in the cost function to the model. These extra variables try to limit the values of the dual multipliers to within a certain box, while still allowing the values to be outside of the box at a certain cost. This cost when a dual variable is outside the box is linear in the size the value of the variables violates the box. In the case of our combined model, we added such slack and surplus variables to all our constraints.

In [5] multiple methods of updating the values for both the cost coefficients δ_- and δ_+, as well as the bounds ϵ_- and ϵ_+ on the values of the surplus and slack variables respectively are suggested. The way we decided to use after some initial tests was to lower the bounds ϵ_- and ϵ_+ every ten iterations. For setting the values of the cost coefficients δ_- and δ_+ we used the approach of setting them to the values of the dual multipliers of the previous iteration.

Solving the ILP. After the LP is solved to optimality by means of column generation, we are not finished yet because this solution might be fractional. In case it is integral, we are finished since we have an integral solution that is optimal. If we do not have an integral solution, we proceed as follows:

1. first we add all unique gate and bus plans that were generated as extra columns while solving the pricing problems.
2. we then add all the unique variables that were taken out during the column generation
3. we reinstate the integrality constraints (10) and (11).

Solving the resulting ILP turned out to be still quite difficult. In order to speed up this solving, we added additional constraints to the problem. These constraints act as a rounding-heuristic. For each stay and for each bus these additional constraints were created in the following way:

1. Determine whether there exists a stay (trip) for which all of the selected gate (bus) plans containing it are of the same type, meaning that in the fractional

solution a stay or a bus trip is always assigned to one particular gate type
or one particular bus shift.
2. Create a constraint that ensures the stay or the trip has to be assigned to
that particular gate type or bus shift in an integral solution.

Although the above constraints might cause the optimal solution of our initial
ILP to be cut off, our experiments did not show any noticeable negative side
effects with respect to the cost of the integral solution compared to the optimal
fractional solution.

3.2   Assigning Gate and Bus Plans to the Actual Gates and Buses

After solving the ILP from the previous section, we have determined the set of
gate and bus plans that provide a (near) optimal solution. For each group of
gates and each group of shifts we have an equal size set of gate plans and bus
plans respectively. The one thing still left to do is to assign each gate plan and
each bus plan to each unique gate and bus respectively.

In case of the bus planning problem, this part is trivial since the buses within
one shift do not have any differences at all; it really does not matter to which of
these buses a particular bus plan is assigned to.

However, for the gate assignment problem it depends on the definition of the
gate types. If each single gate is a separate type, we already have an assignment
of stays to physical gates and this step is also trivial.

If we have grouped the gates with certain equal properties into types, the
individual gates within such a type still might be different on some other, less
important properties. These additional properties can be used for determining
to which physical gate a particular gate plan needs to be assigned.

Since the size of these problems is relatively small (in the order of 5 to 10
gates within one group) it is probably most effective to leave this up to the gate
planner to do this manually.

4   Experimental Results

For testing our model, we wrote a prototype in C++ and ran numerous experiments. All experiments were ran on on Pentium 4 2.8 GHz computer equipped
with 1GB of RAM. The solver we used for solving all (I)LP problems is Cplex 9.1.3
via the Concert Technology interface.

AAS provided us with both data regarding the gate assignment problem,
which consisted of all stay information for three high-season (HS) days and
three low-season (LS) days and data regarding the bus planning problem with
all information regarding buses for one complete month.

From the supplied gate data we created two types of instances. In one type
of instances we aggregate all gates with identical properties (e.g. size, region,
ground handler, pier) into groups of gates. We refer to this type of instances
as Grouped Gates (GG). Furthermore, we constructed instances where every
gate is considered as a group with size one except for the platform gates. Recall

Table 1. Sizes of the provided instances with regard to gates

Instance	Gates	Gate types	Remote stands
Grouped	128	40	34
Single	128	94	34

Table 2. General LP results

Instance	Total time LP (s)			Avg iter	Avg time (s)/iter	
	Average	Min	Max		RMP	Pricing
02-08-GG	1129.6	967.8	1472.0	161.67	2.8	3.9
02-08-SG	2070.1	1752.1	2657.7	171.90	4.8	6.8
03-08-GG	973.9	864.7	1213.2	148.27	2.6	3.7
03-08-SG	1847.4	1627.4	2337.8	163.07	4.4	6.5
04-08-GG	1142.6	1010.4	1641.3	157.50	3.2	4.0
04-08-SG	2575.2	2189.9	3970.3	212.77	4.6	7.2
15-03-GG	658.5	560.3	769.3	165.17	1.1	2.7
15-03-SG	1235.8	1094.6	1472.0	175.17	1.9	4.8
16-03-GG	710.0	623.8	850.4	161.90	1.3	2.8
16-03-SG	1383.4	1144.0	1661.5	175.87	2.5	5.0
17-03-GG	595.0	474.6	775.1	141.37	1.2	2.8
17-03-SG	1125.1	991.3	1422.4	151.70	2.2	4.9

that for these instances our algorithm directly assigns stays to physical gates. We refer to this type of instances as Single Gates (SG). This disaggregation results in over twice the number of gate types, as can be seen in Table 1. This way we created 12 instances with regards to the gate and stay information. The high-season instances contain about 600 stays and about 1000 arrival/departure events for the bus planning. For the low-season instances these numbers are 500 and 900 respectively.

To avoid that our algorithm would get tailored to a small number of instances, we created another set of instances by combining each of the 12 gate assignment instances with the buses and shifts of all 30 of the bus planning problem instances. These instances contain about 60 buses and about 20 types of shifts (of which about 70 percent is long enough to contain a mandatory break). We may expect the set of buses available at each given time of the day should roughly be enough for driving all trips.

In Table 2 we present the general results with regards to solving the LP part of the problem. We combined each instance of the gate assignment problem with the 30 available instances of the bus planning problem and we present the average time over these 30 instances needed for solving each combination, the minimum time, and the maximum time. We also present the average number of iterations needed to solve the LP relaxation and finally, we also present the average time needed in each iteration of the column generation process to solve the pricing problem and the time needed for resolving the Restricted Master Problem (RMP) after we have added the columns found when solving the pricing problem.

Table 3. Improvements with column deletion and stabilization

| Instance | Improvement factor with respect to | | |
	Avg. LP time	Avg. iterations	Avg. time RMP/iter
02-08-GG	8.80	5.20	3.11
02-08-SG	5.29	4.56	1.69
03-08-GG	10.58	5.53	4.31
03-08-SG	6.32	4.76	2.32
04-08-GG	19.49	5.54	7.34
04-08-SG	8.93	3.90	5.20
15-03-GG	3.01	2.87	1.91
15-03-SG	2.56	2.82	1.37
16-03-GG	5.30	5.33	1.54
16-03-SG	4.31	4.46	1.36
17-03-GG	8.75	6.61	2.58
17-03-SG	7.12	6.95	1.55

Our experiments indicate that the LP can be solved within a reasonable amount of time. From Table 2 we can see that a significant amount of the time needed for solving the LP-relaxation is spent in solving all the separate pricing problems. Since all parts of the pricing problem that need to be solved can be solved completely independent from each other, we could easily bring down the influence of the pricing problems on the total time needed for solving the LP-relaxation by making use of parallel programming.

To investigate the effect of the column deletion and the stabilized column generation, we also ran part of the instances without these enhancements. It could be clearly seen that the time needed to solve the LP relaxation to optimality explodes without the use of column deletion and stabilization. One part responsible for this huge increase in time needed is the large increase in the average time needed for solving one iteration of the RMP. This can be explained by the fact that after a couple of iterations, the model quickly becomes very large due to the fact that all columns stay in the model.

We have put the improvement factors in Table 3. It turns out that without column deletion and stabilized column generation, the average number of iterations needed to solve the LP-relaxation to optimality is higher than when both are enabled, while the average time needed for solving the pricing problems is lower. The increase in number of iterations needed is an example of the so-called tailing-off effect. In the beginning there are big improvements in each iteration, while more and more iterations are needed when closer by the optimum. Using the stabilized column generation has a positive effect on this tailing-off effect, as can be seen by the number of iterations needed.

It turns out that the combination of column deletion and stabilized column generation are responsible for a huge improvement, in our experiments by a factor 2.5 up 19, in the time needed for solving the LP-relaxation to optimality with

Table 4. General results ILP

Instance	Average additional constraints		Average solving time ILP (s)
	Flight constraints	Trip constraints	
02-08-GG	121.4	57.6	43.5
02-08-SG	103.4	57.9	54.1
03-08-GG	117.8	57.1	42.0
03-08-SG	105.4	57.7	103.3
04-08-GG	119.3	57.2	82.7
04-08-SG	108.7	57.5	95.2
15-03-GG	108.9	58.4	86.5
15-03-SG	91.0	59.0	271.0
16-03-GG	107.0	59.1	45.8
16-03-SG	84.2	59.3	170.6
17-03-GG	118.5	59.9	20.6
17-03-SG	105.6	59.6	29.5

column generation. Interesting is the fact that the improvement seems larger when the instances are larger (see HS versus LS).

The results for solving the ILP are given in Table 4. The table shows that we were able to solve the very large ILP within a few minutes. In our experiments the integrality gap turned out to be very small.

As mentioned in Section 3 we added additional constraints to the model before solving the actual ILP. These additional constraints can be considered as a kind of rounding-heuristic in the way that if in the optimal solution for the LP-relaxation a stay is always assigned to a certain type of gate in all selected gate plans, we add a constraint that enforces the stay to be assigned to a gate plan of that type.

The average number of constraints that were added for stays as well as for buses is shown in Table 4. These constraints result in ILP models that are a lot smaller and hence in a much smaller solution time. From our experiments we found that the additional constraints did not have a significant impact on the value of the final ILP solution and did not result in infeasibility of the ILP.

One other way to speed up the process of solving the ILP we used is to first only solve the root node relaxation. We then add a so called cut up limit to the model that is 0.5% above the value of the root node. This cut up limit acts for the ILP solver as if an integral solution with that particular value has already been found, meaning that any node with a relaxation value greater than this cut up value is automatically pruned. Strictly speaking this might result in infeasibility of the ILP (when the optimal ILP solution exceeds the threshold), but this never occurred in our experiments.

Furthermore, when looking at the time needed for solving the various final ILP models, we can see that these times are still within very acceptable ranges, also for the Single Gate Problems. This indicates that it is feasible to assign stays and trip directly to physical gates and buses respectively.

5   Conclusion and Further Research

We have investigated the combination of two assignment problems that in practice are solved in a sequential fashion. We formulated the combined problem in
one large model for which we approximate the optimal solution by means of an
approach based on column generation.

We implemented our algorithm and tested it with instances based on real-
life data provided by AAS. The results show that our approach is capable of
solving these instances within acceptable time, especially given the fact that
this approach solves two problems within about the same time that currently
is available at AAS for the computer to present a solution for only the gate
assignment problem.

We also showed that our approach is still capable of solving the instances
within acceptable running times if we create a single gate type for each separate
gate, except for the remote stands. This different model leads to over twice the
number of gate types which significantly increased the size of the instances.

We are currently performing a simulation study of the platform buses, to
evaluate the robustness of the column generation planning compared to a kind
of first-come-first-served method as used at AAS. We can clearly see that the
column generation schedule is more smooth, in the sense that the idle time
is spread more evenly. Currently, the gate assignment at AAS needs a lot of
replanning during the day of operation. However, comparing the quality of our
resulting schedules to the actual schedules in use at AAS is difficult for a variety
of reasons, the main one being the fact it is not possible to retrieve the schedule
we would like to compare to, namely the initial schedule as produced by the
computer for the upcoming day.

An interesting possibility of further investigation is to start looking at a more
operational type of planning. It would be interesting to see how our suggested
approach performs if we do not let it create a schedule from scratch but we
supply it with a schedule and some disturbances and let the program try to
resolve this updated problem.

One of the main things that would have to be considered for this approach is
the fact that any new solution should not deviate too much from the currently
existing solution. So when solving the problems after some parts are fixed (since
they already happened) and other events have changed properties (e.g. earlier or
later Estimated Time of Arrivals and Departures) the cost function would not
only have to consider the robustness of the schedule, but also the similarity to
the original-day-ahead schedule, since too many changes in a schedule will result
in a lot of confusion for the different parties dependent on the schedule.

References

1. Clarke, L.W., Hane, C.A., Johnson, E.L., Nemhauser, G.L.: Maintenance and crew
   considerations in fleet assignment. Transportation Science 30, 249–260 (1996)
2. Cordeau, J.-F., Stojkovic, G., Soumis, F., Desrosiers, J.: Benders decomposition for
   simultaneous aircraft routing and crew scheduling. Transportation Science 35(4),
   375–388 (2001)

3. Diepen, G.: Column Generation Algorithms for Machine Scheduling and Integrated Airport Planning. PhD thesis, Utrecht University (in preparation, 2008)
4. Diepen, G., van den Akker, J.M., Hoogeveen, J.A., Smeltink, J.W.: Using column generation for gate planning at amsterdam airport schiphol. Technical Report UU-CS-2007-018, Institute of Information and Computing Sciences, Utrecht, the Netherlands (2007)
5. du Merle, O., Villeneuve, D., Desrosiers, J., Hansen, P.: Stabilized column generation. Discrete Math. 194(1-3), 229–237 (1999)
6. Freling, R., Huisman, D., Wagelmans, A.P.M.: Models and algorithms for integration of vehicle and crew scheduling. Journal of Scheduling 6(1), 63–85 (2003)
7. Gao, C.: Airline Integrated Planning and Operations. PhD thesis, Georgia Institute of Technology (August 2007)
8. Lohatepanont, M., Barnhart, C.: Airline schedule planning: Integrated models and algorithms for schedule design and fleet assignment. Transportation Science 38(1), 19–32 (2004)
9. Rexing, B., Barnhart, C., Kniker, T., Jarrah, A., Krishnamurthy, N.: Airline fleet assignment with time windows. Transportation Science 34(1), 1–20 (2000)
10. Sandhu, R., Klabjan, D.: Integrated airline planning. In: AGIFORS Symposium 2004, Singapore (2004)

Mining Railway Delay Dependencies in Large-Scale Real-World Delay Data[*]

Holger Flier[1], Rati Gelashvili[2], Thomas Graffagnino[3], and Marc Nunkesser[1]

[1] Institute of Theoretical Computer Science, ETH Zürich, Switzerland
{holger.flier,marc.nunkesser}@inf.ethz.ch
http://www.pw.inf.ethz.ch/
[2] Tbilisi State University, Georgia
gelash@gmail.com
[3] SBB AG Bern, Infrastruktur/Trassenmanagement, Switzerland
thomas.graffagnino@sbb.ch

Abstract. The propagation of delays between trains has a considerable impact on railway operations. Ideally, planners would like to create timetables that avoid such propagation as much as possible. To improve existing timetables, tools for automatic detection of systematic dependencies of delays among trains would be of great aid. We present efficient algorithms to detect two of the most important types of dependencies, namely dependencies due to resource conflicts and due to maintained connections. We give experimental results on real-world data that demonstrate the practical applicability of our algorithms.

1 Problem Statement

During operations, it is unavoidable that trains get delayed. Reasons for delays are manifold: customers blocking doors, train connections, scarce track capacities, weather, technical problems, etc. From a planner's point of view, some causes for delay just have to be accepted, such as customer behavior, and some have to be dealt with in disruption management, such as power failure due to catastrophic weather conditions. There are, however, also systematic dependencies between the delays of trains, which are inherent to the timetable and can be influenced by careful planning. In this paper, we present algorithmic methods to efficiently detect such dependencies in large-scale, real-world railway delay data. The goal is to support planners in improving timetables by providing them with a list of potentially systematic delay dependencies of the current timetable. These dependencies can then be more closely examined by appropriate statistical methods in a following step. Finally, planners may be able to remove or weaken those dependencies by means of small, local modifications to the timetable.

At Schweizerische Bundesbahnen (SBB), delay data are measured by the interlocking system throughout the whole Swiss railway network and recorded on

---

[*] This work was partially supported by the Future and Emerging Technologies Unit of EC (IST priority - 6th FP), under contract no. FP6-021235-2 (project ARRIVAL).

R.K. Ahuja et al. (Eds.): Robust and Online Large-Scale Optimization, LNCS 5868, pp. 354–368, 2009.

a less detailed level comprising about 2300 operating points. These data describe the arrival and departure times of each train for every operating point along its route for every day the train drove. There are, however, no data on the dependencies between delays of different trains.

Delays are usually classified into primary and secondary delays. Primary delays "occur" at some point in the network, e.g., due to doors blocked by customers, technical problems, or accidents. Secondary delays (also called knock-on delays) are the consequences of primary or secondary delays on other trains. For example, a punctual train may accumulate a secondary delay because it waits for a delayed train to maintain a connection. Another example is a pair of trains that need to leave a station via the same track segment in a fixed order, where the first train leaving the station is late, forcing the second train to wait until the track segment is free. If the delay of one specific train causes a secondary delay of another on a regular basis, e.g., on at least 25% of the days, we speak of a *systematic dependency* between the delays.

In this paper, we suggest models that, given certain parameters, describe the patterns underlying the most important types of dependencies. We present algorithms that efficiently find systematic dependencies in large-scale railway delay data. If a train depends on the delays of several other trains, the most significant dependency for the delay of each day can be identified by our methods. Our approach does not rely on any assumption on the statistical distribution of the data. We show results of our method on real-world data.

The paper is organized as follows. In Section 2, we give a brief summary of related work. Section 3 introduces the models of dependencies along with the algorithms to detect them. We show how the delays of a single train can be explained by several dependencies in Section 4. In Section 5, we suggest modifications of the algorithms to account for errors in the data or exceptions to the model. Finally, we present results of our experiments in Section 6, and give a conclusion and outlook in Section 7.

2   Related Work

In her PhD thesis [1], Conte examines several approaches to identify dependencies among delays. Arrival and departure delays of trains are associated with random variables. Assuming a multivariate normal distribution, the Tri-graph method [10] is applied to construct a graph whose nodes represent the random variables. In such a graph, edges are included on the basis of non-zero (partial) correlation coefficients, hence missing edges represent conditional independence. Conte and Schöbel [2] suggest to use the constructed Tri-graph in combination with linear regression to generate so called virtual constraints for the delay management problem. For the latter, refer to, e.g., [5,9].

In this paper, we present an algorithmic approach that makes no assumptions about the distribution of delays. Furthermore, we give real-world examples of dependencies that have very low correlation coefficients, and yet are important. In contrast to the network-wide approach suggested by Conte, however, we are currently detecting dependencies only within a station. Further, our goal is to support planners in improving timetables, rather than making robust delay management decisions during operations.

The problem of distinguishing between primary and secondary delays is not only of interest for timetabling, but also for determining fines due to performance contracts between governments and train operating companies. Daamen, Goverde and Hansen [3] developed a prototype software to register secondary delays due to conflicts on track sections. Their approach requires detailed delay data at the level of signals and track segments. Further, the approach requires dispatchers to identify incidents leading to primary delay. Secondary delays due to waiting for a connection could also be implemented in the prototype given that scheduled connection times are provided.

In contrast, our approach aims at finding systematic dependencies in the timetable rather than precisely ascribing particular delays to train operators. Our approach works with less detailed data on the level of operating points, requires no incidence records, and recognizes both types of delay dependencies.

For on overview of other delay propagation models, refer to [1,4].

3   Models and Algorithms

Two important types of dependencies between delays of different trains are *waiting* and *blocking*. In this section, we formally characterize such dependencies between a train that is originally delayed, called the *delayer*, and the train to which the delay propagates, called the *victim*. To be more precise, when we speak of a delayed train, we actually mean that some event, i.e., the arrival or departure of a train at a specific station, occurs later than scheduled.

In the following, we denote by x the delayer and by y the victim train, and by x_d and y_d the delay on day d of an event of the (potential) delayer and victim trains, respectively.

3.1   Waiting Dependency

A waiting dependency is given if a train waits for another one to maintain a connection. Hence, the delay of the arrival event of the feeder train may propagate to the departure event of the connecting train at a specific station. In order to find such dependencies in the data, we first formulate an idealized model of a waiting dependency. We remark that models of this kind are already known, e.g., see [6]. Based on this model, we provide an algorithm that finds waiting dependencies in the data.

Ignoring for a moment that the victim may depend on more than one delayer and may also suffer from other sources of delay, we can model an idealized waiting dependency as follows. First, there usually is some buffer time s up to which the feeder train may be delayed without affecting the connecting train. If the feeder train is delayed by more than s, the connecting train will wait to maintain the connection, but only up to a maximal waiting time w, i.e., a maximal delay $e = s + w$ of the feeder. Denoting by x_d the delay of the feeder train on day

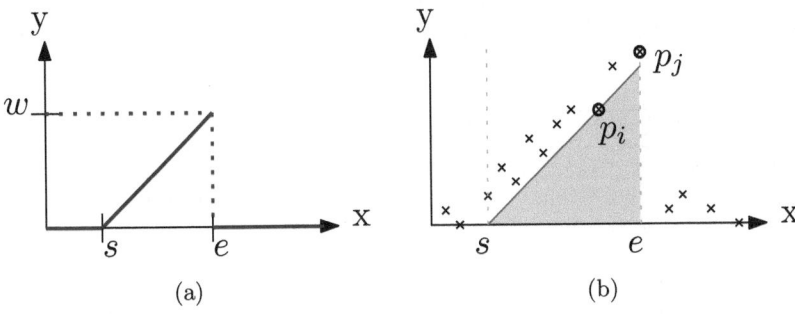

(a)   (b)

Fig. 1. (a) Idealized waiting dependency of the delay of a connecting train y on the delay of a feeder train x within the interval $[s, e]$. The maximum waiting time is $w = e - s$. (b) Hypothetical example data, where each point $p_d = (x_d, y_d)$ corresponds to the observed delays on day d. The interval $[s, e] = [x_i - y_i, x_j]$ is the solution of Problem 2, i.e., it maximizes the number of points $|S|$ above the shaded triangle, subject to the condition that no point may lie within the triangle.

d, and by $\tilde{y}_d$ the (idealized) corresponding delay of the connecting train, the waiting dependency can be formulated as

$$\tilde{y}_d = f(x_d, s, e) = \begin{cases} x_d - s & s \le x_d \le e \\ 0 & \text{otherwise,} \end{cases} \tag{1}$$

as shown in Figure 1(a). The parameters s and e may vary depending on the station and the specific pair of trains involved in the scheduled connection. We remark that planned values for these parameters could be obtained in principle. Unfortunately, it may turn out that during operations, the actual parameters differ from the planned ones. We therefore have to assume that the actual parameters are unknown.

In practice, of course, one victim may depend on several delayers, and furthermore, there may be other sources of delay which we cannot explain. Therefore, $f(x_d, s, e)$ can only be a lower bound on the actual victim's delay within the interval $[s, e]$. As we are interested in systematic dependencies, we want to find an interval $[s, e]$ containing a maximum number of points $p_d := (x_d, y_d)$, for which $f(x_d, s, e)$ is a lower bound on the delay y_d of the victim. Formally, given the delay data x_d and y_d for a set of days $d \in D$ for potential delayer x and victim y, respectively, we get the following problem:

$$\max_{s,e} |S| \tag{2}$$

$$S = \{(x_d, y_d) \mid s \le x_d \le e, \ y_d \ge x_d - s\}$$
$$\emptyset = \{(x_d, y_d) \mid s \le x_d \le e, \ y_d < x_d - s\}$$

Geometrically, we are looking for a rectangular triangle with a maximum number of points above it but none within. An example is given in Figure 1(b).

This problem is solved by Algorithm 1, which sweeps through the points in non-decreasing order of the x-coordinate, i.e., the delay of the incoming train.

Algorithm 1. Detect Waiting Dependency

Input: Delays $p_d = (x_d, y_d)$ of delayer x and victim y on days $d \in \{1, \ldots, n\}$.
1 Sort data according to non-decreasing x_i, breaking ties according to non-increasing y_i ;
2 $p_{n+1} \leftarrow (\infty, 0)$;   // sentinel
3 **for** $i \leftarrow 1$ **to** $n + 1$ **do**
4 | $s_i \leftarrow x_i - y_i$;   // calculate intercepts
5 **end**
6 $k, k^* \leftarrow 0$;   // number of points in current / best solution
7 $s \leftarrow s_1; s^* \leftarrow 0$;   // start of current / best solution
8 $\ell \leftarrow 1$;   // index of last point in current solution
9 **for** $i \leftarrow 1$ **to** $n + 1$ **do**
10 | **if** $s_i > s$ **then**
   | // cannot extend current solution to p_i
11 | **if** $k > k^*$ **then**
   | // update best solution
12 | $k^* \leftarrow k$;
13 | $s^* \leftarrow s$;
14 | $e^* \leftarrow x_{i-1}$;
15 | **end**
   | // initialize new solution
16 | $s \leftarrow s_i$;
   | // find first point in new interval
17 | **while** $x_\ell < s$ **do**
18 | | $\ell \leftarrow \ell + 1$;
19 | **end**
20 | $k \leftarrow i - \ell + 1$;
21 | **else**
22 | | $k \leftarrow k + 1$;
23 | **end**
24 **end**

Output: Number of points k^* in optimal interval $[s^*, e^*]$

Theorem 1. *Algorithm 1 computes a solution to Problem (2) in time $\mathcal{O}(n \log n)$.*

Proof. Every point $p_i = (x_i, y_i)$ defines an interval $[s_i, e_i]$ and a corresponding set of points S as follows: The interval starts at the intercept of the 45 degree line through p_i with the x-axis, namely at $s_i := x_i - y_i$. The interval ends at $e_i = x_j$, the x-coordinate of the rightmost point p_j of the sequence of points above the line, i.e., for all p_k, $k \in \{i, \ldots, j\}$ it holds that $y_k \geq x_k - s_i$.

Notice that in order to maximize $|S|$ it suffices to examine only those intervals $[s_i, e_i]$ which are defined by the points p_i, $i \in \{1, \ldots, n\}$: In any optimal solution there exists one point $p_i^* \in S^*$ with maximal intercept s_i^*. Hence, the start of the optimal interval s^* must be greater or equal to s_i^*, for otherwise p_i^* would not be in S^*. So setting $s^* = s_i^*$ is feasible for all points in S^*, as well as setting the end of the interval $e^* = x_j^*$, with p_j^* being the rightmost point of S^*.

The algorithm sweeps through all points in the order defined on Line 1. Maintaining s as the starting point of the current interval, it maximally extents the interval until the first point below the 45 degree line is met, i.e., the condition on Line 10 is violated. The intervals corresponding to the points above the 45 degree line need not be considered, since they either are infeasible or contain only a subset of the points of the current interval.

Clearly, sorting takes $\mathcal{O}(n \log n)$ time, and the rest of the algorithm runs in time $\mathcal{O}(n)$.   □

To detect all waiting dependencies in the data, Algorithm 1 is run on data of pairs of trains that are scheduled to meet at station within a reasonable time difference, say, up to 15 minutes. Depending on the number of days recorded in the data, we define a minimum number of days that must be in S^* in order to call a dependency *systematic*. For the points in S^* of a systematic dependency, we say that the delay of the victim is *explained* by the dependency, meaning that the delay of the delayer minus the buffer time s^* is a lower bound on the delay of the victim on the days corresponding to the points in S^*.

3.2   Blocking Dependency

If two trains have to use the same infrastructure element, such as a track segment or a platform, then a blocking dependency may exist, since one of them must pass that element first. This dependency can occur between any combination of arrival and departure events, as exemplified in Figure 2.

For reasons of operational safety, a certain headway time must be respected between two consecutive trains accessing the same infrastructure element. If we depict the delay data of two blocking trains as in Figure 3, one can identify a 45 degree line representing all the hypothetical arrival/departure times that would lead to a crash of the two trains. In our model of a blocking dependency, we assume that the headway times are always respected. Hence, there is a stripe around the 45 degree line in which no points may lie. The stripe also partitions

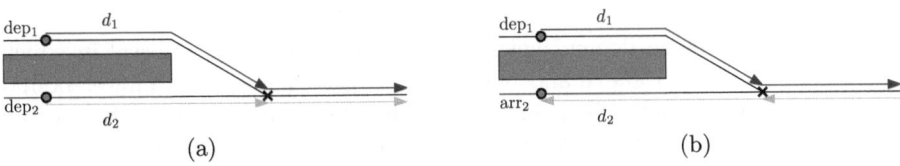

(a)   (b)

Fig. 2. Examples for conflicting trains, driving (a) in the same direction and (b) in opposite directions. From their arrival/departure location, they have to travel distances d_1 and d_2 to their first point of conflict.

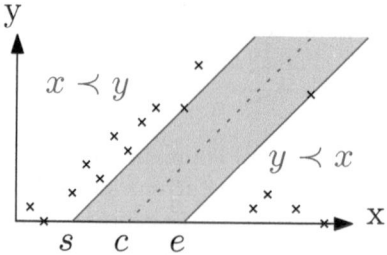

Fig. 3. Example of a blocking relationship between two trains departing in the same direction. The axes denote the delay of the respective train. The intersect c defines a 45 degree line (dotted) on which each point would represent a crash scenario, so $c = (\text{dep}_2 + d_2) - (\text{dep}_1 + d_1)$, where dep_i is the planned departure time of the event and d_i the driving time to the conflict point of train i. Around this line, a stripe (solid lines) represents the headway time that has to be respected, ensuring a safety distance between the trains. Every point above the stripe represents a situation where train x precedes train y, and vice versa for points below the stripe.

the points according to the precedence of the trains. In some cases, such as in Figure 2(b), it may practically not be possible to switch the order of trains, even in case of large delays, such that one region is empty. If the order of trains is fixed, a blocking dependency would also be found by Algorithm 1. We remark that the order of trains on the conflicting track segment is not obvious from the data, since delays are not given at the level of track segments but at the more aggregated level of operating points. Therefore, data about the exact routes of the trains is not available.

As in the case of waiting dependencies, we are searching for a subset of points for which a function of the delay of the delayer is a lower bound on the delay of the victim. Hence, we are interested in all points above the stripe. Formally, given the delay data x_d and y_d for a set of days $d \in D$ for potential delayer x and victim y, respectively, we get the following problem:

$$\max_{s,e} |S| \tag{3}$$
$$S = \{(x_d, y_d) \mid x_d \geq s,\ y_d \geq x_d - s\}$$
$$\emptyset = \{(x_d, y_d) \mid x_d - e < y_d < x_d - s\}$$

A subtlety of blocking dependencies is that there may be an interval in which there are points both above and below the stripe. Therefore, the lower bound on the delay of the victim is defined for a subset S of points, rather than for all points falling in an interval. Algorithm 2 solves Problem (3) using a sweep-line approach:

Algorithm 2. Detect Blocking Dependency

Input: Delays $p_d = (x_d, y_d)$ of delayer x and victim y on days $d \in \{1, \ldots, n\}$.
Minimum width of stripe $\underline{w} > 0$.

```
 1  Sort data according to non-decreasing x_i;
 2  p_{n+1} ← (∞, 0) ;                                   // sentinel
 3  for i ← 1 to n + 1 do
 4  |   s_i ← x_i − y_i ;                                 // calculate intercepts
 5  end
 6  c ← array of non-descending sorted {s_1, . . . , s_{n+1}};
 7  k, k* ← 0 ;                       // number of points in current / best solution
 8  s, s* ← 0 ;                               // left intercept of current solution
 9  r, r* ← 0 ;                              // right intercept of current solution
10  ℓ ← 1 ;                                  // index of first point above stripe
11  for j ← 2 to n + 1 do
12  |   s ← c[j − 1] ;
13  |   e ← c[j] ;
14  |   if e − s ≥ w then
15  |   |   while x_ℓ < s do
16  |   |   |   ℓ ← ℓ + 1 ;                               // first point above stripe
17  |   |   end
18  |   |   k ← j − ℓ ;                                  // number of points above stripe
19  |   |   if k > k* then
    |   |   |   // update best solution
20  |   |   |   k* ← k;
21  |   |   |   s* ← s;
22  |   |   |   e* ← e;
23  |   |   end
24  |   end
25  end
```

Output: Number of points k^* above the optimal stripe defined by s^* and e^*.

Theorem 2. *Algorithm 2 computes a solution to Problem (3) in time $\mathcal{O}(n \log n)$.*

Proof. Every point $p_i = (x_i, y_i)$ defines a 45 degree line through the intercept $(s_i, 0)$ and itself. W.l.o.g., we consider only those stripes whose left and right intercepts are defined by consecutive intercepts of the points (and hence, there are no points in the stripe). The condition on Line 14 ensures that only stripes respecting the minimum width $\underline{w} > 0$ are considered. When k is computed on Line 18, it holds that $e > s$ and that the $n - j + 1$ points below the stripe defined by s and e all have an intercept greater or equal to e. Thus, k is computed correctly, and the algorithm computes a solution to Problem (3).

Clearly, sorting takes $\mathcal{O}(n \log n)$ time, and the rest of the algorithm runs in time $\mathcal{O}(n)$. □

As for the waiting dependency, we consider only those dependencies to be systematic which hold on at least the minimum number of days required. To account for reasonable headway times, we further require an appropriate minimum width of the stripes. The detection of pathological cases can be prevented by computing reasonable bounds on the location of the center of the stripe from the timetable.

4 Multiple Dependencies

It is straightforward to generalize the lower bound obtained from a single dependency to the case where a train is the victim of several dependencies. In the following, we assume that for a victim train several such dependencies have been found. Thus, we may get several lower bounds on the delay of the victim on a particular day, namely from those dependencies that can explain it on that day. We make the usual assumption that the victim is delayed by the worst cause, i.e., the delayer providing the maximum lower bound for the victim's delay.

Formally, we are given a train y that is the victim of k dependencies with delayers x_i, $i \in \{1, \ldots, k\}$. Generalizing from the lower bound for the victim's delay from above, we observe that for each day d

$$g\left(f_1(x_d^1), \ldots, f_k(x_d^k)\right) = \max\left\{f_1(x_d^1), \ldots, f_k(x_d^k)\right\} \tag{4}$$

is a lower bound on the victim's delay y_d, where f_i is the function of the corresponding waiting or blocking dependency with delayer x^i. For a given day d, we call the delayer x^i which assumes the maximum in $g\left(f_1(x_d^1), \ldots, f_k(x_d^k)\right)$ the *best explanation* for y_d. Note that for a given day d, there may be no delayer explaining y_d, yielding only a trivial lower bound as the best explanation.

To visualize the quality of a certain multiple dependency, we plot the victim's delays against the best explanations, see Figures 6, 7, and 8.

5 Extensions

The recorded delay data are subject to inaccuracies, because the measurements on the tracks are aggregated to the level of operating points. SBB requires from their systems that such errors in the data be less than 20 seconds. Furthermore, it may well be the case that on a few days, the operational waiting rule described by Model (1) is violated. Such exceptions are unavoidable during operations. They may be caused by human mistake or as an intentional reaction to an exceptional situation.

For these reasons, there may be points p_i in the data that one would like to ignore, because otherwise, they may prevent a dependency from being detected. A similar problem in statistics is known as the least trimmed squares estimator for linear regression as surveyed in [7], where one seeks to find a subset of points minimizing the squared residuals for that subset. In our case, however, we are restricted to subsets corresponding to intervals, and have a fixed slope for the line we would like to "fix".

Exceptional points may prevent Algorithm 1 from detecting a dependency completely or worsen the resulting lower bounds (by increasing s). It is possible to extend the algorithms to allow for a maximal number $\bar{r}$ of allowed exceptional points. Clearly, practical values of $\bar{r}$ are very small. In the case of waiting dependencies, we want to solve the problem

$$\max_{s,e} |S| \tag{5}$$

$$S = \{(x_d, y_d) \mid s \leq x_d \leq e,\ y_d \geq f(x_d, s, e)\}$$

$$\bar{r} \geq |\{(x_d, y_d) \mid s \leq x_d \leq e,\ y_d < f(x_d, s, e)\}|$$

We sketch the necessary modifications of Algorithm 1 in order to solve Problem (5): We introduce a variable r that keeps track of the number of exceptional points in the current solution's interval $[s, e]$. Further, we keep a priority queue of these r points ordering them by their intersects s_i. We need the queue because points will leave the solution in the order of their intersects, whereas they enter the solution in the order of their x-coordinate. Now, we modify the criterion in Line 10, such that a solution is only extended if less than $\bar{r}$ exceptional points are in the current solution. If the solution is extended and the current point is exceptional, we add it to the priority queue. If the solution is not extended, i.e., there are already $\bar{r}$ exceptional points in the current solution, we remove the point p_l with smallest intercept s_l from the queue and increase s to s_l.

During execution of the algorithm, no more than $\bar{r} < n$ points are in the priority queue, and each point may only be inserted and removed once, at a cost of $\mathcal{O}(\log \bar{r})$. Hence, Problem 5 can still be solved in time $\mathcal{O}(n \log n)$.

6   Experiments

In this section, we present some of the dependencies which can be found in real-world data. The data comprised several important operating points of the SBB network during two months of the 2008 timetable. We required a minimum number of 15 explained days, a minimum interval width of 90 seconds for waiting dependencies, as well as a minimum interval width of 120 seconds for blocking dependencies. The following plots were created with R [8], as well as the correlation statistics (based on Pearson's product moment correlation coefficient). Some of the examples presented here were specifically selected to demonstrate that there are important dependencies which are hard to detect by means of correlation. Even if only the explained points, i.e., those which lie in the interval of a waiting dependency, are taken into account, the correlation of those delays can be very low, see Figure 4(b).

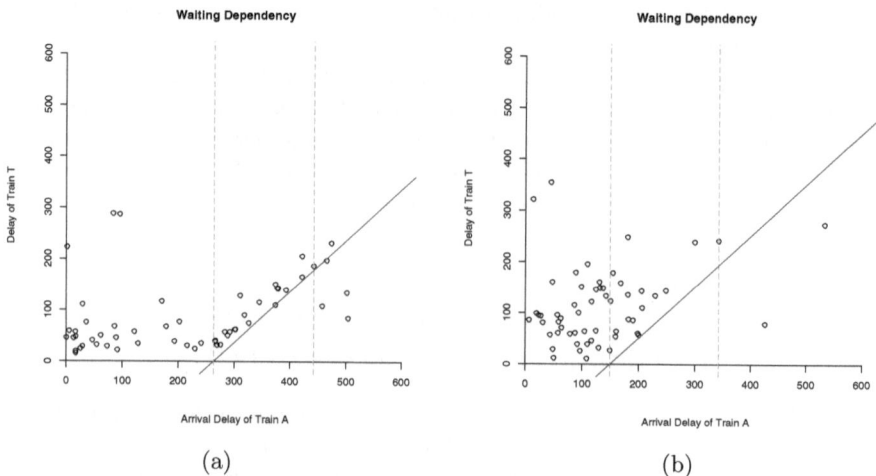

Fig. 4. Two examples of waiting dependencies in Basel. (a) The correlation between arrival and departure over all days is as low as 0.1602 (with a p-value of 0.2215). In the explained interval (between the dashed lines), the correlation is 0.9513 (p-value 3.652e-11). (b) In this example, the correlation over all days is 0.2151 (p-value 0.0959), higher than the correlation over the explained interval, which is 0.0998 (p-value 0.6845).

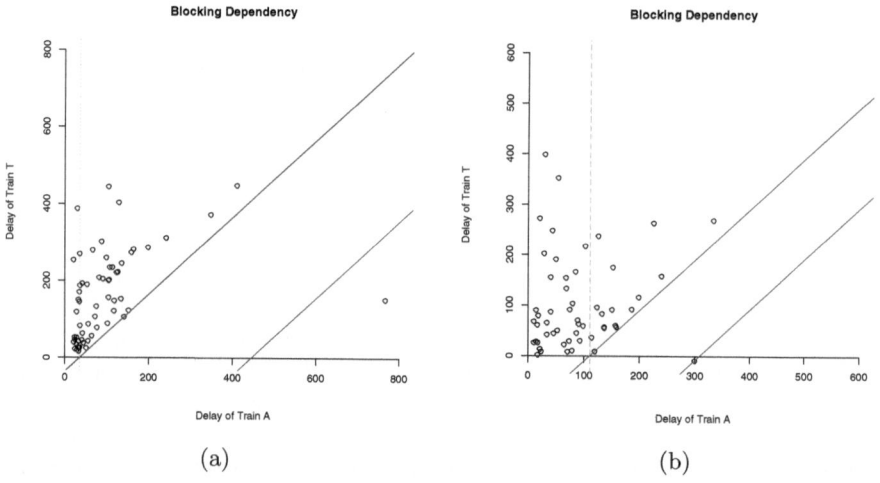

Fig. 5. Examples of blocking dependencies. (a) Two departures in Bern, blocking each other; the correlation is 0.3969 (p-value 0.0015). (b) Blocking dependency in Basel, with correlation 0.2040 (p-value 0.1147).

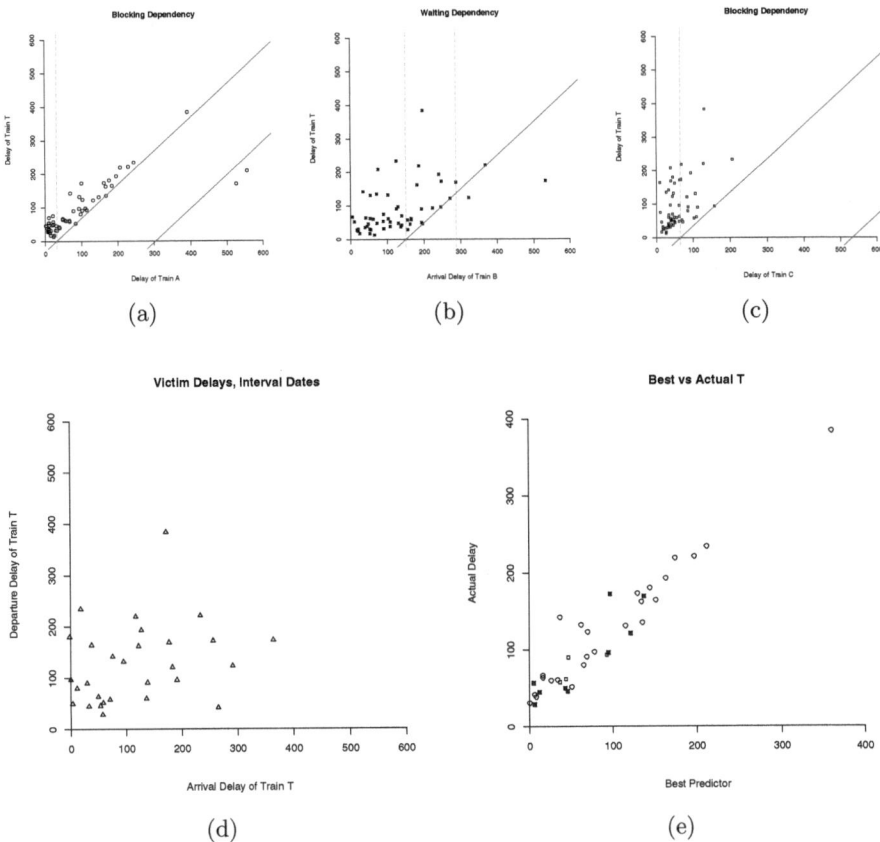

Fig. 6. Multiple dependency of a victim train T with delayers A, B, and C. (a) A blocking T on 34 explained days. (b) Waiting dependency with B, 17 days explained. (c) C blocking T on 16 days. (d) The arrival and departure delays of T have no obvious pattern. (e) Best explanation by delayers. A (circles), B (filled squares), and C (hollow squares) are the best explanation on 65.8%, 23.7%, and 10.5% of the explained days, respectively.

The dependencies in Figures 4 and 5 are single waiting or blocking dependencies for different victims. Multiple dependencies are given in Figures 6, 7, and 8, each showing a victim for which several dependencies could be found. We included a plot of the victim's own arrival delay, where available, and plots showing the best explanation for its delay upon departure for each day.

Notice that in the examples of Figures 6 and 7, arrival and departure delays of the victim do not follow an obvious pattern. Looking at the best explanation

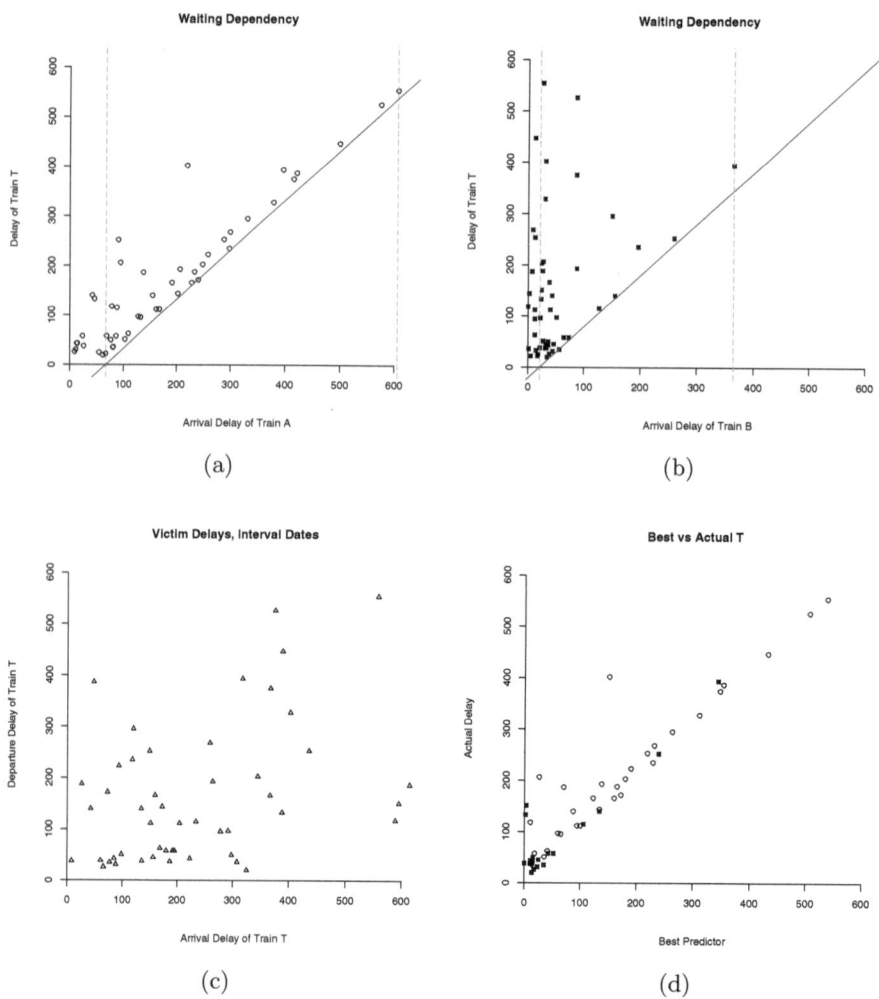

Fig. 7. Multiple dependency of a victim train T with waiting dependencies only. (a),(b) Waiting dependencies with trains A, B, on 17 days, each. (c) The arrival and departure delays of T have no obvious pattern. (d) Best explanation by delayers. A (circles) and B (filled squares) are the best explanation on 60.8% and 39.2% of the explained days, respectively.

plot, however, there is an almost linear dependency of the departure delay of the victim on the respective best delayer. Notice that a perfect explanation would have all points on a 45 degree line through the origin.

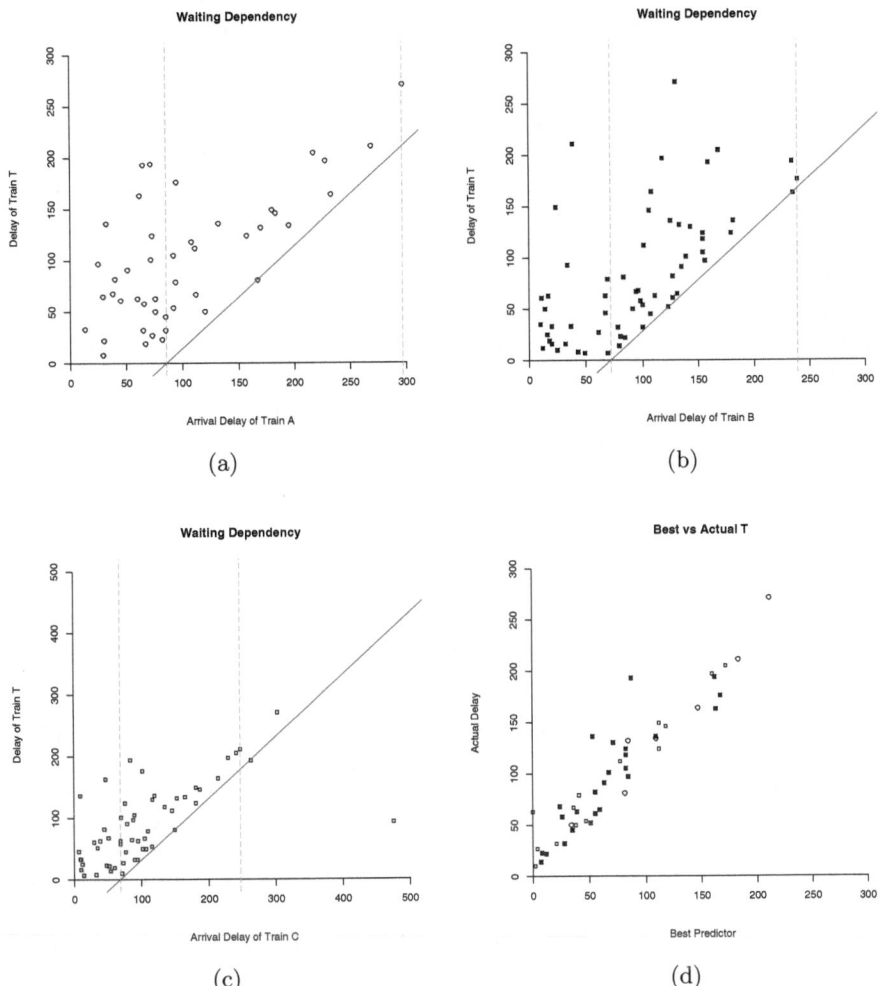

Fig. 8. Multiple dependency of a train T on three trains at the first station of T's trip, for which no arrival delay is available in the data. (a),(b),(c) Waiting dependencies with trains A, B, and C, explaining 20, 38, and 35 days, respectively. (d) Best explanation: A (circles), B (filled squares), and C (hollow squares) are the best explanation on 15.2%, 54.3%, and 30.4% of the explained days, respectively.

7   Conclusion

The results of our experiments are very encouraging. We analyzed real-world delay data from SBB. Using the approach presented in this paper, we were able to find dependencies which have significant impact on daily operations. Allowing for exceptional points as described in Section 5 turned out to be helpful.

We implemented a prototype that is connected to the SBB database, enabling planners to search for dependencies at all operating points of the Swiss railway network.

Our approach is useful to quickly find candidate dependencies from large real-world data sets that provide lower bounds on the delay of trains. In a second step, a statistical examination of the dependencies could be useful to assess their significance, especially in the presence of exceptional data points. It would be interesting to collect additional data that allows the distinction between primary and secondary delay, and to compare these with the dependencies found by our algorithms.

A third step would be the extension of our approach to global dependencies, i.e., to trace back the propagation of delays along the route of trains, possibly yielding a network of delay propagations. Ideally, it would be possible to estimate the effect of a small local change of the timetable, say, by adding a small buffer time, on a network of delay propagations.

References

1. Conte, C.: Identifying Dependencies Among Delays. PhD thesis, Georg-August-Universität zu Göttingen (2007), http://webdoc.sub.gwdg.de/diss/2008/conte/
2. Conte, C., Schöbel, A.: Identifying dependencies among delays. In: Proceedings of IAROR 2007 (2007)
3. Daamen, W., Goverde, R.M.P., Hansen, I.A.: Non-discriminatory automatic registration of knock-on train delays. Networks and Spatial Economics 9(1), 47–69 (2009)
4. D'Ariano, A.: Improving Real-Time Train Dispatching: Models, Algorithms and Applications. PhD thesis, TU Delft (2008), http://repository.tudelft.nl/
5. Gatto, M.: On the Impact of Uncertainty on some Optimization Problems: Combinatorial Aspects of Delay Management and Robust Online Scheduling. PhD thesis, ETH Zürich, No. 17452 (2007)
6. Goverde, R.M.P.: Optimal scheduling of connections in railway systems. Technical report, TRAIL, Delft, The Netherlands (1998)
7. Hubert, M., Rousseeuw, P.J., Aelst, S.V.: High-breakdown robust multivariate methods. Statistical Science 23(1), 92–119 (2008)
8. R Development Core Team. R: A Language and Environment for Statistical Computing. R Foundation for Statistical Computing, Vienna, Austria (2008), http://www.R-project.org
9. Schöbel, A.: Optimization in Public Transportation. Optimization and Its Applications, vol. 3. Springer, Heidelberg (2006)
10. Wille, A., Bühlmann, P.: Tri-graph: a novel graphical model with application to genetic regulatory networks. Technical report, ETH Zürich (2004), ftp://ftp.stat.math.ethz.ch/Research-Reports/Other-Manuscripts/buhlmann/trigraph.pdf

Rescheduling Dense Train Traffic over Complex Station Interlocking Areas

Francesco Corman[1], Rob M.P. Goverde[1], and Andrea D'Ariano[2,1]

[1] Transport & Planning Department, Delft University of Technology,
Stevinweg 1, 2628 CN Delft, The Netherlands
f.corman@tudelft.nl, r.m.p.goverde@tudelft.nl
[2] Dipartimento di Informatica e Automazione, Università degli Studi Roma Tre,
via della Vasca Navale 79, 00146 Roma, Italy
a.dariano@dia.uniroma3.it

Abstract. Railway rescheduling is the task of restoring feasibility in case of disturbances and limiting the propagation of delays through a railway network. This task becomes more difficult when dealing with complex interlocking areas, since operational rules constrain the passage of trains through short track sections. This paper presents a detailed microscopic representation of the railway network that is able to tackle the complexity of a station area with multiple conflicting routes and high service frequency. Two alternative graph formulations are presented to model the incompatibility between routes: one based on track sections and another based on the aggregation of track sections into station routes. An extensive computational study gives useful information on the performance of the two formulations for different disturbance scenarios.

Keywords: Train Rescheduling, Alternative Graph, Incompatibility Graph.

1 Introduction

Railway traffic is usually operated according to an existing plan of operations (off-line timetabling) that specifies train departure and arrival times at station platforms and other relevant points on the network where trains interact, e.g. at the merging and crossing points of lines and routes. During operations, disturbances cause deviations from the original plan which therefore has to be adjusted to resolve route and timetable conflicts as quickly as possible (real-time dispatching).

The real-time process needs accurate modeling of the evolution of train traffic and propagation of delays, starting from an initial disturbed status of the network, with the aim to compute new scheduling plans that should result in feasible train movements. In this context, high density of traffic, severe disturbances, and complex interlocking areas, are factors causing multiple interrelations between train services and increasing the complexity to predict the traffic flow in the network. At present, there is no advanced decision support tool available to traffic

R.K. Ahuja et al. (Eds.): Robust and Online Large-Scale Optimization, LNCS 5868, pp. 369–386, 2009.

controllers who thus have to rely solely on their experience and rules-of-thumb to deal with real-time disturbances. The complexity, time constraint, and limited decision support often leads to sub-optimal dispatching solutions. For these reasons, there is a need for Conflict Detection and Resolution (CDR) systems that are able to (i) model running and headway times using the signaling and safety systems in use, (ii) forecast the delay propagation in a large network, and (iii) propose good dispatching measures in a short time.

The recent literature related to advanced CDR systems can be classified according to two levels of approximation. Macroscopic models represent the railway network as a simplified series of links connecting stations. A fixed running time is used for train runs between two stations, and a fixed headway time is imposed between consecutive trains on the same link or at stations. These models are usually adopted in the planning stage for designing plan intentions or draft timetables, in order to keep complexity manageable. On the other hand, microscopic models use detailed infrastructure and rolling stock characteristics to simulate accurately individual speed profiles and running times taking into account the dynamic constraints of the signaling systems. The latter approach is required when dealing with real-time dispatching of train operations.

For the Dutch railways, DONS is a macroscopic timetable design tool consisting of a user interface and two main computation modules. The network scheduler module, CADANS [1], computes a feasible periodic network timetable, based on periodic interval constraints on e.g. running, dwell, transfer, and headway times, while neglecting capacity constraints at (large) stations that are considered black boxes. If the problem instance is infeasible CADANS returns a minimal set of conflicting constraints that has to be relaxed by the user before a feasible solution can be found. The other module, STATIONS (see e.g. [2]), computes feasible routes in station areas given the arrival and departure times computed by CADANS and train preferences for platforms. For each train a set of possible routes is derived including the blocking time for each track section on the route given the fixed platform arrival and departure times, train characteristics, and interlocking constraints (e.g. sectional-release route locking). From this, conflicting train/route pairs are derived with overlapping infrastructure claims, which results in a conflict graph where each node represents a train/route pair and edges connect nodes associated with conflicting routes or with the same train (each train is assigned to only one route). In this formulation, a complete set of compatible routes translates into a node packing problem, i.e., find a maximum set of nodes in the conflict graph such that no edge connects two nodes. A branch and cut algorithm has been implemented that computes a solution with an average computation time of one minute per station. Solving a larger station like Utrecht takes a few minutes [2].

Caimi et al. [3] also consider the problem of routing trains with a given timetable through a station using the conflict graph. The problem is modeled as an independent set problem over the conflict graph, which is solved by a fixed-point iteration heuristic. The fixed-point iteration algorithm finds feasible routings within seconds or minutes depending on the traffic density, although the algorithm also failed for particular dense instances [4]. Moreover, Caimi et al. [3] describe a local search algorithm to find an improved routing with increased buffer times between the most tight train paths. However, this optimization is much more time consuming (hours). Fuchsberger [4] describes two extensions of the conflict graph: the tree conflict graph and the resource tree conflict graph. In these graphs, all train routes are given by a list of route nodes in the station network topology with associated blocking time intervals. For each train, all routes are combined in a directed (routing) tree with a common entrance point as root. A conflict occurs when two train routes of distinct trains would block a resource at the same time. In the tree conflict graph, these conflicts are modeled by adding a conflict edge between the associated route nodes in both routing trees. In the resource tree conflict graph, groups of conflicting intervals for all trains simultaneously are modeled. Instead of conflict edges between pairs of trains as in the tree conflict graph, resource vertices are added to the graph which are connected to all associated nodes with overlapping intervals. The big advantage of the (resource) tree conflict graphs is that the conflicting route nodes are identified explicitly together with all routes branching from this node. A feasible routing is now obtained by finding directed paths from each root node of the trees to one of the leafs such that the directed paths are not interconnected by a conflict edge or conflict group, respectively. This problem can be formulated as a multi-commodity flow problem. Experiments by Fuchsberger [4] showed that the tree conflict graph model can be solved faster than the conflict graph model, however the construction time of the tree conflict graph is much higher than that of the conflict graph, so that overall the computation time may still take minutes. In contrast, constructing and solving the resource tree conflict graph model took only up to a few seconds for the same test instances.

Caimi et al. [5] solve the joint scheduling and routing problem in large station areas using a time discretization and a conflict graph, where a scheduling solution is represented by an independent set over the conflict graph. Multiple routes are considered for each train in order to find route sets that have some degree of robustness against delays. The running time of a train is assumed the same for all possible routes as a result of the time discretization. The fixed point iteration algorithm of Caimi et al. [3] was adjusted to incorporate the timetabling problem. Together with a heuristic to reduce the number of routings, the approach is able to generate timetables with feasible routings for large stations and dense traffic in less than a minute. The model must be combined with another timetabling problem in the so-called compensation zones between the main stations (or condensation zones) and a coordination procedure that sets the passing times and tracks at the portals between the compensation and condensation zones. A global feasible timetable is obtained by solving the local timetabling problems in the compensation and condensation zones with matching boundary conditions at the portals.

Rodriguez [6] uses a job shop scheduling model with state resource constraints in order to detect and solve train conflicts in a short computation time. Synchronization constraints are formulated to keep trains running with sufficient headway distances, even in case of yellow or red signal aspects. Constraint programming is adopted to dispatch trains in a station area with up to 12 trains.

D'Ariano et al. [7] propose an alternative graph model of the train scheduling problem to compute optimal train orders. A detailed problem formulation is presented that considers the route of each train running in stations and corridors as a set of consecutive block sections. Each train reserves one block section at a time. The problem is solved by a branch and bound algorithm truncated after a given time of computation. Results on a Dutch railway area and a two-hour timetable with 108 trains are reported to assess the effectiveness of the proposed approach. The branch and bound algorithm outperforms simple dispatching rules within a few seconds of computation.

Most of the previous work on train rescheduling investigated simplified railway networks and an implicit representation of route incompatibilities in large station interlocking areas, while this is a challenging issue when dealing with multiple interactions between inbound and outbound routes [8].

This paper presents microscopic representations of railway networks to tackle the complexity of busy station areas with multiple conflicting routes and high service frequencies. The next section describes basic terminology and the conflict detection and resolution problem. Section 3 presents two formulations to model the incompatibility between station routes: one based on track sections and another based on the aggregation of track sections into station routes. Section 4 describes briefly the algorithms used to solve the CDR problem in complex interlocking areas. Section 5 reports the computational results for the two formulations in terms of delay minimization and computation time. The last section discusses the main achievements and gives further research directions.

2   Basic Terminology and Problem Formulation

A railway network can be partitioned into interlocking areas and open tracks. Open tracks are railway lines without diverging or converging tracks. In modern railways safe operation on open tracks is guaranteed by automatic block systems. In fixed block systems a safe separation distance between successive trains is obtained by dividing the open track into *block sections* which may occupy at most one train at a time. A train is only allowed to enter a block section after the train ahead has completely left the block. Blocks are protected by block signals that can show stop (red) or proceed (green) aspects. Figure 1 shows a typical speed profile of a train facing a red signal aspect while traversing a line with five signals. The first signal aspect is green, which enables the train to traverse the subsequent block section at its scheduled speed SS. The aspect of the second signal is yellow, therefore the train decreases its speed to a prescribed approaching speed AS until the next signal aspect, which happens to be red. The train must stop until the signal aspect turns to yellow, and when this happens it increases its speed up to AS. When arriving at the sight distance from the next signal, it can accelerate up to SS if the fourth signal aspect is green.

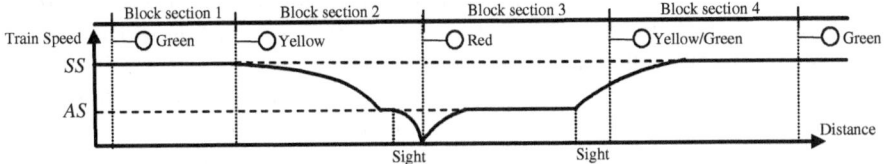

Fig. 1. Three aspects Dutch signaling system

The minimal separation time between two successive trains is thus given by a sight and reaction time of the block signal, the running time over the block, the clearing time until the entire train has left the block, and a switching time until the block signal is released. However, a train must be warned for a red signal over at least the braking distance before the signal which generally exceeds the sighting distance. In three-aspect signaling the block signals can also show warning aspects (yellow) indicating that the next signal shows red and a braking action is required to be able to stop in front of the next (red) signal. Hence, the minimal train separation distance for successive unhindered train movements is two free blocks. The associated blocking time is the minimum signal headway time for unhindered following consisting of the above mentioned time components and the approaching time from the approach signal (previous block signal) to the block signal. Block signals operate automatically based on track-free detection. For this, the railway line consists of *track sections* that are able to detect the presence of a train. A block section contains one or more track sections. When a train enters a block this is detected by the first track section causing the block signal to show a stop aspect. When a train clears the last track section of a block the block signal at the beginning is released and shows a warning aspect until also the next block is cleared.

In contrast, interlocking areas contain merging, diverging, and/or crossing railway lines. Interlocking areas include station layouts with platform tracks for scheduled stops and junctions for merging and crossing railway lines without platforms. Safety in interlocking areas is guaranteed by interlocking systems that prevent simultaneously authorizing conflicting routes. A route in an interlocking area is a sequence of tracks sections and switches between two signals. Routes are set by signalers or automatic route setting systems. After a route calling the interlocking system checks if all track sections are available, switches are in the correct position or free to move, and no opposing routes are called. If the route is proved available the switches move to the required positions and are locked, and finally the route is locked. After a proved route locking the interlocking (or controlled) signal at the beginning of the route can be cleared. The route is released when a train traverses the route. Modern interlocking systems have a sectional-release route locking, where the sections are released one-by-one after track-free detection. This way, a switch becomes available to another route as soon as it is released. Older interlocking systems have a route-release route locking, where the locked switches are released simultaneously after a common release point (behind the last switch) is released. Thus, in interlocking areas each track section has its own blocking time: all track sections on a route are blocked at the same time and released one-by-one according to the passage of a train. Note the difference with a block section on the open track which is released as one section. Also note that the route behind an interlocking signal and the next signal can be different based on the set switch positions. For more information on interlocking and block systems, we refer the interested reader to Hansen and Pachl [8] or Pachl [9].

A station route can be partitioned into an inbound route from an interlocking signal to the platform track(s) and an outbound route from the exit signal at the platform to a block signal on the open track or an intermediate interlocking signal. An inbound route thus connects an interlocking signal at the end of an open track (the home signal) or an intermediate interlocking signal to an exit signal after a platform. The blocking times of the platform track section(s) include the dwell time at the platform and are thus typically much larger than the preceding track sections on the route. Two station routes are called compatible if they can be used at the same time by different trains, and they are incompatible if they have a track section in common.

We consider a timetable which describes the movement of all trains running in the network. Each train has a scheduled arrival/through time at a set of relevant points along its route (such as station platforms, railway junctions, and exit points of the network). Furthermore, a train is not allowed to depart from a platform before its scheduled departure time. A train is considered to be late when arriving at the platform later than the scheduled arrival time.

During operations, process time variations and disturbances cause delays that require the timetable to be adjusted to prevent an accumulation of delays. Examples of perturbations are temporary speed limitations due to technical failures or track maintenance work and extended dwell times at scheduled stops. Other serious disturbances are disruptions corresponding to a track blockage. In this case, an alternative route has to be provided for each train scheduled over the unavailable track. A route conflict occurs when two or more trains claim the same track section at the same time. In this case, a movement authority is given to only one of the trains involved, while the others must wait until the route becomes available or are rerouted. A conflict is called a deadlock if at least one train has to be moved backwards to allow other train movements.

We next introduce some definitions of delays. A delay is the positive difference between the actual arrival time and the scheduled arrival time, and is defined for a set of relevant points in the network such as scheduled station stops and signal passages at the boundaries of the area under study. The total delay is the sum of all delays at the relevant points. A primary (or original) delay is directly caused by process time variations, failures, or disturbances and can only be recovered by exploiting available running time and dwell time supplements, i.e., by running trains at maximum speed and minimum dwell times. A consecutive (or secondary) delay is caused by train interactions and can be minimized by pro-actively managing the railway traffic.

The conflict detection and resolution problem can be defined as follows: given a railway network, a time horizon of operations, a set of train routes and scheduled event times at the relevant points in the network, and the actual position and speed of each train at an initial time t_0, find a conflict-free and deadlock-free schedule for the trains in the network, with feasible speed profiles respecting the signaling system, no early departures, and trains arriving at the relevant points with the smallest consecutive delay.

3   Alternative Graph Modeling of the CDR Problem

This section presents two alternative graph formulations of the CDR problem in complex station areas. The *disaggregated formulation* is a straightforward extension of existing CDR models [7] that takes into account incompatibility of station routes at the level of track sections, and results in optimistic minimum headway times. On the other hand, the *aggregated formulation* aggregates track sections into station routes, leading to slightly pessimistic minimum headway times when dealing with sectional-release route locking.

3.1   Disaggregated Formulation

The combinatorial structure of the CDR problem is similar to that of a job shop scheduling problem with several additional constraints. In job shop scheduling, a job must be processed by a prescribed sequence of servers, *machines*. Each machine is characterized by the ability to process at most one job at a time. The processing of a job on a machine is an *operation*. The job shop scheduling problem therefore consists of defining starting times of all the operations such that each operation starts after the completion of its predecessor and no machine processes two operations simultaneously.

In the disaggregated formulation, the passage of a train over a track section represents an operation. Here, block sections on the open track are assumed to consist of a single track section, but station routes in an interlocking area are separated in the track sections making up the route. The starting time of an operation corresponds to the time t_i a train starts running over the associated track section. In interlocking areas, the sectional-release approach is adopted, i.e., every track section becomes available as soon as it is released by the current train. A machine represents a track section, since it cannot host two or more trains simultaneously. The setup time of a machine consists of the clearing time and switching time. A CDR solution is feasible if the running and setup time constraints are satisfied for each pair of operations associated to the same track section, and there is no deadlock in the network.

Mascis and Pacciarelli [10] introduce alternative graphs to model variants of job shop scheduling problems. An alternative graph is a triple $\mathcal{G} = (N, F, A)$ with N a set of nodes, F a set of fixed arcs, and A a set of pairs of so-called alternative arcs. A graph selection S is a set of alternative arcs chosen from A such that at most one arc is selected for each pair. A solution to the scheduling

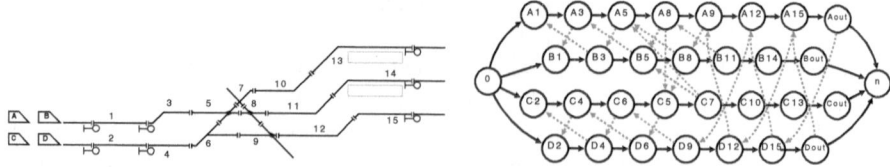

Fig. 2. Infrastructure layout and the corresponding alternative graph when modeling the network at the level of track sections

problem is a complete (exactly one arc from each pair is chosen) and consistent (there are no positive length cycles) graph selection.

The alternative graph model is able to include all operational time constraints of running trains. This can be done by using blocking time theory that describes the minimum required headway times between trains (see e.g. [8]). Specifically, a node of the alternative graph represents the passage of a train into a track section, which starts an operation. The fixed arcs connect the successive nodes of the alternative graph associated to a train route and are weighted with the running time. The alternative arcs are used to model choice of orders between operations on the same machine and are weighted with the setup time. A complete and consistent selection in the alternative graph $G(S)$ therefore corresponds to a conflict-free and deadlock-free schedule.

Figure 2 presents an example of the disaggregated formulation of the CDR problem. The left-hand side shows a complicated interlocking area layout with 15 track sections numbered 1 to 15. Four trains, A, B, C, and D, are running from left to right, and their destination is, respectively, platform track section 15, 14, 13, and 15. The routes of trains A and C cross, over track sections 5 or 8, all other routes in the interlocking area, while the routes of trains B and D do not interfere with the other routes. The right-hand side of Figure 2 shows the resulting alternative graph. A node in the alternative graph corresponds to the passage of a train over a track section. For reason of simplicity, the weight of fixed and alternative arcs is not depicted.

As shown in the example, the disaggregated formulation is able to model the incompatibility between station routes at the microscopic level of track sections but does not take into account the following operational constraints: (i) unscheduled stops are generally not allowed along consecutive track sections of the interlocking area, and (ii) the headway time between the running trains cannot be computed on the basis of short track sections. This formulation thus leads to a larger perceived capacity when scheduling trains in complex interlocking areas. In the next subsection, we cope with these limitations by introducing an aggregated formulation of the CDR problem.

3.2 Aggregated Formulation

Accurate management of complex interlocking areas requires an aggregate formulation of the CDR problem. The idea is to group together the track sections

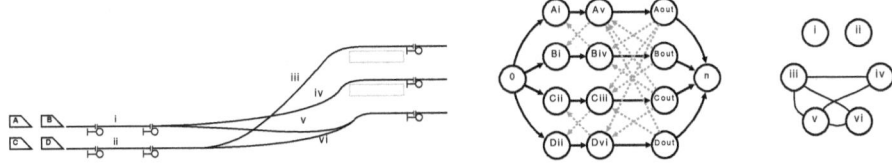

Fig. 3. Infrastructure layout, alternative graph and incompatibility graph resulting when track sections are aggregated into station routes

of each station route, so that all the required operational constraints can be taken into account explicitly. The headway time between trains with incompatible station routes is approximated according to a route-release route locking as follows. When the train releases the last switch section (the release point) on a station route, all track sections up to that point are released simultaneously, and thus become available to other trains. In this way, each train runs over its station route without being delayed by other trains. The setup time computed for the release point of the station route is used as the headway time needed before another conflicting route can be operated.

The aggregated formulation needs the complementary information of routes that are incompatible with each other, since the detailed routes of trains are lost when aggregating the track sections into station routes. The route incompatibility can be represented by means of an *incompatibility graph*, in which each node corresponds to a station route and edges connect every two nodes corresponding to incompatible station routes. Note the difference with the conflict graph [2,5] that defines timing conflicts between train/route pairs and thus depends on arrival/departure times. The incompatibility graph is a characteristic of an interlocking area and used to determine the corresponding alternative graph.

We now show the aggregated formulation of the CDR problem for the example of Section 3.1. The infrastructure layout is depicted in Figure 3 (left-hand), in which the detail of each track section is lost in favor of station routes.

Figure 3 (center) presents the resulting alternative graph. Since only station routes are modeled in the graph, the number of nodes and fixed arcs is decreased considerably. The alternative pairs are also less (7 against 12 for the disaggregated formulation), so the scheduling problem becomes easier to solve, even if there can be alternative arcs implied by the choice of other arcs [7].

The four trains running in the proposed network have the following station routes: i and v for train A, i and iv for train B, ii and iii for train C, ii and vi for train D. Table 1 reports the link between the microscopic train routes of Figure 2 and the station routes of Figure 3.

The incompatibility graph is shown in Figure 3 (right-hand). The incompatible pairs of station routes are the following: (iii, iv) ,(iii, v), (iii, vi), (iv, v) and (v, vi). The routes i and ii are compatible with the others.

A compact representation of the information necessary to model the incompatibility graph is as follows. To determine whether two station routes are

Table 1. Station routes and the corresponding list of track sections

Station routes	List of track sections
i	1
ii	2
iii	4,6,5,7,10,13
iv	3,5,8,11,14
v	3,5,8,9,12,15
vi	4,6,9,12,15

incompatible, it is sufficient to calculate the intersection between the list of track sections associated to each station route. This procedure can be implemented in the alternative graph model by associating *virtual machines* to station routes such that two station routes are incompatible if and only if they are associated to at least a shared virtual machine. In other words, virtual machines represent all the incompatibilities between conflicting routes in a complex interlocking area.

The number of virtual machines is given by the number of nodes of the incompatibility graph that are not connected with all the other nodes. The procedure adopted in this paper is to scan the incompatibility graph and search for the virtual machines needed to model all the incompatibilities between conflicting routes. The virtual machines are then introduced in the alternative graph formulation of the CDR problem, allowing to translate the characteristics of the non-aggregated model into the aggregated one.

3.3   Sectional-Release Route Locking Principle

We now discuss a further sophistication of the aggregated formulation of the CDR problem that models exactly the sectional-release route locking principle. The resulting CDR problem can be represented as a special job shop problem with a careful modeling of the setup times.

The aggregated formulation described in the previous subsection makes the assumption that for each train running on a station route all track sections up to the platform track (if the train stops) or to the end of the station route are released simultaneously. With this assumption, the setup time of each station route only depends on the current train that is running on that route. However, the headway time between consecutive trains, considered by the aggregated formulation, can be safely shortened. To achieve this result, the setup time of each train running on a station route should also depend on the successive trains traversing the same route.

If two trains follow exactly the same station route, the sectional release results in the same setup time as for the aggregated formulation. On the other hand, if two consecutive trains have station routes that diverge at some intermediate point of the interlocking area, the route of the preceding train would be released when the last shared track section has been cleared. This can be achieved by a slight modified formulation of the CDR problem with sequence-dependent setup times, i.e., the setup time between job i and job j depends on both job i and

Table 2. Aggregated formulation with sequence-dependent setup times

Node of origin	Node of destination	Last track section	Arc weight
A_{out}	B_{iv}	8	$S_8 - R_9^A - R_{12}^A$
A_{out}	C_{iii}	5	$S_5 - R_8^A - R_9^A - R_{12}^A$
A_{out}	D_{vi}	12	S_{12}
B_{out}	A_v	8	$S_8 - R_{11}^B$
B_{out}	C_{iii}	5	$S_5 - R_8^B - R_{11}^B$
C_{out}	A_v	5	$S_5 - R_7^C - R_{10}^C$
C_{out}	B_{iv}	5	$S_5 - R_7^C - R_{10}^C$
C_{out}	D_{vi}	6	$S_6 - R_{10}^C$
D_{out}	A_v	12	S_{12}
D_{out}	C_{iii}	6	$S_6 - R_9^D - R_{12}^D$

job j. A good state-of-the-art overview concerning the job shop problem with sequence-dependent setup times can be found e.g. in Artigues and Feillet [11].

In the aggregated formulation with sequence-dependent setup times, the alternative arcs modeling the setup time in a station interlocking area have to consider both the preceding and following trains running in that area. Table 2 shows how to model the interlocking area of the illustrative example of the previous subsections. The first two columns present the origin and destination nodes for each alternative arc of the interlocking area. The third column indicates the last track section in common between the routes of the two involved trains, e.g. in the first row of this table trains A and B follow different paths that diverge on track section 8. The fourth column shows how to compute the weight of each alternative arc.

The idea is first to compute the setup time as for the aggregated formulation with no sequence-dependent setup time, and then to subtract from this value the difference between the running time to traverse the complete station route in the interlocking area and the running time to traverse the part of the station route that is shared between the preceding and following trains. For example, the third row reports the same value of setup time (S_{12}, i.e., the setup time on track section 12) for the two formulations since both trains A and D follow the same station route up to track section 12, while the fifth row reports different values of setup time (S_5 versus $S_5 - R_8^B - R_{11}^B$, where R_8^B and R_{11}^B are the running times of train B on track sections 8 and 11) for the two formulations since the station routes of trains B and C differ after track section 5. Precisely, the station route of train B is $B3, B5, B8, B11$ while the station route of train C is $C4, C6, C5, C7, C10$. Clearly, in this alternative graph formulation some alternative arc can assume a negative weight, as e.g. the weight of the arc between nodes B_{out} and C_{iii} if $S_5 < R_8^B + R_{11}^B$.

3.4 A Qualitative Comparison

In this section we compare the two interlocking approximation formulations with the actual sectional-release route locking principle using an illustrative example.

Figure 4 shows a simple illustrative example with two trains (depicted in different gray colors) running from an open track (at the left) to a station interlocking area. The block section on the open track contains two track sections which are however reserved and released simultaneously according to the fixed block system principles. The station interlocking area contains two platform tracks and two station routes from the open track to the platform track sections. The two routes diverge at the second switch section and lead via an intermediate track section to one of the platform track sections. The routes thus have the first two (switch) track sections in common. The first (light-gray) train traverses the station area over the upper platform track without stopping, while the second (dark-gray) train has a scheduled stop at the lower platform track. The three diagrams (a)–(c) represent the blocking time diagrams corresponding to three-aspect signaling and the three interlocking variants, where the second train follows the first train as close as possible without being hindered, i.e., at the minimum headway distance facing only green signals at sight distance all the way. The blocking time for each track section starts at (the sight time before) the time that the section is reserved and ends after it is released [8,9].

Figure 4(a) shows the blocking time diagram where the second train follows the first at the minimum headway according to the sectional-release route locking principle. The critical section for following at the minimum headway time is the second (switch) track section on the station route. As soon as this section is released, the route to the other platform track can be set, locked, and cleared for the second train. In the blocking time diagram this is visualized by the touching blocking times of both trains in the second shared track section of the station routes. Note that the blocking times of the last two sections seem to overlap (shown by the dotted line), but these sections correspond to a different route for each train which can be used in parallel.

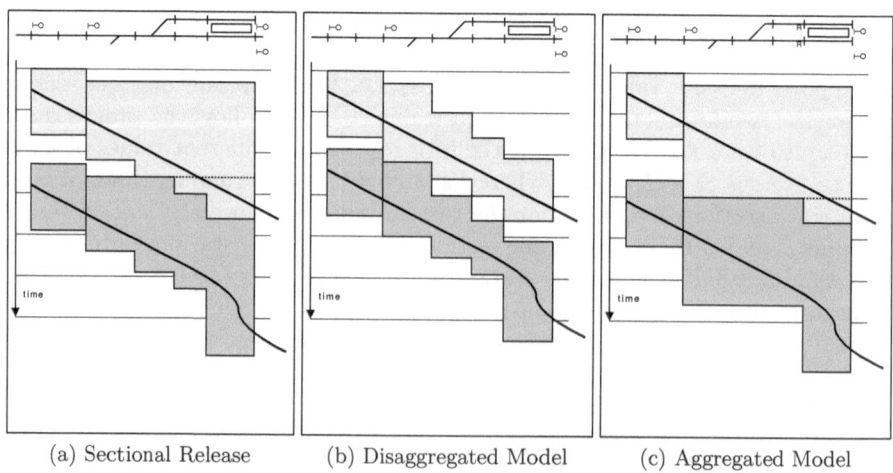

(a) Sectional Release   (b) Disaggregated Model   (c) Aggregated Model

Fig. 4. Blocking time diagrams illustrating the effect of three interlocking principles

Figure 4(b) shows the blocking time diagram when train traffic in the interlocking area is modeled with the disaggregated formulation. Here, each track section in an interlocking route is treated as a separate block. In comparison to Figure 4(a) the blocking times are much shorter and a smaller minimum headway time is obtained corresponding to a larger perceived capacity. The critical section is now given by the first (switch) track section of the station routes. Note that this model leads to an infeasible solution since the second switch section is still occupied by the first train when the second train claims the route after the release of the first switch section. In practice, the interlocking signal will still be red and therefore the block signal at the block section on the open track will be yellow so that the second train has to brake and is hindered.

Figure 4(c) shows the blocking time diagram using the aggregated formulation with non sequence-dependent setup times. Here, the route is released in two steps: first all switch and track sections before the common release point (the last section before the platform track section) indicated by the symbol 'R', and second all platform track sections. Clearly, this solution is more conservative than the one of Figure 4(a) since all track sections up to the platform track are released simultaneously. The critical section is now the third section of the first train's route corresponding to the release point. The solution of the aggregated formulation with non sequence-dependent setup times is compatible with the sectional-release route locking principle giving a slightly larger headway time than the minimum required.

In conclusion, the disaggregated model is too optimistic about station capacity utilization and may give infeasible solutions. On the other hand, aggregated formulation with non sequence-dependent setup times is pessimistic about the station capacity utilization but leads to feasible solutions. The headway time increase depends on the distance of the critical section in the sectional-release route locking solution to the common release point of the route-release route locking solution. This time difference will generally be small and can be considered as a practical buffer time, i.e., a slightly larger-than-necessary headway time. The discussed disaggregated and aggregated formulations lead to a lower and upper bound, respectively, on the maximum consecutive delay with respect to sectional-release route locking. Since the aggregated formulation with non sequence-dependent setup times yields feasible solutions, a small gap between the two bounds implies that the aggregated model solution is a good approximation of the optimal solution.

4   Algorithms for Train Scheduling and Speed Coordination

The scheduling procedure used in this paper is the Branch and Bound (BB) algorithm described in [7]. This algorithm is able to compute near-optimal solutions to practical sized CDR problems with fixed speed profiles. The objective function is the minimization of the maximum consecutive delay at each relevant point of the studied area. For the aggregated formulation with non sequence-dependent setup times, the lower bound procedure adopted for the BB algorithm is the Jackson Preemptive Schedule [12], adapted to deal with the virtual machines.

The scheduling solutions obtained by the BB algorithm do not consider possible speed adjustments needed to satisfy the Dutch three-aspect signaling system in case of conflicts between trains. Therefore, after the train orders have been computed by the BB algorithm, the train speed coordination procedure of D'Ariano et al. [13] is adopted to increase the traversing times for all trains facing yellow and red signal aspects. A typical drivers' behavior is implemented in which the trains proceed at their scheduled speed in case of green signal aspects, decrease speed to an approaching speed (usually 40 km/h in the Netherlands) in case of yellow signal aspects, stop in case of red signal aspects, and re-accelerate after a signal aspect improves. The traversing times for all the trains facing yellow (and red) signal aspects may therefore increase. This simple procedure enables us to compute feasible speed profiles and to predict the practical effects of the proposed train schedules. A more detailed discussion on speed adjustment and driver behavior can be found in [14].

5   Computational Experiments

This section presents the experiments performed to evaluate the two interlocking approximation formulations of the CDR problem, over a large sample of real-life instances. Algorithms have been implemented in C++ and run on a PC equipped with a processor Intel Pentium D (3 Ghz), 1 GB Ram and Linux operating system.

The test case is Utrecht Central station, which is one of the most complex station areas in the Netherlands. This station area has 20 platforms, more than 100 switches and 200 track sections, leading to a large number of possible inbound and outbound routes. The network topology is similar to a star with 5 main traveling directions (Figure 5). In total, the diameter of the entire dispatching area under study is around 20 km and includes more than 600 track sections. We consider one hour of traffic prediction for the 2008 timetable, up to 80 (passenger and freight) trains running in the station area.

The alternative graphs used to model the CDR problem in the case of the disaggregated formulation have 4067 nodes and 12235 alternative pairs. In the aggregated formulation, the alternative graphs have 1847 nodes and 4773 alternative pairs. Note that the smaller the alternative graph, the faster the computation of the CDR solutions.

We test a set of 450 timetable perturbations by combining 30 delay scenarios at the entrance of the network and 15 delay scenarios at Utrecht Central station. The latter delay scenarios are dwell time extensions for trains stopping at station platforms. Realization data were collected and made available by the Dutch infrastructure manager ProRail and random disturbances are generated according to Weibull distributions, as in [15].

We also consider three infrastructure scenarios (case 0: all infrastructure available; case 1: platform 2 of Utrecht Central unavailable; case 2: platform 15 of Utrecht Central unavailable), resulting respectively in 0%, 2% and 5% of the trains having to follow an alternative route to perform their scheduled trip.

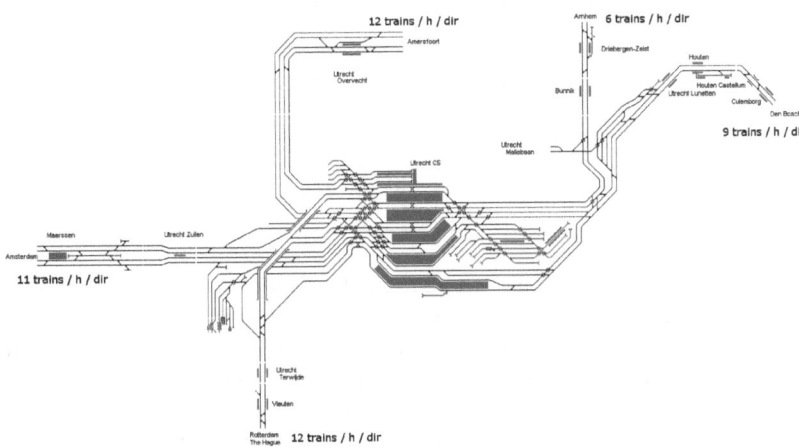

Fig. 5. Utrecht Central station in the center of the diagram; the scheduled hourly traffic per direction is given for each line of the dispatching area

The proposed perturbations are combined with the three infrastructure scenarios, leading to a total amount of 1350 disturbance scenarios with a maximum entrance delay for the trains of around 700 seconds and an average entrance delay experienced by every train of around 30 seconds.

Table 3 reports the average results over all the tested perturbations and disruptions scenarios for the CDR problem with fixed speed profiles (no train speed coordination algorithm is used). The disturbances are divided in the perturbation scenario (case 0) and in the two disruption scenarios, separately (case 1 and 2). For the disaggregated formulation, the maximum consecutive delay (in seconds), the average consecutive delay (in seconds), and the total computation time (in seconds) are shown in Columns 2, 3 and 4, respectively. For the aggregated formulation with non sequence-dependent setup times, the same type of information is presented in Columns 5, 6 and 7.

The fixed-speed CDR solutions presented in Table 3 are computed by the BB algorithm with no time limit of computation. So, the average value of the maximum consecutive delay reported in Column 2 (Column 5) is optimal for the disaggregated (aggregated) formulation. However, since the disaggregated

Table 3. CDR solutions without train speed coordination

Infra	Disaggregated formulation			Aggregated formulation		
Scenario	Max delay (s)	Avg delay (s)	Comput time (s)	Max delay (s)	Avg delay (s)	Comput time (s)
Case 0	91.8	1.5	1.7	108.6	2.6	0.9
Case 1	309.4	4.9	1.7	309.4	6.3	0.4
Case 2	199.4	3.3	1.8	200.7	4.3	1.3

Table 4. CDR solutions with train speed coordination

Infra Scenario	Disaggregated formulation				Aggregated formulation			
	Max delay (s)	Avg delay (s)	Comput time (s)	Coord iter	Max delay (s)	Avg delay (s)	Comput time (s)	Coord iter
Case 0	125.7	2.6	18.0	47	243.6	10.3	3.7	62
Case 1	309.4	6.2	18.8	49	317.4	14.1	3.8	63
Case 2	204.1	4.5	19.5	51	235.5	10.7	3.7	63

formulation does not take into account all the operational constraints in the station interlocking areas, the average value reported in Column 2 is a lower bound on the maximum consecutive delay at all the relevant points. But there is a small gap between the lower bound and the corresponding average value obtained for the aggregated formulation with non sequence-dependent setup times, that is an upper bound for the exact CDR formulation of the sectional-release route locking principle, leading to the conclusion that relaxing or tightening some operational constraints at these interlocking areas has a limited impact on the minimization of train delays.

When comparing the disturbance cases in Table 3, the delay figures are larger when dealing with disruption case 1, since trains have to be rerouted to very busy platforms nearby platform 2 of Utrecht Central.

Table 4 reports the average results over all the tested perturbations and disruptions scenarios after the train speed coordination procedure. In addition to the information presented in Table 3, Column 5 (Column 9) reports the average number of train speed coordination iterations required by the disaggregated (aggregated) formulation in order to satisfy the signaling system constraints when dealing with train conflict situations. The solutions of the disaggregated formulation are better than the ones of the aggregated formulation with non sequence-dependent setup times in terms of delay minimization. The difference between the solutions of the two formulations is more evident when dealing with the perturbation scenarios. However, a longer computation time is required to compute the former solutions since the corresponding alternative graph has a large number of operations (nodes) to be managed.

A comparison between the solutions obtained with and without train speed coordination underlines the impact of varying the speed profiles of the trains facing yellow and red signal aspects. The CDR solutions with fixed speed profiles are found, on average, in less than one third of the time to compute the CDR solutions with variable speed profiles. However, the latter solutions give more precise information on the delay propagation since the speed profiles of the trains involved in the conflicts are managed more accurately.

Another interesting point is to study how the output delays depend on the magnitude of the input delays. Table 5 shows the average results obtained for three group of instances of increasing entrance disturbances. For each group of instances, Column 1 and 2 report the maximum and average entrance delays while the other columns report the maximum and average consecutive delays for the CDR solutions computed with the disaggregated and aggregated formulations

Table 5. Increasing delays versus the two formulations with train speed coordination

Entrance disturbance		Disaggregated formulation		Aggregated formulation	
Max delay (s)	Avg delay (s)	Max delay (s)	Avg delay (s)	Max delay (s)	Avg delay (s)
467	20.2	197	3.2	267	10.3
706	31.9	216	4.7	267	11.9
925	35.8	209	4.3	247	12.6

and train speed coordination. For the disaggregated formulation, increasing entrance disturbances do not result necessarily in more conflicts and larger consecutive delays. In fact, this formulation does not include all the relevant constraints in station interlocking areas. On the other hand, the aggregated formulation is more accurate and presents consecutive delays that increase in a rather regular way compared to the different groups of entrance disturbances.

6   Conclusions

The development of advanced conflict detection and resolution systems is an important direction of research since there is a clear need to improve railway traffic management in case of disturbed operations. This paper proposes disaggregated and aggregated formulations, based on alternative graphs, to model large train scheduling instances with high accuracy. We focus on solving conflicts between consecutive trains at network level and modeling feasible headway distances in complex station interlocking areas. Computational experiments on a main dispatching area of the Dutch railway network, with up to 80 trains per hour, have been presented using existing scheduling and train speed coordination algorithms. The two formulations are compared in terms of delay minimization and computation time for different disturbance scenarios. The aggregated formulation with non sequence-dependent setup times has shown to be a good approximation of sectional-release route locking operations resulting in small extra buffer times.

A number of other issues remain that need further development.

- It would be interesting to design more robust conflict solutions by addressing the question on where and how much extra buffer time should be placed between consecutive train paths in presence of disturbances.
- The potential of the aggregated formulation with sequence-dependent setup times has not been quantified since the train scheduling algorithms need to be adapted to deal with this additional level of model sophistication.
- It remains to be solved the challenging problem of studying the full benefits of dispatching trains in larger networks and for heavily disturbed operations.

Acknowledgements

We thank the Dutch infrastructure manager ProRail (specially D. Middelkoop and L. Lodder) for providing the instances, and Prof. Ingo A. Hansen for his helpful suggestions. This work is partially supported by the programs "Towards Reliable Mobility" of the Transport Research Centre Delft and by the Italian Ministry of Research, Grant number RBIP06BZW8, project FIRB "Advanced tracking system in intermodal freight transportation".

References

1. Schrijver, A., Steenbeek, A.: Dienstregelingontwikkeling voor Railned: Rapport CADANS 1.0. Technical report, Centrum voor Wiskunde en Informatica, Amsterdam, the Netherlands (1994) (in Dutch)
2. Zwaneveld, P.J., Kroon, L.G., Van Hoesel, S.P.M.: Routing trains through a railway station based on a node packing model. European Journal of Operational Research 128(1), 14–33 (2001)
3. Caimi, G., Burkolter, D., Herrmann, T.: Finding delay-tolerant train routings through stations. In: Fleuren, H., Den Hertog, D., Kort, P. (eds.) Operations Research Proceedings 2004, pp. 136–143 (2005)
4. Fuchsberger, M.: Solving the train scheduling problem in a main station area via a resource constrained space-time integer multi-commodity flow. Master's thesis, ETH Zurich (2007)
5. Caimi, G., Burkolter, D., Herrmann, T., Chudak, F., Laumanns, M.: Design of a railway scheduling model for dense services. Networks and Spatial Economics 9(1), 25–46 (2009)
6. Rodriguez, J.: A study of the use of state resources in a constraint-based model for routing and scheduling trains. In: Hansen, I.A., Radtke, A., Pachl, J., Wendler, E. (eds.) Proceedings of the 2nd International Seminar on Railway Operations Modelling and Analysis, Hannover, Germany (2007)
7. D'Ariano, A., Pacciarelli, D., Pranzo, M.: A branch and bound algorithm for scheduling trains in a railway network. European Journal of Operational Research 183(2), 643–657 (2007)
8. Hansen, I.A., Pachl, J. (eds.): Railway Timetable and Traffic: Analysis, Modelling and Simulation. Eurailpress, Hamburg (2008)
9. Pachl, J.: Railway Operation and Control. VTD Rail Publishing, Mountlake Terrace (2002)
10. Mascis, A., Pacciarelli, D.: Job shop scheduling with blocking and no-wait constraints. European Journal of Operational Research 143(3), 498–517 (2002)
11. Artigues, C., Feillet, D.: A branch and bound method for the job-shop problem with sequence-dependent setup times. Annals of Operations Research 159(1), 135–159 (2008)
12. Jackson, J.R.: Scheduling a production line to minimize maximum tardiness. Technical Report 43, University of California, Los Angeles, Management Science Research Project (1955)
13. D'Ariano, A., Pranzo, M., Hansen, I.A.: Conflict resolution and train speed coordination for solving real-time timetable perturbations. IEEE Transactions on Intelligent Transportation Systems 8(2), 208–222 (2007)
14. D'Ariano, A.: Improving Real-Time Train Dispatching: Models, Algorithms and Applications. PhD Thesis, TRAIL Thesis Series T2008/6, The Netherlands (2008)
15. Yuan, J.: Stochastic Modelling of Train Delays and Delay Propagation in Stations. PhD Thesis, TRAIL Thesis Series T2006/6, The Netherlands (2006)

Online Train Disposition: To Wait or Not to Wait?[*]

Luzi Anderegg[1], Paolo Penna[2], and Peter Widmayer[1]

[1] Institute for Theoretical Computer Science, ETH Zentrum, CH-8092 Zürich,
Switzerland
lastname@inf.ethz.ch
[2] Dipartimento di Informatica ed Applicazioni "R.M. Capocelli", Università di
Salerno, via S. Allende 2, I-84081 Baronissi (SA), Italy
penna@dia.unisa.it

Abstract. We deal with an online problem arising from bus/tram/train disposition problems. In particular, we look at the case in which the delay is *unknown* and the vehicle can only wait in a station so as to minimize the passengers' waiting time.

We present deterministic polynomial-time optimal algorithms and matching lower bounds for several problem versions. In addition, all lower bounds also apply to randomized algorithms, thus implying that using randomization does not help.

1 The Setting

While many of the optimization problems encountered in transportation have already been studied in the early days of operations research [3, 5, 7] and have even stimulated the development of the field, this is not the case for *disposition* problems. Disposition (also known as operations control) deals with the *real time reaction* against the negative effects of *unexpected events*. For railways, the goal is to maintain high service quality in spite of events such as delays due to disturbances. Problems of this sort have been attacked mostly by computer simulations [14, 20] (see also [16] for a survey). In this paper, we pursue a different approach: we aim at an understanding of the fundamental algorithmic nature of these problems. In particular, we will look at *worst case* analysis of algorithms that must work with *partial information* (e.g., we know that a vehicle has been delayed, but we do not exactly know by how many time units). To this aim, we will show how these questions can be treated as an *online problem* [4, 8]. Then, we will characterize the performance of algorithms depending on several factors like (i) number of vehicles, (ii) whether the algorithm has some estimation of the delays, (iii) whether it can use randomization, etc. This problem is closely related to the *delay management problem*: a schedule and a delay for one or more vehicles is given and good decisions (e.g., to wait or to depart) for the consecutive trains must be taken. In contrast to our problem, most delay management studies

---

[*] A preliminary version of this paper appeared in [2].

R.K. Ahuja et al. (Eds.): Robust and Online Large-Scale Optimization, LNCS 5868, pp. 387–398, 2009.

388   L. Anderegg, P. Penna, and P. Widmayer

assume that the exact delay is known to the algorithm. In [1, 19] simulation systems are used for analyzing the delay management problem. Other authors, such as [18] (see also [12] for a survey), formulate an optimization problem in which the goal is to minimize the total waiting time subject to different constraints (slack times available at stations and tracks, connections can be dropped, etc.). For the same objective function, [9] described polynomial time algorithms for special cases, such as a limited number of transfers, or a railway network with a path topology. In a follow-up paper [10], a more general variant of the delay management problem was shown to be NP-complete both with and without slack times (or buffer times) in the timetable. More work on the delay management problem has been done in [6, 13]. The online delay management problem has been addressed further in [11] where the authors again considered also the case of a single line. The main difference between the model in that paper and the one here is on the delay incurred by the passengers. In [11], passengers' delay is exclusively due to the fact that a train at a station does not wait and thus a connection is missed. In our model, instead, passengers arrive at the station and we consider how much each of them should wait before their train leaves.

1.1   The Disposition Problem

Consider the following scenario from high-frequency bus (or tram or train) systems: we are given a station with $r > 0$ passengers arriving at each time unit on average (i.e., r is the *arrival rate of passengers at the station*). Buses reach the station regularly every t time units if no delay occurs. Whenever a bus reaches the station, it picks up all waiting passengers (i.e., the seating capacities of the bus are not our concern). We assume that picking up the passengers is instantaneous, i. e., requires no time. This implies that the overall passenger *waiting time at the station* (sum of all individual waiting times) is $r \cdot t^2/2$ per t time units between any two consecutive buses. Now consider the case in which one bus is currently at some station and the next bus is delayed by some amount of $\delta > 0$ time units. Assume the only action we can take is to make a bus wait in a station (this is sometimes referred to as *holding* [17]). The problem now is to decide how long the bus in the station should wait, if at all.

A convenient way of looking at this problem is to consider a snapshot of only three buses B_3, B_2 and B_1 as in Figure 1 that are travelling from left to right (we will drop the limitation to three buses later and consider more buses). Then, from the point of view of the passengers in the station, making B_2 wait for w time units is equivalent to "shift" B_2 leftwards by the distance that a bus travels in w time units. The overall waiting time (denoted as cost) can be computed according to which bus passengers get in (Figure 2 shows the case $w = 0$), as

$$cost = \underbrace{r(t + w)^2/2}_{B_2} + \underbrace{r(t + \delta - w)^2/2}_{B_3}. \tag{1}$$

Clearly, *knowing* δ, the best choice (i.e., the choice that minimizes the value in Equation 1) is $w = \delta/2$. However, for our problem of interest, we only know that

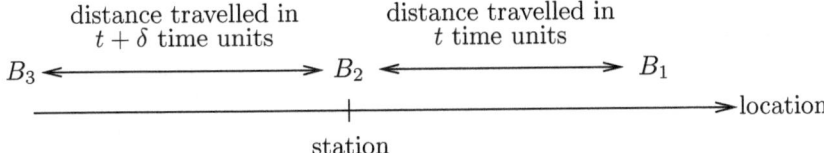

Fig. 1. The case of three buses

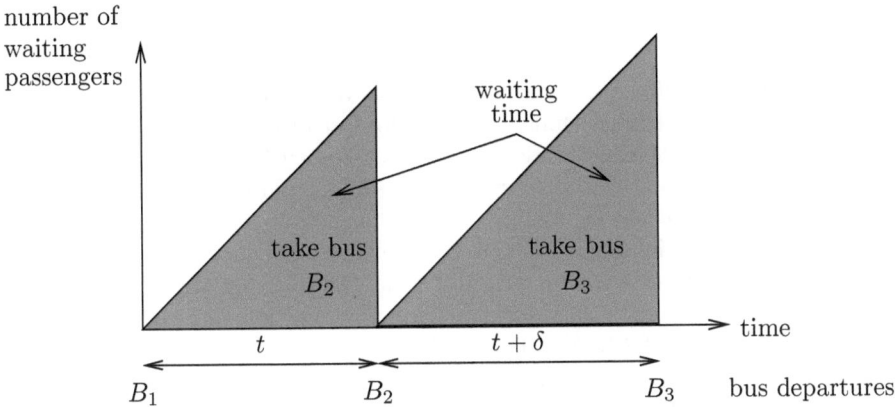

Fig. 2. Passengers waiting time when bus B_3 has a delay δ

B_3 is delayed (e. g., because of a traffic jam), but we do not exactly know δ. In this case, should B_2 leave immediately or wait for a while? In the latter case, how much should it wait for?

1.2   An Algorithmic Perspective

We view this question as an *online* problem in which we have to choose a good w *without knowing* δ (ideally, w should be good for *all* possible delays δ). Because the value of t is purely a matter of scaling time units, we will assume from now on that $t = 1$. Then, the actual waiting time, denoted as cost, is $r(1 + w)^2/2 + r(1 + \delta - w)^2/2$, and the optimum waiting time is $r(1 + \delta/2)^2$. For the competitive ratio, that is, the ratio between the actual and the optimum waiting times, the arrival rate r cancels, and therefore we assume for simplicity of the presentation from now on that $r = 2$. That is, we get the actual waiting time $\mathsf{cost}(w, \delta) = (1+w)^2 + (1+\delta-w)^2$, and the competitive ratio within the interval of arrival times from bus B_1's departure to B_3's arrival is

$$\rho(w, \delta) = \frac{\mathsf{cost}(w, \delta)}{\mathsf{opt}(\delta)} = \frac{(1 + w)^2 + (1 + \delta - w)^2}{2(1 + \delta/2)^2}. \tag{2}$$

We are interested in online algorithms that minimize the above ratio without knowing δ, that is, algorithms that decide w in such a way that $\max_{\delta \geq 0} \rho(w, \delta)$

is as small as possible. This is clearly equivalent to find $\min_{w \geq 0} \max_{\delta \geq 0} \rho(w, \delta)$. Note that if we choose a waiting time $w > 1$ and the adversary chooses not to delay bus B_3 then the situation is as if there is no bus B_2. The competitive ratio becomes strictly greater than two. As choosing not to wait gives a competitive ratio of less than two (compare also Section 2 and the proof of Theorem 3) we only need to consider w in $[0, 1]$ respectively find $\min_{w \in [0,1]} \max_{\delta \geq 0} \rho(w, \delta)$.

We consider two versions of this problem: (a) the *unbounded* case in which δ can be any positive integer; (b) the *bounded* case in which $\delta \leq \Delta$, where Δ is a positive integer *known to the algorithm*, that is, an upper bound on the maximum delay that can occur.

Remark 1 (Competitive measures). Notice that we are adopting the definition of *strictly c-competitive* algorithms. However, for our problem(s) this is equivalent to that of *c-competitiveness*. Indeed, we have assumed $t = 1$ only for the sake of simplicity, but *we do not consider t constant* (or as a parameter of the problem), since we want to derive algorithms that perform well *for any t known to the algorithm*. Under this assumption, we can always construct an instance whose cost is arbitrarily large by increasing t. This allows to apply any lower bound on the strict competitiveness also to the (weaker) definition of *c*-competitiveness. On the other hand, if t is a *constant of the problem*, then the definition of *c*-competitiveness is meaningless: the worst solution we can get has cost at most $\text{opt} + \Delta r = \text{opt} + O(1)$. This would imply a 1-competitive algorithm, *regardless of what we do*, while the (strictly) competitive ratio tells us whether the strategy is good or not.

1.3   Our Contribution

We consider the above mentioned online problem and its natural extension in which a set of $n + 2$ buses (instead of three) is given: bus B_1 already left the station, bus B_{n+2} has a delay δ and we have to decide the waiting time w_i for each B_i, for $i = 2, \ldots, n + 1$. This provides a family of basic disposition problems that capture some fundamental aspects of the real situations. For these problems, we completely characterize the competitive ratio of both deterministic and randomized algorithms, depending on n and Δ. In particular, we prove the *tight* bounds shown in Figure 3.

Interestingly, all the upper bounds are given via *deterministic algorithms*, while the lower bounds also apply to *randomized* ones. Indeed, we show that the

Problem version	Lower bound	Upper bound
Unbounded delays	$n + 1$	$n + 1$
Bounded delays $(\delta \leq \Delta)$	$1 + n\left(\frac{\Delta}{2 + 2n + \Delta}\right)^2$	$1 + n\left(\frac{\Delta}{2 + 2n + \Delta}\right)^2$

Fig. 3. Our results on the competitive ratio of online algorithms. All upper bounds are obtained via deterministic algorithms, while lower bounds also apply to randomized ones.

competitive ratio attained by our deterministic algorithms cannot be improved even when considering randomized algorithms against an *oblivious* adversary [4, 8]. In other words, randomization is useless for our disposition problems.

Paper organization. For the sake of clarity, we first present our results for the case $n = 1$. In particular, Sections 2 and 3 deal with the unbounded and the bounded case, respectively. We then extend the results to the case $n > 1$ in Section 4. Finally, in Section 5 we discuss further extensions and open questions.

2   Unbounded Delays

We first observe that two strategies are always possible:

No wait. In this case $w = 0$ and $\rho(w, \delta) = \frac{1+(1+\delta)^2}{2(1+\delta/2)^2}$. For $\delta \to \infty$, this ratio tends to 2 from below.

Wait "forever". This means that B_2 waits until B_3 arrives in the station. Then, $\rho(w, \delta) = 2$.

The above two strategies seem quite inefficient. Indeed, a better choice might be a compromise of them (i.e. wait, but not too much). Unfortunately, the following result shows that finding such a compromise is impossible:

Theorem 1. *No (randomized) algorithm can be better than 2-competitive in the case of unbounded delays.*

Proof. Every (randomized) algorithm ALG chooses an upper bound $W \in \mathcal{R}^+ \cup \{\infty\}$ on the waiting time according to some probability distribution independent of δ (this value is chosen by the adversary and is not known to the algorithm). For every δ, the waiting time is $\min(W, 1 + \delta)$ because we never wait more than the time the delayed bus arrives at the station (at that point δ is disclosed to the algorithm and the optimal decision is to have the bus to leave). In particular $W = \infty$ corresponds to the "wait forever" strategy meaning that, for every δ, the waiting time w is equal to $1 + \delta$. Observe that the "no wait" strategy ($W = 0$) strictly dominates the "wait forever" strategy ($W = \infty$) because $\rho(0, \delta) = \frac{1+(1+\delta)^2}{2(1+\delta/2)^2} = \frac{2+\delta^2+2\delta}{2+\delta^2/2+2\delta} \leq 2 = \rho(1 + \delta, \delta)$, for every $\delta \geq 0$ (see Equation 2). Therefore, for every ALG that chooses $W = \infty$ (the "wait forever" strategy) with nonzero probability, there is another algorithm ALG′ which has the same or a better competitive ratio and that chooses $W = \infty$ with probability zero.

We can thus focus on algorithms that choose always a *finite* upper bound W on the waiting time according to some probability distribution. This implies that, for any $p \in (0, 1]$, there exists $\overline{w}$ such that $Pr[W \leq \overline{w}] \geq 1 - p$. Since $\rho(w, \delta)$ is decreasing for $w \in [0, \delta/2]$ (see Equation 2) we have that $\rho(W, \delta) \geq \rho(\overline{w}, \delta)$ for all $W \leq \overline{w}$ and $\delta \geq 2\overline{w}$. Therefore, the competitive ratio is at least $(1-p) \cdot \rho(\overline{w}, \delta)$ for every $\delta \geq 2\overline{w}$. Since p can be arbitrarily small and since the adversary can choose δ arbitrarily large to make $\rho(\overline{w}, \delta)$ close to 2 (note that, for any fixed $\overline{w}$, $\rho(\overline{w}, \cdot)$ tends to 2 for $\delta \to \infty$), the lower bound $(1-p) \cdot \rho(\overline{w}, \delta)$ on the competitive ratio can be made arbitrarily close to 2. Hence the theorem follows.   □

Although the above theorem implies that both strategies above are optimal for large delays, it is clear that "no wait" is always better than "wait forever". Moreover, the former performs quite well whenever δ is small. In the subsequent section we investigate this version of the problem.

3   Bounded Delays

In this section we consider the version of the problem in which $\delta \leq \Delta$, where Δ is a positive constant *known to the algorithm*. The purpose of this is twofold: on the one hand we want to study whether this additional information allows for improved competitive ratios; on the other hand, we are interested in finding tight bounds that show how fast the competitive ratio tends to 2 as Δ increases. The "no wait" strategy provides a first upper bound. However, the reader can easily check that choosing $w = \Delta/2$ gives already an improvement. In the next section we give a tight bound for deterministic algorithms.

3.1   Deterministic Algorithms

Our algorithm DET should choose a good value of w based solely on the information that $\delta \leq \Delta$. To this aim, we first restrict ourselves to a *weaker adversary* that chooses only $\delta = 0$ or $\delta = \Delta$. Therefore, our goal will become

$$\min_w \max\{\rho(w,0), \rho(w,\Delta)\}. \tag{3}$$

In order to determine the best value for w according to Equation 3, we look for which values of w the adversary would give us $\delta = 0$, that is $\rho(w,0) \geq \rho(w,\Delta)$. The latter condition is equivalent to

$$\frac{(1+w)^2 + (1-w)^2}{2} \geq \frac{(1+w)^2 + (1+\Delta-w)^2}{2(1+\Delta/2)^2},$$

which corresponds to $w \geq \Delta/(4+\Delta) =: w_0(\Delta)$. Since $\rho(w,\Delta)$ is monotonically decreasing in $[0, w_0(\Delta)]$ and $\rho(w,0)$ is monotonically increasing in $[w_0(\Delta), \Delta]$, we have (see also Figure 4)

$$\min_w \max\{\rho(w,0), \rho(w,\Delta)\} = \rho(w_0(\Delta),0) = 1 + \left(\frac{\Delta}{4+\Delta}\right)^2. \tag{4}$$

The following lemma is used to show that DET performs well also against an adversary choosing *any* $\delta \in [0, \Delta]$.

Lemma 1. *For any $w \geq 0$, $\max_{0 \leq \delta \leq \Delta} \rho(w,\delta) \leq \max_{\delta \in \{0,\Delta\}} \rho(w,\delta)$.*

Proof. We distinguish the two cases $w \geq w_0(\Delta)$ and $w < w_0(\Delta)$. For $w \geq w_0(\Delta)$, we show that $\rho(w,\delta) \leq \rho(w,0)$ whereas for $w < w_0(\Delta)$, we show that $\rho(w,\delta) \leq \rho(w,\Delta)$. So, let us assume that $w \geq w_0(\Delta)$. Then

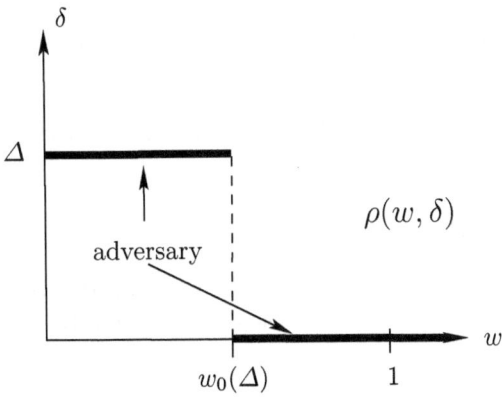

Fig. 4. The worst cases for the deterministic algorithm

$$
\begin{aligned}
\rho(w,\delta) - \rho(w,0) &= -\left(\frac{\delta(1+w)(w(\delta+4)-\delta))}{(2+\delta)^2}\right) \\
&\leq -\left(\frac{\delta(1+w)(w_0(\Delta)(\delta+4)-\delta)}{(2+\delta)^2}\right) \\
&= -\left(\frac{\delta(1+w)(\delta(w_0(\Delta)-1)+4w_0(\Delta))}{(2+\delta)^2}\right) \\
&\leq -\left(\frac{\delta(1+w)(\Delta(\frac{\Delta}{\Delta+4}-1)+4\frac{\Delta}{\Delta+4})}{(2+\delta)^2}\right) = 0
\end{aligned}
$$

Hence, in the first case $\rho(w,\delta)$ is maximized for δ equal zero.

For the case $w < w_0(\Delta)$, we consider the difference $\rho(w,\delta) - \rho(w,\Delta)$. Since

$$
\frac{d^2}{dw^2}[\rho(w,\delta) - \rho(w,\Delta)] = \frac{8}{(2+\delta)^2} - \frac{8}{(2+\Delta)^2} \geq 0 \qquad (\text{for } \delta \leq \Delta),
$$

the difference is convex between 0 and Δ with respect to w. Hence over the region $w \in [0, w_0(\Delta)]$ the maximum of $\rho(w,\delta) - \rho(w,\Delta)$ must be either at $w = 0$ or at $w = w_0(\Delta)$. But

$$
\rho(0,\delta) - \rho(0,\Delta) = \frac{4(\delta-\Delta)(\delta+\Delta+\delta\Delta)}{(2+\delta)^2(2+\Delta)^2} \leq 0
$$

and

$$
\rho(w_0(\Delta),\delta) - \rho(w_0(\Delta),\Delta) = \frac{8\delta(\delta-\Delta)(2+\Delta)}{(2+\delta)^2(4+\Delta)^2} \leq 0.
$$

Hence, in the second case $\rho(w,\delta)$ is maximized for δ equal Δ. $\square$

Because of the above lemma and the definition of $w_0(\Delta)$, we obtain the following:

Theorem 2. *No deterministic algorithm can be strictly better than* $1 + w_0(\Delta)^2$ *competitive, where* $w_0(\Delta) = \Delta/(4 + \Delta)$. *Therefore,* DET *is optimal for any* $\Delta \geq 0$.

As expected, this bound tends to 2 when Δ goes to infinity (which corresponds to the case of unbounded delays).

3.2   Lower Bound for Randomized Algorithms

In this section, we show that no randomized algorithm RAND can achieve an expected competitive ratio smaller than the competitive ratio of the deterministic algorithm DET.

Theorem 3. *For any* $\Delta > 0$, *no randomized algorithm* RAND *can be better than* DET.

Proof. Suppose w is chosen randomly with corresponding random variable W. Then, for any W, the adversary chooses δ such that $E[\rho(W, \delta)]$ is maximized. Because $\rho(w, \delta)$ is a convex function in w (see Equation 2), we can apply Jensen's inequality [15], and reduce the randomized case to the deterministic one. In particular, Jensen's inequality implies

$$\max_{\delta} E[\rho(W, \delta)] \geq \max_{\delta} \rho(E[W], \delta).$$

The latter quantity is the competitive ratio of the deterministic algorithm choosing $w = E[W]$. Hence, the result is implied by Theorem 2. □

4   Many Buses

Consider a set of $n + 2$ buses $\{B_1, \ldots, B_{n+2}\}$ such that: (i) B_1 already left the station, (ii) B_{n+2} has been delayed by δ, and (iii) the set $\{B_2, \ldots, B_{n+1}\}$ corresponds to the *control set* of n buses that we can delay in order to minimize the overall waiting time. Let $w = (w_2, w_3, \ldots, w_{n+1})$ represent such waiting times, i.e., bus B_i is delayed by w_i, $i = 2, \ldots, n + 1$. The cost is clearly

$$\text{cost}(w, \delta, n) = \underbrace{(1 + w_2)^2}_{B_2} + \sum_{i=2}^{n} \underbrace{(1 + w_{i+1} - w_i)^2}_{B_{i+1}} + \underbrace{(1 + \delta - w_{n+1})^2}_{B_{n+2}}. \quad (5)$$

As we do assume that B_i always precedes B_{i+1}, for $i = 1, \ldots, n + 1$, we restrict ourselves to those w for which $w_i \leq w_{i+1}$, $i = 2, \ldots, n$. In fact, we next show that without loss of generality it is enough to consider certain "balanced" waiting times:

Definition 1 (Balanced vector). *Let the* balanced vector $u(x)$ *be the vector assigning waiting time* $w_i = (i - 1)x/n$ *to bus* B_i *for* $i = 2, \ldots, n + 1$.

Observe that for a balanced vector $u(x)$ the distance between two any consecutive buses is identical, namely equal to $1 + x/n$ (except between buses B_{n+1} and B_{n+2} where it is $1 + \delta - x$). The following fact then follows from Equation 5 and Definition 1.

Fact 4. *For every w that is not balanced it holds that*

$$\forall \delta : cost(w, \delta, n) > cost(u(w_{n+1}), \delta, n).$$

Because of the above fact, we have to choose an optimal waiting time for bus B_{n+1} to compensate the delay δ. The buses B_2 to B_n are then evenly distributed. That is, we have to set $w_{n+1} = x$ so that

$$\max_{\delta} \rho(u(x), \delta) = \max_{\delta} \text{cost}(u(x), \delta)/\text{opt}(\delta)$$

is as small as possible. In the sequel, we let

$$\rho(x, \delta, n) := \rho(u(x), \delta) = \frac{n(1 + x/n)^2 + (1 + \delta - x)^2}{(n + 1)(1 + \delta/(n + 1))^2}. \tag{6}$$

Observe that, for all $x > 1$, $\rho(x, 0, n) \geq \max_{\delta} \rho(0, \delta, n)$. Hence, we will restrict ourselves to $x \in [0, 1]$. The following result is a simple generalization of Theorem 1.

Theorem 5. *For the case of $n+2$ buses and unbounded delays, no (randomized) algorithm can be better than $(n + 1)$-competitive.*

4.1 Bounded Delays

We first consider an adversary that always picks $\delta \in \{0, \Delta\}$, as in the case $n = 1$. Then, we observe that $\rho(w, 0, n)$ is equal to $\rho(w, \Delta, n)$ for

$$w = n\Delta/(2 + 2n + \Delta) =: w_0(\Delta).$$

Further, $\rho(w, \Delta, n)$ is greater than $\rho(w, 0, n)$ and monotonically decreasing in $[0, w_0(\Delta)]$. In $[w_0(\Delta), \Delta]$, $\rho(w, 0, n)$ is greater than $\rho(w, \Delta, n)$ and monotonically increasing. Therefore, the best deterministic algorithm DET against the restricted adversary is given by the value $w_0(\Delta)$ with competitive ratio equal to $\rho(w_0(\Delta), 0, n)$. From Equation 6, it follows that

$$\forall w, \quad \rho(w, 0, n) = 1 + w^2/n,$$

thus implying a competitive ratio of $1 + n \left(\frac{\Delta}{2 + 2n + \Delta} \right)^2$.

The next lemma shows that the adversary cannot profit from choosing δ in $[0, \Delta]$.

Lemma 2. *For any $w \geq 0$, $\max_{0 \leq \delta \leq \Delta} \rho(w, \delta, n) \leq \max_{\delta \in 0, \Delta} \rho(w, \delta, n)$.*

Proof. The proof follows the same steps as Lemma 1.
For $w \geq w_0(\Delta)$:

$$
\begin{aligned}
\rho(w, \delta, n) - \rho(w, 0, n) &= \frac{\delta(n+w)(\delta n - (\delta + 2n + 2)w)}{n(\delta + n + 1)^2} \\
&\leq \frac{\delta(n+w)(\delta n - (\delta + 2n + 2)w_0(\Delta))}{n(\delta + n + 1)^2} \\
&= \frac{\delta(n+w)(\delta(n - w_0(\Delta)) - (2n + 2)w_0(\Delta))}{n(\delta + n + 1)^2} \\
&\leq \frac{\delta(n+w)(\Delta(n - w_0(\Delta)) - (2n + 2)w_0(\Delta))}{n(\delta + n + 1)^2} = 0
\end{aligned}
$$

For $w < w_0(\Delta)$: Since for $\delta \leq \Delta$

$$
\frac{d^2}{dw^2}[\rho(w, \delta, n) - \rho(w, \Delta, n)] = (n+1)\left(\frac{2 + \frac{2}{n}}{(\delta + n + 1)^2} - \frac{2 + \frac{2}{n}}{(\Delta + n + 1)^2}\right)
$$

is greater equal zero, the maximum of this difference is either at $w = 0$ or at $w = w_0(\Delta)$. But

$$
\rho(0, \delta, n) - \rho(0, \Delta, n) = (n+1)\left(\frac{n(\delta - \Delta)(n\delta + 2\Delta\delta + \delta + n\Delta + \Delta)}{(\delta + n + 1)^2(\Delta + n + 1)^2}\right) \leq 0
$$

and

$$
\rho(w_0(\Delta), \delta, n) - \rho(w_0(\Delta), \Delta, n) = \frac{4\delta n(n+1)(\delta - \Delta)(n + \Delta + 1)}{(\delta + n + 1)^2(2n + \Delta + 2)^2} \leq 0. \qquad \square
$$

Lemma 2 together with the definition of $w_0(\Delta)$ implies the following:

Theorem 6. *For any $n \geq 1$ and for any $\Delta > 0$, no deterministic online algorithm can have competitive ratio better than $1 + w_0(\Delta)^2/n$, where $w_0(\Delta) = n\Delta/(2 + 2n + \Delta)$. Therefore,* DET *is optimal.*

Similar to Section 3 we can extend this result to any randomized algorithm RAND.

Theorem 7. *For any $n \geq 1$ and for any $\Delta > 0$, no randomized algorithm* RAND *can be better than* DET.

Proof. We first observe that the function $\rho(\cdot, \delta, n)$ is convex, that is, for every $\delta \geq 0$ and for any two vectors $\boldsymbol{w}, \boldsymbol{z}$ it holds that

$$
\lambda \cdot \rho(\boldsymbol{w}, \delta, n) + (1 - \lambda) \cdot \rho(\boldsymbol{z}, \delta, n) \geq \rho(\lambda \boldsymbol{w} + (1 - \lambda)\boldsymbol{z}, \delta, n), \tag{7}
$$

for $\lambda \in [0, 1]$. This follows from Equation 5 and from the fact that the function $(1 + x)^2$ is convex. In particular, for $\boldsymbol{e} = (E[W_2], \ldots, E[W_{n+1}])$ being the expected vector of $\boldsymbol{W} = (W_2, \ldots, W_{n+1})$, we obtain

$$
\max_\delta E[\rho(\boldsymbol{W}, \delta)] \geq \max_\delta \rho(\boldsymbol{e}, \delta) \qquad \text{(Jensen's inequality)}
$$

$$
\geq \max_\delta \rho(\boldsymbol{u}(e_{n+1}), \delta) \quad \text{(apply Fact 4 with } e_{n+1} = E[W_{n+1}])
$$

The latter quantity is the competitive ratio of the deterministic algorithm corresponding to the vector $u(e_{n+1})$. Theorem 6 thus implies the desired result.   □

5 Conclusion

Disposition in a transportation system is critical for customer satisfaction. In this paper we look at a disposition problem arising in high-frequency bus (or tram or train) systems from an algorithmic point of view. We formulate the problem as an online problem and prove tight bounds on the competitive ratio of the problem.

This work provides a basis for competitive analysis of disposition in more complex high-frequency transportation systems. We prove our results for a basic setting which captures some of the main aspects of more complex situations. In particular, our model focuses on the waiting time experienced by passengers waiting at a station. Our results characterize the loss of efficiency (w.r.t. the total waiting time) depending on the amount of resources available (i.e., the number of buses) and on the amount of information about the delay we have (i.e., an upper bound Δ on the delay δ).

We have shown that optimal solutions can be obtained from very simple deterministic algorithms which fix the waiting time of the bus B_{n+1} preceding the delayed bus to a value $w_0(\Delta) \cdot t$, where t is the time two consecutive buses reach the station if not delayed. The optimal waiting time for the other buses is always uniquely determined by the waiting time chosen for B_{n+1}. This simple strategy outperforms any possible randomized choice of buses waiting times.

As future research, our setting can be extended in several ways to reflect real world aspects. For instance, several stations with different arrival rates, several bus lines sharing some stations, or other cost functions can be considered.

Moreover, the formulation as an online problem can be applied to other disposition problems in transportation systems. In this context, the model should be adapted to the different types of possible reactions in the case of an unexpected event and the appropriate cost function.

Acknowledgments. The authors wish to thank an anonymous reviewer for suggesting the use of Jensen's inequality argument in the analysis of randomized algorithms and for suggesting a simpler proof of Lemma 1. This work has been partially supported by the Swiss National Science Foundation under Project no. 200021-107685 (Algorithmic Methods for Delay Management). Most of this work was done while the second author was working at ETH Zürich.

References

1. Adenso-Díaz, B., Gonzáles, M.O., Gonzáles-Torre, P.: On-line timetable rescheduling in regional train services. Transportation Research 33B, 287–398 (1999)
2. Anderegg, L., Penna, P., Widmayer, P.: Online train disposition: to wait or not to wait? In: Proc. ATMOS (2002); also available in the Electronic Notes in Theoretical Computer Science 66(6), 32–41 (2002)

3. Bertossi, A., Carraresi, P., Gallo, G.: On some matching problems arising in vehicle scheduling models. Networks 17, 271–281 (1987)
4. Borodin, A., El-Yaniv, R.: Online Computation and Competitive Analysis. Cambridge University Press, Cambridge (1998)
5. Brucker, P., Hurink, J.L., Rolfes, T.: Routing of railway carriages: A case study. In: Memorandum No. 1498, Fac. of Mathematical Sciences. Univ. of Twente, Fac. of Math. Sciences (1999)
6. Cicerone, S., D'Angelo, G., Di Stefano, G., Frigioni, D., Navarra, A.: Delay management problem: Complexity results and robust algorithms. In: Yang, B., Du, D.-Z., Wang, C.A. (eds.) COCOA 2008. LNCS, vol. 5165, pp. 458–468. Springer, Heidelberg (2008)
7. Dantzig, G., Fulkerson, D.: Minimizing the number of tankers to meet a fixed schedule. Nav. Res. Logistics Q 1, 217–222 (1954)
8. Fiat, A., Woeginger, G. (eds.): Online Algorithms: The State of the Art. LNCS, vol. 1442. Springer, Heidelberg (1998)
9. Gatto, M., Glaus, B., Jacob, R., Peeters, L., Widmayer, P.: Railway delay management: Exploring its algorithmic complexity. In: Hagerup, T., Katajainen, J. (eds.) SWAT 2004. LNCS, vol. 3111, pp. 199–211. Springer, Heidelberg (2004)
10. Gatto, M., Jacob, R., Peeters, L., Schöbel, A.: The computational complexity of delay management. In: Kratsch, D. (ed.) WG 2005. LNCS, vol. 3787, pp. 227–238. Springer, Heidelberg (2005)
11. Gatto, M., Jacob, R., Peeters, L., Widmayer, P.: Online delay management on a single train line. In: Geraets, F., Kroon, L.G., Schoebel, A., Wagner, D., Zaroliagis, C.D. (eds.) Railway Optimization 2004. LNCS, vol. 4359, pp. 306–320. Springer, Heidelberg (2007)
12. Ginkel, A.: Event-activity networks in delay management. Master's thesis, University of Kaiserslautern (2001)
13. Ginkel, A., Schöbel, A.: To Wait or Not to Wait? The Bicriteria Delay Management Problem in Public Transportation. Transportation Science 41(4), 527 (2007)
14. Heimburger, D.E., Herzenberg, A.J., Wilson, N.H.M.: Using simple simulation models in operational analysis of rail transit lines: Case of study of boston's red line. Transportation Research Record 1677, 21–30 (1999)
15. Jensen, J.L.W.V.: Sur les fonctions convexes et les inégalités entre les valeurs moyennes. Acta Mathematica 30(1), 175–193 (1906)
16. Mansilla, S.: Report on disposition of trains. Technical report, ETH Zürich (2001)
17. O'Dell, S.W., Wilson, N.H.M.: Optimal real-time control strategies for rail transit operations during disruptions. In: Computer-Aided Transit Scheduling. Lecture Notes in Economics and Math. Sys., pp. 299–323 (1999)
18. Schöbel, A.: A model for the delay management problem based on mixed-integer programming. In: Proc. ATMOS 2001 (2001); also available in the Electronic Notes in Theoretical Computer Science 50(1), 1–10 (2001)
19. Suhl, L., Biederbick, C., Kliewer, N.: Design of customer-oriented dispatching support for railways. In: Computer-Aided Scheduling of Public Transport. Lecture Notes in Economics and Mathematical Systems, vol. 505, pp. 365–386. Springer, Heidelberg (2001)
20. Zhu, P., Schnieder, E.: Determining traffic delays through simulation. In: Computer-Aided Scheduling of Public Transport. Lecture Notes in Economics and Math. Sys., pp. 387–398 (2001)

Disruption Management in Passenger Railway Transportation

Julie Jespersen-Groth[1,2], Daniel Potthoff[3,5], Jens Clausen[1,2],
Dennis Huisman[3,5,6], Leo Kroon[4,5,6], Gábor Maróti[4,5],
and Morten Nyhave Nielsen[1,2]

[1] DSB S-tog, Denmark
[2] Department of Informatics and Mathematical Modelling, Technical University of
Denmark, DK-2800 Kongens Lyngby, Denmark
[3] Econometric Institute, Erasmus University Rotterdam, P.O. Box 1738, NL-3000
DR Rotterdam, The Netherlands
[4] Rotterdam School of Management, Erasmus University Rotterdam, P.O. Box 1738,
NL-3000 DR Rotterdam, The Netherlands
[5] Erasmus Center for Optimization in Public Transport (ECOPT)
[6] Department of Logistics, Netherlands Railways, P.O. Box 2025, NL-3500 HA
Utrecht, The Netherlands

Abstract. This paper deals with disruption management in passenger railway transportation. In the disruption management process, many actors belonging to different organizations play a role. In this paper we therefore describe the process itself and the roles of the different actors.

Furthermore, we discuss the three main subproblems in railway disruption management: timetable adjustment, and rolling stock and crew re-scheduling. Next to a general description of these problems, we give an overview of the existing literature and we present some details of the specific situations at DSB S-tog and NS. These are the railway operators in the suburban area of Copenhagen, Denmark, and on the main railway lines in The Netherlands, respectively.

Finally, we address the integration of the re-scheduling processes of the timetable, and the resources rolling stock and crew.

1 Introduction

Many Europeans travel frequently by train, either to commute or in their leisure time. Therefore, the operational performance of railway systems is often discussed in the public debate. Travelers expect to arrive at a specific time at their destination. If they travel by rail, they expect to arrive more or less at the time published in the timetable. However, unforeseen events often take place, which cause delays or even cancelations of trains. As a result, passengers arrive later than expected at their final destinations. Due to missed connections, the delay of a passenger can be even much larger than the delays of his individual trains.

Due to the importance for the public on one hand and the deregulation of the railway market on the other, railway operators now put more emphasis on their

R.K. Ahuja et al. (Eds.): Robust and Online Large-Scale Optimization, LNCS 5868, pp. 399–421, 2009.

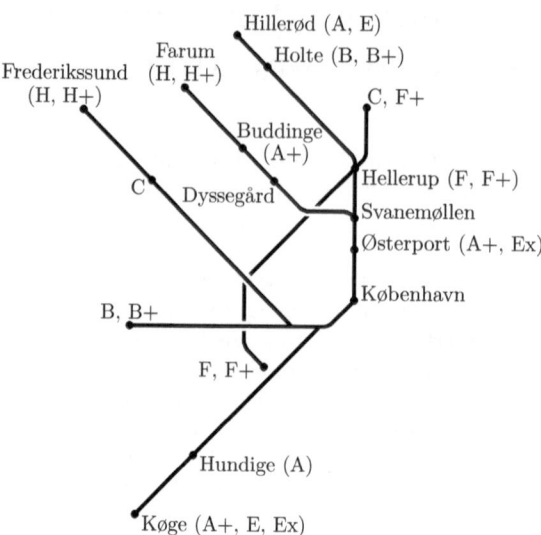

Fig. 1. The S-tog railway network

operational performance than in the past. Furthermore, due to the separation of the management of the infrastructure and the operations in many European countries (including Denmark and The Netherlands), several organizations are responsible for the performance of the railway system.

This paper deals with passenger railway transport only. However, in addition to the passenger railway operator itself, the infrastructure manager and other (also cargo) operators have a strong influence on the performance of the railway services of that single operator. Therefore, the role and the objectives of the infrastructure manager and of the operators are discussed.

We consider two passenger railway operators in more detail: DSB S-tog and NS. DSB S-tog is the operator of local train services in the greater Copenhagen area, see Figure 1. NS is the main operator in The Netherlands, having the exclusive right to operate passenger trains on the so-called Dutch Main Railway Network until 2015, see Figure 2. Both companies operate a set of lines on their network, where a *line* is defined as a route between two stations, sometimes called terminals, operated with a certain frequency, e.g. line A of S-tog runs between Hillerød and Hundige every 20 minutes.

Unfortunately, trains do not always run on time due to unexpected events. Examples are infrastructure malfunctions, rolling stock break downs, accidents, and weather conditions. Such events are called *disruptions*. The Dutch railway network has approximately 17 disruptions related to the infrastructure per day with an average duration of 1.8 hours. About 35% of these infrastructure related disruptions are related to technical failures, while another 35% is related to third parties (e.g. accidents with other traffic). Next to the disruptions to infrastructure failures, there are also disruptions caused by the operators. The three main causes for delays contributed to DSB S-tog are delays due to passengers (45%),

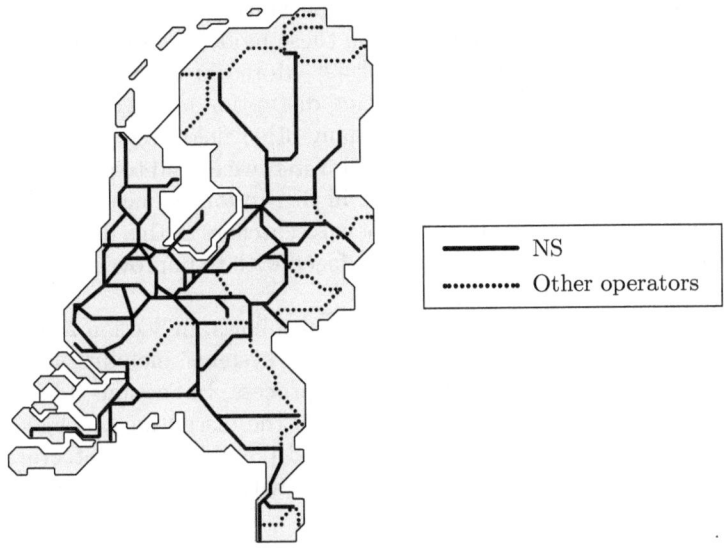

Fig. 2. The Dutch railway network (in 2008)

rolling stock problems (30%) and drivers (15%). The proportion between the disruptions caused by the operators and the infrastructure is roughly 50-50 in The Netherlands as well as in Denmark.

Of course, infrastructure managers and operators try to avoid disruptions. Unfortunately, many of them are hard to influence. Therefore, it is very important to limit the consequences of these disruptions. A very common problem in railways is that, due to the strong interdependencies in the railway network and due to cost efficient resource schedules, disruptions are very likely to spread over the network in space and time. This well-known phenomenon is called *knock-on effect*. The key to a good performance of railways is to limit the knock-on effect and thereby to limit the impact of single disruptions. Therefore, operating plans should be robust and effective disruption management is required. In this paper, we will only look at the second problem. In addition, note that the consequences for passengers can be limited by delaying connecting trains such that passengers can still have their connection even if their arriving train has a delay. This latter problem is known as delay management ([21,22]), however this topic falls outside the scope of the current paper.

So far, Operations Research (OR) models have hardly been applied in practice for disruption management in railway systems. Nevertheless, it is our strong belief that OR models can play an important role to limit the impact of disruptions and thereby to improve the performance of railway systems. This belief is supported by the fact that nowadays OR models and techniques play a major role in several railway companies during the *planning* phase, where the focus is on a good balance of the service level offered to the passengers and the efficiency of the resources rolling stock and crew. The best example is probably the introduction of the new Dutch timetable, for which NS received the 2008 Franz Edelman Award, [14]. For an overview on these models and techniques, we refer to surveys of [3,5,7,11], and to the book [9]. Moreover OR models have proven to be quite effective already for supporting disruption management processes in the airline context, see e.g. [26] and in many other fields, see [25].

The objectives of this paper are twofold. First, we intend to give a comprehensive description of the problems arising in disruption management for railway systems. Second, we aim at attracting new researchers to this field by describing the challenges that railway companies are faced with to improve their operational performance.

The remainder of this paper is organized as follows. In Section 2 we give a description of disruption management for railway systems, including a description of organizations and actors involved in this process. In Sections 3-5, we discuss timetabling, rolling stock and crew aspects of the disruption management process. Section 6 deals with the advantages and possibilities of integrating some of these processes. Finally, we finish the paper with some concluding remarks in Section 7.

2   Description of Disruption Management

Clausen et al. [6] give the following definition of a *disruption* in relation to airline operations: "An event or a series of events that renders the planned schedules for aircraft, crew, etc. infeasible." By definition, a disruption is hence a cause rather than a consequence. In this paper we use the same definition for railway operations, substituting "aircraft" with "rolling stock".

A disruption does not necessarily have immediate influence on the timetable - some disruptions like a track blockage renders the planned timetable immediately infeasible, while others as e.g. shortage of crew due to sickness may lead to cancelations either immediately, in the long run or not at all, depending on the number of stand-by crews. Note that a disruption leads to a *disrupted situation*. Even though this is a slight abuse of terms, we will occasionally refer to the disrupted situation as the disruption itself.

Accordingly, we define *railway disruption management* as the joint approach of the involved organizations to deal with the impact of disruptions in order to ensure the best possible service for the passengers. This is done by modifying the timetable, and the rolling stock and crew schedules during and after the disruption. The involved organizations are the infrastructure manager and the operators.

Of course, one first has to answer the question if the situation is disrupted, i.e. if the deviation from the original plan is sufficiently large or not. Similar to the airline world (see [13]), this question is normally answered by dispatchers monitoring the operations. Although this is a difficult task, it is not considered any further in the remainder of this paper.

The Sections 2.1 to 2.3 introduces a framework of organizations, actors and processes in disruption management, which is valid for several European railway systems. In Section 2.4 we discuss the organizational context of the disruption management process.

2.1 Organizations

The organizations directly involved in disruption management are the infrastructure manager and the railway operators. These organizations usually have contracts with the involved government. Moreover, there are direct appointments between the infrastructure manager and the operators. These issues are described below.

The infrastructure manager has a contract with the government that obliges it to provide the railway operators with a railway network of a certain infrastructure capacity and reliability. The infrastructure manager has also the responsibility of maintaining the railway network as efficiently as possible.

A passenger railway operator obtains a license to operate passenger trains on the network from the government. The operator is contractually bound to provide a performance that exceeds certain specified thresholds on certain key performance indicators. For example, there may be thresholds for the number of train departures per station, for the (arrival) punctuality at certain stations, for the percentage of realized connections, for the seating probability, etc. Here, the punctuality is the percentage of trains arriving within for example 3 or 5 minutes of their scheduled arrival time at certain stations. The realization figures on these performance indicators have to be reported to the government periodically. If an operator does not reach one of the thresholds, it has to pay a certain penalty to the government. If the performance is very poor, another operator may be given the license to operate trains on the network.

As a consequence, usually the main objective of the railway operator is to meet all thresholds set in the contract with the government at minimum cost. The latter is due to the fact that the railway operators are commercially operating companies. Thus, the number of rolling stock units on each train must match with the expected number of passengers. Deadheading of rolling stock units between depots and to and from maintenance facilities must be minimized. Furthermore, the number of crews needed to run the operations and to cover unforeseen demand must be minimized as well.

In more detail, an important objective of the operators in the disruption management process is to minimize the number of passengers affected by the disruption, and to minimize the inconvenience for the affected passengers. Indeed, small delays of trains are usually not considered as a bad service by the passengers, but large disruptions are. If passengers are too often confronted with large disruptions, which usually lead to long extensions of travel times and, even worse, to a lot of *uncertainty* about travel options and travel times, they may decide to switch to a different mode of transport. In relation to this, passenger operators usually prefer to return to the original timetable as soon as possible after a disruption. Indeed, the original timetable is recognizable for the passengers. Therefore, the original timetable provides a better service than a temporary ad hoc timetable during a disruption.

The passengers are the direct customers of the railway operators, and they are only indirect customers of the infrastructure manager. This may imply that the manager has less knowledge of the expected passenger demand on each train and of the real-time passenger locations in the operations. The latter may prohibit a passenger focused dispatching, and may instead lead to a network capacity focused dispatching, i.e. dispatching focusing on supplying sufficient buffer times in the network to recover from disruptions.

Furthermore, each delay of a train may be attributed either to a railway operator or to the infrastructure manager (where the latter one is usually also responsible for delays caused by external factors), depending on the nature of the disruption. However, this creates a natural conflict between the organizations that may prohibit an effective communication and co-operation in the operations. The latter may be counter-productive for the operational performance of the railway system. Thus, although the infrastructure manager and the railway operators have the same general objective of providing railway services to the passengers of a high quality level, there are also conflicting elements in their objectives.

2.2 Actors

In railway disruption management, the actors are the dispatcher of the infrastructure manager and those of the railway operators. The major tasks to be carried out are *timetable adjustment, rolling stock re-scheduling*, and *crew rescheduling*. Figure 3 shows how the responsibilities for the different elements are shared among the actors.

The infrastructure manager controls and monitors all train movements in the railway network. *Network Traffic Control* (NTC) covers all tasks corresponding to the synchronization of the timetables of the different operators. NTC has

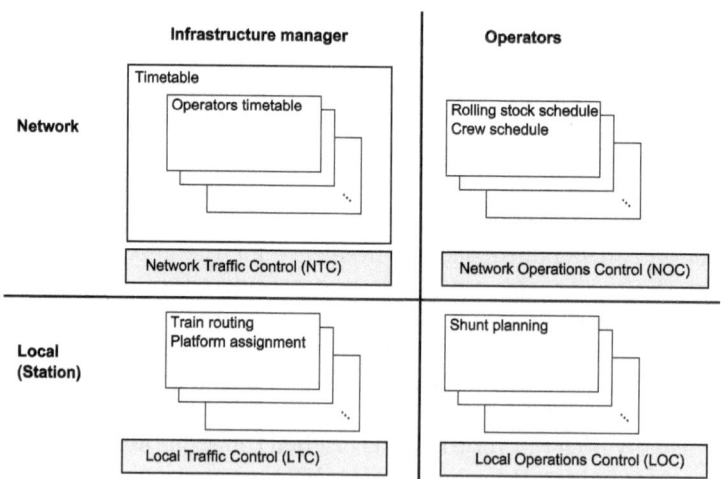

Fig. 3. Schematic view of actors, timetables and resource schedules

to manage overtaking, re-routing, short turning, or canceling trains in order to prevent them from queueing up. The latter is a permanent threat at the basically one-dimensional railway infrastructure. Queueing up of trains immediately leads to extensions of travel times.

On a local level, the process is managed by the *Local Traffic Control* (LTC). For example, LTC is responsible for routing trains through railway stations and for platform assignments. Safety is ensured by headways and automatic track occupancy detection systems.

The *Network Operations Control* (NOC) of each passenger operator keeps track of the operations of the operator on a network level. The dispatchers of NOC are acting as decision makers for the operator in the disruption management process. Depending on the size of the operator, there are one or more dispatchers for rolling stock and crew, respectively. These dispatchers monitor and modify the rolling stock and crew movements. NOC dispatchers are the counterparts of the dispatchers of NTC.

Dispatchers of the *Local Operations Control* (LOC) of the railway operators are responsible for coordinating several local activities at the stations, such as shunting processes. They support NOC by evaluating whether changes to the rolling stock schedules can be implemented locally.

Train drivers and *conductors* are also important elements in the disruption management process. They are usually the first ones that are confronted with passengers that are affected by a disruption. If train drivers and conductors work on different lines, they may carry a delay from one line to another. In order to avoid this situation, the crew dispatchers may have to modify several duties. Besides making the decisions, the dispatchers also have to instruct and sometimes to convince the crew members to carry out the modifications, see Section 5.

2.3   Processes

NTC dispatchers constantly monitor the operations and have to decide if an actual situation is a disruption or will lead to a disruption in the near future. When this is the case, they start the disruption management process. Within this process, the original timetable may need to be changed. This is done by carrying out a dispatching plan. Figure 4 displays the information flows between the different actors in this process.

First, NTC determines all trains that are affected by the disruption. NOC of the corresponding operators must then be informed about the disruption and its direct consequences. In the next step, the dispatchers have to find out to which extent it is still possible to run traffic on the involved route. Some pre-defined emergency scenarios give an indication about which trains should be overtaken, re-routed, short turned, or canceled. Using this information, an initial dispatching plan can be constructed. This dispatching plan must be evaluated by LTC. Almost simultaneously, the proposed dispatching plan is communicated

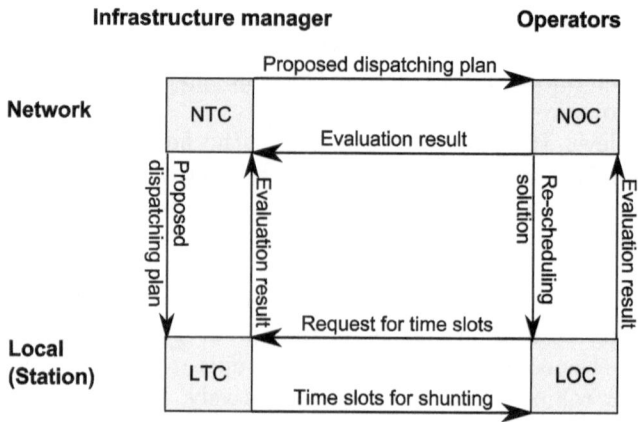

Fig. 4. Information flow during the dispatching plan development

to NOC of the operators. A complicating factor is the uncertainty about the duration of the disruption, for example NTC can only *estimate* how long it will take to repair a broken switch or signal.

The dispatching plan may correspond to changes in the planned operations of several operators. As a whole, these changes are compatible with respect to the safety regulations. However, for the operators it may be impossible to operate the dispatching plan due to their resource schedules for rolling stock or crew. Therefore, the decision about the dispatching plan is taken in consultation between the infrastructure manager and the operators.

Hence, NOC dispatchers have to check whether it is possible for them to operate the proposed dispatching plan. In particular, they have to check whether they can adapt their resource schedules to the proposed dispatching plan. Furthermore, LOC has to verify that the modified timetable and the adapted resource schedules can be carried out locally. Because the resource scheduling problems are NP-hard and the available time is limited, not all re-scheduling options can be evaluated. The re-scheduling solutions represent a trade-off between the available time and the quality of the solution.

This evaluation procedure can basically have three different outcomes. First, NOC and LOC may find a re-scheduling solution to the proposed dispatching plan where no additional cancelations or delays are needed. Second, they may find an initial solution, but trains have to be canceled in a second stage because rolling stock and/or crews are unavailable. A cancelation of a train has, however, a strong negative impact on the service level. Finally, NOC may come up with a request for changes to the proposed dispatching plan if this enables them to construct a much better solution.

Of course, not only one but several operators may ask for changes in the proposed dispatching plan. When these requests are conflicting, it is the responsibility of NTC to make a fair decision. This may involve another iteration of proposal and evaluation between NTC and the operators.

After the final decision about the dispatching plan has been taken by NTC, it is communicated to LTC and to the operators. LTC has to implement the new train routes and change platform assignments. NOC has to inform the train drivers and conductors whose duties have been changed. LOC has to generate new shunting plans. LOC communicates directly with LTC to ask for time slots for shunting movements in the station area. Furthermore, passengers need to be informed in trains, at stations, and via internet and teletext about the changes in the timetable and alternative travel routes.

2.4   Organizational Issues

The description in Section 2.2 of the actors in the disruption management process is a functional description, and not an organizational. For example, it suggests that all dispatchers of each of the mentioned actors are located in the same office. However, this need not be the case.

For example, in the Danish case, NTC, LTC and the timetable and rolling stock dispatcher of the NOC of S-tog are located in the same room, but the crew dispatcher of NOC is located at the crew depot of S-tog. This division was made on request of the train drivers. In practice, it creates some challenges regarding effective communication between the different dispatchers.

In The Netherlands, the situation is even more complex: The Netherlands have been split up into 4 regions, and each region has its own NTC office and its own NOC office of NS. Moreover, there is a central NOC office of NS for coordinating the rolling stock re-scheduling process. Similarly, there are 13 LTC offices and 13 LOC offices of NS. Obviously, this organizational split leads to a lot of additional communication within NTC and within NOC, which is counter-productive in the disruption management process. Therefore, there are currently plans to bring all offices of NTC together, and to do the same with the NOC offices. Moreover, it is investigated how the separation between the infrastructure manager and the operators can be reduced.

3   Timetable Adjustments

3.1   Problem Description

NTC has the overall responsibility of the railway operations and coordinates the disruption management process. When a disruption is recorded, NTC evaluates its effect and, if it is considered as severe, NTC tries to re-schedule the events of the timetable affected by the disruption.

The severeness of a disruption is not easily assessed. It is described as a combination of how much time will pass until the operations are according to plan again and how many trains will be affected. The number of passengers that get delayed because of a disruption also contributes to its degree of severeness. Finally, it makes a large difference to the severeness whether the time intervals between trains on the same track (headways) are small or large. The effect caused by a blockage will be less on sections of the network with much time between the trains than on sections with little time between the trains.

Timetables are constructed with included buffer time. Therefore, a timetable is able to absorb some disruptions. Buffer times are included in the dwell times, the running times, and the headways. When a disruption occurs, the buffer times in the timetable are used to gain time whenever possible. Thus they enable recovery from a disruption.

The state of the daily operation of a train operator at some point in time is influenced by a number of factors, including the current state of the infrastructure (the rail network), and the state of all resources necessary in the operational phase, most notably rolling stock and crew. In the following we introduce the concepts of infrastructural capacity, operational capacity, utilization, and residual operational capacity.

The infrastructural capacity of a rail network in a particular state is the maximum amount of traffic (number of trains per time unit) which is continously able to flow through the network in this state. The infrastructural capacity is independent of the current amount of traffic. The operational capacity of the network is the maximum amount of traffic which is continously able to flow through the network given the current states of network and resources in terms of crew and rolling stock. Note that this is always less than or equal to the infrastructural capacity. The utilization of the network at the current point in time is the amount of continuously flowing traffic. It depends on both the network state and the state of each resource and is always less than or equal to the operational capacity. The residual operational capacity or just the residual capacity is the difference between the operational network capacity and the utilization at the current point in time.

When a severe disruption occurs and it cannot be absorbed by the buffers in the timetable, the utilization of the network decreases, and trains may queue up. In that case, NTC aims to increase the operational capacity and thereby the residual capacity in the network by e.g. moving trains faster through the network, allowing overtaking at relevant stations, turning trains earlier, canceling departures, etc. Residual capacity is in general maintained by controlling the traffic flowing in the network and by preventing blocking situations to occur.

In Sections 3.2 and 3.3 we distinguish between disruptions with low and high impact on the timetable. Low level impact disruptions are those where recovery to the originally planned timetable is possible by using so-called dispatching rules. High level impact disruptions are those where recovery in this way is not possible, for example, if a complete blockage occurs at some part of the network. In such a case, more significant recovery measures are needed.

3.2   Dispatching Rules at S-tog and NS

Dispatching rules are used for disruptions that have a lower level of impact on the railway system. Dispatching rules are further divided into three subgroups according to the level of severeness of the disruption that invoked them. For disruptions with the lowest level of impact, where no substantial decrease in utilization has yet emerged, it is sufficient to make few modifications to the timetable. At the next level, where the traffic is more affected by the disruption, it is necessary to increase the utilization of the network. This may require that the operational capacity is increased, for example through changes in the timetable in stopping patterns. The severest of the low impact disruptions need an increase in residual network capacity through a decrease in utilization before recovery to a state with larger utilization (corresponding to the original timetable) is possible. We give examples of dispatching rules at the two latter levels: Overtaking and changing stopping patterns, and cancellation of single trains and entire train lines.

The different rules have different abilities to relieve disruptions and they have different effects for the passengers. From the passengers' point of view, a rule may affect the number of train departures per station or it may force the passengers to change their routes. The effect of a dispatching rule on the delays of trains and its effect on the passengers can be conflicting. Increasing the residual capacity often implies a decrease in utilization through a reduction in the number of train departures, which is undesirable from the passengers' point of view. However, not increasing the residual capacity will make it very hard to absorb a delay, and this is also undesirable for the passengers.

Overtaking and Changing Stopping Patterns. Handling operations is less complex if there is a predetermined order of train lines, which in case of a disruption can be broken on stations with multiple platforms in the same direction i.e. where overtaking between trains is possible. This is, for example, used when a fast train reaches a delayed stop train at a station with two platforms available in the same direction.

If a stop train, S, is delayed and a fast train, F, catches up with it, and no overtaking is possible, another option is to change the stopping patterns of S and F provided that the two trains are of the same rolling stock type. This rule is specifically used at S-tog. The passengers on train S are informed that after the next stop S becomes a fast train. This enables passengers to get off in time, if their destinations are stations, where fast trains do not stop. These passengers just enter the immediately following train F (the former fast train) and are hence not delayed further. The passengers in S for the fast train stops experience a catch up of the delay due to the switch of stopping patterns. The passengers on F are similarly informed that their train becomes a stop train. Since overtaking is not possible, they experience hardly any additional delay compared to the initial situation, where train F is blocked by the stop train S. Hence, no passenger is delayed more than what was caused by the initial disruption. Note that the operation does not require a platform long enough to accommodate two trains, and that only part of the passengers from the stop train S have to change train.

Increasing Residual Capacity. Residual capacity is increased when departures are canceled. Canceling a departure from a terminal will increase the residual capacity along the entire route of the train. However, from the point of view of NOC, it leaves a number of train units at the departure terminal. This might enforce the cancelation of a departure at the terminal at the other end of the line and may create parking capacity problems at the shunting areas.

It is also possible to cancel an entire train line. An example of how this dispatching rule is used in practice is the cancelation of line B+, which is a line in the present S-tog timetable, cf. Figure 1. Suppose that there are signaling problems between Hellerup and Holte and the trains must run slower than indicated by the timetable. The lines operated on this route are lines A and E running from Hillerød and lines B and B+ running from Holte. To enable better absorbtion of the ongoing disruption, NTC decides to increase the residual capacity by cancelling line B+. This decreases the network utilization thereby allowing an increase in the headways between the remaining trains. The cancellation is implemented by shunting trains on line B+ to shunting areas as these are reached along the route of line B+.

The advantages of the described dispatching rules are that they all increase the residual capacity for absorbing delays in the disrupted situation. The passengers, however, will experience that there are less departures, which for some passengers will obstruct their travel plans. Also, the seat capacity of the trains still in operation is most likely insufficient. Customer questionnaires show that, like delays and canceled departures, this is also considered as poor quality of service.

3.3   Larger Disruptions

For high impact disruptions, a set of emergency scenarios may exist, e.g. when tracks in one or both directions are completely blocked. These emergency scenarios describe for each section in the network and each direction an alternative timetable.

The immediate reaction to a high impact disruption is to apply an appropriate emergency scenario. On heavily utilized networks, the headways are so tight that the system will queue up immediately if no adequate measures are taken after a high impact disruption has occurred. Therefore, almost all railway traffic is canceled around the disrupted area. Trains may be turned as closely as possible to this location. Otherwise, trains may be rerouted, but this requires sufficient capacity on the detour route. Finally, some lines may be canceled completely.

As an example, consider a situation in which the tracks in both directions between stations Dyssegård and Buddinge near Copenhagen are blocked, see Figure 1. The lines crossing this section in a normal situation are the lines A+, H, and H+. Line A+ is operated between Køge and Buddinge, and lines H and H+ are operated between Frederikssund and Farum. The emergency scenario for this blockage is presented in Tables 1 and 2.

Table 1 shows how the lines are changed and whether they are canceled partly or fully. Unless other disruptions occur, only the lines directly involved in the blockage are included in the emergency scenario.

Table 2 specifies how many trains are necessary and which turnaround times must be used for them. Each line is changed according to its stopping pattern. Lines A+ and H+ are shortened, and therefore they can be run by 6 and 8

Table 1. Changes of the lines on the section Dyssegård to Buddinge

Line	Changed from and to	Canceled from and to
A+	Køge to Østerport	Østerport to Buddinge
H	Frederikssund to Dyssegård Buddinge to Farum	Dyssegård to Buddinge
H+	Frederikssund to Svanemøllen	Svanemøllen to Farum

Table 2. Turnaround times and necessary numbers of trains

Line	Traffic south of blockage	Traffic north of blockage
A+	Køge to Østerport Turnaround time: 10 min. Trains necessary: 6	Canceled
H	Frederikssund-Dyssegård Turnaround time: 19 min. Trains necessary: 8	Farum-Buddinge Turnaround time: 13 min. Trains necessary: 3
H+	Frederikssund-Svanemøllen Turnaround time: 16 min. Trains necessary: 8	Canceled

trains, respectively, whereas 8 and 10 trains are necessary normally. Line H is split into two parts and needs 8 plus 3 trains in the disrupted situation, whereas 10 trains are necessary normally.

Given the information in Tables 1 and 2, NTC knows which lines to cancel, where to launch bus-services, how many trains to use for each line, and how many train units to shunt to shunting areas.

4 Rolling Stock Re-scheduling

4.1 Problem Description

This section describes rolling stock re-scheduling in a disrupted situation. Here the assumption is that, whenever this is necessary, the timetable has already been adjusted to the disrupted situation. The main goal is to decide how the rolling stock schedules can be adjusted to this new timetable at reasonable cost and with a minimum amount of passenger inconvenience.

The most characteristic feature of rolling stock is that it is bound to the tracks: rolling stock units cannot overtake one another, except at locations with parallel pairs of tracks. A broken rolling stock unit may entirely block the traffic – actually, this is a frequent cause of disruptions. Moreover, the operational rules of rolling stock units are largely determined by the shunting possibilities at the stations. Unfortunately, shunting is a challenging problem in itself, even for a medium-size station. Therefore, NOC must constantly keep contact with

LOC and check whether or not their intended measures can be implemented in practice. The modifications may be impossible due to lack of shunting drivers or infrastructure capacity.

In the re-scheduling process, the timetable services must be provided with rolling stock which is usually available in several different types. The assignment must fulfill some elementary requirements. For example, electrical rolling stock units cannot run on lines without catenary wires, and no train should be longer than the shortest platform on its route. The shunting capacity of most stations is severely restricted in a disrupted situation. Therefore the re-scheduling process aims at reducing the number of modification of previously planned shunting operations, without introducing new ones at locations or points in time where they do not occur in the original schedules.

Railway operators usually keep rolling stock on stand-by. These units can be used only in case of disruptions. Moreover, many of the rolling stock units are idle between the peak hours, since the available rolling stock capacity is usually larger than than the off-peak passenger demand. If a disruption takes place during off-peak hours, these idle units can act as stand-by units.

In case of a disruption, the first dispatching task is to assign the available rolling stock units to train tasks. These decisions are taken under high time pressure, often guided by the emergency scenarios which tell how the trains have to turn. Whenever there is room for changes, the planners try to cover the seat demand as well as possible. In some cases, however, they are forced to cancel trains due to lacking rolling stock.

After a disruption, it is preferable for the rolling stock schedules to return to the originally planned schedules as quickly as possible, since the feasibility of the originally planned schedules has been checked in detail. As a consequence of all these measures, the rolling stock units will not finish their daily duties at the locations where they were planned prior to the disruption. This is not a problem if two units of the same type get switched: rolling stock units of the same type can usually take each other's duty for the rest of the day. It is more likely that, at the end of the day, some stations have more units of a type than originally planned, and others have less. Thus, unless expensive deadheading trips are used, the traffic on the next day is influenced by the disruption. Modifications of the schedules for the busy peak hours of the next morning are highly undesirable. Therefore additional measures are to be taken so that the rolling stock balance at night is as close to the planned balance as possible.

A further important element in rolling stock re-scheduling is maintenance of rolling stock. Train units need preventive maintenance regularly. The railway operators limit the maximum number of kilometers a unit can serve between two maintenance checks. The train units undergo these large-scale inspections typically once a month. Due to efficiency reasons, units are usually in service just until they reach the maintenance limit. Units that are close to this limit and have to undergo a maintenance check in the forthcoming couple of days are monitored permanently. The latter is particularly important during and after a disruption which may have cut the planned route of the units towards a maintenance facility. NOC has to make sure that these units reach a maintenance facility in time. Usually, only a small number of rolling stock units is involved in planned maintenance routings. Other units of a given type are interchangeable, both in the planning and in the operations.

4.2   Current Practice at S-tog and NS

Both companies operate a dense railway system. This basically allows for many alternative rolling stock schedules through exchanges of train units. However, usually trains have short turn-around times, which rules out complex shunting operations at end points. Also, the shunting capacity (shunting area and crews) of stations is often a bottleneck. NS and S-tog operate rolling stock units of several types. Moreover, a train may contain units of different types, and then the order of the train units in the train is important. Notice that combining different types allows adjusting the seat capacity well to the passenger demand. In case of disruptions, however, the dispatchers have the additional task of monitoring and re-balancing exchanged rolling stock types.

Both NS and S-tog do use sophisticated computer systems for rolling stock management. NS uses an automated tracking and tracing system for real-time monitoring the individual units. At S-tog, the tracking and tracing data is updated manually. These systems, however, lack algorithmic decision support tools; nearly all decisions have to be taken and to be fed to the system manually. As a consequence of the lack of decision support, the dispatchers focus on the immediately forthcoming time period only. Dispatchers identify possible conflicts, and handle them in order of urgency. This approach is commonly known as *rolling horizon*. Note that even if there were decision support tools available, the rolling horizon approach is still a natural candidate for handling disruptions since it can deal with the intrinsic uncertainty of future events.

As mentioned in Section 3.3, a recovery method employed by S-tog for large disruptions is canceling train lines. NOC at S-tog has the responsibility of determining a plan how to re-insert the lines after the disruption. A model has been constructed for finding an optimized re-insertion plan, see [12]. Based on the given number of trains that must be re-inserted from each depot along the line and the start time of the re-insertion, the model calculates which trains must be re-inserted from which depots, and how the drivers for these trains can get to these depots. The automatic decision support system for re-inserting train lines is used in the operations. Moreover, in an ongoing project, the problem of re-allocating rolling stock units to trains in the operations is addressed.

4.3   New Developments

Compared to medium-term planning, there is a very scarce literature on real-time rolling stock re-scheduling. In the recent years intensive research has been conducted to develop methods for the real-time problems as well.

Budai et al. [4] study the Rolling Stock Balancing Problem. It is assumed that the timetable and a feasible rolling stock schedule are given. Moreover, the target rolling stock balance is given. This target is equal to the number of units per type that were originally supposed to arrive at the stations at the end of the planning horizon. The Rolling Stock Balancing Problem aims at modifying the input schedule in such a way that the realized end-of-day balance is as close to the target as possible.

Although the problem was first studied for the operational planning phase, it is also relevant in real-time re-scheduling after a disruption when all immediate conflicts have been resolved (that is, there is a feasible schedule) but the realized end-of-day rolling stock balance differs from the target balance.

Budai et al. [4] prove that an off-balance of a single train unit leads to an NP-hard optimization problem. Also, two heuristic algorithms are developed and compared to exact optimization methods. The computational results on real-life problem instances of NS indicate that the heuristic algorithms provide solutions of promising quality very quickly, within a few seconds.

Another track of research aims at applying an existing rolling stock circulation model of [8] for real-time planning.

Nielsen [16] measures the overall performance of the rolling stock re-scheduling algorithm by three objective criteria: (i) cancellation of trips; (ii) deviation from the originally planned shunting process; and (iii) deviation from the originally planned end-of-day balance.

Criterion (i) is related to keeping a high service quality, while criteria (ii) and (iii) measure the deviation from the undisrupted schedule. In particular, minimizing criterion (ii) enhances the chance that the found solution can be implemented in practice. Indeed, new, non-planned shunting operations can turn out to be impossible due to lacking shunting capacity. Finally, minimizing criterion (iii) tries to reduce the disruption's consequences for the next day.

The algorithm of [16] is based on the flexible linear integer programming model of [8]; this model has been used by NS since 2004 for medium term planning. To find a schedule for the disrupted timetable, the model of [8] is extended in such a way that the criteria (i), (ii) and (iii) are taken into account. However, the model cannot deal with uncertainties of the input data, and solving it by commercial MIP software can take several hours. Therefore [16] developed a rolling horizon based solution approach for dealing with real-time re-scheduling problems of NS. The main idea of the rolling horizon algorithm is to consider at any moment the forthcoming, say, 3 hours only. Based on the latest forecasts on the duration of the disruption, the updated timetable is created. Then the extended MIP model is solved for the considered time horizon. This optimization can indeed be performed in a few seconds. An hour later, or whenever new, relevant information arrives, the model is solved again for the forthcoming 3 hours. This process is repeated until the end of the day. The algorithm is inspired by the current rolling stock disruption management process in practice.

While the criteria (i) and (ii) are easily incorporated in a rolling horizon framework, the deviation from the target rolling stock balance is conceptually more difficult: The end of the day is not visible before the very last iteration. Nielsen [16] proposes a heuristic way to guide the rolling horizon algorithm to the originally planned end-of-day balance.

Nielsen [16] reports computational results on several realistic problem instances of NS. These include disruptions on the so-called "Noord-Oost" case, a particularly complex rolling stock scheduling instance. The rolling horizon based algorithm found solutions with very little deviations from the undisrupted schedule, both in terms of shunting and in terms of rolling stock balance. On-going research focuses on making the algorithm fully comply with the restrictions of railway practice. This includes fine-tuning the algorithm as well as some extensions such as dealing with maintenance of rolling stock units.

5 Crew Re-scheduling

5.1 Problem Description

Recall that the recovery of the timetable, the rolling stock schedule, and the crew schedule is usually done in a sequential fashion. For an estimated duration of the disruption, a modified rolling stock schedule has been determined for a modified timetable. Both are input for the *crew re-scheduling problem*, in which the crew schedule needs to be modified in order to have a driver and an appropriate number of conductors for each task of the modified timetable. Tasks can be either passenger train movements, empty train movements, or shunting activities.

The modified timetable contains the unchanged tasks from the original timetable which have not yet started and additional tasks which were created as reaction to the disrupted situation. A duty becomes infeasible due to a time or a location conflict. The latter may occur, e.g. when one of its tasks has been canceled, and hence the corresponding driver cannot perform the remaining part of his duty.

In Figure 5, we show an example of an infeasible duty. Because of a disruption, the train containing task t_3 is canceled. Driver d has already finished task t_1 and is at station B. He can perform the next task in his duty, but since t_3 is canceled he cannot go from station C to D. Hence, he will not be able to perform the two last tasks of his duty. Furthermore, this means that, if no action is taken, these two tasks need to be canceled as well. Moreover, driver d has to get back to his crew depot at station A in an appropriate way and at a reasonable time.

The crew re-scheduling problem tries to re-assign tasks, such that the new crew schedule covers all tasks of the modified timetable. Stand-by crews located at major stations play an important role as they can be utilized in the new crew schedule.

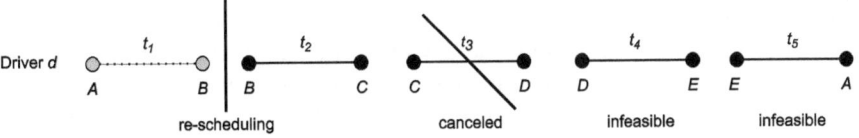

Fig. 5. An infeasible duty

The possibilities for changing duties on the day of operation are based on rules and agreements between the railway company and labor unions. These possibilities usually vary from company to company. For example, the driver's route knowledge has to be taken into account as well as his license for certain rolling stock types. In order to increase the flexibility of the crews, they can be repositioned to another station by traveling on trains as passengers. This option is called crew deadheading.

The objective of the crew re-scheduling problem is a combination of different aspects, namely feasibility, operational costs, and stability. The feasibility aspect is by far the most important, since infeasibility usually implies that trains have to be canceled. It is the decision of the operator how to balance the aspects operational costs and stability.

First of all, there is the feasibility aspect. It is not evident that all tasks can be covered by a solution. Given two solutions with different uncovered tasks, there may exist a preference for one of them, depending on the urgency and the expected numbers of passengers of the uncovered tasks. If a task cannot be covered, canceling it will lead to a feasible crew re-scheduling solution. An additional cancelation, however, leads to more inconvenience for the passengers, which is against the general aim of disruption management. Moreover, such a cancelation has to be approved by the rolling stock dispatchers and the local planners, since it disturbs the rolling stock circulation. Because a cancelation is a change of the timetable, it has to be approved by NTC.

Operational costs are the second aspect in the objective. In the railway context, the crew payments are often based on fixed salaries. Nevertheless, some parts of a re-scheduling solution influence the operational costs. Crew deadheading on trains can be considered to have no costs other than time, whereas using other transport options for repositioning and taking home stranded crews is not free. Also, operator specific compensations for extra work due to modified duties need to be considered.

The third aspect in the objective is stability. Humans are involved in the implementation of every re-scheduling solution and can cause its failure. A crew dispatcher may, for example, forget to call a driver and inform him about the modifications in his duty. Therefore, a solution is considered to be more stable if the number of modified duties is smaller.

5.2   Current Practice at S-tog and NS

A closely related problem is crew re-scheduling in short term planning. This occurs for instance due to timetable changes based on maintenance work on tracks. The resulting crew schedule is called a *special plan*. For the construction of special plans additional rules have to be taken into account. If a special plan is made prior than 72 hours before the day of operation, duties may start and end up to 30 minutes earlier (respectively later) compared to the planned schedule. Within the last 72 hours before the day of operation duties may start earlier or end later only if this is accepted by the crew member.

Both railway operators S-tog and NS nowadays use optimization software to construct special plans. The optimization software TURNI, described e.g. in [2], is used at S-tog. TURNI was designed for and has been used successfully for solving 'classical' crew scheduling problems in long term planning. NS is using a dedicated approach ([10]) which has been integrated into the CREWS planning system ([15]). Both applications rely on a combination of column generation and Lagrangian relaxation.

For crew re-scheduling on the day of operation neither of the two companies is using a decision support system. The crew dispatchers use an interactive software system. This provides them with information about the actually planned duties, and enables them to store their duty modifications in the system. The system informs them about delays of trains and about modifications in the timetable and rolling stock schedules. The system also indicates time and location conflicts in the duties. Recovery options, however, have to be found manually without algorithmic support. In the manual procedure, conflicts are resolved one at a time in order of urgency.

As mentioned earlier several agreements exist about the way duties may be modified on the day of operations. However, if a dispatcher finds an option outside these rules he might ask the affected drivers if they are willing to accept the changes to their duties. Experiments were carried out to inform crew members automatically via SMS about duty modifications. However, communicating modifications via telephone is still common practice.

5.3   New Developments

Crew re-scheduling within disruption management was subject to several research projects within S-tog and NS. Experience from short term planning made already clear that is is not possible to consider all duties and tasks in the re-scheduling problem due to too long computation times. Therefore all studies on re-scheduling on the day of operations consider only a small part of the crew schedule, given by a subset of the duties and a time window.

In a first study at S-tog, the standard version of TURNI was tested. The tests have shown that in general these kinds of re-scheduling problems can be tackled with column generation based optimization methods. However, tailored systems are needed in order to reduce computation time.

One such tailored solution method to solve the crew re-scheduling problem was developed by [20] and [19]. The problem is formulated as a set partitioning problem and possesses strong integer properties. The proposed solution approach is therefore a depth-first search in a branch-and-price tree. The LP-relaxation of the problem is solved with a column and constraint generation algorithm. The problem is first initialized with a very small disruption neighborhood, which contains only duties that cover delayed, canceled or re-routed tasks and is limited by a recovery period. As long as tasks are uncovered, while solving the LP-relaxation, the disruption neighborhood is extended by either adding more duties to the problem or by extending the recovery period. In order to deal with new information becoming available the author(s) propose to use the crew re-scheduling algorithm in a rolling time horizon approach similar to the one proposed by [16] for rolling stock re-scheduling. The algorithm was tested on instances based on historic disruptions using real-life crew schedules from S-tog. The obtained results are very good in terms of solution quality as well as in terms of computation time.

Potthoff et al. [18] extended the approach of [10] to crew re-scheduling during disruptions. A column generation based heuristic computes solutions for core problems containing a small subset of duties and tasks. Lagrangian relaxation and subgradient optimization is used to obtain dual solutions for the restricted master problems. A greedy Lagrangian heuristic constructs feasible solutions from the generated columns. First an initial core problem is made of the affected duties and a small number of duties that cover tasks on the same train lines as canceled or re-routed trains within a specified time interval. If tasks remain uncovered in the solution, new core problems are generated containing duties in the "neighborhood" of an uncovered task. The main idea is similar to the approach by [20] and [19]. Start with a small core problem and consider more possibilities when tasks cannot be covered. While in [20] and [19] the initial problem gets extended, [18] propose to consider different core problems. They present a rule for the construction of a core problem based on a "neighborhood" of an uncovered task. The idea behind the "neighborhood" is to find a set of duties such that all tasks can be covered. The paper contains experiments on disruptions that took place at NS and provides very promising results.

Finally, there are some experiments at NS with multi-agent technology. In this approach each driver is represented by a driver-agent. If due to the disruption a driver-agent can no longer perform a certain task, this driver-agent starts a negotiation process with other driver-agents to transfer the task to another driver-agent. For more details, we refer to [1].

6   Integrated Recovery

The integrated recovery approach has received little attention up till now. To the best of our knowledge [24] is the only paper presenting a model that manipulates the timetable and the crew schedule at the same time. The objective is to simultaneously minimize the deviation of the new timetable from the original one, and the cost of the crew schedule. One part of the model represents the timetable adjustment, a second part corresponds to a set partitioning model for the crew schedules. In addition, there are some constraints linking the timetabling and crew scheduling part. It should be mentioned that the railway system addressed in the research is of a relatively simple structure.

The benefits of such an approach compared to the sequential approach may, however, be large in terms of quality of service, and the field is expected to become an active research field in the future.

7   Conclusions

Railway operators pay much attention to improve their operational performance. One of the key issues is to limit the number of delays by reducing the knock-on effect of single disruptions. To achieve this goal, effective disruption management is required. In this paper, we have explained the role of the different organizations and actors in the disruption management process. An important issue here is that next to the operator itself, the infrastructure manager plays a major role in the disruption management process. The different objectives of both organizations on one hand and difficult communication schemes on the other hand, complicate the disruption management process a lot.

After the description of disruption management, we discussed the three sub-problems arising in railway disruption management: timetable adjustment, and rolling stock and crew re-scheduling. To adjust the timetable, several dispatching rules are applied in practice. Unfortunately, no optimization techniques are involved to solve this problem currently. For the re-scheduling of rolling stock and crew some first attempts have been made in the literature to come up with OR models and solution techniques. Most of these have been derived from similar problems in the airline world. However, most of these ideas are in an early stage and have not been applied in practice yet.

In other words, there is a major challenge for the OR community to develop new models and come up with new solution approaches to tackle these problems. Therefore, we hope and expect that another review paper on railway disruption management in about 5 years contains much more models and solution approaches than this one, and moreover that many of them have been applied in practice.

Acknowledgments

This research was partially sponsored by the Future and Emerging Technologies Unit of EC (IST priority 6th FP), under contract no. FP6-021235-2 (ARRIVAL). Furthermore, this research is made possible with support of TRANSUMO. TRANSUMO (TRANsition SUstainable MObility) is a Dutch platform for companies, governments and knowledge institutes that cooperate in the development of knowledge with regard to sustainable mobility.

References

1. Abbink, E., Mobach, D., Fioole, P.-J., Kroon, L.: Actor-agent based train driver rescheduling. In: Proceedings of BNAIC (2008)
2. Abbink, E.J., Fischetti, M., Kroon, L.G., Timmer, G., Vromans, M.J.C.M.: Reinventing crew scheduling at Netherlands Railways. Interfaces 35(5), 393–401 (2005)
3. Assad, A.A.: Models for rail transportation. Transportation Research A 14, 205–220 (1980)

4. Budai, G., Maróti, G., Dekker, R., Huisman, D., Kroon, L.: Re-scheduling in railways: The rolling stock balancing problem. Technical report, Econometric Institute, Erasmus University Rotterdam (2007)
5. Caprara, A., Kroon, L.G., Monaci, M., Peeters, M., Toth, P.: Passenger Railway Optimization. In: Barnhart, C., Laporte, G. (eds.) Transportation. Handbooks in Operations Research and Management Science, vol. 14, pp. 129–187. Elsevier, Amsterdam (2007)
6. Clausen, J., Larsen, A., Larsen, J., Rezanova, N.J.: Disruption management in the airline industry - Concepts, models and methods. Computers & Operations Research (in press, 2009), doi:10.1016/j.cor.2009.03.027
7. Cordeau, J.-F., Toth, P., Vigo, D.: A survey of optimization models for train routing and scheduling. Transportation Science 32, 380–404 (1998)
8. Fioole, P.-J., Kroon, L., Maróti, G., Schrijver, A.: A rolling stock circulation model for combining and splitting of passenger trains. European Journal of Operational Research 174, 1281–1297 (2006)
9. Geraets, F., Kroon, L.G., Schoebel, A., Wagner, D., Zaroliagis, C.D. (eds.): Railway Optimization 2004. LNCS, vol. 4359. Springer, Heidelberg (2007)
10. Huisman, D.: A column generation approach to solve the crew re-scheduling problem. European Journal of Operational Research 180, 163–173 (2007)
11. Huisman, D., Kroon, L.G., Lentink, R.M., Vromans, M.J.C.M.: Operations Research in passenger railway transportation. Statistica Neerlandica 59(4), 467–497 (2005)
12. Jespersen-Groth, J., Clausen, J.: Optimal reinsertion of cancelled train lines. Technical Report Report-2006-13, Informatics and Mathematical Modelling, Technical University of Denmark (August 2006)
13. Kohl, N., Larsen, A., Larsen, J., Ross, A., Tiourine, S.: Airline disruption management - perspectives, experiences and outlook. Journal of Air Transport Management 13, 147–162 (2007)
14. Kroon, L.G., Huisman, D., Abbink, E., Fioole, P.-J., Fischetti, M., Maróti, G., Schrijver, L., Steenbeek, A., Ybema, R.: The new Dutch timetable: The OR revolution. Interfaces 39, 6–17 (2009)
15. Morgado, E., Martins, J.: CREWS_NS - scheduling train crews in the Netherlands. AI Magazine 19, 25–38 (1998)
16. Nielsen, L.K.: A decision support framework for rolling stock rescheduling. Technical report, ARRIVAL Project (2008)
17. Nissen, R., Haase, K.: Duty-period-based network model for crew rescheduling in European airlines. Journal of Scheduling 9, 255–278 (2006)
18. Potthoff, D., Huisman, D., Desaulniers, G.: Column generation with dynamic duty selection for railway crew rescheduling. Technical report, Econometric Institute, Erasmus University Rotterdam (2008)
19. Rezanova, N.J.: The Train Driver Recovery Problem - Solution Method and Decision Support System Framework. Ph.D. Thesis, Technical University of Denmark (2009)
20. Rezanova, N.J., Ryan, D.N.: The train driver recovery problem - A set partitioning based model and solution method. Computers & Operations Research (in press, 2009), doi:10.1016/j.cor.2009.03.023
21. Schöbel, A.: Optimization in Public Transportation. Springer Optimization and Its Applications, vol. 3. Springer, Berlin (2006)

22. Schöbel, A.: Integer programming approaches for solving the delay management problem. In: Geraets, F., Kroon, L.G., Schöbel, A., Wagner, D., Zaroliagis, C. (eds.) Railway Optimization 2004. LNCS, vol. 4359, pp. 145–170. Springer, Heidelberg (2007)
23. Törnquist, J.: Railway traffic disturbance management. PhD thesis, Blekinge Institute of Technology, Karlskrona, Sweden (2006)
24. Walker, C.G., Snowdon, J.N., Ryan, D.N.: Simultaneous disruption recovery of a train timetable and crew roster in real time. Computers & Operations Research 32, 2077–2094 (2005)
25. Yu, G., Qi, X.: Disruption management: Framework, models and applications. World Scientific Publishing Company, Singapore (2004)
26. Yu, G., Argüello, M., Gao, S., McCowan, S.M., White, A.: A new ara for crew recovery at Continental Airlines. Interfaces 33(1), 5–22 (2003)

Author Index